Advances in Cancer Detection, Prediction, and Prognosis Using Artificial Intelligence and Machine Learning

Rajeev Nema • Ashok Kumar
Dinesh Kumar Saini
Editors

Advances in Cancer Detection, Prediction, and Prognosis Using Artificial Intelligence and Machine Learning

Springer

Editors
Rajeev Nema
Department of Biosciences
Manipal University Jaipur
Jaipur, Rajasthan, India

Ashok Kumar
Biochemistry
AIIMS Bhopal
Bhopal, Madhya Pradesh, India

Dinesh Kumar Saini
Department of IoT and Intelligent Systems
Manipal University Jaipur
Jaipur, Rajasthan, India

ISBN 978-981-96-9345-0 ISBN 978-981-96-9346-7 (eBook)
https://doi.org/10.1007/978-981-96-9346-7

This Springer imprint is published by the registered company Springer Nature Singapore Pte Ltd.
The registered company address is: 152 Beach Road, #21-01/04 Gateway East, Singapore 189721, Singapore

Preface

Advances in cancer detection, prediction, and prognosis using artificial intelligence and machine learning continue to evolve. Genetic alterations in cancer, while being fundamental drivers of tumorigenesis, can also lead to various metabolic adaptations that enable cancer cells to survive and proliferate in diverse tumor microenvironments. This flexibility in AI and ML differs from normal cellular metabolic processes and results in heterogeneity in cancer metabolism, both within the same cancer type and within the same tumor. In this book, the authors explore the complexity and diversity of cancer metabolism, emphasizing that understanding this heterogeneity is essential for developing effective metabolism-based therapies for cancer treatment. Deciphering how cancer cells utilize a variety of nutrient resources will enable researchers and clinicians to match specific chemotherapeutic agents with patients who are most likely to respond positively, facilitating more cost-effective and personalized cancer treatment. This book is the second edition and includes all the content from the first edition, along with new information that makes it distinct. It consists of four major parts: Basic AI and Advances in Cancer Detection, Prediction, and Prognosis; Heterogeneity of Cancer Metabolism; Relationship Between Cancer Cells and Cancer-Associated Artificial Intelligence and Machine Learning; and The Network Interplay Between Cancer and Other Diseases. This book is aimed at cancer artificial intelligence and machine learning researchers, cancer biologists, other researchers, physicians, epidemiologists, healthcare professionals from various disciplines, policymakers, and marketing and economic strategists. It is also designed for teaching undergraduate and graduate students. The metabolic pathways and their regulations discussed in this book serve as illustrations of the heterogeneity of cancer metabolism and are non-inclusive.

Jaipur, Rajasthan, India — Rajeev Nema
Bhopal, Madhya Pradesh, India — Ashok Kumar
Jaipur, Rajasthan, India — Dinesh Kumar Saini

Acknowledgments

This book was made possible thanks to the contributions of the members of the Nema Cancer Research Laboratory at Manipal University Jaipur. We extend our heartfelt gratitude to the outstanding support of the Associate Editor at Springer Nature Publisher. A special thank you goes to Dr. Ashok Kumar Cooper for his editing and valuable suggestions. The figures included in this book were created using BioRender.com.

Introduction

Artificial intelligence (AI) and machine learning (ML) are closely related fields, with ML being a subset of AI that enables machines to learn from data without being explicitly programmed (Sarker, 2021). The integration of AI and ML in oncology promises to revolutionize the way cancer is diagnosed, treated, and managed. By leveraging these technologies, healthcare providers can not only detect cancer earlier but also personalize treatment plans based on individual patient data. This tailored approach has the potential to enhance patient outcomes and quality of life, ultimately reshaping the landscape of cancer care. AI algorithms have demonstrated significant potential in improving the accuracy and efficiency of cancer detection by analyzing medical imaging data, such as mammograms, MRIs, and CT scans (Khalifa & Albadawy, 2024). These algorithms can quickly and accurately identify suspicious areas that may be overlooked by human radiologists, leading to earlier diagnosis and treatment. AI-powered diagnostic tools for identifying cancerous cells and predicting tumor growth can transform the field of oncology by utilizing machine learning algorithms to analyze complex imaging data (Dlamini et al., 2020). Incorporating genetic data and lifestyle information into predictive models will further refine the accuracy of cancer risk assessments, allowing healthcare providers to identify high- risk individuals even before symptoms manifest. By adopting a holistic approach to patient care, medical professionals can create personalized prevention plans that consider individual genetic predispositions and lifestyle factors. This customized approach not only improves the chances of early detection and successful treatment but also empowers patients to take an active role in their health and well-being. As AI continues to evolve and incorporate more data sources, the potential for early detection and prevention will only increase, offering new hope in the battle against this devastating disease. In conclusion, the application of AI and ML in cancer diagnosis and treatment is reshaping the landscape of cancer prognosis. By analyzing patient data, healthcare providers can more accurately predict how a patient's cancer will progress and respond to treatment, leading to more tailored treatment plans. Predicting treatment outcomes based on AI-generated prognostic models has the potential to revolutionize cancer care by providing tailored and targeted therapies for each individual patient. By incorporating real-time data and continually updating algorithms, healthcare providers can stay ahead of disease progression and adjust treatment plans as needed. This personalized approach has the potential to save lives and enhance the quality of life for cancer patients worldwide. The impact of AI-powered prognostic tools on patient care and decision-making is substantial. With the integration of AI-powered tools, medical professionals can better understand the unique needs of each individual and tailor their treatment plans accordingly. This personalized approach not only enhances patient care but also empowers healthcare providers to make more precise and effective decisions, ultimately leading to improved patient outcomes and an overall enhancement in the quality of care. However, ethical considerations are crucial in the healthcare industry. Ensuring patient privacy and data security and addressing potential biases in AI algorithms are key challenges that must be carefully managed. Moving forward, it is vital for healthcare professionals to collaborate closely with AI developers and ethicists to establish guidelines and standards for the ethical use of AI in cancer care (Mennella et al., 2024). Limited AI systems in handling complex cancer cases highlight the need for continuous training and validation of AI algorithms to ensure accuracy and effectiveness. Furthermore, the lack of transparency regarding how AI makes decisions presents a challenge in gaining trust from both healthcare providers and patients (Eke & Shuib, 2025). Collaboration between healthcare professionals, AI developers, and regulatory bodies is essential to address these limitations and ensure that AI technology is employed responsibly in cancer care. Potential advancements in AI technology to enhance accuracy and efficiency in cancer detection and treatment are promising, but it is crucial for all stakeholders to work together to ensure that these advancements are implemented ethically and effectively. By uniting to overcome these challenges, AI has the potential to significantly improve the quality of cancer care and outcomes for patients across the globe.

Dr. Rajeev Nema, PhD
Associate Professor,
Department of Biosciences Manipal University Jaipur,
Jaipur, Rajasthan, India
E-mail: rajeev.nema@jaipur.manipal.edu

References

Dlamini, Z., & Francies, F. Z., Hull, R., & Marima, R. (2020). Artificial intelligence (AI) and big data in cancer and precision oncology. *Computational and Structural Biotechnology Journal, 18*, 2300–2311.

Eke, C. I., & Shuib, L. (2025). The role of explainability and transparency in fostering trust in AI healthcare systems: a systematic literature review, open issues and potential solutions. *Neural Computing & Applications, 37*, 1999–2034.

Khalifa, M., & Albadawy, M. (2024). AI in diagnostic imaging: Revolutionising accuracy and efficiency. *Computer Methods and Programs in Biomedicine Update, 5*, 100146.

Mennella, C., Maniscalco, U., Pietro, G. D., & Esposito, M. (2024). Ethical and regulatory challenges of AI technologies in healthcare: A narrative review. Heliyon [Internet]. [cited 2025 Apr 21];10. Available from: https://www.cell.com/heliyon/abstract/S2405-8440(24)02328-4

Sarker, I. H. (2021). Machine learning: Algorithms, real-world applications and research directions. *SN Computer Science, 2*, 160.

Contents

About the Editors

Rajeev Nema is currently serving as an Associate Professor at Manipal University Jaipur, Rajasthan. With over 9 years of post-doctoral experience, his research has been deeply rooted in diverse aspects of molecular cancer biology. He earned his PhD from the Department of Life Science at Barkatullah University, where he utilized natural product chemistry to investigate the key chemical activities of three distinct cancer cell lines: A549 (lung cancer), MCF-7 (breast cancer), and PA-1 (pancreatic cancer). From 2014 to 2017, he worked as a research associate at the All India Institute of Medical Sciences (AIIMS), Bhopal, under the Indian Council of Medical Research (ICMR). His work in the Department of Biochemistry focused on deciphering the crosstalk between ROR1/2 and S1P signaling pathways, with an emphasis on their roles in regulating cell proliferation, apoptosis, and invasion in lung cancer cells. Recognized for his potential as a young scientist, he was awarded the Ministry of Health and Family Welfare (MoHF) fellowship, which allowed him to continue his research on these pathways from 2018 to 2021. During this period, he developed extensive expertise in lung and breast cancer biology, mastering various cellular and molecular biology techniques, including primary and established cancer cell culture methods. Between 2021 and 2023, he held the position of senior research scientist in the Department of Gene Expression (Oncology Development) at 3B BlackBio Biotech India Ltd., Bhopal. Most recently, from February to August 2023, he worked as a postdoctoral researcher in the Department of Oncological Sciences at the Icahn School of Medicine at Mount Sinai, New York, further expanding his international research exposure.

Ashok Kumar is currently working as a Professor in the Department of Biochemistry of All India Institute of Medical Sciences Bhopal. His current research interest understands the role of non-coding RNAs and sphingosine-1-phosphate signaling in Head and Neck Cancer. He received his PhD degree from Sanjay Gandhi Postgraduate Institute of Medical Sciences (SGPGIMS) Lucknow, India. He received his postdoctoral training from UCSF-Benioff's Children's Hospital Oakland & Research Centre, Oakland, CA, USA. His major areas of research work are Cell and Molecular Biology, Cancer Biology, Immunology, and Sphingolipid signaling. He has more than 15 years of research experience. He has published more than 50 research articles and 10 book chapters, and has edited 2 books. He is a member of several international professional societies including American Association of Cancer Research and Fellow of Royal Society of Biology.

Dinesh Kumar Saini is currently working as a Professor in the Department of Computer and Communication Engineering at the School of Computing and Intelligent Systems at Manipal University Jaipur, Jaipur, India. His current research interests are computer science and engineering, information technology, health science and AI, software systems, business administration, and human resource management. He received his PhD degree from Birla Institute of Technology and Science, Pilani, Rajasthan, India. While serving as Dean of the Faculty of Computing and Information Technology in Oman, he maintained a strong academic record. He has administrative experience as a university senator, university research council member, university library committee member, faculty research coordinator, etc. He has conceived, planned, developed, organized, and implemented numerous seminars, conferences. He has published more than 140 research articles and 10 book chapters, as well as edited 2 books. He is a member of several international professional societies.

Cancer Diagnosis: An Overview

Mradu Bhadauriya, Rupali Yadav, Dinesh Ahirwar,
Deepti Joshi, Rajeev Nema, and Ashok Kumar

Abstract

Cancer is a progressive disease characterized by the uncontrolled cell growth and its metastasis to other part of the body. It is multifactorial disease and one of the leading causes of death worldwide. Based on its progression and complexity, cancers are categorized into four stages, referred to as stage I, II, III, and IV. Unfortunately, approximately 80% of cancer patients are diagnosed in the advanced stages. Therefore, early detection is the main difficulty for cancer diagnosis, and a single test is insufficient. Thus, a holistic strategy that includes physical, pathological, radiographic, and laboratory investigations for tumor biomarkers is necessary for early cancer diagnosis. It necessitates the interdisciplinary collaboration among radiologists, pathologists, and clinicians to expedite these diagnostic processes and arrive at a definitive diagnosis and rational treatment strategy. The search for new tumor biomarkers is still ongoing, and traditional methods are being improved in response to demand by utilizing the most recent technological advancements, such as MRI, SPECT, and immunohisto-

M. Bhadauriya · R. Yadav · A. Kumar (✉)
Department of Biochemistry, All India Institute of Medical Sciences (AIIMS) Bhopal, Bhopal, India
e-mail: ashok.biochemistry@aiimsbhopal.edu.in

D. Ahirwar
Department of Bioscience and Bioengineering, Indian Institute of Technology Jodhpur, Jodhpur, Rajasthan, India

D. Joshi
Department of Pathology & Laboratory Medicine, All India Institute of Medical Sciences (AIIMS) Bhopal, Bhopal, India

R. Nema
Department of Biosciences, Manipal University Jaipur, Jaipur, Rajasthan, India

R. Nema et al. (eds.), *Advances in Cancer Detection, Prediction, and Prognosis Using Artificial Intelligence and Machine Learning*,
https://doi.org/10.1007/978-981-96-9346-7_1

chemistry. Even with the tremendous advances in our understanding of its molecular properties, cancer remains a challenging disease for patients, researchers, and doctors due to its complicated and adaptive nature. Early detection, predicting aggressiveness, metastasis, and recurrence, assessing tumor margins during surgery, monitoring tumor growth and treatment resistance, and differentiating between preneoplastic and neoplastic lesions are some of the challenges in managing cancer. In the current chapter, we provide a comprehensive review on the conventional methods and technologies used for the screening and diagnosis of common tumors. Further, we also put forth the challenges in the early diagnosis of cancer.

Keywords

Cancer · Tumor biomarkers · Cancer diagnosis

1 Introduction

Cancer is one of the leading causes of death worldwide. According to the data from the Global Cancer Observatory (GLOBOCAN), in 2022, approximately 20 million new cases of cancer and 9.7 million fatalities occurred worldwide (Bray et al., 2024). According to the recent GLOBOCAN data, the most common cancers worldwide are breast, prostate, lung, colon and rectum, melanoma of the skin, bladder, kidney and renal pelvis, non-Hodgkin lymphoma, endometrial, pancreatic, leukemia, thyroid cancer, and liver cancer (Cao et al., 2024). Approximately 50% of common cancers in males are prostate, lung, and colorectal malignancies (Didkowska et al., 2022). On the other hand, breast, lung, and colorectal cancers account for half of the cancer diagnosed in women (Greene et al., 2022).

Most solid cancers are categorized into four stages, referred to as stage I, II, III, and IV. Approximately 80% of cancer patients are diagnosed in the advanced stages. In order to promote the early detection of cancer, screening for common cancers is recommended. For example, mammography, colonoscopy, PAP smear, and prostate-specific antigen have been recommended for the breast, colorectal, cervix, and prostate cancer, respectively (Giaccone, 2011). Early detection rates have been made possible using currently available cancer diagnostic techniques, such as positron emission tomography (PET), X-ray computed tomography (CT), magnetic resonance imaging (MRI), and molecular diagnostic techniques. These techniques are crucial for the early cancer detection and therapeutic management of cancer patients. The rate of accurate precursor lesion identification is a critical component of these cancer screening programs' efficacy; a higher identification rate in the early stages enables earlier onset treatment, which lowers the long-term incidence of invasive cancer and improves overall prognosis. In the current chapter, we will discuss the conventional methods and technologies used for the screening and diagnosis of common solid tumors. Further, we also put forth the challenges in the early diagnosis of cancer.

2 Conventional Tools

Malignancies cannot be correctly diagnosed by a single test. A comprehensive clinical history, physical examination, and diagnostic testing are typically necessary for a patient's full evaluation. To find out if someone has cancer or if their symptoms are being mimicked by another illness (such as an infection), numerous tests are required. Efficient diagnostic testing is used to monitor the progression of the disease, confirm or rule out its presence, and plan and assess the efficacy of treatment. Repeat testing may be required in certain situations, such as when a person's health has changed. Medical imaging, laboratory testing (including tumor marker testing), tumor biopsy, endoscopic examination, surgery, or genetic testing are some of the diagnostic methods used to diagnose malignancies (Fig. 1).

3 Cancer Screening Methods

"Cancer screening" refers to tests that search for signs of cancer before symptoms manifest. Statistics show that several screening tests help identify cancer early on, when it is easier to treat and possibly even curable. They are a way to find out if a patient might have cancerous cells or early signs of cancer. Physical examinations, laboratory studies, and imaging tests are all part of the cancer screening process. In addition to routine check-ups during the visit, physical examinations serve as a baseline and may reveal any alterations, such as enlarged organs, strange tumors, or changes in skin color. The staging, future treatment, and prognosis can all be drastically altered by a single, mildly aberrant physical exam finding that prompts additional testing for metastatic illness (Rosenzweig et al., 2014). Laboratory investigations such as complete blood counts (CBC), tumor marker tests, flow cytometry, circulating tumor cell tests, urine tests, and biopsies are examples of laboratory testing for cancer that can be used to detect abnormalities or cancer cells, which can help in the identification of patients who are at risk of developing cancer or already have cancer (Fig. 1).

3.1 Radiological Investigations to Detect Neoplasia

Several imaging technologies, including CT, MRI and ultrasound, are used to diagnose malignancies.

Ultrasonography An image of the internal organs can be created by using a transducer to send sound waves into the body, which are then reflected from tissues and organs. Ultrasonography (USG) allows the detailed imaging of the solid organs and the identification of possible morphologic changes that may be considered as a sign of malignancy. In order to provide a clear diagnosis, healthcare professionals require certain data, such as the presence of an abnormality in target tissue, the size of the organ, blood flow, and blood vessels, which guides the radiologists in narrowing

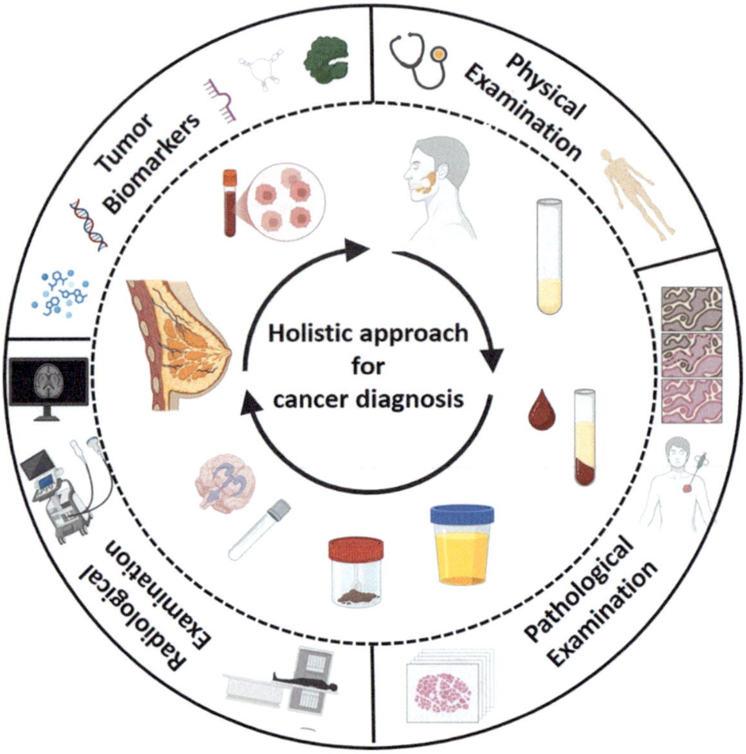

Fig. 1 Holistic approaches being used for the screening and diagnosis of cancer. It includes physical examinations, pathological examination (biopsy followed by histopathology and immunohistochemistry), radiographic examination (X-Rays, USG, CT scan, PET scan), and laboratory tests for tumor biomarkers. These biomarkers can be assessed in biofluids such as blood, serum, plasma, urine, saliva, cerebrospinal fluid, stool, nipple discharge, circulatory tumor cells, and in cancer tissues. (Created using Biorender)

down the possibility of a tumor being malignant. In addition to showing tumors, USG can direct medical professionals doing ultrasound guided fine needle aspiration cytology and biopsies. Furthermore, contrast-enhanced ultrasound (CEUS) has shown utility in vascular ultrasound, characterization of hepatic masses and breast lesions (Koutras et al., 2023). In addition, USG is often routinely performed along with mammography to increase the sensitivity of screening in women with non-dense breasts.

Computed Tomography (CT) SCAN Computer-controlled X-rays are used in computed tomography (CT) scans, often known as CAT scans, to produce three-dimensional images of the body, but a CT scan and a radiograph display distinct kinds of data. A plain radiograph is typically two-dimensional, but a skilled radiologist can use it to acquire an idea of the approximate three-dimensional location of a tumor. A radiograph of the arm or chest looks all over the body, but it cannot determine how deep anything is. Through imaging and examining multiple three-dimensional slices of the body, such as bread slices, a physician can determine not only whether a tumor is there but also its approximate depth within the body. Because the data regarding how many X-rays are going through a body is gathered on a computer rather than just a flat piece of film, a CT scan can be three-dimensional. Compared to a simple radiograph, the data from a CT scan can be improved to be more colorful. To make it easier to see the boundaries between organs or between organs and tumors on plain radiographs and CT scans, the patient may be given a contrast agent to drink or by injection.

Low doses of radioactive materials associated with chemicals that are utilized by body cells or that adhere to tumor cells are employed in molecular and nuclear imaging. It is possible to track the radioactive materials in the body to determine their concentration and location by using specialized detecting equipment. Positron emission tomography (PET) scans produce digital pictures of chemical alterations in tissue, including the metabolism of sugar (Galieni et al., 2021). Usually, a mixture of modified glucose that has been radioactively labeled is injected into the patient. Because cancer cells absorb glucose more readily than normal bodily tissues, the radioactive sugar can aid in tumor location. The distribution of the radiolabeled glucose in the tumor and body is found using a PET scanner. The ability to distinguish between normal and diseased tissues is enhanced when a CT scan and PET pictures are matched. This data is converted by a computer into the pictures that a radiologist interprets. When assessing if a mass is malignant, PET scans may be useful. PET scans, however, are more accurate in identifying larger and more aggressive tumors than they are in identifying less aggressive malignancies or tumors smaller than 8 mm, or half of a thumbnail. The smallest tumor mass detected by PET is continuously getting smaller. When other imaging methods yield normal results, they might also be able to identify malignancy. When assessing and staging recurrent disease—cancer that has returned—PET scans may be useful. PET scans are starting to be widely used to determine whether a treatment is effective, namely whether tumor cells are dying and need less sugar as a result. PET-CT shows better accuracy in the detection of malignancies in the thyroid, head and neck, lung, breast, esophageal, colorectal, melanoma, lymphoma, and solitary pulmonary nodules. Single photon emission computed tomography (SPECT) is like PET in that it records data using radioactive tracers and a scanner, which a computer then forms into two- or three-dimensional images. A little quantity of a radioactive substance is injected into a vein, and precise pictures of the parts of the body where the radioactive material is absorbed by the cells are created using a scanner. SPECT can provide details on the body's chemical reactions (metabolism) and blood flow to tissues. This process involves attaching a radioactive material to antibodies, which are proteins that identify and adhere to tumor cells. The antibodies will adhere to any tumor that may be present. The location of the tumor and the detection of the radioactive material can then be determined by a SPECT scan.

Magnetic Resonance Imaging To get tissues to release radio waves of their own, MRI uses radio waves in the presence of a strong magnetic field that surrounds the MRI machine's opening where the patient rests. A picture of the body's organs can be shown on a computer screen because different tissues, including tumors, send signals that vary in intensity depending on their chemical composition. Although MRI is sometimes more sensitive than CT scans at differentiating soft tissues, both may create three-dimensional images of parts of the body.

4 Pathological Examinations

Histopathology Histopathology is vital to the diagnosis, prognosis, and treatment of illness. These days, it is enhanced by molecular methods that enable the identification of the molecular "lesions" that underlie the onset of illness. Developing targeted therapeutics for various cancer types is crucial; early detection and effective therapy can be facilitated by a deeper comprehension of the pathophysiology and clinical characteristics. During a workup for several issues, including cancer, histopathology is a crucial diagnostic tool. It is widely acknowledged as the most reliable method for neoplasia diagnosis (Tseng et al., 2023). Taking a sample of tissue for microscopic examination is known as a biopsy. The only way to know for sure if an atypical spot is cancer or not is to do this. To learn more about the cancer, one might potentially get a biopsy. This includes the kind of cancer and its rate of growth.

Immunohistochemistry One of the most potent tools in the diagnostic surgical pathologist's toolbox is immunohistochemistry (IHC), which uses monoclonal and polyclonal antibodies to identify antigens in tissue sections. To ascertain the tissue distribution of an antigen of interest in both health and illness, immunohistochemistry (IHC) is a crucial use of both monoclonal and polyclonal antibodies (Mebratie & Dagnaw, 2024). Because certain tumor antigens are expressed de novo or are upregulated in some tumors, it is frequently utilized for cancer diagnosis. In pathology, IHC is crucial, especially in the subspecialties of hematopathology, neuropathology, and oncologic pathology. The availability of biopsies is necessary for IHC; these are cut into sections using a microtome and subsequently treated with the proper antibody. A marker that is directly attached to the primary antibody or to a suitable secondary antibody, such as a fluorescent dye, enzyme, radioactive element, or colloidal gold, makes the site of antibody binding visible under a standard or fluorescent microscope (Ghahremani et al., 2022).

5 Tumor Biomarkers

Tumor biomarkers are the measurable biomolecules that can clearly differentiate between the normal and dysregulated processes involved in the development of the tumor thereby help in the diagnosis of cancer. They are important for screening,

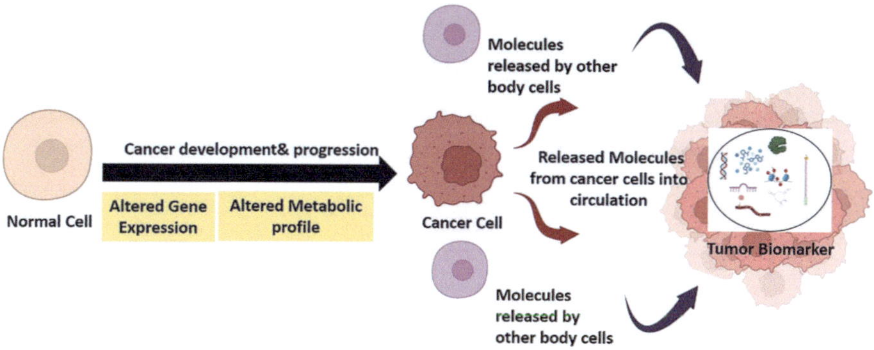

Fig. 2 Cancer cells exhibit differential gene expression and metabolic profile compared to normal healthy cells. Tumor biomarkers are therefore molecules released into the circulation by cancer cells or other body cells in response to malignancy. (Figure is created using Biorender)

early detection, diagnostic confirmation, prognosis, therapeutic response prediction, disease progression, and recurrence monitoring (Sharma, 2009).

There is multiple evidence suggesting the altered gene expression, metabolic and proteomic profile during the onset and progression of cancer cells compared to normal or healthy cells. Thus, the molecules released by cancer cells or other body cells in response to cancer are secreted into the circulation and serve as tumor biomarkers (Fig. 2). For example, alpha-fetoprotein (AFP), carcinoembryonic antigen (CEA), and human chorionic gonadotropin (hCG), which are oncofetal antigens present in the normal fetus and expressed in minute concentrations in healthy tissues, but their expression is upregulated in some malignancies and are known to serves as a tumor biomarker. These markers can be proteins (like an enzyme or receptor), nucleic acids (like a microRNA or other non-coding RNA), antibodies, and peptides (Mahadevarao Premnath & Zubair, 2025). Depending on their chemical structures, tumor biomarkers may be simple molecules, such as catecholamines or well-characterized proteins, such as hormones, enzymes, or gene products; or more heterogeneous glycoproteins or mucins, such as carbohydrate antigen 125 (CA 125).

Biomarkers can be assessed either in biofluids such as blood, serum, plasma, urine, saliva, cerebrospinal fluid, stool, or nipple discharge or can be derived from tissue, which necessitates biopsy or specialized imaging (Fig. 3). Recently, there has been interest in using molecular biomarkers in biofluids to diagnose, predict a drug response in a cancer patient and for clinical research due to their non-invasive features. For example, salivary mRNA such as KRAS, MBD3L2, ACRV1, and DPM1 have been well documented in studies to be highly specific biomarkers for the detection of pancreatic cancer (Zhang et al., 2010). Additionally, a few salivary proteins, such as Calprotectin, AZGP1, and HP, have been reported to be highly sensitive and specific for lung cancer detection. In mouth and throat malignancies, salivary DNA can also identify mutations in the genes PI3K, CDKN2A, FBXW7, HRAS, and KRAS (Liu et al., 2017). Since proteomic indicators are more relevant to the development and progression of tumors because they alter molecular

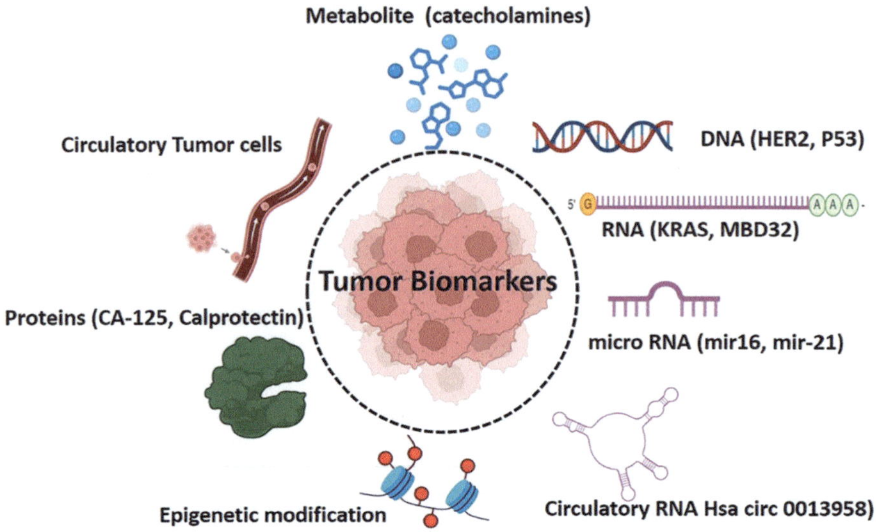

Fig. 3 Types of tumor biomarkers. Depending on the chemical nature, these markers can be DNA, RNA, miRNA, small circular RNA, proteins, epigenetic modifications, circulatory tumor cells, or cellular metabolites. (Created using Biorender)

pathways and activities in both malignant and normal cells, serum-derived single proteins make up most FDA-approved cancer biomarkers, which are now being used in clinical settings (Fig. 3). For example, testicular cancer is staged using HCG, AFP, and LDH. Cytokeratins and HER2/NEU expression is being used to enhance the prognosis of breast cancer (Wang & Xu, 2019). The markers, tetraspanin 8, CD171, and CD151, can clearly distinguish lung cancer patients of all subgroups from healthy people (Sandfeld-Paulsen et al., 2016). This field is growing enormously, and new emerging biomarkers are being investigated; however, there is still a pressing need to identify a suitable marker, which should be tissue-specific and can clearly demonstrate the onset and progression of the cancer in stage-specific manner. Recently plasma biomarker proteins have been discovered that could help in bladder cancer early detection. Compared to the healthy group, patients with low-grade bladder cancer have been shown to have a greater level of haptoglobin, which raises the possibility that this protein could play an important role in the early phases of bladder carcinogenesis (Nedjadi et al., 2020). For pancreatic cancer detection, several body fluids, such as bile, pancreatic juice, urine, and fluid from pancreatic cysts, have been shown to include proteins from pancreatic cancer, namely, high-mobility group box 1, CA19-9, osteopontin, hepatocyte growth factor, etc. which hold enormous promise as practical biomarkers (Afify et al., 2022; Al-Shaheri, 2025; Jin et al., 2025).

Apart from molecular markers, the proteins that are overexpressed or mutated and have differential gene expression during the different stages of cancer are the genetic reason for cancer development and are also being used as genetic markers for cancer diagnosis. For example, EGFR in lung, colorectal, and head and neck cancers; KRAS in lung, pancreatic, and colorectal adenocarcinoma; BRAF in melanoma and colorectal cancers; ALK in some NSCLC instances; HER2 in breast and gastric cancers; and PSA as a prostate cancer (Passaro et al., 2024). Single nucleotide polymorphisms (SNPs) in several genes, such as XRCC1, ATM, and p53 in lung, head, and neck cancers; CYP1A1, RAD1, BRCA1 and BRCA2 in breast cancer; and PGS2 in lung cancer, are also being used as genetic markers (Gudur et al., 2023; Karaağaç et al., 2023). Thus, differential expression of these genetic markers, mutations in tumor suppressor genes and mismatch-repair genesf and oncogene alterations have been linked to cancer, particularly metastatic cancer (Herceg & Hainaut, 2007). Mutations in the tumor suppressor gene, p53 has been reported in over half of sporadic malignancies. Moroever, alterations in the KRAS and APC oncogenes indicate metastatic progression (Aghagolzadeh & Radpour, 2016). A hereditary TP53 mutation (Li-Fraumeni syndrome) increases the risk of many malignancies. Recently, mutations in mitochondrial DNA have also been suggested as emerging diagnostic markers for several types of cancer by several scientists (Chatterjee et al., 2011; Vikramdeo et al., 2022). Haplotype analysis is currently being used to investigate the mitochondrial inheritance pattern in cancer patients (Vikramdeo et al., 2023). For screening of such genes, it can be evaluated using IHC (Immunohistochemistry), FISH (Fluorescence in situ hybridization), gene-specific PCR (Polymerase chain reaction), and NGS (Next Generation sequencing). These biomarkers are essential for the diagnosis, categorization, and choice of cancer treatment.

Additionally, the circulating tumor cells (CTCs) can be simple yet powerful indicators in the field of cancer diagnosis (Fig. 3). Numerous clinical study findings indicate that CTC is a significant prognostic factor in patients with metastatic lung, prostate, and breast cancer (Strati et al., 2021). The molecular diagnosis of CTCs can be done using chromatin immunoprecipitation sequencing, RNA in situ hybridization, DNA sequencing, and RNA sequencing (Lin et al., 2021). CTC provides an early and precise indication of the disease's progression and survival for patients receiving systemic therapy for metastatic breast cancer (Kotsifaki et al., 2024). Compared to conventional tumor markers (such as CA27-29), cancer treatment targets (CTTs) are more accurate prognostic indicators (Das et al., 2024). The selection of an effective treatment regimen can also be influenced by the prevalence of therapeutic targets on CTCs, and the results of treatment can be evaluated following the first medication cycle. According to reports, the frequency of CTCs can predict a patient's survival at different stages of treatment for metastatic breast cancer. In patients with metastatic prostate cancer, CTC counts have been shown to be a reliable predictor of prognosis and responsiveness to treatment. EpCAM-positive CTCs have a prognostic relevance to the detection of hepatocellular carcinoma, according to Schulze et al.

Tumor-infiltrating lymphoid cell histopathology analysis has also been validated as a reliable and prognostically valuable biomarker. By distinguishing between self-antigens and foreign antigens, the immune system can maintain immunological tolerance and trigger defensive immunity against foreign antigens. The immune cell count in scanned tissue has already been used to find trustworthy and clinically valuable biomarkers across a number of tumor entities, including colorectal cancer and liver metastases. The most common immune cells associated with clinical consequences at the tumor site are T lymphocytes and macrophages (Greene Carbone et al., 2024). T cells maintain self-tolerance, which helps prevent immunological diseases (Peña-Romero & Orenes-Piñero, 2022). As per the literature, cancer patients' inadequate immunological responses to tumor antigens were associated with elevated production of regulatory T-cells (T-regs), suggesting that this may encourage immune dysregulation and tumor growth (Huppert et al., 2022). Numerous T-regs have already been found in the circulation or within the tumor in patients with liver, cutaneous, breast, pancreatic, and lung malignancies (Ohue & Nishikawa, 2019; Togashi et al. 2019).

Tumor microRNA profiles can also be used as a marker for cancer diagnosis. It is feasible to detect cancer-associated microRNA markers in physiological fluids, enabling less intrusive techniques for cancer microRNA monitoring. In 2002, the first research on microRNA dysregulation in cancer was published. Chronic lymphocytic leukemia patients have been shown to have high levels of two microRNAs, miR-16 and miR-15 (Duroux-Richard et al., 2022). Another study revealed the higher expression of miRNA-483-3p in pancreatic ductal adenocarcinoma compared to healthy controls (Shao et al., 2021). Higher expressions of miRNA-483-3p and miRNA-21 expression have been reported in plasma in the case of intraductal papillary mucinous neoplasms and in liver and lymph node metastases, respectively (Das et al., 2024). miRNAs control a variety of targets and can be either tumor suppressors or oncogenes. MiR-18a, miR-205-5p, and miR-155 inhibit the proliferation, invasion, and migration of colorectal cancer cells, in contrast to miR494, miR-17-3p, and miR-598, which stimulate these processes (Pal et al., 2021; Vychytilova-Faltejskova & Slaby, 2022).

Circular RNAs, non-coding RNAs with a closed-loop structure that play a crucial role in gene regulation, have also been upregulated in cancer. All lung adenocarcinomas displayed higher amounts of Hsa circ 0013958, with 20 circRNAs downregulated and 39 upregulated, according to Zhu et al. (Zhu et al., 2017). The findings suggest that Hsa circ 0013958 could be a potent non-invasive marker for the early diagnosis of lung cancer. Using real-time PCR, researchers searched for Hsa circ 0013958 levels in lung adenocarcinoma (LAC) (Das et al., 2024). The LAC cell line was found to have higher amounts of Hsa circ 0013958 than the healthy human bronchus epithelial cell line. Zhu et al. discovered a positive association between the expression level of Hsa circRNA 101996 and tumor size, TNM staging, and lymphovascular invasion in cervical cancer. Moreover, overexpression of Hsa circRNA 101996 is linked to a poorer prognosis. They discovered that miR-8075, which is controlled by Hsa circRNA 101996, reduces TPX2 expression and promotes the growth and metastasis of cervical cancer. Epigenetic modifications are also being considered as potent cancer biomarkers because they are prevalent for specific genes, durable, and detectable with minimal invasiveness. DNA methyltransferases, which add methyl groups to the cytosine groups of DNA, have been discovered to be changed in cancer cells in a number of studies. The local CpG island promoter is hypermethylated, which silences the tumor suppressor genes and results in gene alterations in those genes. The NKX2-6, SPAG6, PER1, and ITIH5 genes were shown to be methylated in serum from breast cancer patients. For instance, hypermethylation of the promoter p16 in blood DNA has been linked to recurrent colorectal cancer. Methylation of the RASSF1A and p16Ink4 genes has been associated with a 15-fold increase in the comparative risk of lung cancer. The methylation status of several genes in clinical specimens may be a potential non-invasive technique for identifying smokers at risk of lung cancer. Ibanez de Caceres et al. found that the RASSFIA or BRCA1 gene's DNA promoter was hypermethylated in 68% of ovarian cancer tissue. Because it inhibits tumor suppressor genes and activates oncogenes, histone acetylation plays a role in the growth and metastasis of malignancies. Demonstrated that, in comparison to samples of healthy tissue, prostate cancer cell lines exhibit higher histone deacetylase activity after lower levels of histone H3 acetylation at specific sites. Several genetic biomarkers are displayed and contrasted.

6 Limitations

To achieve effective and accurate diagnosis, its early detection is very important and thus cancer diagnostic procedures are continuously being developed and altered. It requires multidisciplinary cooperation between radiologists, pathologists, and clinicians to reduce the time required to complete these diagnostic procedures as well as to reach a conclusive diagnosis and logical treatment plan. Typically, imaging is the first step in the cancer diagnostic process, followed by a tissue biopsy. Because of this two-step process, radiologists and pathologists must constantly communicate to make sure that the biopsy results directly match the images collected. The picture archiving and communication system (PACS) and other systems incorporated into the various imaging modalities are another area of cancer diagnostics that might be improved to be more optimally efficient. It is advantageous because it substitutes traditional film with digital images, makes it easy to save and access images from numerous imaging modalities, and enables the simultaneous viewing of many images. Since PACS is such a helpful system, it can store radiology reports in a digitally organized format for convenient access, improve image visualization and interpretation because images can be rotated and enlarged, and it will be more economic as it will eliminate the need to print films.

Furthermore, nanotechnology has emerged as a compelling method for cancer diagnosis and treatment because of its capacity to distribute many ligands, target receptors, and other biological elements. Liposomes, iron oxide, dendrimers, quantum dots, gold nanoparticles, and carbon nanotubes are among the many kinds of nano-formulations used for diagnostic purposes in optical, MRI, PET, CT, SPECT, and X-ray procedures. These nanoparticles can be employed as contrast agents in MRI and photoacoustic tomography imaging, and they can be actively or passively targeted into the tumor. Using SPECT imaging, DTPA-labeled pegylated liposomes have been utilized to image many cancer types, including lung, head and neck, and breast cancer. PET imaging uses F-liposomes, while MRI imaging uses gadolinium-loaded nanoparticles. Rapid advancements in nanotechnology could be crucial for tumor imaging and treatment. A potential future roadmap that expands the field of early cancer detection and diagnostics by creating technologies that target essential cellular processes like hypoxia, angiogenesis, and apoptosis and are implicated in the treatment of refractory tumors will require a great deal of interdisciplinary information sharing between clinical and basic research.

7 Applications of AI and ML in Cancer Diagnosis

Based on the limitations, advances in computer engineering and statistics over the years have encouraged many scientists to apply computational methods like multivariate statistical analysis to analyse the prognosis of the disease. As discussed above, cancer is an aggressive disease with a low median survival rate, and ironically, the treatment process is lengthy and very costly due to its high recurrence and mortality rates. Accurate early diagnosis and prognosis prediction of cancer are essential to enhance the patient's survival rate. However, every procedure has its significant characteristics and limitations. Furthermore, cancer prediction performance has increased to unprecedented levels as artificial intelligence (AI), particularly machine learning, and has been increasingly popular in clinical cancer research in recent years. The scientific community has taken notice of AI due to its deep learning and problem-solving capabilities. In his study "Computing Machinery and Intelligence" from the nineteenth century, Alan Turing introduced the concept of artificial intelligence (Turing, 1950). The Al-based technology era is just getting started. Ten years ago, there were very few publications on AI and medical imaging, but in the years that followed, that number skyrocketed (Pesapane et al., 2018). Numerous dynamic applications of AI exist, including picture interpretation and categorization, data organization, information storage, information mining, and many more. AI is anticipated to greatly help the radiologist in enhancing the diagnosis specificity because of its broad application in biomedical imaging technology (Recht & Bryan, 2017). In summary, USG, X-ray, CT, MRI, PET, histopathology, biopsy, and immunohistochemistry will undoubtedly continue to be the most common methods of diagnosing cancer for many years to come. However, the scientific community is becoming increasingly interested in utilizing computational input and support resulting in a major technological revolution that will enable real-time diagnosis and prediction of human health-related problems improving, the current clinical approaches to cancer treatment (Iqbal et al., 2021).

Acknowledgement AK would like to thank the funding support from the Indian Council of Medical Research (ICMR 5/13/93/2020/NCD-III), and RN would like to thank the funding support from Manipal University Jaipur for the Enhanced Seed Grant under the Endowment Fund (No. E3/2023-24/QE-04-05) and DST-FIST project (DST/2022/1012) from Govt. of India to Department of Biosciences, Manipal University Jaipur.

References

Afify, S. M., Hassan, G., Ishii, H., Monzur, S., Nawara, H. M., Osman, A., Abu Quora, H. A., Sheta, M., Zahra, M. H., Seno, A., & Seno, M. (2022). Functional and molecular characters of cancer stem cells through development to establishment. In S. Tanabe (Ed.), *Cancer stem cell markers and related network pathways* (pp. 83–101). Springer. https://doi.org/10.1007/978-3-031-12974-2_4

Aghagolzadeh, P., & Radpour, R. (2016). New trends in molecular and cellular biomarker discovery for colorectal cancer. *World Journal of Gastroenterology, 22*(25), 5678–5693.

Al-Shaheri, F. N. S. (2025). *Early detection of tumour recurrence in pancreatic cancer patients by the detection of specific microRNA and protein signatures in peripheral blood* [Dissertation]. https://doi.org/10.11588/heidok.00036169

Bray, F., Laversanne, M., Sung, H., Ferlay, J., Siegel, R. L., Soerjomataram, I., & Jemal, A. (2024). Global cancer statistics 2022: GLOBOCAN estimates of incidence and mortality worldwide for 36 cancers in 185 countries. *CA: A Cancer Journal for Clinicians, 74*(3), 229–263. https://doi.org/10.3322/caac.21834. Epub 2024 Apr 4. PMID: 38572751.

Cao, W., Qin, K., Li, F., & Chen, W. (2024). Comparative study of cancer profiles between 2020 and 2022 using global cancer statistics (GLOBOCAN). *Journal of the National Cancer Center, 4*(2), 128–134. https://doi.org/10.1016/j.jncc.2024.05.001

Chatterjee, A., Dasgupta, S., & Sidransky, D. (2011). Mitochondrial subversion in cancer. *Cancer Prevention Research (Philadelphia, Pa.), 4*(5), 638–654. https://doi.org/10.1158/1940-6207.CAPR-10-0326. PMID: 21543342; PMCID: PMC3298745.

Das, S., Dey, M. K., Devireddy, R., & Gartia, M. R. (2024). Biomarkers in cancer detection, diagnosis, and prognosis. *Sensors, 24*(1), Article 1. https://doi.org/10.3390/s24010037

Didkowska, J., Wojciechowska, U., Michalek, I. M., & Caetano dos Santos, F. L. (2022). Cancer incidence and mortality in Poland in 2019. *Scientific Reports, 12*(1), 10875. https://doi.org/10.1038/s41598-022-14779-6

Duroux-Richard, I., Gagez, A.-L., Alaterre, E., Letestu, R., Khalifa, O., Jorgensen, C., Leprêtre, S., Tchernonog, E., Moreaux, J., Cartron, G., & Apparailly, F. (2022). miRNA profile at diagnosis predicts treatment outcome in patients with B-chronic lymphocytic leukemia: A FILO study. *Frontiers in Immunology, 13*. https://doi.org/10.3389/fimmu.2022.983771

Galieni, A., D'Ascenzo, N., Stagnari, F., Pagnani, G., Xie, Q., & Pisante, M. (2021). Past and future of plant stress detection: An overview from remote sensing to positron emission tomography. *Frontiers in Plant Science, 11*. https://doi.org/10.3389/fpls.2020.609155

Ghahremani, P., Li, Y., Kaufman, A., Vanguri, R., Greenwald, N., Angelo, M., Hollmann, T. J., & Nadeem, S. (2022). Deep learning-inferred multiplex immunofluorescence for immunohistochemical image quantification. *Nature Machine Intelligence, 4*(4), 401–412. https://doi.org/10.1038/s42256-022-00471-x

Giaccone, G. (2011). Grand challenges in oncology. *Frontiers in Oncology, 1*, 26. https://doi.org/10.3389/fonc.2011.00026. PMID: 22649758; PMCID: PMC3355910.

Greene Carbone, F., Russo, C., Colamatteo, A., La Rocca, C., Fusco, C., Matarese, A., Procaccini, C., & Matarese, G. (2024). Cellular and molecular signaling towards T cell immunological self-tolerance. *Journal of Biological Chemistry, 300*(4), 107134. https://doi.org/10.1016/j.jbc.2024.107134

Greene, G., Griffiths, R., Han, J., Akbari, A., Jones, M., Lyons, J., Lyons, R. A., Rolles, M., Torabi, F., Warlow, J., Morris, E. R. A., Lawler, M., & Huws, D. W. (2022). Impact of the SARS-CoV-2 pandemic on female breast, colorectal and non-small cell lung cancer incidence, stage and healthcare pathway to diagnosis during 2020 in Wales, UK, using a national cancer clinical record system. *British Journal of Cancer, 127*(3), 558–568. https://doi.org/10.1038/s41416-022-01830-6

Gudur, A. K., Gudur, R. A., Bhosale, S. J., Kale, S. R., & Datkhile, K. D. (2023). Single nucleotide polymorphisms in DNA repair genes (XRCC1, XRCC2, XRCC3) and their association with radiotherapy toxicity among head and neck cancer patients: A study from South-Western Maharashtra. *Asian Pacific Journal of Cancer Prevention, 24*(9), 3049–3057. https://doi.org/10.31557/APJCP.2023.24.9.3049. PMID: 37774056; PMCID: PMC10762732.

Herceg, Z., & Hainaut, P. (2007). Genetic and epigenetic alterations as biomarkers for cancer detection, diagnosis and prognosis. *Molecular Oncology, 1*(1), 26–41. https://doi.org/10.1016/j.molonc.2007.01.004. Epub 2007 Mar 14. PMID: 19383285; PMCID: PMC5543860.

Huppert, L. A., Green, M. D., Kim, L., Chow, C., Leyfman, Y., Daud, A. I., & Lee, J. C. (2022). Tissue-specific Tregs in cancer metastasis: Opportunities for precision immunotherapy. *Cellular & Molecular Immunology, 19*(1), 33–45. https://doi.org/10.1038/s41423-021-00742-4

Iqbal, M. J., Javed, Z., Sadia, H., et al. (2021). Clinical applications of artificial intelligence and machine learning in cancer diagnosis: Looking into the future. *Cancer Cell International, 21*, 270. https://doi.org/10.1186/s12935-021-01981-1

Jin, D., Khan, N., Gu, W., Lei, H., Goel, A., & Chen, T. (2025). Informatics strategies for early detection and risk mitigation in pancreatic cancer patients. *Neoplasia (New York, N.Y.), 60*, 101129. https://doi.org/10.1016/j.neo.2025.101129

Karaağaç, M., Geredeli, Ç., Yıldırım, M. S., Altınok, T., Dede, İ., İnal, A., Zamani, A. G., Kaya, B., Demirkazık, A., & Artaç, M. (2023). The XRCC1 and TP53 gene polymorphisms are associated with advanced-stage disease and early distant metastasis at diagnosis in non-small cell lung cancer. *Journal of Cancer Research and Therapeutics, 19*(5), 1248–1254. https://doi.org/10.4103/jcrt.jcrt_1657_21. PMID: 37787291.

Kotsifaki, A., Maroulaki, S., & Armakolas, A. (2024). Exploring the immunological profile in breast cancer: Recent advances in diagnosis and prognosis through circulating tumor cells. *International Journal of Molecular Sciences, 25*(9), Article 9. https://doi.org/10.3390/ijms25094832

Koutras, A., Perros, P., Prokopakis, I., Ntounis, T., Fasoulakis, Z., Pittokopitou, S., Samara, A. A., Valsamaki, A., Douligeris, A., Mortaki, A., Sapantzoglou, I., Katrachouras, A., Pagkalos, A., Symeonidis, P., Palios, V. C., Psarris, A., Theodora, M., Antsaklis, P., Makrydimas, G., Chionis, A., Daskalakis, G., & Kontomanolis, E. N. (2023). Advantages and limitations of ultrasound as a screening test for ovarian cancer. *Diagnostics (Basel), 13*(12), 2078. https://doi.org/10.3390/diagnostics13122078. PMID: 37370973; PMCID: PMC10297553.

Lin, D., Shen, L., Luo, M., et al. (2021). Circulating tumor cells: Biology and clinical significance. *Signal Transduction and Targeted Therapy, 6*, 404. https://doi.org/10.1038/s41392-021-00817-8

Liu, H.-J., Guo, Y.-Y., & Li, D.-J. (2017). Predicting novel salivary biomarkers for the detection of pancreatic cancer using biological feature-based classification. *Pathology – Research and Practice, 213*(4), 394–399. https://doi.org/10.1016/j.prp.2016.09.017

Mahadevarao Premnath, S., & Zubair, M. (2025). Laboratory evaluation of tumor biomarkers. [Updated 2023 Oct 28]. In *StatPearls [Internet]*. StatPearls Publishing. Available from: https://www.ncbi.nlm.nih.gov/books/NBK597378/

Mebratie, D. Y., & Dagnaw, G. G. (2024). Review of immunohistochemistry techniques: Applications, current status, and future perspectives. *Seminars in Diagnostic Pathology, 41*(3), 154–160. https://doi.org/10.1053/j.semdp.2024.05.001

Nedjadi, T., Benabdelkamal, H., Albarakati, N., Masood, A., Al-Sayyad, A., Alfadda, A. A., Alanazi, I. O., Al-Ammari, A., & Al-Maghrabi, J. (2020). Circulating proteomic signature for detection of biomarkers in bladder cancer patients. *Scientific Reports, 10*(1), 10999. https://doi.org/10.1038/s41598-020-67929-z. PMID: 32620920; PMCID: PMC7335182.

Ohue, Y., & Nishikawa, H. (2019). Regulatory T (Treg) cells in cancer: Can Treg cells be a new therapeutic target? *Cancer Science, 110*(7), 2080–2089. https://doi.org/10.1111/cas.14069. Epub 2019 Jun 18. PMID: 31102428; PMCID: PMC6609813.

Pal, S., Garg, M., & Pandey, A. K. (2021). Biomarkers as putative therapeutic targets in colorectal cancer. In G. P. Nagaraju, D. Shukla, & N. K. Vishvakarma (Eds.), *Colon cancer diagnosis and therapy: Volume 1* (pp. 123–177). Springer. https://doi.org/10.1007/978-3-030-63369-1_8

Passaro, A., Al Bakir, M., Hamilton, E. G., Diehn, M., André, F., Roy-Chowdhuri, S., Mountzios, G., Wistuba, I. I., Swanton, C., & Peters, S. (2024). Cancer biomarkers: Emerging trends and clinical implications for personalized treatment. *Cell, 187*(7), 1617–1635. https://doi.org/10.1016/j.cell.2024.02.041. PMID: 38552610; PMCID: PMC7616034.

Peña-Romero, A. C., & Orenes-Piñero, E. (2022). Dual effect of immune cells within tumour microenvironment: Pro- and anti-tumour effects and their triggers. *Cancers, 14*(7), Article 7. https://doi.org/10.3390/cancers14071681

Pesapane, F., Codari, M., & Sardanelli, F. (2018). Artificial intelligence in medical imaging: Threat or opportunity? Radiologists again at the forefront of innovation in medicine. *European Radiology Experimental, 2*(1), 35.

Recht, M., & Bryan, R. N. (2017). Artificial intelligence: threat or boon to radiologists? *Journal of the American College of Radiology, 14*(11), 1476–1480.

Rosenzweig, M. Q., Gardner, D., & Griffith, B. (2014). The history and physical in cancer care: a primer for the oncology advanced practitioner. *Journal of the Advanced Practitioner in Oncology, 5*(4), 262–268. https://doi.org/10.6004/jadpro.2014.5.4.3. PMID: 26110070; PMCID: PMC4457181.

Sandfeld-Paulsen, B., Jakobsen, K. R., Bæk, R., Folkersen, B. H., Rasmussen, T. R., Meldgaard, P., et al. (2016). Exosomal proteins as diagnostic biomarkers in lung cancer. *Journal of Thoracic Oncology, 11*(10), 1701–1710. https://doi.org/10.1016/j.jtho.2016.05.034

Shao, H., Zhang, Y., Yan, J., Ban, X., Fan, X., Chang, X., Lu, Z., Wu, Y., Zong, L., Mo, S., Yu, S., & Chen, J. (2021). Upregulated microRNA-483-3p is an early event in pancreatic ductal adenocarcinoma (PDAC) and as a powerful liquid biopsy biomarker in PDAC. *Oncotargets and Therapy, 14*, 2163–2175. https://doi.org/10.2147/OTT.S288936

Sharma, S. (2009). Tumor markers in clinical practice: General principles and guidelines. *Indian Journal of Medical and Paediatric Oncology, 30*(1), 1–8. https://doi.org/10.4103/0971-5851.56328. PMID: 20668599; PMCID: PMC2902207.

Strati, A., Nikolaou, M., Georgoulias, V., & Lianidou, E. S. (2021). RNA-based CTC analysis provides prognostic information in metastatic breast cancer. *Diagnostics, 11*(3), Article 3. https://doi.org/10.3390/diagnostics11030513

Tseng, L. J., Matsuyama, A., & MacDonald-Dickinson, V. (2023). Histology: The gold standard for diagnosis? *Canadian Veterinary Journal, 64*(4), 389–391. PMID: 37008634; PMCID: PMC10031787.

Togashi Y., Shitara K. & Nishikawa H. (2019). Regulatory T cells in cancer immunosuppression - implications for anticancer therapy. Nature Reviews Clinical Oncology, 16(6),356-371. PMID: 30705439.

Turing, A. M. (1950). Computing machinery and intelligence. In E. A. Feigenbaum (Ed.), *Computers and Thought*. McGraw-Hill.

Vikramdeo, K. S., Anand, S., Khan, M. A., et al. (2022). Detection of mitochondrial DNA mutations in circulating mitochondria-originated extracellular vesicles for potential diagnostic applications in pancreatic adenocarcinoma. *Science Reports, 12*, 18455. https://doi.org/10.1038/s41598-022-22006-5

Vikramdeo, K. S., Anand, S., Sudan, S. K., Pramanik, P., Singh, S., Godwin, A. K., Singh, A. P., & Dasgupta, S. (2023). Profiling mitochondrial DNA mutations in tumors and circulating extracellular vesicles of triple-negative breast cancer patients for potential biomarker development. *FASEB Bioadvance., 5*(10), 412–426. https://doi.org/10.1096/fba.2023-00070. PMID: 37810173; PMCID: PMC10551276.

Vychytilova-Faltejskova, P., & Slaby, O. (2022). Chapter 17—Pathophysiology roles and translational opportunities of miRNAs in colorectal cancer. In M. Negrini, G. A. Calin, & C. M. Croce (Eds.), *MicroRNA in human malignancies* (pp. 203–244). Academic. https://doi.org/10.1016/B978-0-12-822287-4.00007-4

Wang, J., & Xu, B. (2019). Targeted therapeutic options and future perspectives for HER2-positive breast cancer. *Signal Transduction and Targeted Therapy, 4*, 34. https://doi.org/10.1038/s41392-019-0069-2

Zhang, L., Farrell, J. J., Zhou, H., Elashoff, D., Akin, D., Park, N. H., Chia, D., & Wong, D. T. (2010). Salivary transcriptomic biomarkers for detection of resectable pancreatic cancer. *Gastroenterology, 138*(3), 949–57.e1–7. https://doi.org/10.1053/j.gastro.2009.11.010. Epub 2009 Nov 18. PMID: 19931263; PMCID: PMC2831159.

Zhu, X., Wang, X., Wei, S., Chen, Y., Chen, Y., Fan, X., Han, S., & Wu, G. (2017). hsa_circ_0013958: A circular RNA and potential novel biomarker for lung adenocarcinoma. *The FEBS Journal, 284*, 2170–2182. https://doi.org/10.1111/febs.14132

Computational Methods in Oncology

B. Aishwarya, J. Priyanka, Mummadi Shyam Kumar Reddy, R. Mohan, Rohit Saluja, Divya Biligere Shivanna, and Roopa S. Rao

Abstract

According to the World Health Organization, cancer is one of the leading causes of death globally, accounting for approximately ten million fatalities annually (2020). Cancer encompasses more than 100 distinct types, including breast, lung, skin, and blood cancers. These diseases are characterized by the uncontrolled growth of abnormal cells, which can invade nearby tissues and metastasize to other parts of the body. The complexity of cancer arises from the interplay of genetic, environmental, and lifestyle factors, such as smoking, diet, lack of physical activity, and exposure to carcinogens. Both inherited and acquired genetic mutations play significant roles in the development and progression of cancer, resulting in diverse clinical presentations and varying responses to treatment. Given this complexity, there is an urgent need for advancements in prevention strategies, early detection methods, and personalized interventions. Recent developments in artificial intelligence (AI), particularly machine learning (ML) and deep learning (DL), have revolutionized cancer care. These technologies facilitate earlier and more accurate cancer detection, prediction, and prognosis, thereby improving patient outcomes. This chapter explores the potential of AI-driven automation to address the complexities of cancer diagnosis and treatment, emphasizing its transformative role in enhancing global cancer care.

B. Aishwarya · J. Priyanka · M. S. K. Reddy · R. Mohan · D. B. Shivanna
Department of Computer Science and Engineering, Faculty of Engineering and Technology, MS Ramaiah University of Applied Sciences, Bangalore, India

R. Saluja (✉)
All India Institute of Medical Sciences, Hyderabad, India

R. S. Rao
Department of Oral Pathology, Faculty of Dental Sciences, MS Ramaiah University of Applied Sciences, Bangalore, India

R. Nema et al. (eds.), *Advances in Cancer Detection, Prediction, and Prognosis Using Artificial Intelligence and Machine Learning*,
https://doi.org/10.1007/978-981-96-9346-7_2

Keywords

Artificial intelligence · Cancer · Diagnosis · Machine learning

1 Introduction

Cancer is a significant global health challenge and remains one of the top causes of death worldwide. According to the World Health Organization (WHO), it was responsible for nearly ten million deaths in 2020, accounting for approximately one-sixth of all global fatalities. Rather than being a single disease, cancer encompasses over 100 distinct types, including breast, lung, skin, and blood cancers. These conditions are defined by the uncontrolled growth of abnormal cells, which can invade nearby tissues and spread to other parts of the body through a process known as metastasis. Extensive metastasis is a major factor contributing to cancer-related deaths (https://www.who.int/news-room/fact-sheets/detail/cancer).

Cancer involves a sequential progression in increasingly abnormal states of cellular condition, caused mainly by mutations in genes that constitute driver mutations for the cancers (Martínez-Jiménez et al., 2020). Figure 1 simplifies this pathway from normal cells to hyperplasia, dysplasia, and ultimately, malignant cancer. Hyperplasia is defined as an inappropriate increase in number of cells; however tissue organization has remained essentially normal (Kurman & Norris, 1987). Further accumulation of genetic mutations results in dysplasia, where atypical cellular morphology and disrupted tissue architecture become the hallmark (Steinbeck, 2004). Unchecked cell growth, invasiveness, and a loss of tissue integrity are characteristics of precancerous stages that culminate in cancer. In this process, driver

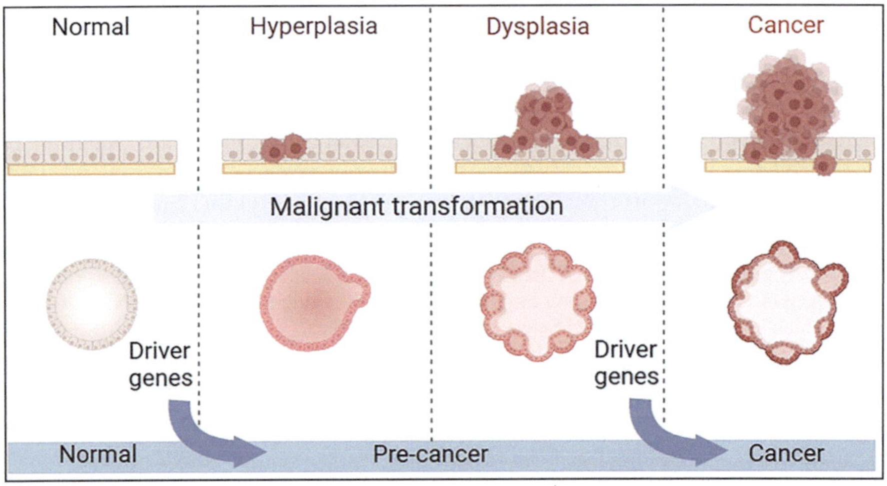

Fig. 1 Malignant transformation: The figure depicts the sequential progression of cells from a normal state to hyperplasia, dysplasia, and ultimately malignant cancer, highlighting the role of driver mutations in promoting uncontrolled growth and invasive behavior

genes are crucial because they provide the altered cells with growth benefits, which results in malignant transformation (Mareel & Leroy, 2003). Understanding transitions is crucial for diagnosing and intervening in cancer. Identifying precancerous alterations can prevent carcinogenesis and provide early detection (González & Agudo, 2012).

The complexity of cancer arises from its interwoven factors that are influenced by a combination of genetic predisposition, environmental exposures, and lifestyle choices (Korn et al., 2022). Factors like use of tobacco, dietary habits, physical inactivity, and over exposure to carcinogens have been indicated as the primary cause of development and the progression of various cancers. Also, genetic mutations which include inherited and acquired play a major role in the onset and progression of cancer in the body. Since the disease in itself is so distinct, it leads to the development of diverse clinical outcomes and responses to treatments (Irigaray et al., 2007).

Therefore, understanding the intricacy of the disease and developing effective prevention strategies, improving the early detection methods has become a matter of utmost importance. Advances in automation through machine learning (ML) and deep learning (DL) have become instrumental in revolutionizing cancer care. These automation techniques enhance early detection and accurate diagnoses of patients (Saba, 2020).

2 Oncology

Oncology is the branch of medicine that deals with the study, diagnosis, treatment, and prevention of cancer. The field is constantly evolving as new therapies and diagnostic techniques are being developed through ongoing research. Oncology includes multiple subfields like ***Medical oncology*** that deals with the treatment of cancer with medications including chemotherapy, immunotherapy, and targeted therapies. ***Radiation oncology*** deals with the use of radiation therapy to destroy cancer cells or shrink tumours. ***Surgical oncology*** deals with removal of tumors and surrounding tissues and this can sometimes include lymph nodes, thereby helping prevent the spread of cancer within the body (Sharma et al., 2023).

2.1 How Does Cancer Develop?

Cancer develops due to a combination of genetic mutations and cellular abnormalities that cause the disruption of the cell cycle, thereby causing the cells to grow uncontrollably and form tumors (Hanahan, 2022).This process is influenced by many factors like genetics, environmental exposures, and lifestyle choices.

2.2 What Are the Visible Symptoms of Cancer?

Visible symptoms in cancer can vary widely based on the site where the tumor cell arises from. However, few of these symptoms include lumps, unexplained weight loss, persistent fatigue, skin changes, prolonged painful cough etc. (Sung et al., 2021). Being aware of such symptoms and seeking medical evaluation is essential as early-stage detection can significantly improve the chances of successful treatment (Crosby et al., 2022).

When symptoms are noticed early, it makes it potentially easier for clinicians to handle the situation and also enables less aggressive treatment with better prognosis. However, neglecting these early visible symptoms can limit treatment options and potentially reduce survival outcomes (Siegel et al., 2022).

3 Types of Cancer

3.1 Breast Cancer

Breast cancer is the most frequently diagnosed cancer worldwide. Its occurrence is influenced by various risk factors, including genetic and hereditary predispositions. Female breast cancer has surpassed lung cancer as the most commonly diagnosed cancer globally. In 2020, an estimated 2.3 million cases of breast cancer were reported, accounting for about 11.7% of all newly diagnosed cancers. The overall five-year survival rate for breast cancer stands at 91.2%, meaning that 91 out of 100 individuals diagnosed with the disease are still alive five years later (Hong & Xu, 2022; Rommasi, 2023).

Various methods are employed to diagnose breast cancer, each with its unique approach. ***Mammogram***: This X-ray of the breast is often the first line of screening and diagnosis for breast cancer, especially effective for detecting abnormalities in dense breast tissue (Tan et al., 2017). ***Ultrasound***: Used to further investigate areas of concern found on a mammogram, ultrasound helps to differentiate between solid masses (which may be cancerous) and fluid-filled cysts (usually non-cancerous) (Geisel et al., 2018). ***Breast MRI (Magnetic Resonance Imaging)***: An MRI provides highly detailed images using magnetic fields, often recommended for high-risk patients or to assess the extent of cancer after diagnosis (Mann et al., 2019). **Biopsy**: A definitive way to diagnose breast cancer, a biopsy involves removing a small sample of breast tissue for examination under a microscope (Alimirzaie et al., 2019). Different types include the following: ***Fine Needle Aspiration (FNA)***: A thin needle removes a small sample. ***Core Needle Biopsy***: A larger needle removes a tissue core. ***Surgical Biopsy***: A surgeon removes a portion or the entire suspicious area.

Digital Breast Tomosynthesis (3D Mammography): This modern mammography technique produces a three-dimensional image of the breast, allowing for more precise detection of small or hidden cancers (Tagliafico et al., 2018). ***Molecular Breast Imaging (MBI)*** is a specialized imaging technique that uses a radioactive tracer to identify areas of aberrant cell activity in the breast. It is frequently utilized when other approaches are inconclusive (Huppe et al., 2018). ***PET Scan (Positron Emission Tomography)***: PET scans are used in cases of suspected advanced breast cancer to assess whether the cancer has spread to other parts of the body (Zangheri et al., 2004). Genetic Testing: In cases of family history or genetic susceptibility, testing for mutations in BRCA1, BRCA2, and other genes can assist assess risk, but it is not a diagnostic tool for active cancer (Schlich-Bakker et al., 2006).

Artificial intelligence (AI) and machine learning have led to major advances in breast cancer diagnosis, prediction, and prognosis. AI algorithms can analyze enormous datasets and uncover complicated patterns that will provide insights to revolutionize cancer care by employing deep learning techniques such as convolutional neural networks (CNNs) and other neural networks (Sharafaddini et al., 2024). AI-powered diagnostic systems for mammography, ultrasound, and MRI have demonstrated increased accuracy in breast cancer diagnosis. CNNs are a common deep learning model that has proven to be useful in image analysis, allowing for early cancer detection by finding very small and subtle variations in tissues that human eyes cannot detect. However, to integrate these AI-powered models for breast cancer detection and diagnosis, there is a need to demonstrate their capabilities across diverse test samples and prove their potential to improve clinical workflow for breast cancer (Yala et al., 2021).

3.2 Oral Cancer

Oral squamous cell carcinoma (OSCC) is a malignant tumor that arises from the squamous epithelial cells lining of the oral cavity. It is the most common type of oral cancer and can occur in various sites, like the lips, tongue, floor of mouth, and gums (Oral Squamous Cell Carcinoma > Clinical Keywords > Yale Medicine). It has a significant challenging prognosis, thereby prompting ongoing research into prognostic factors and novel diagnostic methods that could potentially influence disease outcomes (Oral Squamous Cell Carcinoma – Merck Manual Professional Edition). OSCC is the seventh most prevalent cancer worldwide and constituting about 4.5% of all cancer diagnosis globally. Also, it leads to around 4,50,000 deaths annually, thereby representing about 4.6% of global cancer related deaths. The global incidence of OSCC has been increasing in numerous countries, particularly among younger individuals, with a projected 30% rise in cases annually by 2030 (Fatima et al., 2024).

OSCC is more prevalent in men than in women and is common in adults. OSCC incidence is highly observed in South and Southeast Asia, mainly due to high and widespread consumption of the carcinogenic Arecanut and betel quid. It is mainly prevalent in these regions, as these regions comprise of developing nations where more than 60% of the population lives below poverty line. Due to this, the people are unable to afford basic healthy lifestyle that forces them to resort to such harmful practices. This will consequently help them to sustain their hunger, that lowers their expenditure on food and related products (Omar, 2013).

The integration of AI and ML algorithms in the detection, treatment, and prognosis of OSCC is capable of a transformative potential by significantly reducing both inter and intra observer bias among pathologists. Due to the heterogenous nature of OSCC tissue characteristics, even experienced pathologists can sometimes encounter challenges in evaluating the nature of these tissues. AI driven systems especially the ones using deep learning models can analyze vast quantities of histopathological data and imaging and identify features specific to OSCC that maybe missed in traditional evaluations. Therefore, we can say that the integration of AI models with the work of pathologists could build the OSCC care and improve survival rates (Pereira-Prado et al., 2023).

3.3 Lung Cancer

Lung cancer remains to be one of the most prevalent and deadliest cancers worldwide. Lung cancer contributes to a major part of cancer deaths with an estimated 1.8 million deaths, i.e., 18% in 2020. The common types of lung cancer are non-small carcinoma (NSCLC) and small cell carcinoma (SCLC). NSCLC is more common and grows slower, while SCLC is less common and grows more quickly. Lung cancer remains asymptomatic in early stages and that contributes to the high mortality rate. Common symptoms of lung cancer include persistent cough, chest pain, shortness of breath, unexplained weight loss, and coughing up blood. These symptoms are often misdiagnosed with other respiratory conditions, thus arising the need for more advanced and accurate diagnostic tools (Thandra et al., 2021).

Artificial Intelligence (AI) powered algorithms can facilitate the classification of lung cancer subtypes based on imaging patterns, genetic markers, and histopathological data. DL models can distinguish between NSCLC and SCLC by analyzing the tissue morphology thereby supporting faster and more accurate treatment decisions. Therefore, these automation techniques can help oncologists and pathologists to make the quality of cancer care better and provide personalized patient care (Li et al., 2021).

3.4 Colorectal Cancer

Colorectal cancer is the third most frequent cancer in men and women globally, contributing to 9.4% cancer deaths, thereby, becoming the second leading cause of death from cancer. It is also the primary cause of mortality in gastrointestinal cancer. The chance of acquiring this cancer is linked to poor eating habits, smoking, intestinal inflammatory illness, polyps, genetics, and aging. Of the individuals diagnosed with colorectal cancer, 90% are over the age of 50, with a median age of 64 years; nevertheless, the illness is more aggressive in younger patients. The American Cancer Association reported that it caused more than 49,700 deaths in 2015. Survival rate is currently employed in prediction for patient age (Granados-Romero et al., 2017).

Colorectal cancer is not always symptomatic and may go undetected in its early stages. If these symptoms develop, they may include changes in bowel habits that persist several days, such as diarrhea, constipation, or the sensation that the gut is not emptying properly. The stool is smaller or shaped differently than usual. The presence of blood in the stool, whether bright red or extremely dark, recurrent gas pains, bloating, fullness, or cramps, weight loss without apparent explanation, and exhaustion (Adelstein et al., 2011).

In using AI in colorectal cancer, early detection and diagnosis can be made; hence tailored treatment approaches but patient progress can also be monitored. Machine learning may be combined with deep learning that is convolutional neural networks in the analysis of images of colonoscopy by artificial intelligence systems where improved detection rates of polyps thus early identification of cancer can be made possible. Computer vision and bioinformatics help in the characterization of subtypes of cancer using predictions of genetic mutations within pathology. Natural language processing retrieves crucial information from clinical records; predictive analytics can help personalize the treatment plan and predict the progression of the disease for proper care. Generally, AI is revolutionizing colorectal cancer care in terms of increased precision and efficiency and personalization in patient outcomes (Yin et al., 2023).

3.5 Prostate Cancer

Prostate cancer is the second most prevalent cancer diagnosis among men and the fifth greatest cause of death worldwide. Prostate cancer may not cause symptoms in the early stages and frequently progresses, necessitating regular monitoring. According to GLOBOCAN 2018, 1,276,106 new cases of prostate cancer were reported worldwide in 2018, with an incidence rate of 7.3% globally. The differing rates of prostate cancer around the world reflect the amount of diagnostic tests used in each location. Prostate cancer incidence and mortality rates are strongly linked to age, with older men (over 65) at the highest risk (Rawla, 2019).

Prostate cancer may beget or show no signs or symptoms in its early stages. Prostate cancer which is more advanced may beget signs and symptoms similar as trouble urinating, dropped force in the sluice of urine, blood in the urine. The more advanced stages may beget signs and symptoms similar as trouble urinating, reduced force in the sluice of urine, blood in the urine, bone pain, blood in the semen, losing weight without trying and erectile dysfunction (https://www.mayoclinic.org/diseases-conditions/prostate-cancer/symptoms-causes/syc-20353087).

Doctors diagnose and manage prostate cancer by looking at data from different tools. These include PSA levels, MRI-guided biopsies, genomic biomarkers, and Gleason grading. They use this info to figure out the diagnosis, assess risk, and keep an eye on patients, but the process can be pretty subjective. AI can lend a hand to doctors in handling huge sets of data. It spots complex connections and boosts productivity and correctness. AI algorithms cut down on subjectivity, which leads to better use of resources and more accurate diagnoses. This review looks at how

AI-based artificial neural networks (ANN) are making strides in improving how we diagnose and manage prostate cancer (Van Booven et al., 2021).

3.6 Stomach Cancer

Despite dramatic reductions in both incidence and mortality rates over the last few decades, stomach cancer remains the fourth most common cancer and the second major cause of cancer-related deaths with a rate of 5.6% cancer-related deaths worldwide. There is a tenfold variation in incidence rates between populations at high and low risk. The rates are especially high in East Asia, Eastern Europe, and certain parts of Central and South America, with men affected at about twice the incidence of women. Unfortunately, the prognosis is often dismal, with a 5-year relative survival rate of less than 30% in most nations. The most well-established risk factors for stomach cancer include Helicobacter pylori infection, which is by far the strongest known risk factor for distal stomach cancer, along with male gender, a family history of the disease and smoking (Zhai et al., 2022).

In the initial stages of stomach cancer, most patients do not exhibit any signs or symptoms. This is usually the reason why such common indications as weight loss without any apparent cause and pain in the region of the stomach occur late in the disease. Other symptoms include decreased desire for food, pain when consuming food, tiredness and lack of energy or strength getting sick or feeling sick, excessive salivation or difficulty with gas coming up or out, as well as normal feelings of discomfort and slight pain in the abdomen, uncomfortable gas or pressure in the stomach after meals. Patients may also complain of abdominal discomfort, most pronounced at the level of the umbilicus, as well as a sense of bloating or a full feeling after eating very little food or even snacks (Smyth et al., 2020).

While managing stomach cancer, AI incorporates techniques like convolutional neural networks and computer vision for improved early detection and diagnosis. These systems evaluate endoscopic or CT images for the occurrence of metastatic lesions at a very early stage, which helps in increasing the benefit of diagnosis. In pathology, CNN's are useful when examining biopsy specimens, and the types of gastric carcinoma cells are detected, rendering faster and more accurate diagnoses to the pathologists. Human language technology makes it possible to retrieve relevant information from patient's files, which allows for informed quantitative risk estimation. In addition, AI and simulation techniques are also used for understanding the effectiveness of treatment interventions and evaluating disease progress for targeted treatment strategies in the management of stomach cancers (Ali et al., 2024).

3.7 Liver Cancer

Liver cancer has become a significant and evolving health concern worldwide, with increasing cases with an incidence of 4.7% globally. Global estimates for liver cancer show that it was responsible for nearly 905,700 new cases and 830,200 deaths in

2020. The primary causes of liver cancer remain hepatitis B and C infections, yet in developed countries, cases linked to non-alcoholic fatty liver disease (NAFLD) and alcohol consumption are rising due to shifts in lifestyle and aging populations. These trends highlight an urgent need for specific strategies to screen and manage risk in populations affected by these factors. From a regional perspective, East Asia has historically reported high rates of liver cancer due to hepatitis prevalence, but recent decades have shown a decline in liver cancer mortality due to widespread hepatitis B vaccination efforts. Contrarily, in North America and high-income Asia-Pacific countries, the burden of liver cancer is increasing. The World Health Organization predicts that without additional prevention measures, liver cancer cases will rise by around 55% by 2040. Prevention strategies like hepatitis B vaccination and lifestyle interventions against NAFLD are essential as these could prevent many cases. Understanding the shifting causes and the impact of prevention methods is key to addressing liver cancer's burden. Targeted interventions tailored to the unique needs of each region such as vaccinations in high-risk areas and lifestyle changes in regions with increasing non-viral causes are essential to curb the rising incidence globally (Liu et al., 2024a). Figure 2 illustrates the progression of liver cancer, showing a transition from a normal liver to a cancerous liver, alongside cellular-level changes, beginning with DNA damage in a normal cell, leading to the formation and multiplication of defective cells, and culminating in cancer.

Women and men who suffer from liver cancer usually show no signs and symptoms in the initial stages of the disease. However, with the progression of the illness, they may start developing symptoms associated with liver dysfunction. Early warnings can be in the form of unintentional weight loss, absence of desire to eat, pain in the upper stomach, dry heaving and emesis, lethargy and overall weakness. Other

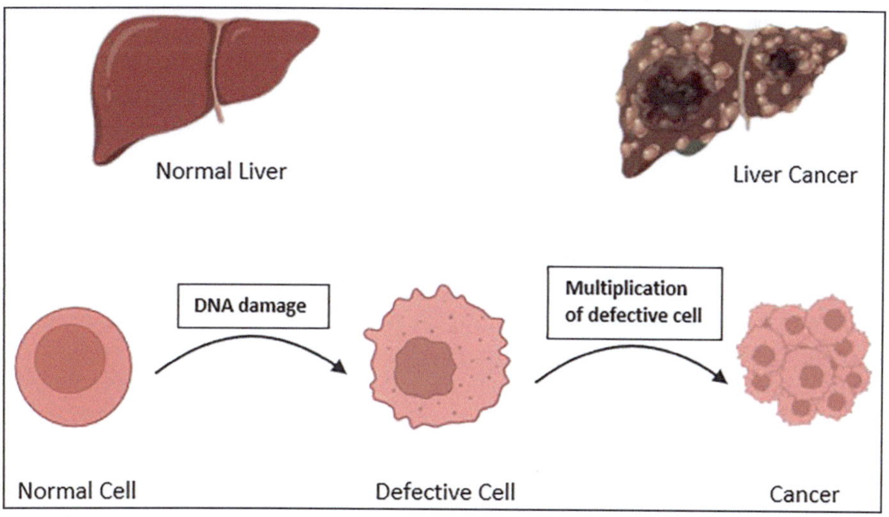

Fig. 2 The figure depicts the progression of liver cancer, from a normal liver and cell to DNA damage, the formation of defective cells, their multiplication, and the development of cancer

signs include bloody vomiting, hypoglycemia, abdominal bloating and pain, hypercalcemia, polycythemia, yellowing of eyes, and grayish or whitish stool (Israelsen et al., 2024).

The use of Artificial Intelligence in the management of liver cancer has been made better by introducing different sophisticated means that seek to improve early detection, diagnosis, and treatment planning. For example, by employing machine learning techniques, it is possible to utilize imaging modalities such as MRI and CT scans to find liver tumors with great precision resulting in early diagnosis. Algorithms adapted to Convolutional Neural Networks are largely used in medical imaging to explore the borders of a tumor and distinguish its appearance from healthy tissue as well as from other lesions of a different pathology. Moreover, it considers patient information such as demographic data, medical history, and genomic data to evaluate personal risks and forecast the progression with the help of predictive analytic powers of artificial intelligence. Natural language processing technology saves doctors a lot of time by pulling out important data from electronic health records for treatment purposes. All these Artificial Intelligence approaches mean better diagnosing accuracies with treatment protocols, higher efficacy in the management of liver cancer, and better outlooks for the patients (Calderaro et al., 2024).

3.8 Esophageal Cancer

Esophageal cancer (EsC), including both squamous cell carcinoma and adenocarcinoma, remains one of the most challenging cancers due to high mortality rates and poor prognosis. According to recent estimates from GLOBOCAN, over 600,000 new EsC cases were reported globally in 2020, contributing to 3% of cancer cases. The subtype distribution varies by region; squamous cell carcinoma is more common in Asia and Africa due to risk factors like tobacco and alcohol, while adenocarcinoma has surged in North America and Europe, partly linked to obesity. This cancer burden is projected to grow by 50% by 2040, emphasizing the need for preventive measures and early detection initiatives to address its rising impact globally (Morgan et al., 2022).

Difficulty in swallowing is often the first noticeable symptom of esophageal cancer, followed by other signs such as pain in the throat or back, discomfort behind the breastbone or between the shoulder blades. Patients may also experience vomiting or coughing up blood, persistent heartburn, hoarseness, chronic cough, and unintentional weight loss (CureScience Institute, 2024).

In the regions where populations have a high predisposition to esophageal cancer, treatment of cancer of the esophagus is not only complicated, but also cost inefficient. However, the advent of artificial intelligence can help in improving the accuracy of diagnosis and even prognosis of esophageal cancer based on clinical, radiological, and genomic data. For instance, recent enhancements in deep learning techniques have positively impacted the identification of early esophageal squamous cell carcinoma and adenocarcinoma in Barrett's esophagus. Such an effect on artificial intelligence is sought in relation to aid to medical professionals towards advanced cancer cases, and towards analyzing endoscopic images depicting precancerous transformations (Zhang et al., 2020).

3.9 Cervical Cancer

Approximately 60,400 new cases and 34,200 deaths were reported in 2020, in addition to an incidence rate of 3.1% globally, making cervical cancer one of the frequent cancers among women. Besides, it is the only form of cancer that has a possibility of being prevented completely through primary strategies that include a 9-valent human papillomavirus (HPV) vaccine that is fully effective, treatment of secondary prevention where patient's especially women's are detected for the disease and treated at their early stages (Hou et al., 2022).

AI is enhancing substantially the diagnosis and management of cervical cancer by a variety of techniques. There is CNN application for Pap smear and cervical biopsy image interpretation which enhances the accuracy of pre cancer and cervical cancer detection. Information from electronic health records is utilized through natural language processing (NLP) to improve patient stratification and follow-up care. In addition, machine learning algorithms analyze risk factors and extrapolate the course of disease from the patients' demographics and clinical data. Furthermore, such tools can incorporate treatment planning optimization through analysis of the patients' responses to the therapies enhancing their management in cervical cancer (Mathivanan et al., 2024).

3.10 Bladder Cancer

Every year in the USA, there are about 74,000 new bladder cancer cases diagnosed, with about 16,000 deaths occurring, with a global incidence rate of 3%. There are approximately 578,000 people living with the disease following diagnosis, thanks to the positive survival rate trends. Bladder cancer is ranked as the most expensive cancer due to its tendency to recur, making it necessary for most patients to be under constant evaluation for the rest of their lives. The median age of diagnosis has increased from 67 years, and in 2008–2012, it was 73 years, whereby the incidence rates show a progressive rise with age, suggesting that it poses an increasing burden with time. There are about 1.3 million diagnosed with the disease each year, worldwide. About 65% of the population is currently believed to be responsible for bladder cancer, and these days, the risk for the active smokers has risen to three, four or five times in relation to the non-smokers in the last two decades (Silverman et al., 2017).

Hematuria or blood in the urine is one of the signs and symptoms of bladder cancer. It can make the urine appear bright red or cola colored, although it can also occasionally appear normal with blood only found in laboratory tests. Back pain, uncomfortable urination, and frequent urine are other symptoms (Leslie et al., 2024).

Through a number of important applications, AI technologies enhance the diagnosis and treatment of bladder cancer. Convolutional neural networks help pathologists grade tumors from biopsy samples, while machine learning algorithms evaluate cystoscopy images to improve the accuracy of tumor diagnosis. Natural language processing helps with risk assessment and treatment planning by extracting pertinent information from electronic health records. Personalized therapy techniques are made possible by predictive analytics models that predict treatment responses and recurrence risks. When combined, these AI tools improve patient outcomes in bladder cancer care by streamlining diagnostics and enabling customized treatment plans (Bashkami et al., 2024).

3.11 Non-Hodgkin Lymphoma

B-cell lymphomas make up over 90% of non-Hodgkin lymphoma cases, which are a variety of lymphoid system malignancies that are mainly divided into B-cell and T-cell neoplasms. In the United States between 55,000 and 60,000 new instances of follicular lymphoma and diffuse large B-cell lymphoma are diagnosed each year, while being the cause of 2.6% of cancer-related deaths worldwide. Although they may not be immediately apparent, swollen lymph glands, persistent fatigue, fever, night sweats, shortness of breath, inexplicable weight loss, and skin irritation or itching are typical signs and symptoms of lymphoma (Saragea, 2024; Sharafaddini et al., 2024).

Treatment varies according to histology and stage; radiation is usually used for early-stage follicular lymphoma, while chemotherapy, immunotherapy, or radioimmunotherapy are needed for more advanced instances (Kumar & Parikh, 2025).

Certain artificial intelligence (AI) technologies are significantly improving the management of non-Hodgkin lymphoma (NHL). Clinical and genetic data are analyzed by machine learning algorithms to find biomarkers that forecast treatment outcomes and patient outcomes. To improve patient stratification and individualized treatment plans, natural language processing (NLP) is used to extract pertinent information from unstructured clinical notes. Additionally, imaging investigations like PET and CT scans are analyzed using deep learning techniques, especially convolutional neural networks which increase the precision of lymphoma staging and identification. In the end predictive analytics models help with better clinical decision-making and individualized patient care by using previous patient data to forecast illness progression and treatment efficacy. When combined these AI tools improve non-Hodgkin lymphoma diagnosis, treatment planning, and monitoring (Ansell & Armitage, 2005).

4 How Computational Methods Are Revolutionizing Oncology?

Techniques like Artificial Intelligence (AI), Machine Learning (ML), Deep Learning (DL), and Computer Vision are now making it possible to make breakthroughs in analyzing large-scale image datasets, thereby improving cancer detection in terms of accuracy and speed (Gaur & Jagtap, 2022). These AI models excel in capturing deviations from normal patterns, or anomalies, which enables them to assist clinicians, radiologists, and pathologists in detecting subtle signs of cancer that might be missed by the human eye (Rai et al., 2024). For instance, convolutional neural networks (CNNs) can analyze mammograms, CT scans, and histopathology slides, identifying patterns indicative of early-stage tumours with high precision (Sharafaddini et al., 2024).

By reducing human errors and inter-observer variability, these technologies not only provide a second opinion but also help standardize diagnostics, leading to faster and more reliable cancer detection (Schüßler et al., 2024). This ability to recognize minute details allows AI-powered tools to improve the efficiency of the diagnostic workflow and enable early intervention, which is critical for successful cancer treatment. Moreover, as AI models are continuously trained on diverse datasets, their predictive accuracy improves over time, adapting to various cancer types and patient demographics (Zeb et al., 2024). Consequently, AI and ML are shaping the future of oncology by supporting healthcare professionals, enhancing diagnostic accuracy, and ultimately, improving patient outcomes (Lotter et al., 2024).

5 How Patient Can Be Improved Through Computational Methods?

Computational methods in oncology enhance the patient outcomes by enabling earlier diagnosis and more accurate prognosis. This further enables tailored treatments. AI and ML models can analyze a patient's medical history, imaging, and genetic data to predict disease progression and treatment response since these models are very good and pattern and anomaly detection (Khaligh et al., 2024). For example, ML models are being used to predict chemotherapy effectiveness and side effects, improving personalized treatment decisions and minimizing adverse effects (Sri Chandana & Balakrishna, 2022). There is also scope to explore natural language processing to extract relevant information from clinical notes to identify patient specific risk factors and enhance personalized care. Also, in the recent past, computational advancements have offered real-time health data that can lead to timely alerts and interventions (Tavabi et al., 2024).

6 Methods

There are three main groups of the advancing methods of cancer treatment. These include conventional, modern, and future techniques (Raghani et al., 2024). Conventional methods constitute the traditional approaches easily recognized to be used in controlling and removing cancer cells. The approaches include surgery, chemotherapy, and radiotherapy (Liu et al., 2024b). Modern methods are accurate treatments like photodynamic therapy, immunotherapy, targeted therapy, and hormonal therapy, and it involves specific disease types with higher efficiency (Hong & Chang, 2024) (Fig. 3).

Another way of identifying the severity of cancer is by using Grading Systems. Some of the grading systems include such as Histologic Grade for prostate cancer (Agosti & Munari, 2024), Bloom-Richardson grading system for breast cancer (Yılmaz & Karagöz, 2023) and Fuhrman grading system for renal cell cancer (Gupta et al., 2024) to name a few, where the degree of malignancy and aggressiveness becomes a basis in issuing prognosis and finding a treatment option. However, these grading systems prove difficult and nonuniform. Hence, using machine learning techniques, this can be automated and standardized to ensure consistency in grading processes (Koh et al., 2022).

Use of AI and ML for automation of most processes will push cancer care forward into the future. Through automation, new avenues for cancer care will come through diagnostics and personal treatment plans and elevated patient care (Shiwlani et al., 2024).

Advances in computer-based approaches for cancer treatments enabled personalized and precision oncology to be transformed into customized treatments matched according to each patient's genetic and unique characteristics of their cancer (Shams, 2024). With the help of computational approaches, including machine

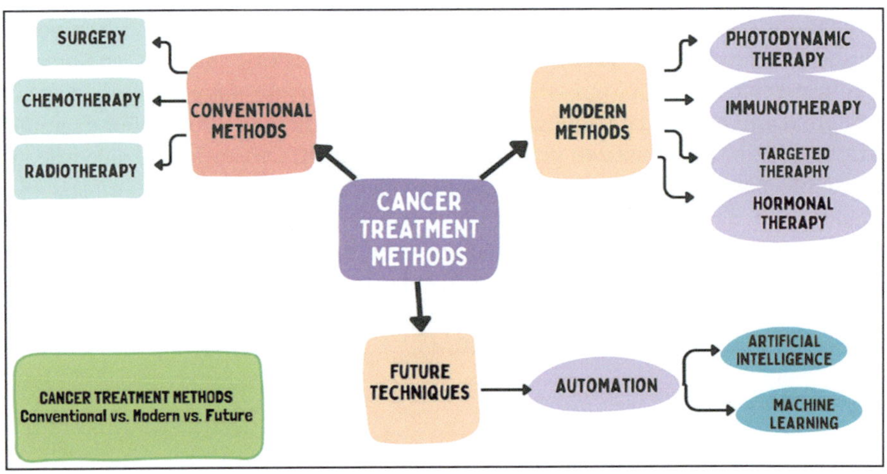

Fig. 3 Cancer treatment methods

learning and artificial intelligence, genomics, proteomics, and other omics technologies provide data that are vastly analyzed from large datasets for detecting particular cancer mutations and predicting patient responses to treatment options, thereby, leading to the development of optimized plans for therapy. These approaches allow the tailoring of treatments to the unique biology of every patient, leading to better efficacy as well as unwanted side effects (Chandrashekar et al., 2024). Precision oncology, driven by computational tools, underpins biomarkers as well as drug discovery aids in predicting cancers' progress and leads eventually to more targeted and effective treatments (Saraf et al., 2024).

7 Further Directions

The future of cancer detection, prediction, and prognosis with AI and ML holds a transformative potential for healthcare by focusing on precision, personalization, and predictive accuracy (Zeb et al., 2024). Future research can be directed towards enhancing early detection models by refining AI algorithms that can identify intricate molecular, genetic, and imaging biomarkers with high sensitivity (Khandakar et al., 2024). Emerging fields like multi-omics integration and the incorporation of real-world data from wearable devices will be AI tools closer to real-time, non-invasive diagnostics (Chenel et al., 2024). Additionally, advances in federated learning will enable broader collaboration between institutions while safeguarding patient privacy and this could revolutionize data-driven cancer research (Chenel et al., 2024).

Further studies could also be directed towards developing explainable AI models, thereby helping clinicians interpret predictions with greater transparency. This will help build trust in AI-driven recommendations (Nasarian et al., 2024). As personalized oncology becomes more prominent, AI models will need to integrate a wide array of patient data, ranging from genetic profiles to environmental influences allowing for more personalized treatment (Sherani et al., 2024). The integration of continuous learning mechanisms will enable AI systems to evolve alongside new medical discoveries thereby ensuring their relevance and efficacy in dynamic clinical settings. Collectively, these advancements promise a future in which AI not only augments the expertise of healthcare professionals but also will potentially bring in unprecedented precision and hope to patients worldwide (Aminizadeh et al., 2024).

8 Conclusion

Cancer is still one of the main medical health issues in the world, causing nearly ten million deaths every year and one-sixth of all deaths on the planet. In this chapter, we described the most widespread cancer diseases according to their features and high mortality cases: breast, lung, colorectal, prostate, and stomach cancers. Conventionally, the method of diagnosis has been performed through biopsy and imaging. However, with these methods come limits in early detection as well as precision, thus, necessitating a need to advance oncology further.

There is a rigorous research going on where, the medical field has intertwined with the AI specialists to create several seamless and precise techniques to aid in early detection, diagnosis, as well as in the treatment of cancer. Computational methods, AI, ML, and DL aim to transform the field of cancer care. Algorithms for AI power most diagnostic models that lead to better accuracy in diagnostics and find minute abnormalities and support the precise grading of the severity of cancer. From applying the capability of CNN based imaging systems to natural language processing on patient data, these tools accelerate efficiency, reduce the risk of human errors, and support individualized treatment plans. Publicly available apps, automated diagnostic devices, and vast datasets can now serve clinicians and researchers across the globe through real time access to sophisticated diagnostics. Altogether, these computational methods promise a future in oncology wherein medical care becomes personalized, accurate, and perfect in bringing about better patient outcomes.

References

Adelstein, B. A., Macaskill, P., Chan, S. F., Katelaris, P. H., & Irwig, L. (2011). Most bowel cancer symptoms do not indicate colorectal cancer and polyps: A systematic review. *BMC Gastroenterology, 11*, 65. https://doi.org/10.1186/1471-230X-11-65. PMID: 21624112; PMCID: PMC3120795.

Agosti, V., & Munari, E. (2024). Histopathological evaluation and grading for prostate cancer: Current issues and crucial aspects. *Asian Journal of Andrology, 26*(6), 575–581.

Ali, H., Muzammil, M. A., Dahiya, D. S., Ali, F., Yasin, S., Hanif, W., Gangwani, M. K., Aziz, M., Khalaf, M., Basuli, D., & Al-Haddad, M. (2024). Artificial intelligence in gastrointestinal endoscopy: A comprehensive review. *Annals of Gastroenterology, 37*(2), 133.

Alimirzaie, S., Bagherzadeh, M., & Akbari, M. R. (2019). Liquid biopsy in breast cancer: A comprehensive review. *Clinical Genetics, 95*, 643. https://doi.org/10.1111/cge.13514

Aminizadeh, S., Heidari, A., Dehghan, M., Toumaj, S., Rezaei, M., Navimipour, N. J., Stroppa, F., & Unal, M. (2024). Opportunities and challenges of artificial intelligence and distributed systems to improve the quality of healthcare service. *Artificial Intelligence in Medicine, 149*, 102779.

Ansell, S. M., & Armitage, J. (2005). Non-Hodgkin lymphoma: Diagnosis and treatment. In *Mayo clinic proceedings* (Vol. 80, No. 8, pp. 1087–1097). Elsevier.

Bashkami, A., Nasayreh, A., Makhadmeh, S. N., Gharaibeh, H., Alzahrani, A. I., Alwadain, A., Heming, J., Ezugwu, A. E., & Abualigah, L. (2024). A review of Artificial Intelligence methods in bladder cancer: Segmentation, classification, and detection. *Artificial Intelligence Review, 57*(12), 339.

Calderaro, J., Žigutytė, L., Truhn, D., Jaffe, A., & Kather, J. N. (2024). Artificial intelligence in liver cancer—New tools for research and patient management. *Nature Reviews Gastroenterology & Hepatology, 21*, 1–15.

Chandrashekar, K., Niranjan, V., Vishal, A., & Setlur, A. S. (2024). Integration of artificial intelligence, machine learning and deep learning techniques in genomics: Review on computational perspectives for NGS analysis of DNA and RNA seq data. *Current Bioinformatics, 19*(9), 825–844.

Chenel, H., Marku, M., James, T., Zinovyev, A., & Pancaldi, V. (2024). Multilayer network approaches to omics data integration in Digital Twins for cancer research. arXiv preprint arXiv:2410.07252.

Crosby, D., Bhatia, S., Brindle, K. M., Coussens, L. M., Dive, C., Emberton, M., Esener, S., Fitzgerald, R. C., Gambhir, S. S., Kuhn, P., & Rebbeck, T. R. (2022). Early detection of cancer. *Science, 375*(6586), eaay9040.

CureScience Institute. (2024). Esophageal cancer. In *Disease review* (pp. 1–22).

Fatima, J., Fatima, E., Mehmood, F., Ishtiaq, I., Khan, M. A., Khurshid, H. M. S., & Kashif, M. (2024). Comprehensive analysis of oral squamous cell carcinomas: Clinical, epidemiological, and histopathological insights with a focus on prognostic factors and survival time. *Cureus, 16*(2).

Gaur, K., & Jagtap, M. M. (2022). Role of artificial intelligence and machine learning in prediction, diagnosis, and prognosis of cancer. *Cureus, 14*(11).

Geisel, J., Raghu, M., & Hooley, R. (2018). The role of ultrasound in breast cancer screening: The case for and against ultrasound. *Seminars in Ultrasound, CT and MRI, 39*(1), 25–34., ISSN 0887-2171,. https://doi.org/10.1053/j.sult.2017.09.006

González, C. A., & Agudo, A. (2012). Carcinogenesis, prevention and early detection of gastric cancer: Where we are and where we should go. *International Journal of Cancer, 130*(4), 745–753.

Granados-Romero, J. J., Valderrama-Treviño, A. I., Contreras-Flores, E. H., Barrera-Mera, B., Herrera Enríquez, M., Uriarte-Ruíz, K., Ceballos-Villalba, J. C., Estrada-Mata, A. G., Alvarado Rodríguez, C., & Arauz-Peña, G. (2017). Colorectal cancer: A review. *International Journal of Research in Medical Sciences, 5*(11), 4667.

Gupta, A., Garg, S., Yadav, N., Dhanakshirur, R. R., Jain, K., Nayyar, R., Kaushal, S., & Das, C. J. (2024). Predicting renal cell carcinoma subtypes and Fuhrman grading using multiphasic CT-based texture analysis and machine learning techniques. *Indian Journal of Radiology and Imaging., 35*, 306.

Hanahan, D. (2022). Hallmarks of cancer: New dimensions. *Cancer Discovery, 12*(1), 31–46.

Hong, G., & Chang, J. E. (2024). Enhancing cancer treatment through combined approaches: Photodynamic therapy in concert with other modalities. *Pharmaceutics, 16*(11), 1420.

Hong, R., & Xu, B. (2022). Breast cancer: An up-to-date review and future perspectives. *Cancer Communications, 42*(10), 913–936.

Hou, X., Shen, G., Zhou, L., Li, Y., Wang, T., & Ma, X. (2022). Artificial intelligence in cervical cancer screening and diagnosis. *Frontiers in Oncology, 12*, 851367.

https://www.mayoclinic.org/diseases-conditions/prostate-cancer/symptoms-causes/syc-20353087

https://www.who.int/news-room/fact-sheets/detail/cancer

Huppe, A. I., Mehta, A. K., & Brem, R. F. (2018). Molecular breast imaging: A comprehensive review. *Seminars in Ultrasound, CT and MRI, 39*(1), 60–69., ISSN 0887-2171,. https://doi.org/10.1053/j.sult.2017.10.001

Irigaray, P., Newby, J. A., Clapp, R., Hardell, L., Howard, V., Montagnier, L., Epstein, S., & Belpomme, D. (2007). Lifestyle-related factors and environmental agents causing cancer: An overview. *Biomedicine & Pharmacotherapy, 61*(10), 640–658.

Israelsen, M., Francque, S., Tsochatzis, E. A., & Krag, A. (2024). Steatotic liver disease. *The Lancet, 404*(10464), 1761–1778.

Khalighi, S., Reddy, K., Midya, A., Pandav, K. B., Madabhushi, A., & Abedalthagafi, M. (2024). Artificial intelligence in neuro-oncology: Advances and challenges in brain tumor diagnosis, prognosis, and precision treatment. *npj Precision Oncology, 8*(1), 80.

Khandakar, S., Al Mamun, M. A., Islam, M. M., Hossain, K., Melon, M. M. H., & Javed, M. S. (2024). Unveiling early detection and prevention of cancer: Machine learning and deep learning approaches. *Educational Administration: Theory and Practice, 30*(5), 14614–14628.

Koh, D. M., Papanikolaou, N., Bick, U., Illing, R., Kahn, C. E., Jr., Kalpathi-Cramer, J., Matos, C., Martí-Bonmatí, L., Miles, A., Mun, S. K., & Napel, S. (2022). Artificial intelligence and machine learning in cancer imaging. *Communications Medicine, 2*(1), 133.

Korn, A. R., Walsh-Bailey, C., Pilar, M., Sandler, B., Bhattacharjee, P., Moore, W. T., Brownson, R. C., Emmons, K. M., & Oh, A. Y. (2022). Social determinants of health and cancer screening implementation and outcomes in the USA: A systematic review protocol. *Systematic Reviews, 11*(1), 117.

Kumar, R., & Parikh, R. R. (2025). Role of radiotherapy in the management of elderly patients with lymphoma. In *Seminars in radiation oncology* (Vol. 35, No 1, pp. 57–66). WB Saunders.
Kurman, R. J., & Norris, H. J. (1987). Endometrial hyperplasia and metaplasia. In *Blaustein's pathology of the female genital tract* (pp. 322–337). Springer.
Leslie, S., Soon-Sutton, T., & Aeddula, N. (2024). *Bladder cancer*. StatPearls.
Li, Y., Chen, D., Wu, X., Yang, W., & Chen, Y. (2021). A narrative review of artificial intelligence-assisted histopathologic diagnosis and decision-making for non-small cell lung cancer: Achievements and limitations. *Journal of Thoracic Disease, 13*(12), 7006.
Liu, J., Luo, R., Zhang, Y., & Li, X. (2024a). Current status and perspective on molecular targets and therapeutic intervention strategy in hepatic ischemia-reperfusion injury. *Clinical and Molecular Hepatology, 30*, 585.
Liu, B., Zhou, H., Tan, L., Siu, K. T. H., & Guan, X. Y. (2024b). Exploring treatment options in cancer: Tumor treatment strategies. *Signal Transduction and Targeted Therapy, 9*(1), 175.
Lotter, W., Hassett, M. J., Schultz, N., Kehl, K. L., Van Allen, E. M., & Cerami, E. (2024). Artificial Intelligence in oncology: Current landscape, challenges, and future directions. *Cancer Discovery, 14*(5), 711–726.
Mann, R. M., Cho, N., & Moy, L. (2019). Breast MRI: State of the art. *Radiology, 292*(3), 520.
Mareel, M., & Leroy, A. (2003). Clinical, cellular, and molecular aspects of cancer invasion. *Physiological Reviews, 83*(2), 337–376.
Martínez-Jiménez, F., Muiños, F., Sentís, I., Deu-Pons, J., Reyes-Salazar, I., Arnedo-Pac, C., Mularoni, L., Pich, O., Bonet, J., Kranas, H., & Gonzalez-Perez, A. (2020). A compendium of mutational cancer driver genes. *Nature Reviews Cancer, 20*(10), 555–572.
Mathivanan, S. K., Francis, D., Srinivasan, S., Khatavkar, V., & P, K. and Shah, M.A. (2024). Enhancing cervical cancer detection and robust classification through a fusion of deep learning models. *Scientific Reports, 14*(1), 10812.
Morgan, E., Soerjomataram, I., Rumgay, H., Coleman, H. G., Thrift, A. P., Vignat, J., Laversanne, M., Ferlay, J., & Arnold, M. (2022). The global landscape of esophageal squamous cell carcinoma and esophageal adenocarcinoma incidence and mortality in 2020 and projections to 2040: New estimates from GLOBOCAN 2020. *Gastroenterology, 163*(3), 649–658.
Nasarian, E., Alizadehsani, R., Acharya, U. R., & Tsui, K. L. (2024). Designing interpretable ML system to enhance trust in healthcare: A systematic review to proposed responsible clinician-AI-collaboration framework. *Information Fusion, 108*, 102412.
Omar, E. A. (2013). The outline of prognosis and new advances in diagnosis of oral squamous cell carcinoma (OSCC): Review of the literature. *Journal of Oral Oncology, 2013*(1), 519312.
Oral Squamous Cell Carcinoma – Merck Manual Professional Edition
Oral Squamous Cell Carcinoma > Clinical Keywords > Yale Medicine
Pereira-Prado, V., Martins-Silveira, F., Sicco, E., Hochmann, J., Isiordia-Espinoza, M. A., González, R. G., Pandiar, D., & Bologna-Molina, R. (2023). Artificial intelligence for image analysis in oral squamous cell carcinoma: A review. *Diagnostics, 13*(14), 2416.
Raghani, N. R., Chorawala, M. R., Mahadik, M., Patel, R. B., Prajapati, B. G., & Parekh, P. S. (2024). Revolutionizing cancer treatment: Comprehensive insights into immunotherapeutic strategies. *Medical Oncology, 41*(2), 51.
Rai, H. M., Yoo, J., & Razaque, A. (2024). Comparative analysis of machine learning and deep learning models for improved cancer detection: A comprehensive review of recent advancements in diagnostic techniques. *Expert Systems with Applications, 255*, 124838.
Rawla, P. (2019). Epidemiology of prostate cancer. *World Journal of Oncology, 10*(2), 63.
Rommasi, F. (2023). Identification, characterization, and prognosis investigation of pivotal genes shared in different stages of breast cancer. *Scientific Reports, 13*(1), 8447.
Saba, T. (2020). Recent advancement in cancer detection using machine learning: Systematic survey of decades, comparisons and challenges. *Journal of Infection and Public Health, 13*(9), 1274–1289.
Saraf, S., De, A., & Tripathy, B. K. (2024). Effective use of computational biology and artificial intelligence in the domain of medical oncology. In *Computational intelligence for oncology and neurological disorders* (pp. 228–252). CRC Press.

Saragea, P. D. (2024). A comprehensive overview of malignant lymphomas: Classification, epidemiology, and clinical insights.
Schlich-Bakker, K. J., ten Kroode, H. F. J., & Ausems, M. G. E. M. (2006). A literature review of the psychological impact of genetic testing on breast cancer patients. *Patient Education and Counseling, 62*(1), 13–20, ISSN 0738-3991,. https://doi.org/10.1016/j.pec.2005.08.012
Schüßler, D., Blanco, M. B., Guthrie, N. K., Sgarlata, G. M., Dammhahn, M., Ernest, R., Evasoa, M. R., Hasiniaina, A., Hending, D., Jan, F., & Le Pors, B. (2024). Morphological variability or inter-observer bias? A methodological toolkit to improve data quality of multi-researcher datasets for the analysis of morphological variation. *American Journal of Biological Anthropology, 183*(1), 60–78.
Shams, A. (2024). Leveraging state-of-the-art AI algorithms in personalized oncology: From transcriptomics to treatment. *Diagnostics, 14*(19), 2174.
Sharafaddini, A. M., Esfahani, K. K., & Mansouri, N. (2024). Deep learning approaches to detect breast cancer: A comprehensive review. In *Multimedia tools and applications* (pp. 1–112).
Sharma, I., Sharma, A., Tomer, R., Negi, N., & Sobti, R. C. (2023). History, evolution, milestones in cancer research and treatment. In *Handbook of Oncobiology: From basic to clinical sciences* (pp. 1–29). Springer.
Sherani, A. M. K., Khan, M., Qayyum, M. U., & Hussain, H. K. (2024). Synergizing AI and healthcare: Pioneering advances in cancer medicine for personalized treatment. *International Journal of Multidisciplinary Sciences and Arts, 3*(1), 270–277.
Shiwlani, A., Khan, M., Sherani, A. M. K., Qayyum, M. U., & Hussain, H. K. (2024). Revolutionizing healthcare: The impact of artificial intelligence on patient care, diagnosis, and treatment. *JURIHUM: Jurnal Inovasi dan Humaniora, 1*(5), 779–790.
Siegel, R. L., Miller, K. D., Fuchs, H. E., & Jemal, A. (2022). Cancer statistics, 2022. *CA: A Cancer Journal for Clinicians, 72*(1), 7–33.
Silverman, D. T., Koutros, S., Figueroa, J. D., Prokunina-Olsson, L., & Rothman, N. (2017). Bladder cancer. In *Schottenfeld and fraumeni cancer epidemiology and prevention* (4th ed., pp. 977–996). Oxford University Press.
Smyth, E. C., Nilsson, M., Grabsch, H. I., van Grieken, N. C., & Lordick, F. (2020). Gastric cancer. *The Lancet, 396*(10251), 635–648.
Sri Chandana, K., & Balakrishna, S. (2022). Cancer treatment recommendation using machine learning technologies. *Cancer, 7*(1), 83–88.
Steinbeck, R. G. (2004). Dysplasia in view of the cell cycle. *European Journal of Histochemistry, 48*(3), 203–212.
Sung, H., Ferlay, J., Siegel, R. L., Laversanne, M., Soerjomataram, I., Jemal, A., & Bray, F. (2021). Global cancer statistics 2020: GLOBOCAN estimates of incidence and mortality worldwide for 36 cancers in 185 countries. *CA: A Cancer Journal for Clinicians, 71*(3), 209–249.
Phi, XA., Tagliafico, A., Houssami, N. *et al.* Digital breast tomosynthesis for breast cancer screening and diagnosis in women with dense breasts—A systematic review and meta-analysis. BMC Cancer 18, 380 (2018).https://doi.org/10.1186/s12885-018-4263-3.
Tan, Y. J., Sim, K. S., & Ting, F. F. (2017). Breast cancer detection using convolutional neural networks for mammogram imaging system. In *2017 International Conference on Robotics, Automation and Sciences (ICORAS), Melaka, Malaysia* (pp. 1–5). https://doi.org/10.1109/ICORAS.2017.8308076
Tavabi, N., Singh, M., Pruneski, J., & Kiapour, A. M. (2024). Systematic evaluation of common natural language processing techniques to codify clinical notes. *PLoS One, 19*(3), e0298892.
Thandra, K. C., Barsouk, A., Saginala, K., Aluru, J. S., & Barsouk, A. (2021). Epidemiology of lung cancer. *Contemporary Oncology/Współczesna Onkologia, 25*(1), 45–52.
Van Booven, D. J., Kuchakulla, M., Pai, R., Frech, F. S., Ramasahayam, R., Reddy, P., Parmar, M., Ramasamy, R., & Arora, H. (2021). A systematic review of artificial intelligence in prostate cancer. *Research and Reports in Urology, 13*, 31–39.
Yala, A., Mikhael, P. G., Strand, F., Lin, G., Smith, K., Wan, Y. L., Lamb, L., Hughes, K., Lehman, C., & Barzilay, R. (2021). Toward robust mammography-based models for breast cancer risk. *Science Translational Medicine, 13*(578), eaba4373.

Yılmaz, D., & Karagöz, F. (2023). Concurrent expression of immunhistochemical parameters in breast cancer patients; clinical implications and consistency with Bloom-Richardson grading system. *Journal of Experimental and Clinical Medicine, 40*(2), 294–299.

Yin, Z., Yao, C., Zhang, L., & Qi, S. (2023). Application of artificial intelligence in diagnosis and treatment of colorectal cancer: A novel Prospect. *Frontiers in Medicine (Lausanne)., 10,* 1128084. https://doi.org/10.3389/fmed.2023.1128084. PMID: 36968824; PMCID: PMC10030915.

Zangheri, B., Messa, C., Picchio, M., et al. (2004). PET/CT and breast cancer. *European Journal of Nuclear Medicine and Molecular Imaging, 31*(Suppl 1), S135–S142. https://doi.org/10.1007/s00259-004-1536-7

Zeb, S., Nizamullah, F. N. U., Abbasi, N., & Fahad, M. (2024). AI in healthcare: Revolutionizing diagnosis and therapy. *International Journal of Multidisciplinary Sciences and Arts, 3*(3), 118–128.

Zhai, S., Yu, L., & Li, J. (2022). *Construction of stomach cancer lesion detection combined with drug therapy based on artificial intelligence*.

Zhang, Y. H., Guo, L. J., Yuan, X. L., & Hu, B. (2020). Artificial intelligence-assisted esophageal cancer management: Now and future. *World Journal of Gastroenterology, 26*(35), 5256–5271.

Overview of Computational Approaches for Cancer Diagnosis

Geeta Chhabra Gandhi, Roshmeet Chakraborty, Dinesh Kumar Saini, Sumit Kalra, and Jabar Yousif

Abstract

Cancer detection has witnessed a paradigm shift with the advent of artificial intelligence (AI) and machine learning (ML), revolutionizing traditional diagnostic approaches. This research explores advanced computational techniques, including Convolutional Neural Networks (CNNs), Generative Adversarial Networks (GANs), Long Short-Term Memory (LSTM) networks, XGBoost, and 3D-CNNs, applied across a spectrum of cancers such as lung, liver, ovarian, brain, and thyroid. These methods leverage medical imaging and cytology data to enhance accuracy, sensitivity, and early-stage identification, addressing limitations in conventional methods. By utilizing deep learning architectures and hybrid models, the study highlights their potential in improving diagnostic precision, reducing variability, and facilitating personalized treatment strategies. However, challenges like data heterogeneity, interpretability, and clinical integration persist. The paper underscores the need for explainable AI, robust validation frameworks, and interdisciplinary collaborations to overcome these barriers. Concluding with a forward-looking perspective, it emphasizes the transformative role of AI/

G. C. Gandhi · R. Chakraborty
Manipal University Jaipur, Jaipur, India

D. K. Saini
Department of IoT and Intelligent Systems, Manipal University Jaipur, Jaipur, Rajasthan, India

S. Kalra
Indian Institute of Technology Jodhpur, Jodhpur, Rajasthan, India

J. Yousif (✉)
Sohar University, Sohar, Oman
e-mail: jyousif@su.edu.om

R. Nema et al. (eds.), *Advances in Cancer Detection, Prediction, and Prognosis Using Artificial Intelligence and Machine Learning*,
https://doi.org/10.1007/978-981-96-9346-7_3

ML in oncology, advocating for continued innovation and global cooperation to realize the vision of accessible, accurate, and proactive cancer diagnostics.

1 Introduction

Cancer continues to be one of the world's top causes of death, impacting millions of people each year and posing a serious problem for healthcare systems. To improve prognosis, optimize treatment regimens, and ultimately save lives, early and precise identification is essential. However, conventional diagnostic techniques like imaging and biopsies frequently have drawbacks, such as inconsistent interpretation, delayed diagnosis, and reliance on the availability of qualified specialists. The diagnosis of cancer has been transformed by developments in machine learning (ML) and artificial intelligence (AI), providing tools that enhance accuracy, efficiency, and accessibility.

This research paper explores a range of AI-driven methodologies tailored to specific cancer types. From Convolutional Neural Networks (CNNs) applied to CT scans for lung cancer to 3D-CNNs for MRI analysis in brain cancer and Generative Adversarial Networks (GANs) for liver cancer detection, AI has demonstrated its versatility and impact across diverse imaging modalities. Additionally, models like Long Short-Term Memory (LSTM) networks and eXtreme Gradient Boosting (XGBoost) have been effectively used for ultrasound imaging, offering solutions for thyroid and ovarian cancer diagnostics, respectively. Each approach is meticulously designed to address the challenges posed by the unique characteristics of different cancers and imaging datasets, pushing the boundaries of traditional medical diagnostics.

Despite these advancements, challenges remain, including the need for large, labeled datasets, interpretability of AI models, and the integration of these systems into clinical workflows. This paper proposes a hybrid AI-based solution, leveraging the strengths of existing algorithms while incorporating attention mechanisms, explainable AI frameworks, and semi-supervised learning techniques. By uniting these technologies, the proposed solution aims to address existing gaps, enabling early detection, reducing false positives, and improving diagnostic outcomes across a range of cancers.

This work aims to contribute to the ongoing efforts to make cancer detection faster, more reliable, and globally accessible, enhancing patient care and outcomes.

2 Cancer—An Introduction

Cancer is a group of diseases characterized by the uncontrolled growth and spread of abnormal cells, leading to significant morbidity and mortality globally. Despite advances in medical science, cancer remains a major public health challenge. Early detection is critical, as it improves treatment outcomes and increases survival rates. However, traditional diagnostic methods such as biopsies, blood tests, and imaging modalities like X-rays, CT scans, and MRIs are often limited in their ability to detect cancer at early stages. Challenges such as low sensitivity, inter-observer variability, and high false-positive rates hinder accurate diagnoses. Furthermore, cancers like pancreatic and ovarian cancer often present with non-specific symptoms, delaying detection until advanced stages when treatment options are limited. The sheer volume and complexity of medical data, combined with disparities in healthcare infrastructure across regions, exacerbate these challenges (Rennie, 1996).

2.1 Overview of Computational Approaches in Cancer Detection

Cancer detection has undergone significant transformation with the advent of computational approaches, which have enhanced accuracy, reduced human error, and allowed for early-stage identification across various types of cancers. Traditional methods often rely on manual assessments such as biopsies or imaging interpretation, which can be subjective and prone to errors. Computational techniques, powered by artificial intelligence (AI) and machine learning (ML), leverage vast datasets and sophisticated algorithms to automate detection, analyze patterns, and make predictions with remarkable precision (Dimitrakopoulos & Beerenwinkel, 2017). These technologies utilize features extracted from imaging modalities like CT, MRI, ultrasound, and histopathological data, applying algorithms like convolutional neural networks (CNNs) for image analysis (Anwar et al., 2018), support vector machines (SVMs) for classification (Huang et al., 2018), and generative adversarial networks (GANs) for synthetic data augmentation (Frid-Adar et al., 2018).

Advancements in computational approaches have allowed models to not only detect malignancies but also differentiate between aggressive and benign tumors. For instance, 3D-CNNs for brain cancer effectively analyze volumetric data (Ramzan et al., 2020), while XGBoost models process ultrasound images for ovarian cancer detection with high accuracy (Chen et al., 2020). Despite these successes, challenges remain, such as the dependence on labeled datasets, model generalization across populations, and explainability in clinical applications. Researchers are addressing these issues by incorporating techniques like transfer learning, hybrid models, and feature visualization, ensuring these solutions are both robust and interpretable (Fig. 1).

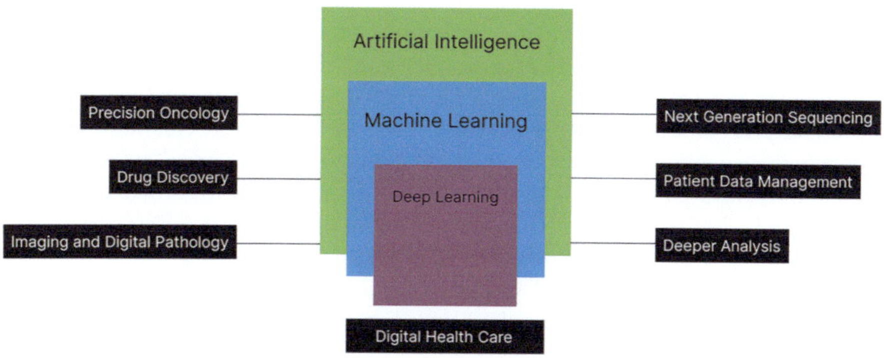

Fig. 1 Applications of AI, ML, and DL in oncology and cancer detection

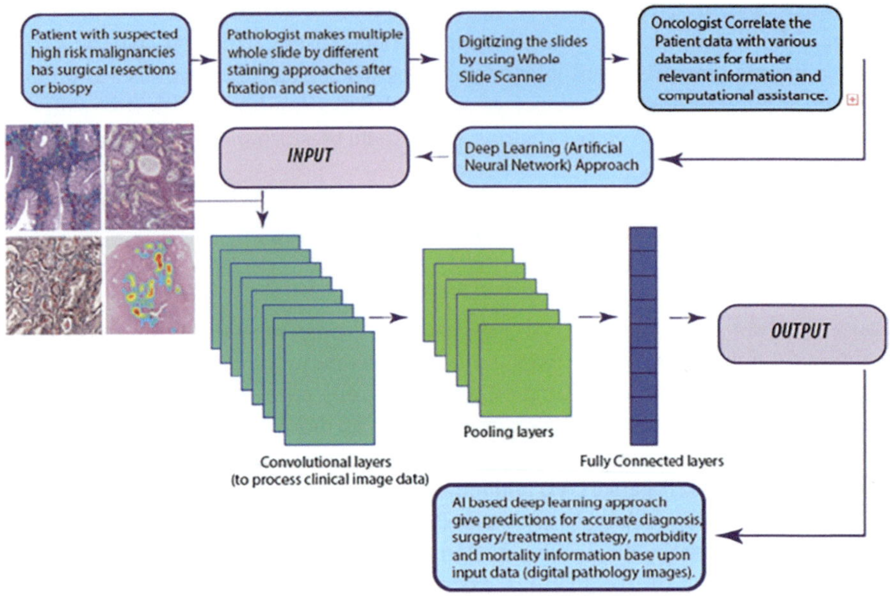

Fig. 2 Artificial Intelligence approached for cancer detection. (Chen et al., 2020)

As AI technologies evolve, the integration of computational approaches into diagnostic workflows is reshaping oncology. These methods have the potential to democratize cancer care by enabling remote diagnostics and tailoring treatments to individual patients based on predictive analytics. Transitioning to the next pivotal area, it is essential to explore the specific roles that AI and ML play in revolutionizing detection processes, bridging the gap between traditional diagnostics and modern computational methodologies (Fig. 2).

2.2 Role of Artificial Intelligence and Machine Learning in Cancer Diagnosis

By overcoming the drawbacks of conventional techniques, artificial intelligence (AI) and machine learning (ML) are transforming the diagnosis of cancer. These technologies use sophisticated algorithms to examine intricate medical data at a speed and precision never before possible. Early and accurate detection of malignant lesions is made possible by deep learning models that are excellent at processing and interpreting imaging data, such as Convolutional Neural Networks (CNNs) (Hunter et al., 2022) and Long Short-Term Memory networks (LSTMs). Similarly, to improve diagnosis accuracy, machine learning methods such as Support Vector Machines (SVMs) and Gradient Boosted Trees are used to examine trends in cytology and radiomics data (Huang et al., 2018) (Fig. 3).

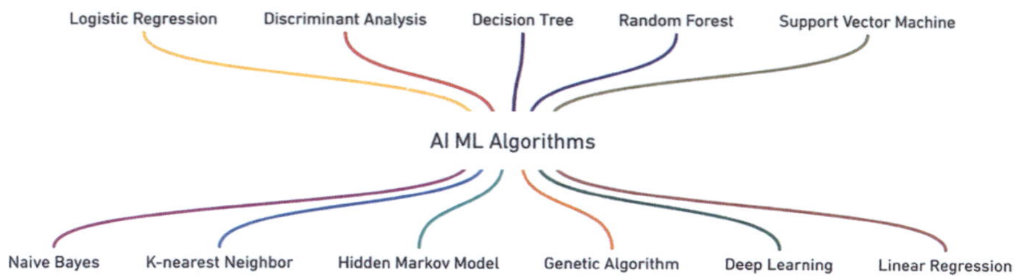

Fig. 3 Types of AI ML algorithm for accurate and precise cancer detection

AI-powered solutions have enabled automated analysis of mammograms, MRIs, and CT scans, reducing human error and increasing diagnostic consistency (Rasool et al., 2024). Moreover, semi-supervised models such as Generative Adversarial Networks (GANs) allow learning from limited labeled datasets, a critical advancement for rare cancers with scarce data. Beyond imaging, AI models analyze genomic, proteomic, and clinical data, aiding in predictive modeling and personalized treatment plans. By integrating AI into clinical workflows, healthcare systems can overcome diagnostic challenges, enabling early detection, reducing variability, and facilitating precision oncology (Alqahtani et al., 2019).

2.3 Ongoing and Existing Research in Cancer Detection

The application of Artificial Intelligence (AI) and Machine Learning (ML) in the field of cancer detection has sparked a revolution in the medical field, offering promising advancements in diagnostic accuracy, speed, and precision. Over the past decade, these technologies have seen remarkable growth, particularly in imaging-based diagnostic techniques, such as mammography, CT scans (Iqbal et al., 2021a), MRIs, and ultrasound. Cancer detection, traditionally reliant on human expertise in interpreting medical images and lab tests, is increasingly supported by AI and ML models that can identify subtle patterns and anomalies often invisible to the human eye. This change is poised to reduce diagnostic errors, speed up diagnosis, and, ultimately, improve patient outcomes (Hunter et al., 2022).

Research in AI-driven cancer detection is diverse and multidisciplinary, incorporating imaging, genomics, clinical data, and even patient history to create more holistic models of diagnosis (Khanna et al., 2020). For instance, deep learning models like Convolutional Neural Networks (CNNs) have made substantial strides in medical imaging by learning hierarchical patterns in pixel data. These models have shown particular success in detecting cancers such as breast, lung, and skin cancers through image analysis, with accuracy rates often surpassing those of human radiologists. In breast cancer, for example, CNNs are employed to analyze mammograms and can distinguish malignant tumors from benign ones, enabling early detection and reducing false positives (Alzubaidi et al., 2021) (Fig. 4).

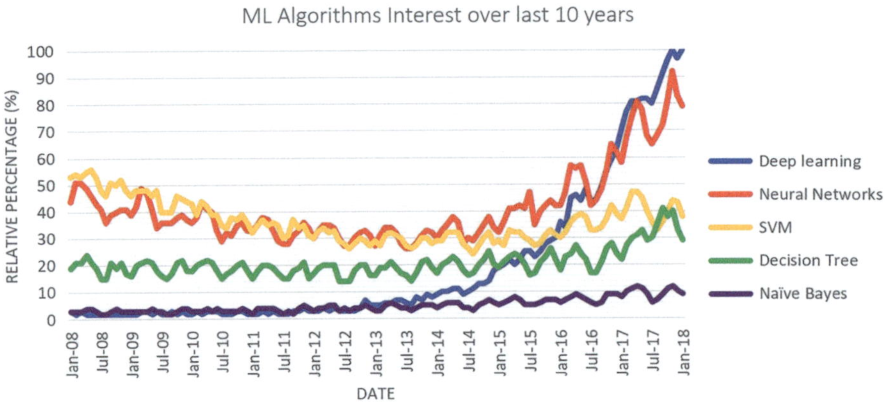

Fig. 4 Overview of ML Algorithms over last 10 years. (Forghani et al., 2019)

The advancement in radiomics—where machine learning algorithms analyze features extracted from medical images—has also contributed to cancer detection. Algorithms like Random Forests (Liu et al., 2021) and Support Vector Machines (SVM) are commonly used in analyzing tumor characteristics such as texture, shape, and size to classify tumors accurately. This approach is frequently applied in prostate and ovarian cancers, where distinguishing between benign and malignant growths can significantly impact treatment decisions (Forghani et al., 2019).

In parallel, generative models like Generative Adversarial Networks (GANs) have been applied to medical imaging, especially in situations where limited labeled data is available. GANs generate synthetic images that mimic real medical data, thereby augmenting small datasets for training machine learning models. In liver cancer detection, for instance, GANs have shown efficacy in detecting lesions in MRI scans, even with a limited amount of labeled training data. This addresses a significant challenge in the medical field—data scarcity, particularly for rare forms of cancer (Dimitriadis et al., 2022).

2.4 Integration of AI with Genomics and Multi-modal Data

Another exciting area of research involves the integration of imaging data with genomic information. By combining AI models that analyze medical images with those that process genomic and molecular data, researchers are creating more precise diagnostic tools. For example, combining MRI scans with genomic data can help not only detect cancer but also predict its progression, identify genetic markers, and personalize treatment plans for individual patients.

One approach that has garnered attention is the use of multi-modal AI systems that integrate medical imaging, genomics, and clinical data. Such models are being developed to handle complex data types, enabling more accurate predictions regarding the type, stage, and aggressiveness of the cancer. These multi-modal approaches are particularly beneficial in cancers like glioblastoma, where MRI scans and genomic analysis are crucial in understanding the tumor's behavior and response to treatment (Quazi, 2022; Bi et al., 2019; Boeken et al., 2023).

2.5 Natural Language Processing (NLP) and Electronic Health Records (EHRs)

While medical imaging has seen the most attention in AI-driven cancer detection, natural language processing (NLP) and machine learning techniques are also being employed to analyze Electronic Health Records (EHRs) for early cancer detection. EHRs contain a wealth of unstructured data, such as doctor's notes, patient histories, and lab reports, which can be difficult to analyze manually. However, NLP algorithms can extract meaningful patterns from this unstructured data to identify patients at higher risk for certain types of cancer (Li et al., 2023).

This is particularly useful for cancers where early detection relies on patient history and symptoms rather than imaging, such as in pancreatic or ovarian cancers. For instance, AI models can analyze patient demographics, family history, symptoms, and clinical test results to identify patterns that might suggest the early onset of cancer, helping to flag at-risk individuals for further testing (Gholipour et al., 2023; Sangariyavanich et al., 2023).

2.6 AI and ML in Surgical Planning and Treatment Monitoring

AI and ML techniques are not just confined to the diagnostic stage of cancer care but are also making a significant impact in treatment planning and monitoring. Radiomics, for instance, can be used to predict how tumors will respond to various treatments by analyzing the texture and morphology of cancerous tissues. This predictive capability is helping oncologists to tailor treatment strategies, ensuring that patients receive the most appropriate therapies based on the unique characteristics of their tumors (Khanna et al., 2020).

AI is also helping in the monitoring of cancer progression, detecting subtle changes in tumor characteristics that may signal recurrence or resistance to treatment. Such predictive models are particularly valuable in cancers that are prone to relapse, such as breast cancer and lung cancer. By continually analyzing patient data, including medical images and test results, AI systems can alert healthcare providers to potential issues before they become clinically apparent, leading to timely interventions (Chen et al., 2021).

3 Techniques Involved in Cancer Diagnosis

Cancer detection techniques combine traditional diagnostics with advanced computational methods to enhance accuracy and early detection. Imaging techniques such as mammography, ultrasound, CT scans, and MRI remain standard, providing insights into anatomical abnormalities. However, manual interpretation often leads to inconsistencies, driving the adoption of computational approaches like AI and ML (Cabral et al., 2023).

Deep learning models, including Convolutional Neural Networks (CNNs), excel in image analysis, detecting subtle patterns in breast and skin cancer (Shah et al., 2023; Fu'adah et al., 2020). For volumetric imaging, 3D-CNNs capture spatial relationships critical for brain cancer diagnosis (Irmak, 2021). Generative Adversarial Networks (GANs) enhance detection through data augmentation, particularly useful in liver cancer diagnosis with limited datasets (Ikeda et al., 2021). Long Short-Term Memory (LSTM) networks analyze sequential data, aiding in thyroid cancer detection (Habchi et al., 2023; Nagalakshmi & Priya, 2023).

Traditional ML techniques also play a role. XGBoost has been utilized for ovarian cancer detection, classifying masses based on ultrasound data (Yuan et al., 2022), while Support Vector Machines (SVMs) integrated with radiomics distinguish tumor aggressiveness for personalized treatment. These methods improve scalability, minimize diagnostic errors, and support clinicians in making informed decisions (Cui et al., 2021).

3.1 Convolutional Neural Networks (CNNs)

In medical image processing, Convolutional Neural Networks (CNNs) have become a key technique, especially for identifying different types of cancer. These networks are very useful for jobs like detecting cancers or nodules in CT scans, MRI pictures, or mammograms since they are made to automatically recognize and categorize information inside images. In order to extract significant characteristics including edges, textures, and patterns suggestive of malignant growths, CNNs process the input picture via several layers of convolutions, pooling, and fully connected layers (Dabeer et al., 2019). The key advantage of CNNs in cancer detection is their ability to learn hierarchical patterns from raw data without needing extensive feature engineering, which enhances accuracy and reduces manual input. Their use in lung, breast, and skin cancer detection has been validated in various research studies, where CNNs have demonstrated high sensitivity and specificity, often surpassing traditional diagnostic methods. The current stage of CNN development in cancer detection is rapidly evolving, with ongoing efforts to refine architectures for greater precision and generalization, particularly through transfer learning and fine-tuning on domain-specific datasets (Huang et al., 2017). Convolutional neural networks, or CNNs, were modeled after the visual cortex of animals, whose cells process small parts of an image. According to Goodfellow et al. (Alzubaidi et al., 2021), CNNs offer three primary advantages: comparable representations, sparse interactions, and parameter sharing. CNNs use shared weights and local connections to process 2D input data, such images, more effectively than fully connected networks. This reduces the number of parameters and makes training easier. A common CNN architecture includes convolutional layers, pooling layers, and fully connected layers, just like a multi-layer perceptron (MLP) (Fig. 5).

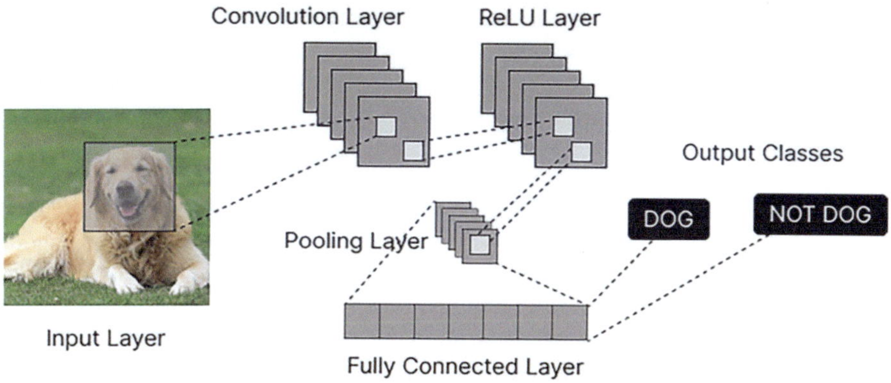

Fig. 5 An example of CNN architecture for image classification

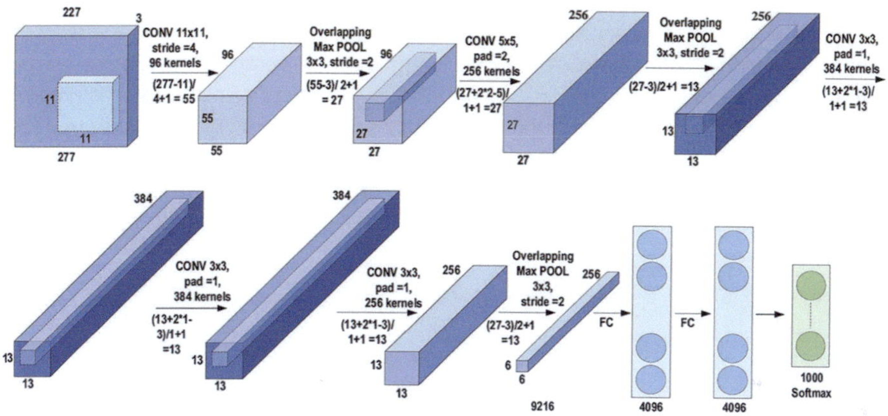

Fig. 6 AlexNet Architecture. (Krizhevsky et al., 2017)

AlexNet is a highly regarded deep CNN architecture (Krizhevsky et al., 2017), known for its groundbreaking achievements in image recognition and classification (Fig. 6).

3.2 Generative Adversarial Networks (GANs)

Generative Adversarial Networks (GANs) have gained significant attention in the healthcare domain, especially for augmenting datasets in medical imaging. A GAN consists of two neural networks: the generator and the discriminator, which work against each other to produce high-quality synthetic images. GANs are used to create artificial medical pictures, such MRI scans, for cancer identification. These images may be used to augment genuine datasets, particularly when there is a lack of annotated data. While the discriminator assesses the produced pictures' validity, the generator produces lifelike visuals that closely match genuine patient data. Better training of diagnostic models is made possible by the GAN's ability to generate increasingly realistic pictures that are indistinguishable from actual medical images through this adversarial process. GANs have shown promise in improving the performance of CNNs by providing them with a broader range of examples during the training phase, which is particularly beneficial in rare cancer types where data scarcity can be a major challenge. The goal of GANs in cancer research at the moment is to enhance the quality of synthetic data and get beyond issues like mode collapse, which occurs when the generator only generates a small range of pictures. In order to guarantee clinical dependability, research is also focused on improving the interpretability of GAN-generated outputs (Hunter et al., 2022; Dimitriadis et al., 2022; Ikeda et al., 2021). GANs come in a variety of forms, each designed to address certain problems or improve performance. High-quality picture production is made possible by Deep Convolutional GANs (DCGANs), which employ convolutional neural networks in both the discriminator and generator. Conditional GANs (cGANs) use extra data, such as textual descriptions or class labels, to regulate the data creation process. CycleGANs learn mappings between two picture domains, which allows them to perform very well in image-to-image translation. ***Progressive Growing of GANs (PGGANs)*** improve stability and resolution by gradually increasing image size during training, enabling high-resolution image generation. **StyleGAN** offers fine-grained control over features and attributes, focusing on the style of generated images (Fig. 7).

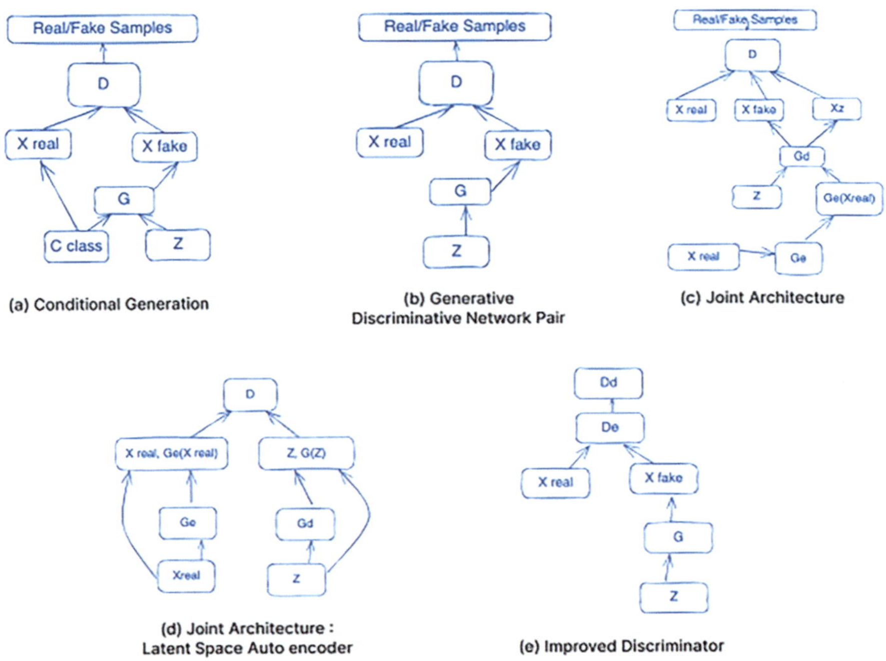

Fig. 7 Types of GAN

3.3 Long Short-Term Memory (LSTM)

To overcome the limitations of traditional Recurrent Neural Networks (RNNs) in handling long-term dependencies in sequential input, a particular type of RNN known as Long Short-Term Memory (LSTM) networks was developed. While traditional RNNs struggle to retain information over extended sequences, LSTMs excel at recognizing and taking advantage of long-range dependencies. Their distinctive architecture, which includes memory cells and a system of gates that control the information flow inside the network, is the source of this capability (Habchi et al., 2023). A memory cell that has the capacity to store data over time and three gates input, output, and forget that regulate the flow of data into and out of the cell make up each LSTM unit. The forget gate chooses which data should be removed from the cell, the output gate regulates which data from the cell should be carried over to the following time step, and the input gate chooses what new data should be recorded in the cell. LSTMs may learn intricate patterns in sequential data with long-range dependencies thanks to this gating mechanism, which lets them selectively keep or dismiss information. Because of its ability to capture long-term relationships, LSTMs are particularly well-suited for a range of applications requiring sequential data, including time-series analysis, machine translation, speech recognition, and natural language processing. In natural language processing, LSTMs have shown useful for tasks including sentiment analysis, language modeling, and question answering. In speech recognition tasks such as acoustic modeling and phoneme recognition, LSTMs have achieved state-of-the-art results.

In conclusion, long-term dependencies in sequential data are well captured by LSTMs, a potent kind of RNN that tackles the vanishing gradient issue. They can learn intricate patterns in data with long-range temporal linkages thanks to their special architecture, which combines memory cells and gates. This makes them an invaluable tool for a variety of sequence-based activities (Nagalakshmi & Priya, 2023) (Fig. 8).

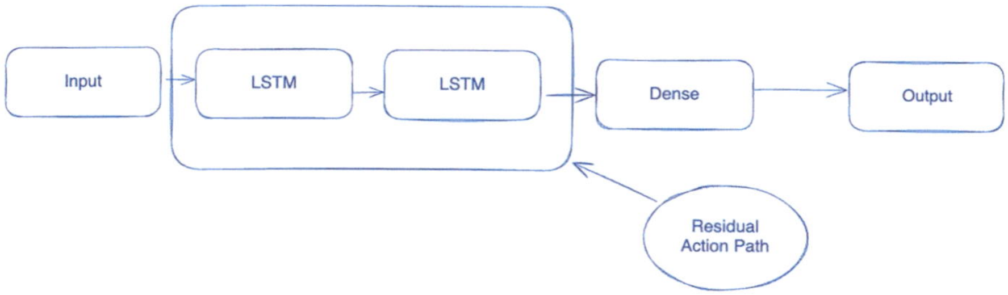

Fig. 8 SR-LSTM structure

3.4 XGBoost

XGBoost (Extreme Gradient Boosting) is a highly effective machine learning algorithm that has been extensively used for classifying cancerous versus non-cancerous samples. XGBoost is a decision tree-based ensemble algorithm that works by creating a series of weak models (usually decision trees) and combining them to form a robust, strong model. The algorithm iteratively refines its predictions by minimizing the residuals (errors) of previous models using gradient descent. In cancer detection, XGBoost is applied to structured datasets, such as those derived from genomic data, pathology reports, or clinical records. One of the key advantages of XGBoost is its ability to handle a mix of numerical and categorical data, making it well-suited for datasets with various features. For ovarian cancer, XGBoost has been used to differentiate between benign and malignant masses, demonstrating high accuracy and reliability. The current development of XGBoost focuses on improving its interpretability and ensuring that the models are explainable, a critical aspect for clinical applications where understanding model decisions is necessary for regulatory and ethical reasons. Research is also focused on optimizing XGBoost for better performance with smaller datasets, which are common in healthcare settings (Yuan et al., 2022; Cui et al., 2021). XGBoost employs "feature subsampling," a technique also utilized in random forests to mitigate overfitting, is more advanced and sophisticated than traditional machine learning algorithms. It uses random seeds to enable repeated operations, enhancing the model without altering parameters. Known for its flexibility and efficiency, XGBoost handles missing data seamlessly and combines weak predictive models to deliver highly accurate results. Moreover, it surpasses other algorithms in computation speed, making it a superior choice for various applications (Yuan et al., 2020). f_1 (X, θ_1) (Fig. 9).

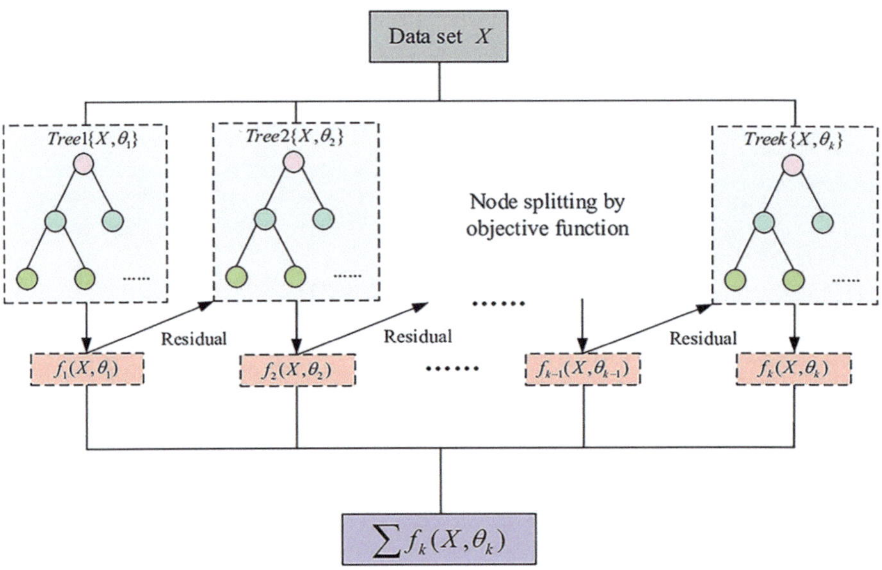

Fig. 9 (Guo et al., 2020): XG Boost Flow chart

Compared to other algorithms, SVMs face challenges with large datasets and numerous variables, while Bayesian networks are fast to train but lack complexity (Kazem, 2017). XGBoost, on the other hand, has proven highly effective in handling large datasets, such as the SEER database. It excels in predicting outcomes like lymph node metastasis in oral squamous cell carcinoma and prostate cancer and survival rates in esophageal cancer, making it a powerful tool for complex medical predictions (Kwak et al., 2021; Gong et al., 2021; Wei et al., 2021).

3.5 ResNet

Residual Networks (ResNet) are a class of deep learning architectures that have revolutionized the field of computer vision, particularly in the medical domain. ResNet addresses the issue of vanishing gradients in very deep neural networks by introducing shortcut connections that allow gradients to flow more easily during training. This makes it possible to build very deep networks, which are more capable of capturing complex features in images. In cancer detection, ResNet has been widely used for analyzing dermoscopic images to detect skin cancers like melanoma. By learning deeper and more intricate features, ResNet-based models outperform traditional CNNs in terms of accuracy and robustness (Targ et al., 2016).

The generalized residual block is a flexible design that can function like a standard CNN layer (by ignoring the residual stream) or as a single-layer ResNet block (by ignoring the main stream). By stacking multiple generalized residual blocks, this architecture can adapt to behave like anything in between, including the standard 2-layer ResNet block (Fig. 10c). This flexibility lets the network decide how many processing steps to apply to the residuals before adding them back, as seen in visualizations. Importantly, this block isn't limited to CNNs—it works with fully connected and other feedforward layers too. By replacing each convolutional layer in a traditional ResNet block (Fig. 1a) with a generalized residual block (Fig. 10b), we create a new architecture called ResNet in ResNet (RiR) (Fig. 10d).

The ResNet algorithm works by stacking convolutional layers with residual connections, where each layer learns to improve the output of the previous layer, enabling it to focus on more abstract image features. Current research on ResNet focuses on optimizing the architecture to handle multi-modal medical data, such as

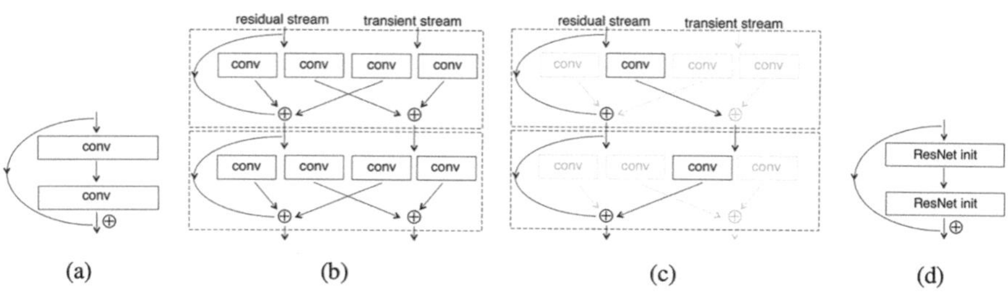

Fig. 10 (Wu et al., 2019): ResNet

combining imaging data with clinical or genomic data for more accurate diagnosis. ResNet's success in skin cancer detection has led to its exploration in other areas of oncology, such as breast and lung cancer (Wu et al., 2019).

3.6 Support Vector Machines (SVM) with Radiomics

Support Vector Machines (SVM) in conjunction with Radiomics is a potent method for obtaining high-dimensional characteristics from medical pictures, which are subsequently applied to classification tasks like cancer diagnosis. Radiomics is the process of extracting a variety of quantitative properties from medical pictures, such as CT scans or MRIs, including texture, shape, and intensity (Sun et al., 2019). SVM, a supervised machine learning technique that divides data points into distinct categories by identifying a hyperplane that maximizes the margin between classes, is then used to assess these features. SVM with radiomics has been used in cancer detection to categorize tumors according to their imaging properties. This makes it useful for distinguishing between aggressive and indolent cancers, such as lung and prostate cancer. This method's strength is its capacity to employ high-dimensional, complex data that human clinicians could find difficult to understand. To increase diagnostic accuracy, SVM with radiomics is now utilized in combination with other machine learning approaches. Research is still being done to improve model resilience and refine feature extraction strategies for clinical usage (Tapani et al., 2022).

Before characteristics are recovered from 16-second epochs, the newborn EEG is first pre-processed. The SVM combines the features into a decision statistic, which is further post-processed (maximum across channels, collaring, median filter in time) to create the final output. The original SDA publication provides a detailed explanation of the post-processing procedures, feature calculation, and selection (Tapani et al., 2019) shown in Fig. 11.

3.7 3D CNNs for Brain Cancer

3D Convolutional Neural Networks (3D CNNs) are specialized variations of traditional CNNs, designed to process volumetric data like MRI scans, making them particularly effective for detecting brain cancers, such as glioblastomas (ırmak, Emrah., 2021). Unlike 2D CNNs, which analyze individual slices of medical

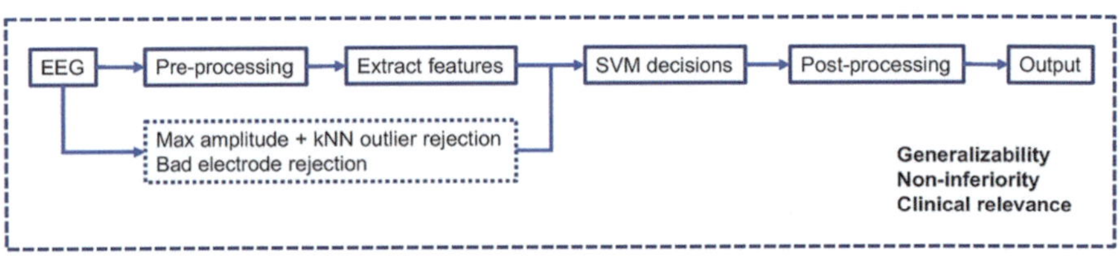

Fig. 11 (Tapani et al., 2022): Flow chart of SVM

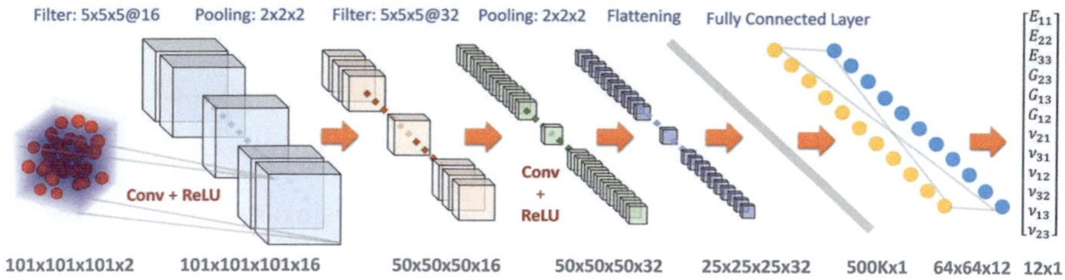

Fig. 12 (Rao & Liu, 2020): 3D CNN Architecture

images, 3D CNNs consider the entire brain volume, capturing spatial dependencies across different slices. This capability is crucial for identifying subtle features and anomalies indicative of glioblastomas or other brain tumors. To prepare the MRI scans for analysis, preprocessing steps are applied, such as reducing noise, enhancing contrast, and adjusting intensity. The images are resized to optimize memory usage, and the dataset is augmented using techniques like flipping or rotating the images. These steps increase the training data size and reduce overfitting, where the model performs well on training data but poorly on new data. A deep multi-scale 3D CNN architecture is often chosen for this task because it effectively captures both local and global contextual information from the 3D MRI scans. The network comprises 3D convolutional layers that extract features in three dimensions, followed by pooling layers and fully connected layers for classification. During training, the preprocessed and augmented MRI data are used to teach the model to identify distinguishing features between cancerous and non-cancerous tissues. The trained model is then evaluated on a separate set of MRI scans using metrics like accuracy to determine its effectiveness in detecting brain cancer. A key advantage of 3D CNNs in this application is their ability to consider contextual information from surrounding tissues, leading to more accurate diagnoses. Current research focuses on optimizing 3D CNNs for faster processing times and higher resolution to support real-time clinical applications. Additionally, researchers are exploring multi-modal approaches, integrating 3D CNNs with data sources like genetic information to further enhance diagnostic performance (Joseph & Dennisan, 2022) (Fig. 12).

3.8 EUS-FNA Cytology for Pancreatic Cancer

Endoscopic Ultrasound with Fine Needle Aspiration (EUS-FNA) cytology is a key diagnostic tool for pancreatic cancer, allowing direct sampling of pancreatic tissue for accurate evaluation. The procedure is valued for its ability to provide high-resolution imaging of the pancreas, enabling precise targeting of lesions for tissue collection. In recent years, machine learning (ML) algorithms, including Support Vector Machines (SVMs) and deep learning techniques, have been integrated with EUS-FNA to assist pathologists in diagnosing malignant cells. These advanced models analyze cytological images obtained during the procedure, improving diagnostic precision, especially in distinguishing benign from malignant lesions, and minimizing human error.

A study conducted on 293 pancreatic specimens from patients undergoing EUS-FNA for solid and cystic pancreatic lesions highlights the utility of this approach. The study excluded FNA specimens from other abdominal areas, such as peripancreatic lesions, lymph nodes, or bile duct masses, focusing solely on pancreatic lesions. Comprehensive clinical data, laboratory test results, and cytopathological and imaging reports were collected for each patient to ensure robust analysis. Imaging techniques such as Ultrasonography (USG), Computed Tomography (CT), and Magnetic Resonance Imaging (MRI) were used to evaluate the size, location, and characteristics of pancreatic lesions before and during the EUS-FNA procedure.

During EUS-FNA, an onsite pathology team, including a trained pathologist and technician, was present to ensure accurate sample collection and immediate preliminary analysis. This setup reduced delays in obtaining results and improved diagnostic reliability. The integration of ML algorithms with EUS-FNA cytology has shown potential in addressing challenges posed by inconclusive histological findings, offering a more robust framework for early detection and classification of pancreatic lesions. These technologies aim to enhance cytology-based diagnoses, especially in complex cases, by refining the differentiation of lesion types, ultimately aiding early diagnosis and improving treatment outcomes for patients with pancreatic cancer (Rastogi et al., 2019) (Fig. 13).

4 Types of Cancer and Its Detection Methodology

4.1 Lung Cancer

The implementation of Convolutional Neural Networks (CNNs) for lung cancer detection from CT scans leverages their ability to automatically analyze and classify medical images, improving diagnostic accuracy and reducing human error. CNNs consist of multiple layers that sequentially process the input data, extracting increasingly complex features at each stage. Initially, raw CT scan images are preprocessed, including steps such as resizing, normalization, and augmentation to ensure uniformity and diversity in training data. The first stage of the CNN, the convolutional layer, applies filters to detect basic features such as edges and textures. As the data moves through the network, deeper layers capture more abstract representations like shapes and patterns, specifically focusing on features indicative of malignant nodules. Pooling layers follow the convolutional layers, reducing the dimensionality of the data and retaining essential features while minimizing computational load. Once the features have been extracted, fully connected layers classify the image into categories, distinguishing between benign and malignant nodules (Hunter et al., 2022; Alzubaidi et al., 2021; Al-Yasriy et al., 2020).

Using backpropagation to reduce prediction errors, the model learns to identify patterns in labeled datasets during the training phase. Optimization methods like the Adam optimizer, which modifies the neural network's weights according to the loss

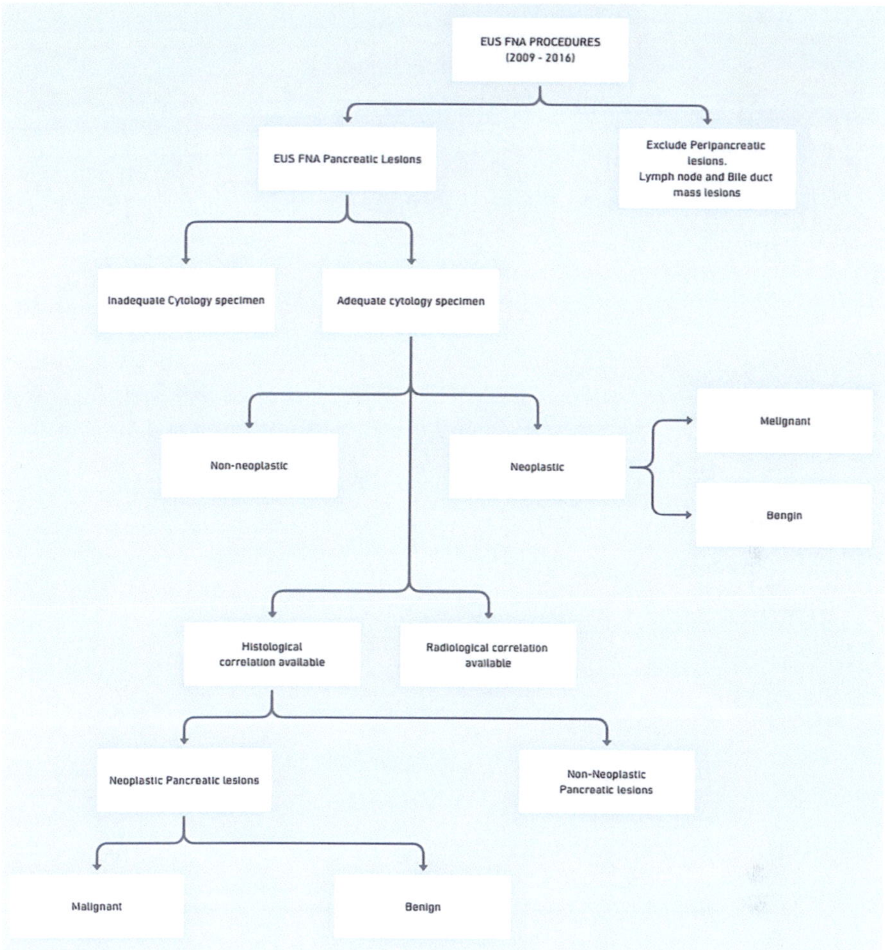

Fig. 13 Flow chart depicting frequency of pancreatic lesions using EUS FNA

function, are used to train the model. To avoid overfitting and enhance the model's capacity to generalize across various datasets, regularization techniques such as batch normalization and dropout are used. The CNN divides the training data into batches and continuously adjusts its parameters according to the performance of each batch (Fig. 14).

Sensitivity, specificity, accuracy, and recall are among the criteria used to assess the CNN model once it has been trained. In medical applications, sensitivity is especially important since it guarantees that a high proportion of cancerous nodules may be accurately identified by the model. To make sure that the model's predictions are not skewed by previous exposure to the data, CNN's performance is evaluated against a testing dataset that it has never seen before. Metrics like the F1 score, which balances precision and recall, are also used to test the model's reliability in real-world settings.

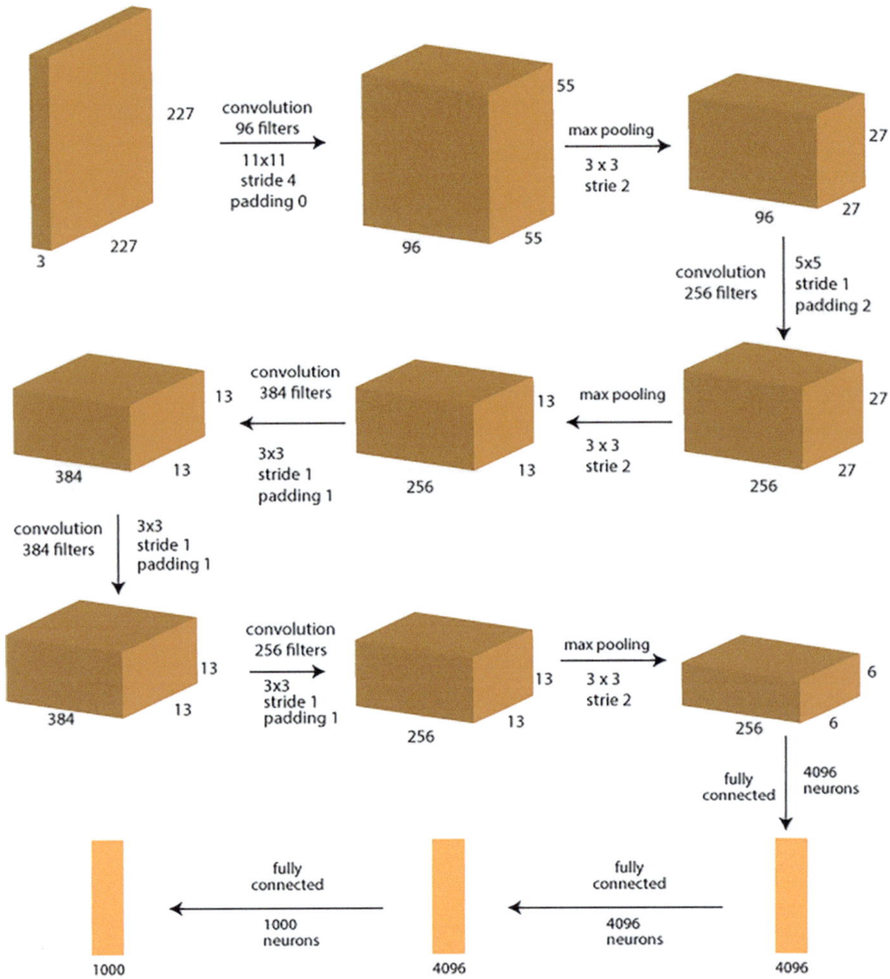

Fig. 14 (Jain, 2016) AlexNet Architecture

We separate the data into datasets for testing, validation, and training. The model is subsequently created and put into use, being trained on the training dataset and verified on the validation dataset. To evaluate the model's performance, this procedure is repeated. To enhance the model during this iteration, we might add more layers with various filter widths, apply average pooling, or change the architecture. Optimizing the model's performance is the aim. The test dataset should not be used for training or model optimization; it is only to be used for final evaluation. At the very end, the test dataset is utilized once to evaluate the actual performance of the model. Lastly, as seen in Fig. 15, which depicts the study design, we use Grad-CAM to display the model (Fig. 16).

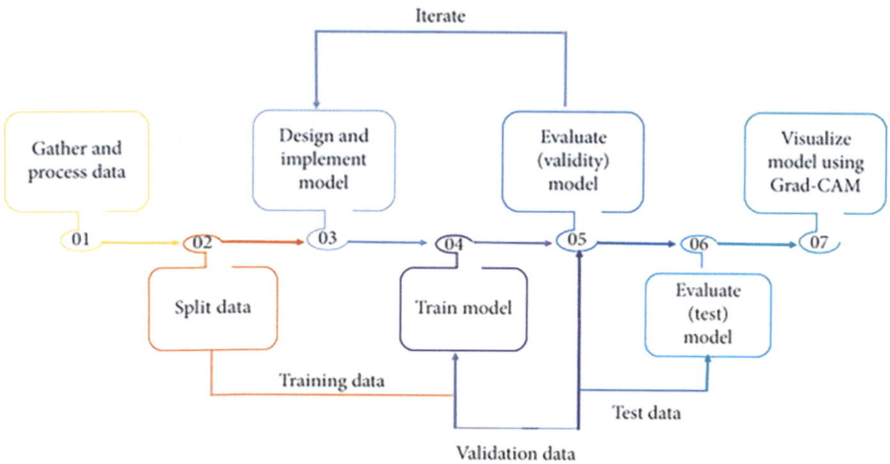

Fig. 15 (Joshua et al., 2021) GradCam Visualize

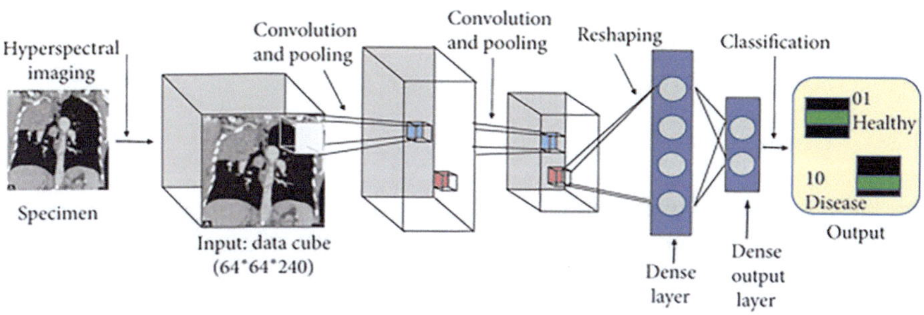

Fig. 16 (Joshua et al., 2021) CNN for liver cancer detection

To enhance the interpretability of the model, techniques like Grad-CAM (Gradient-weighted Class Activation Mapping) are used to visualize which areas of the CT scan the CNN focuses on when making predictions. This transparency is essential for clinicians who need to understand and trust AI-driven diagnoses. By highlighting the areas in the CT scan that contributed to the decision-making process, Grad-CAM helps validate the model's output, ensuring that it aligns with clinical expectations (Joshua et al., 2021) (Fig. 17).

The integration of CNNs into clinical workflows offers significant potential for early-stage lung cancer detection. By automating the detection of malignant nodules in CT scans, CNNs can reduce the workload of radiologists and minimize the risk of missed diagnoses. Moreover, the model's high sensitivity improves the likelihood of detecting cancer at earlier, more treatable stages, ultimately enhancing patient prognosis and survival rates.

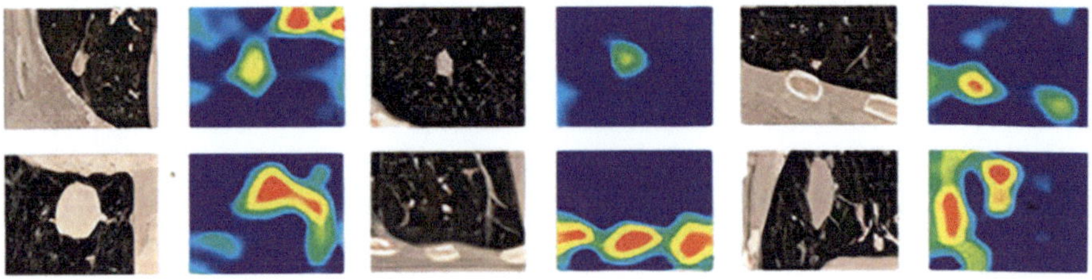

Fig. 17 (Joshua et al., 2021): Grad-CAM generated maps of CT scan

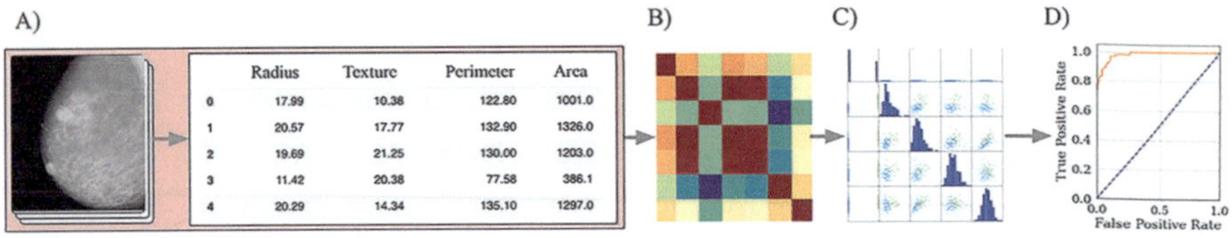

Fig. 18 (Abdelrahman et al., 2021) conventional CAD procedure. (**a**) Create a feature that describes the image by hand. (**b**) Choose pertinent features using correlation matrices and other filtering techniques. (**c**) The cleaned characteristics are then used in a statistical model to (**d**) draw conclusions

4.2 Breast Cancer

The integration of ***Convolutional Neural Networks (CNNs)*** for breast cancer detection from mammograms represents a significant advancement in ***Computer-Aided Detection (CAD)*** systems, offering accurate identification of malignant and benign abnormalities. Mammography is a minimally invasive first-line screening method, often followed by advanced imaging techniques like MRI or histological analysis for high-risk cases or confirmed anomalies. Traditional CAD pipelines relied heavily on handcrafted feature engineering, requiring experts to extract features such as texture, shape, and Gray-level intensity. These features were then used in statistical classifiers like ***Support Vector Machines (SVMs)*** or ***Random Forests (RF)*** to predict disease states. However, the reliance on manually engineered features made these methods labor-intensive and time-consuming, limiting their scalability and accuracy (Abdelrahman et al., 2021) (Fig. 18).

Each layer in the CNN architecture has a distinct function in feature extraction and categorization. Pooling layers minimize computational complexity while maintaining important information by reducing spatial dimensions after convolutional layers apply filters to identify low-level characteristics like edges and textures. The network uses several convolutional-pooling rounds to gradually learn high-level features. These characteristics are combined for classification by fully linked layers at the network's end, which forecast whether malignancy will be present or not.

Before being fed into the CNN, mammograms undergo preprocessing procedures such pixel leveling, image resizing to uniform dimensions, and data augmentation techniques like rotation and flipping to increase robustness. By ensuring consistency in the input data, these methods lessen the impact of different imaging conditions. The model's performance is validated using metrics such as F1-score, sensitivity, specificity, and accuracy. High sensitivity, usually between 94 and 96%, guarantees early breast cancer identification, reduces false negatives, and offers vital support for early therapies.

CNN-based systems are further optimized by sophisticated techniques like transfer learning. For tasks particular to mammograms, transfer learning entails optimizing CNN models that have already been trained, like as those that were trained on massive datasets like ImageNet. This method makes use of pre-trained networks' generic image analysis skills while cutting down on training time. Furthermore, radiologists can validate AI predictions by using interpretability approaches such as Gradient-weighted Class Activation Mapping (Grad-CAM), which highlight regions of relevance in mammograms. CNNs become more dependable and trustworthy by incorporating these techniques, which makes them suitable for clinical operations in the real world (Fig. 19).

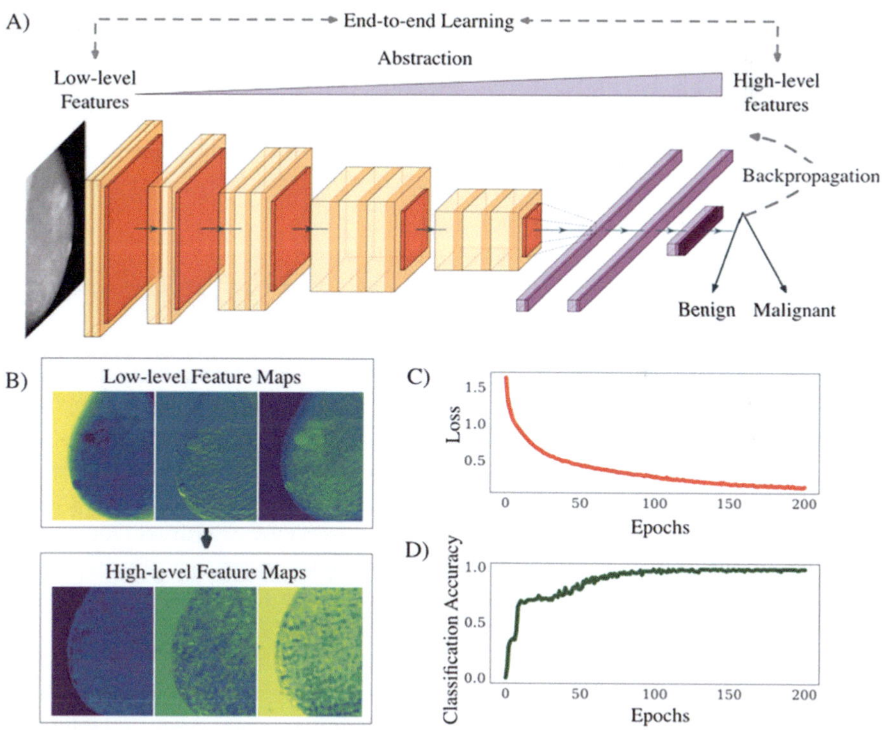

Fig. 19 (**a**) CNN Architecture and Model Learning: After pooling, the output is dark orange. Feature maps (orange stacks) are produced by many filters. Prior to classification, these maps are flattened into purple vectors. Backpropagation is used to update feature map settings depending on the error-measuring loss function. With every x-axis convolution, the feature maps get wider. The PlotNeuralNet GitHub repository was used in part to make this fig. (**b**) Feature Abstraction: Higher-level features are extracted as the model gets deeper. Convolutions reveal intricate patterns in high-level characteristics that cause CAD, whereas radiologists see patterns in low-level features. (**c**) Loss Function: An example plot that illustrates how the model reduces mistakes (loss) throughout training epochs. (**d**) Accuracy Improvement: Classification accuracy rises as the model gains knowledge and reduces loss. (Abdelrahman et al., 2021)

By removing the necessity for intensive feature engineering, CNNs have revolutionized this procedure. Instead, to preserve spatial qualities, CNN designs' convolutional layers use pattern-specific filters to extract low-level data like edges and textures. To maximize computing efficiency while preserving important information, pooling layers minimize the dimensionality of feature maps. Deeper levels of these iterative layers extract abstract, high-level information that are essential for categorizing anomalies. At the network's end, fully connected layers combine these characteristics to forecast malignancy. When a CNN is being trained, labeled mammography datasets are fed into the network. Backpropagation and optimization techniques like Adam or SGD are used to modify weights and reduce prediction errors, which are quantified by loss functions such binary cross-entropy (Shah et al., 2023).

The integration of CNNs into breast cancer detection workflows has the potential to revolutionize diagnostics. These systems provide real-time assistance to radiologists, identifying suspicious regions in mammograms for further investigation. Unlike traditional CAD methods, CNNs offer an end-to-end learning process with minimal manual intervention, significantly improving accuracy and efficiency. This advanced approach facilitates early detection, reduces diagnostic errors, and enhances survival rates through timely intervention. By addressing limitations of earlier CAD systems, CNN-based methodologies position themselves as a critical tool in the fight against breast cancer, seamlessly bridging technology and healthcare to save lives (Singh et al., 2006).

4.3 Prostate Cancer

Radiomics, in conjunction with Support Vector Machines (SVM), offers a powerful methodology for the detection and classification of prostate cancer, especially in distinguishing aggressive tumors from indolent ones. The process begins with acquiring medical imaging data, commonly MRI scans, which provide detailed anatomical and functional information of the prostate. Radiomics focuses on extracting a large number of quantitative features from these images, which include texture, shape, and intensity features that can reflect underlying tumor biology. These features are often not visible to the naked eye but carry significant information regarding the tumor's characteristics, such as its aggressiveness (Varan et al., 2023).

The suggested prostate cancer classification system is shown in Fig. 20, where a fine-tuned ridge regressor is used for all feature selectors. The Support Vector Machine (SVM) classifier and regressor hyperparameters are optimized using the random search algorithm to improve accuracy and performance. Following their extraction from the prostate imaging data, the radiomic characteristics are fed into machine learning techniques, particularly support vector machines (SVMs). SVM is a supervised learning technique that can differentiate between aggressive and indolent prostate cancers because it performs well in binary classification tasks. Finding the hyperplane that best divides data points from two distinct classes is the fundamental concept of support vector machines (SVM). The SVM would be

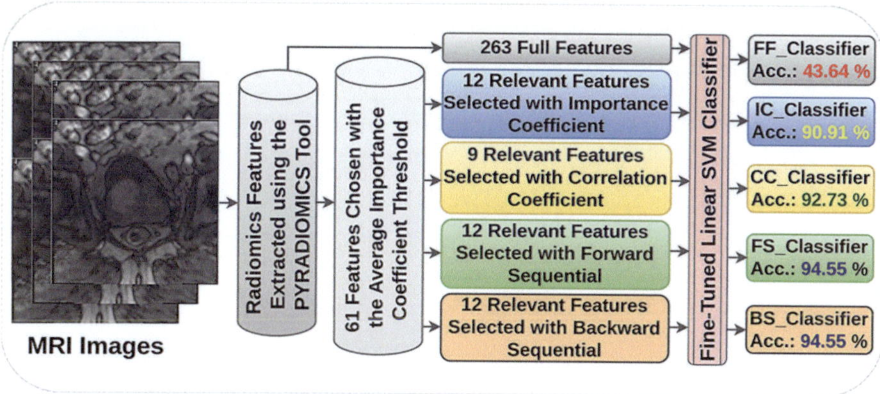

Fig. 20 The suggested categorization schemes for prostate cancer. All feature selections were modelled using the fine-tuned ridge regressor. The random search approach was used to optimize the regressor and SVM classifier's hyperparameters. (Varan et al., 2023)

trained to categorize prostate cancer tumors using the radiomic characteristics that were retrieved. The model is trained using labeled datasets in which clinical specialists have already identified the aggressive and indolent cancers. The SVM model may then accurately classify new, unseen tumors by identifying a border (or hyperplane) that optimizes the margin between these two classes (Akbari et al., 2012) (Fig. 21).

The flowchart for our LS-SVM classification approach is displayed in Fig. 22. For LS-SVM training, we initially segment the tumor on a single set of hyperspectral pictures. A new set of hyperspectral pictures is subjected to automatic tumor categorization using the trained LS-SVM. We employed 251 element input vectors in photos with a spectral resolution of 2 nm. The classification of the image's normal and cancerous tissue is the output of the trained LS-SVM.

Figure 21 depicts a mouse with a tumor that was utilized in the HSI experiment. The in vivo imaging investigation involved 11 mice in all. For every mouse, hyperspectral pictures were obtained. The spectral images of a normal mouse are displayed in Fig. 23. The sensitivity of pictures taken at different wavelengths to identify the same tumor varies. The spectra of cancer and healthy tissue were compared using quantitative analysis of the spectral pictures.

The implementation of this technology begins with data preprocessing, which involves normalizing and standardizing the MRI images to ensure uniformity and remove noise from the data. After preprocessing, a feature extraction step takes place, where algorithms are applied to quantify the textural, morphological, and intensity-based properties of the tumor region. These features may include histogram-based features, fractal dimensions, and higher-order statistical measures like entropy, correlation, and contrast, all of which can help capture the subtle differences between aggressive and indolent tumors. Once the features are extracted, the data is split into training and testing sets (Wu et al., 2015).

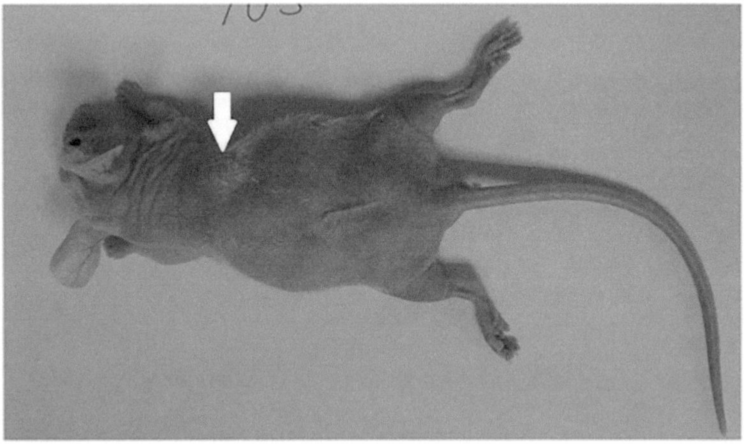

Fig. 21 Rat with prostate tumor (arrow). (Akbari et al., 2012)

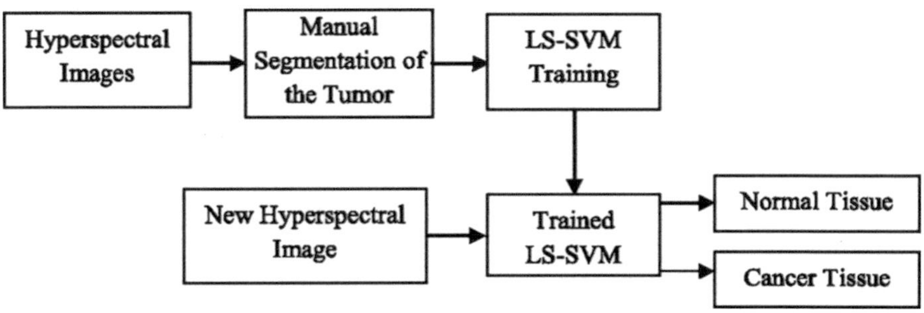

Fig. 22 The flow chart of the tissue classification method for hyperspectral images

Advanced prediction models must be incorporated into the diagnostic process, especially for the "grey zone" of prostate cancer diagnosis, which occurs when PSA values are less than 10 ng/ml. A more thorough knowledge of the efficacy of such models may be found in Figs. 3 and 4 of this work. Figure 24's calibration plots evaluate the degree of agreement between expected probability and actual results. Although neither model achieves full calibration, Chun's nomogram demonstrates greater calibration in the lower probability ranges, indicating that a well-calibrated model nearly corresponds with the 45-degree line. The net benefit of using each model at different threshold probabilities is examined in the decision curve analysis (DCA) shown in Fig. 4 to assess clinical utility. Chun's nomogram and the new PCPT calculator both perform better than the "treat all" (Perdonà et al., 2010) (Fig. 25).

The next step is model training, where the SVM algorithm is fed the training dataset. The SVM uses these features to construct an optimal hyperplane, iteratively refining its parameters to minimize classification errors. Regularization techniques such as kernel functions (e.g., radial basis function) may be employed to handle

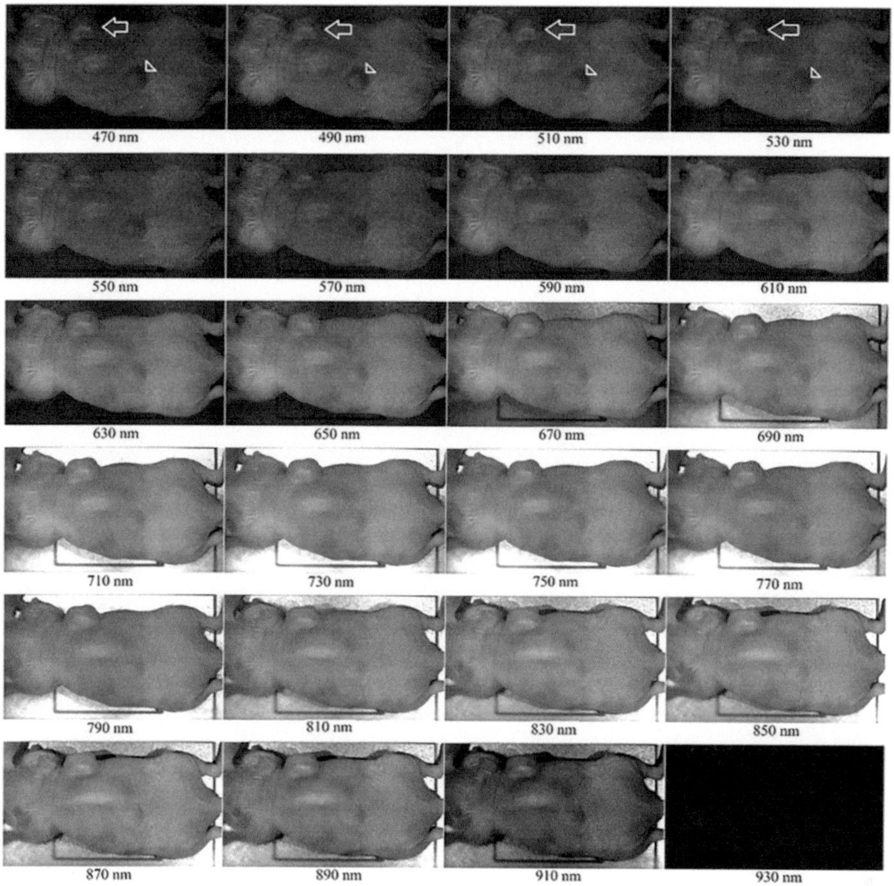

Fig. 23 Spectral images of a tumor-bearing mouse at different wavelengths. The tumor (arrow) and the pellet (arrow head) are visible in images taken at various wavelengths). (Akbari et al., 2012)

non-linear relationships between the features, making the SVM robust enough to deal with the complex nature of medical images (Song et al., 2020). Advanced tools like the updated PCPT calculator, which prioritizes accuracy, and Chun's nomogram, which balances calibration and clinical net benefit, further enhance this analytical framework.

After training, the model is evaluated on a separate testing dataset to assess its performance. Performance metrics such as accuracy, precision, recall, and the area under the receiver operating characteristic curve (AUC-ROC) are used to validate the effectiveness of the model in distinguishing between aggressive and indolent tumors. The updated PCPT calculator, for example, achieves an AUC-ROC of 79.6%, while Chun's nomogram attains 71.5%, reflecting its slightly lower accuracy but stronger calibration and higher net benefit as revealed by the DCA. The higher the accuracy and AUC, the better the model is at correctly classifying new,

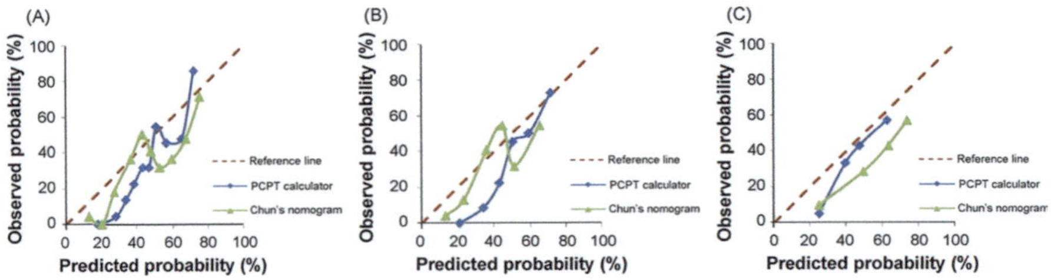

Fig. 24 Receiver operating characteristic (ROC) curves are shown for the PCPT risk calculator, Chun's nomogram, prostate cancer antigen 3 (PCA3), and prostate-specific antigen (PSA) across three scenarios: (**a**) the entire cohort, (**b**) initial biopsies, and (**c**) repeat biopsies. (Perdonà et al., 2010)

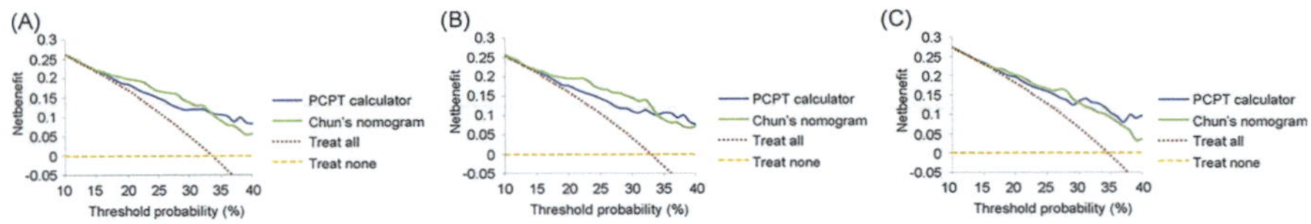

Threshold probability (%)		10	15	20	25	30	35	40
Net Benefit (%)	PCPT	26.2	22.2	18.4	15.0	12.0	10.6	8.3
	Chun's	26.4	22.2	19.8	16.8	13.3	8.9	5.7
	Treat All	26.1	21.7	16.9	11.3	5.0	-2.3	-10.8
Net reduction in avoidable biopsies (%)	PCPT	1	2	6	11	16	24	29
	Chun's	2	3	12	17	21	21	25

Fig. 25 (Perdonà et al., 2010) shows how to use Chun's nomogram and the Prostate Cancer Prevention Trial (PCPT) risk calculator to predict positive prostate biopsies using decision curve analysis (DCA). Within a threshold probability range of 10–40%, the analysis encompasses three scenarios: the complete cohort, first biopsies, and repeat biopsies. The accompanying table shows the whole cohort's net benefits and unnecessary biopsy reductions as compared to the "treat all" approach for different threshold probabilities

unseen tumor cases. Additionally, cross-validation techniques may be employed to ensure that the model is not overfitting to the training data and is generalizable to different patient cohorts (Perdonà et al., 2010).

Radiomics with SVM has the potential to significantly impact prostate cancer diagnosis by enabling personalized treatment plans. By differentiating between aggressive and indolent tumors, the technology allows for more tailored therapeutic decisions. For instance, aggressive tumors may be treated with more intensive therapies like surgery or radiation, while indolent tumors might be managed with active surveillance, avoiding overtreatment and its associated side effects. Furthermore, the integration of radiomics with SVM could help in reducing inter-radiologist variability, ensuring a more consistent and objective analysis of prostate cancer. In clinical settings, this methodology could be implemented in routine imaging workflows, enhancing early detection and improving patient outcomes.

4.4 Skin Cancer

Dermoscopic image analysis using deep learning, particularly with models like ResNet (Residual Networks), has proven to be an effective method for skin cancer detection. Drawing insights from recent advancements in melanoma research, including a novel study proposing Lightweight Convolutional Neural Networks (LWCNN) with transfer learning, this report explores the methodology and impact of such technologies in distinguishing melanoma from benign lesions (Fig. 26).

The process begins with the collection of high-resolution dermoscopic images that provide detailed visual representations of skin lesions. These images undergo rigorous preprocessing to standardize them for deep learning models. Key steps include resizing the images to 224x224 pixels, normalization, and augmentation techniques like rotation and flipping, which help models generalize better across diverse conditions. Image enhancement techniques are also employed to highlight features critical for melanoma detection.

The usefulness of ResNet, a deep convolutional neural network renowned for its residual connections that lessen the vanishing gradient issue, is highlighted in the paper. ResNet is especially well-suited for evaluating intricate dermoscopic pictures because of these connections, which allow it to learn deeper and more abstract properties without losing important information. ResNet gradually learns higher-level patterns, including lesion asymmetry, border abnormalities, and pigmentation changes—important indicators for differentiating melanoma—after extracting low-level information, such edges and textures, from the initial layers (Mohapatra et al., 2021).

Additionally, the study explores the potential of LWCNN, a lightweight model designed to achieve high accuracy with lower computational requirements. This innovation is crucial for resource-constrained settings where access to high-performance computing infrastructure may be limited. LWCNN, alongside established models like GoogleNet, ResNet-18, and MobileNet-v2, leverages transfer

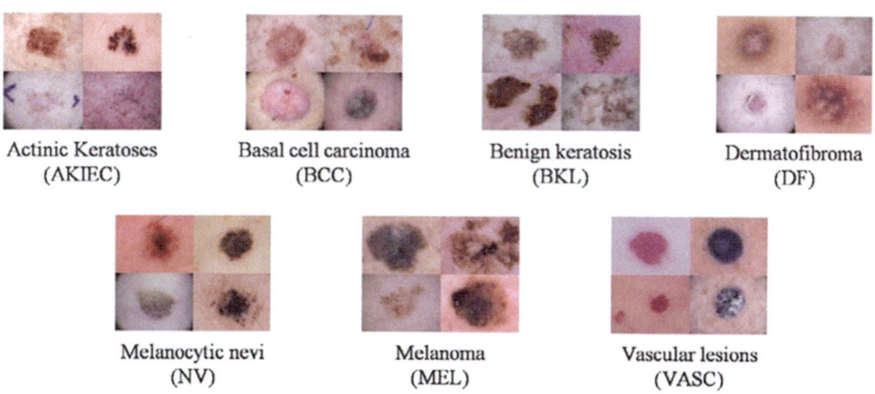

Fig. 26 Sample images of 7 category of diagnostics. (Qin et al., 2020)

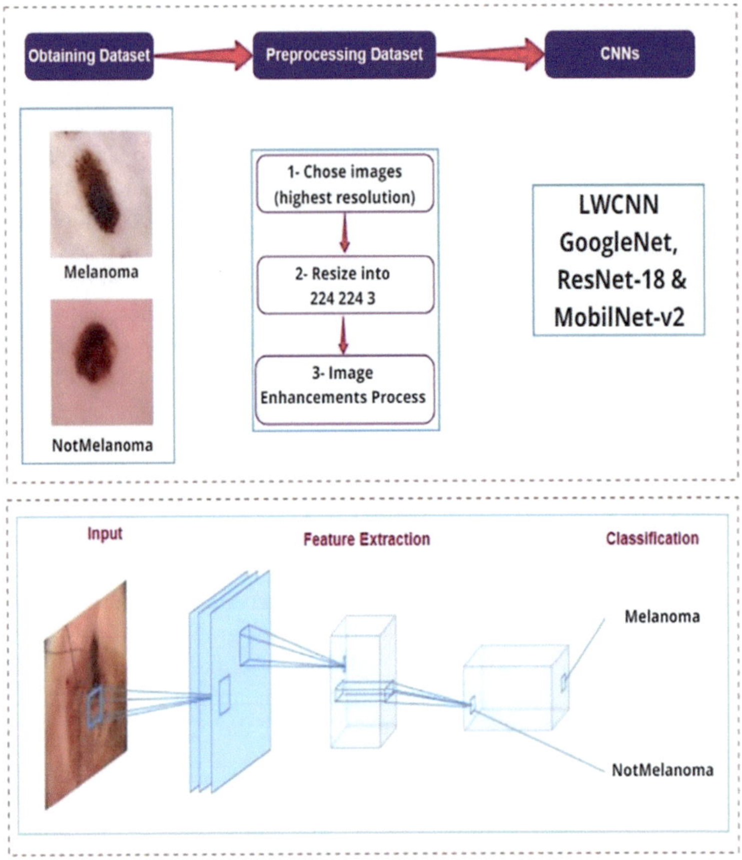

Fig. 27 (Alabdulkreem et al., 2024): Methodology for detection using CNN and MATLAB of Melanoma and NotMelanoma

learning to adapt pre-trained knowledge from other datasets, enhancing its performance on melanoma detection tasks. These models collectively analyze images to extract distinguishing features and classify lesions into benign and malignant categories (Fig. 27).

During the training phase, labeled datasets annotated by dermatologists are used to teach the models. Loss functions measure prediction errors, and backpropagation techniques, supported by optimizers such as Adam, refine the model's weights to improve accuracy. Data augmentation further enriches the dataset, ensuring robustness in handling varied and unseen cases. The integration of lightweight architectures, such as LWCNN, and preprocessing methods emphasizes the importance of efficiency in medical imaging tasks. The reference study highlights that lightweight models can complement traditional deep learning architectures by balancing computational demands and diagnostic precision. This dual approach addresses the challenge of early melanoma detection in regions with limited access to

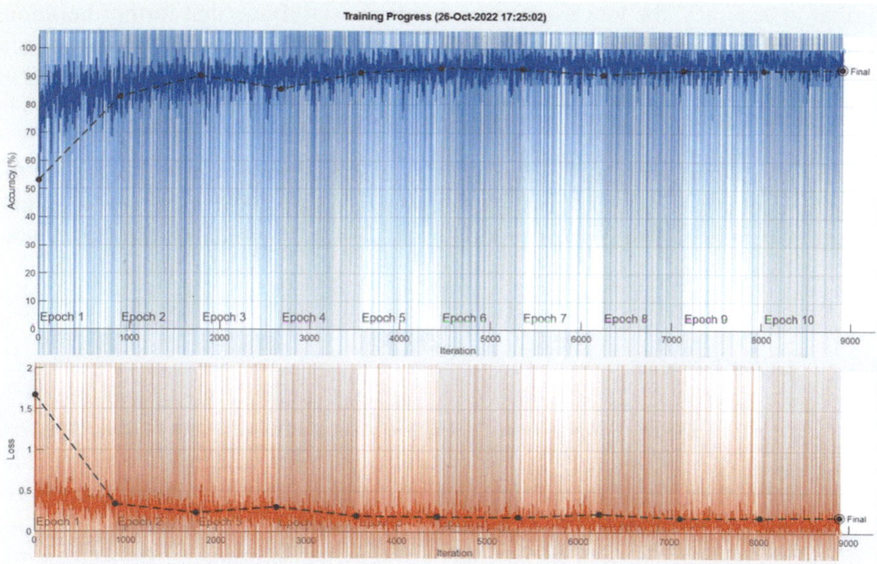

Fig. 28 (Alabdulkreem et al., 2024): ResNet-18 in experiment (1) accuracy and training loss

dermatologists, potentially reducing fatal outcomes through timely intervention. This analysis is depicting the critical role of preprocessing, feature extraction, and transfer learning in advancing melanoma diagnosis. By drawing from state-of-the-art studies and models, it becomes evident that combining innovative architectures like LWCNN with established deep learning frameworks can significantly enhance the accuracy and accessibility of skin cancer detection systems. These methodologies not only improve diagnostic reliability but also provide scalable solutions to bridge the gap in medical expertise, particularly in underserved areas (Alabdulkreem et al., 2024).

Figure 28 provides a comprehensive visual overview of the ResNet-18 model's performance during the training process in the first experiment. The graph tracks the progression of the training iterations along the x-axis, offering insight into how the model evolves as it learns. The y-axis measures two critical metrics: accuracy, which indicates the model's ability to correctly classify skin lesions, and loss, which quantifies the prediction errors made by the model. The trends depicted in the figure reveal a consistent improvement in the model's accuracy as training progresses. With each iteration, the accuracy steadily climbs, reflecting the model's increasing proficiency in distinguishing between melanoma and non-melanoma lesions. This upward trend continues until a saturation point is reached, beyond which additional training yields minimal improvements. This plateau suggests that the model has successfully learned the essential features necessary for effective classification. Conversely, the loss metric displays a downward trajectory over the course of training. A decreasing loss indicates that the model is progressively making fewer errors in its predictions, demonstrating an enhanced understanding of the input data.

Similar to accuracy, the loss eventually stabilizes, signifying that further iterations bring diminishing returns in performance improvements. This stabilization reflects the model's convergence and its readiness for deployment (Mohapatra et al., 2021).

In summary, Fig. 28 effectively illustrates the learning dynamics of the ResNet-18 model, showcasing its ability to refine performance through increased accuracy and reduced prediction errors as training progresses. The trends depicted highlight the model's capability to achieve a balance between learning key features and minimizing errors, culminating in a robust system for skin lesion classification (Alabdulkreem et al., 2024).

4.5 Liver Cancer

Liver cancer detection using a Semi-supervised Generative Adversarial Network (GAN) for MRI analysis represents a cutting-edge approach that addresses challenges in acquiring sufficient labeled data, a common issue in medical image processing. This method makes use of the capabilities of GANs, a kind of machine learning methods made up of two neural networks: a discriminator and a generator. The discriminator attempts to distinguish between created and genuine MRI pictures in the setting of liver cancer, whereas the generator produces artificial images that resemble actual lesions. Both networks get better over time as a result of this adversarial process, with the discriminator getting better at identifying minute distinctions and the generator creating pictures that are more realistic (Zhao et al., 2020) (Figs. 29 and 30).

The semi-supervised aspect of the GAN is particularly valuable in medical imaging, where acquiring labeled data (i.e., MRI scans annotated with specific diagnoses) is time-consuming and expensive. In traditional supervised learning, large quantities of labeled data are required for effective training. However, in the case of liver cancer, where expert annotations are limited, semi-supervised GANs can learn

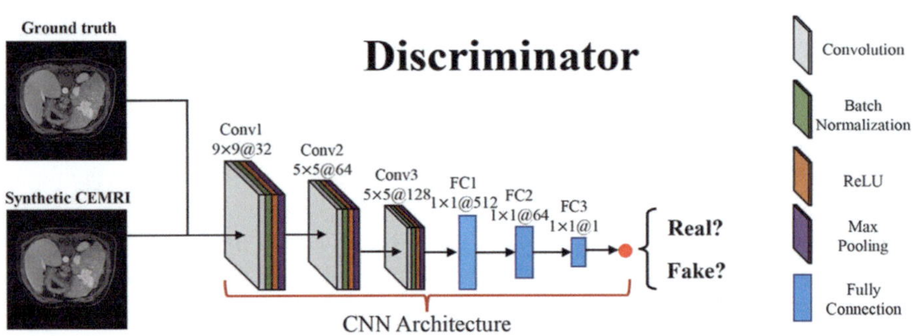

Fig. 29 The discriminative network's CNN architecture, which is intended to analyze both synthetic and ground truth CEMRI inputs. The network determines if the input data is synthetic or genuine by producing discriminative outputs. By employing an adversarial approach, it successfully directs the attention-aware generator to recognize and correct its own errors, enhancing the synthetic CEMRI's realism and authenticity. (Zhao et al., 2020)

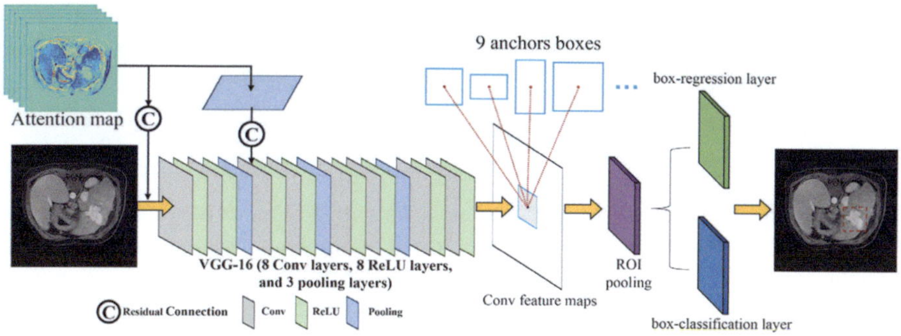

Fig. 30 The structure of the tumor detection network, which uses synthetic CEMRI to precisely locate and categorize the tumor's Region of Interest (RoI). The VGG-16-based convolutional processes are improved by incorporating attention maps from the generator into the detector through residual connections. This makes it possible to extract tumor-specific information more successfully and boosts overall tumor detection performance. Furthermore, the generator is encouraged to narrow its emphasis on the distinctive features that differentiate two types of tumors by the back-propagation of the classification loss Lcls. Through back-propagation, the detector and generator may both improve when Lclsin is included into the Tripartite-GAN architecture. (Zhao et al., 2020)

from both labeled and unlabeled data, dramatically reducing the need for a large, annotated dataset. The model can use a small set of labeled MRI scans to train the generator and discriminator, while simultaneously learning from a much larger pool of unlabeled data. This approach effectively uses the limited labeled data to guide the model in understanding the complex patterns associated with liver lesions, while the unlabeled data further improves the model's generalization ability (Zhao et al., 2020; Qin et al., 2020).

Once trained, the GAN can be used to analyze new MRI scans, detecting liver lesions with high accuracy (91%). The generated synthetic data plays a crucial role in augmenting the original dataset, providing the model with a more diverse set of examples and improving its ability to detect various types of lesions, including those that may be rare or atypical. The generator's synthetic lesions assist the discriminator in identifying these challenging cases, which might otherwise be missed in a traditional machine learning framework (Qin et al., 2020).

Moreover, the semi-supervised GAN approach improves diagnostic efficiency. In clinical practice, radiologists are often faced with large volumes of imaging data, and manually annotating each scan is both time-consuming and error-prone. The GAN-based model automates much of this process, highlighting areas of interest in the MRI scans that may warrant closer inspection. By doing so, it not only reduces the burden on radiologists but also helps in the early detection of liver cancer, potentially improving patient outcomes. This is particularly important for liver cancer, as early-stage detection can lead to more effective treatments and higher survival rates (Adamu et al., 2024).

Overall, semi-supervised GANs provide a powerful tool for liver cancer detection, enhancing diagnostic accuracy while minimizing the need for large labeled datasets. The combination of generative and discriminative learning, along with the ability to handle both labeled and unlabeled data, offers a promising solution for improving liver cancer diagnosis and reducing the time and resources required for manual analysis. By integrating this technology into clinical workflows, the detection of liver lesions can become faster, more accurate, and more efficient, ultimately benefiting patients and healthcare providers alike (Qin et al., 2020; Adamu et al., 2024).

The algorithm used for liver cancer detection with a Semi-supervised Generative Adversarial Network (GAN) involves two main components: the **generator** and the **discriminator**. The **generator** creates synthetic MRI images, while the **discriminator** differentiates between real and fake images. In the semi-supervised version, the discriminator is trained using both labeled and unlabeled data. The GAN's adversarial process helps both networks improve over time, enabling the model to detect liver lesions effectively, even with fewer labeled samples. This enhances diagnostic accuracy and reduces the reliance on large labeled datasets (Gouda et al., 2022).

4.6 Ovarian Cancer

Ultrasound-based machine learning (ML) using XGBoost for ovarian cancer detection leverages advanced computational techniques to analyze ultrasound imaging data and differentiate between benign and malignant ovarian masses. The process begins with collecting a high-quality dataset of labeled ultrasound images, annotated to indicate whether the masses are benign or malignant. These datasets include features such as size, shape, texture, and acoustic properties, extracted through image processing techniques. Popular datasets like the "Ovarian Cancer Dataset" provide a solid foundation for training machine learning models (Bhattacharjee et al., 2017). These kinds of jobs are well suited for the XGBoost method, an improved gradient boosting approach that can handle big, unbalanced datasets and simulate intricate non-linear connections. By building an ensemble of decision trees, XGBoost increases accuracy while reducing overfitting, a significant problem in medical imaging with sparse labeled data, by correcting mistakes from the prior trees (Hsiao et al., 2021).

The XGBoost algorithm, an optimized gradient-boosting method, has proven particularly effective in medical imaging tasks. It builds an ensemble of decision trees where each subsequent tree corrects the errors of the previous one, improving overall accuracy while minimizing common challenges like overfitting. XGBoost is especially suited to handle high-dimensional and imbalanced datasets, as often encountered in medical imaging, where the number of malignant cases is significantly smaller than benign ones (Hsiao et al., 2021). Preprocessing steps, such as resizing images, normalization, and augmentation techniques like flipping and rotation, standardize the data and enhance the diversity of the training set, ensuring that the model can generalize effectively across various conditions. Extracted features, including texture-based attributes, shape descriptors, and statistical properties, are then used to train XGBoost, enabling it to identify patterns that distinguish benign from malignant cases.

GA-XGBoost extends the capabilities of XGBoost by integrating Genetic Algorithms (GA) for feature selection and optimization. While XGBoost excels at refining decision trees for accurate classification, GA enhances the process by identifying the most relevant features. This optimization occurs through iterative processes inspired by natural selection, such as mutation, crossover, and selection, which refine the feature subset to maximize model performance. Additionally, GA-XGBoost employs a bagging technique, training multiple models independently and aggregating their predictions. In identifying high-risk instances, this lowers variation, increases robustness, and improves accuracy. Consequently, GA-XGBoost has outperformed conventional techniques such as LASSO regression in terms of accuracy and sensitivity when it comes to differentiating between benign and malignant tumors (Patil et al., 2023). Genetic algorithms (GAs) have been designed to replicate the concept of natural selection by searching for an available combination of gene expression profiling probes that would provide a prediction model with higher performance (Dabeer et al., 2019; Huang et al., 2017). Therefore, we call this technique GA-XGBoost, which employs a GA to further improve the feature selection of the XGBoost algorithm. Figure 31 shows the five main steps of a GA: starting population, fitness function, selection, crossover, and mutation.

With an accuracy of around 85%, XGBoost-based methodologies have shown significant promise in clinical settings. They are particularly effective at identifying malignant masses that might otherwise be missed by traditional diagnostic methods. The performance of these models is assessed using key evaluation metrics such as accuracy, sensitivity, specificity, and the area under the receiver operating characteristic (ROC) curve. Sensitivity is particularly critical in medical applications to ensure that malignant cases are not overlooked, as missing such diagnoses can have severe consequences for patient outcomes. The ability to detect subtle patterns in ultrasound data that may not be apparent to the human eye makes these models

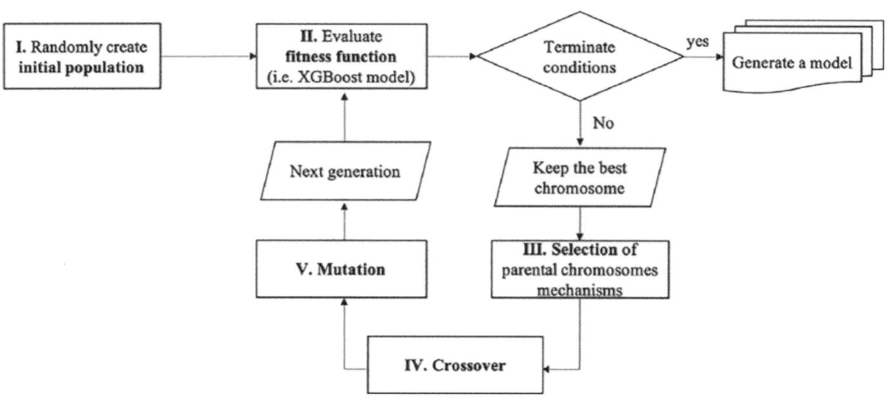

Fig. 31 (Hsiao et al., 2021) Genetic Algorithm flowchart

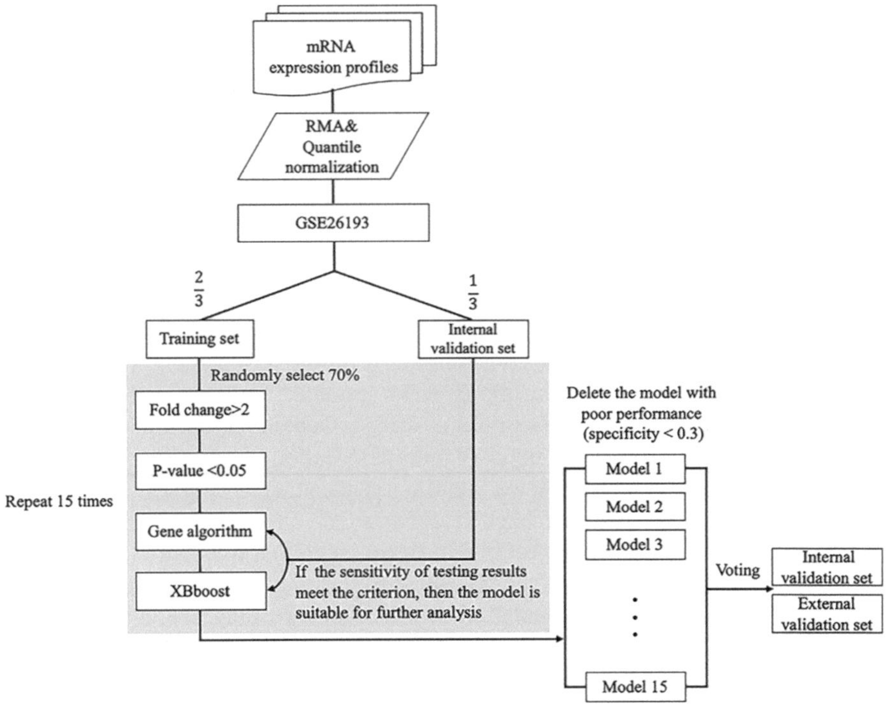

Fig. 32 (Hsiao et al., 2021): Pipeline of GA XGBoost Bagging Algorithm

invaluable tools for early and accurate cancer detection (Yuan et al., 2022). Figure 31 provides an overview of the workflow designed to identify high-risk ovarian cancer (OC) patients. This comprehensive approach includes data preprocessing, dimensionality reduction, a bagging-based algorithm incorporating GA-XGBoost models, and external validation to construct a robust prediction model (Fig. 32).

By integrating XGBoost and GA-XGBoost into clinical workflows, healthcare professionals can benefit from more precise diagnostic tools that reduce the likelihood of diagnostic errors. These models also lessen the burden on medical staff by automating complex analyses, enabling clinicians to focus on more challenging cases. Furthermore, these methodologies support personalized treatment strategies by stratifying patients based on risk, ensuring that high-risk individuals receive timely and targeted interventions while low-risk patients avoid unnecessary procedures. The integration of ultrasound imaging with advanced machine learning algorithms marks a transformative step in ovarian cancer detection, offering efficient, accessible, and highly accurate diagnostic solutions in clinical practice (Fig. 33).

The algorithm used for ovarian cancer detection in ultrasound-based machine learning is XGBoost (Extreme Gradient Boosting). This algorithm is a type of gradient boosting that builds an ensemble of decision trees sequentially. It focuses on reducing prediction errors by correcting the mistakes of previous trees. XGBoost is efficient in handling large datasets, capable of managing high-dimensional features,

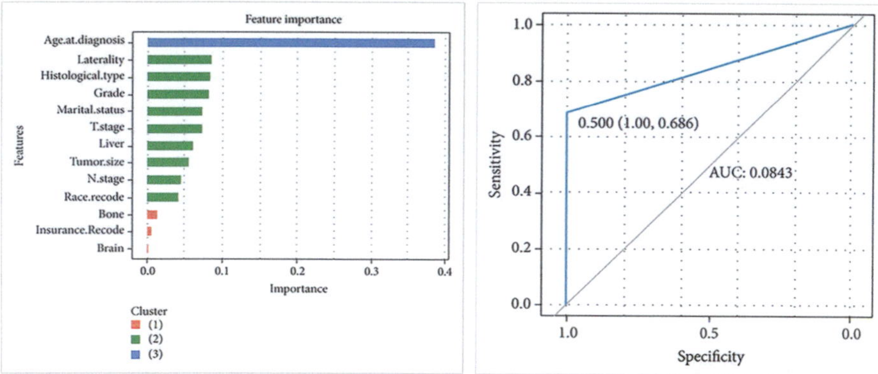

Fig. 33 (Cui et al., 2021): Age, laterality, histology type, grade, and marital status were the top five parameters determined by the gbm (Light Gradient Boosting Machine) method. After generating receiver operating characteristic (ROC) curves, XGBoost showed excellent discriminative performance with an AUC of 0.843. It had a sensitivity of 1.000, a specificity of 0.686, and an accuracy of 0.982

and effective in minimizing overfitting. It is particularly useful for classifying benign and malignant ovarian masses based on extracted features like size, shape, and texture from ultrasound images (Hsiao et al., 2021).

4.7 Thyroid Cancer

The detection of thyroid cancer has significantly advanced with the integration of two robust methods: deep convolutional neural networks (CNNs) for computer-aided diagnosis (CAD) and Long Short-Term Memory (LSTM) networks for ultrasound-based analysis. Each approach independently enhances the precision and reliability of thyroid cancer detection, leveraging distinct strengths tailored to specific clinical and technical challenges.

The CNN-based CAD system focuses on thyroid ultrasound (US) images to detect malignancy and predict the BRAFV600E mutation, a critical genetic marker linked to aggressive papillary thyroid cancer (PTC). The process begins with the acquisition of high-quality US images, where regions of interest (ROIs) are defined and standardized through preprocessing steps like normalization. The CNN extracts essential spatial features such as echogenicity, nodule orientation (e.g., taller-than-wide shape), and margin characteristics, identifying patterns associated with malignancy. The CAD system outputs a binary classification (cancer or not) and a continuous malignancy risk score (0–100%) (Yoon et al., 2020). This risk score is further integrated with clinical parameters like patient age and tumor size into a multivariable logistic regression model, enhancing predictive accuracy. The CAD system achieves an AUC of 0.646, which improves to 0.706 when combined with clinical factors, demonstrating its ability to identify high-risk nodules effectively. Additionally, this approach minimizes inter-observer variability and operator dependency, providing consistent and objective results critical for preoperative decision-making. The model's high sensitivity (85.3%) ensures that malignant nodules are reliably identified, reducing the likelihood of missed diagnoses. This method is particularly effective for early detection, where image features can correlate with genetic markers like the BRAFV600E mutation, streamlining risk stratification and optimizing surgical strategies (Yoon et al., 2020).

The process of the suggested framework is depicted in Fig. 34. Prior to being divided into several smaller patches (e.g., 256 × 256 pixels) inside each WSI, each slide is first divided into its foreground region (i). (iii) All foreground patches are transformed into sets of low-dimensional feature embeddings for the attention network using feature extraction. (iv) The diagnostic prediction is then generated by the attention network by integrating patch-wise data into slide-level representations. (v) The attention network scores each region on the slide and produces an attention score based on how crucial each region is to the slide-wise diagnosis after the slide-wise representations have been obtained. Attention pooling then compresses patch-level information into slide-level data representations by balancing patches

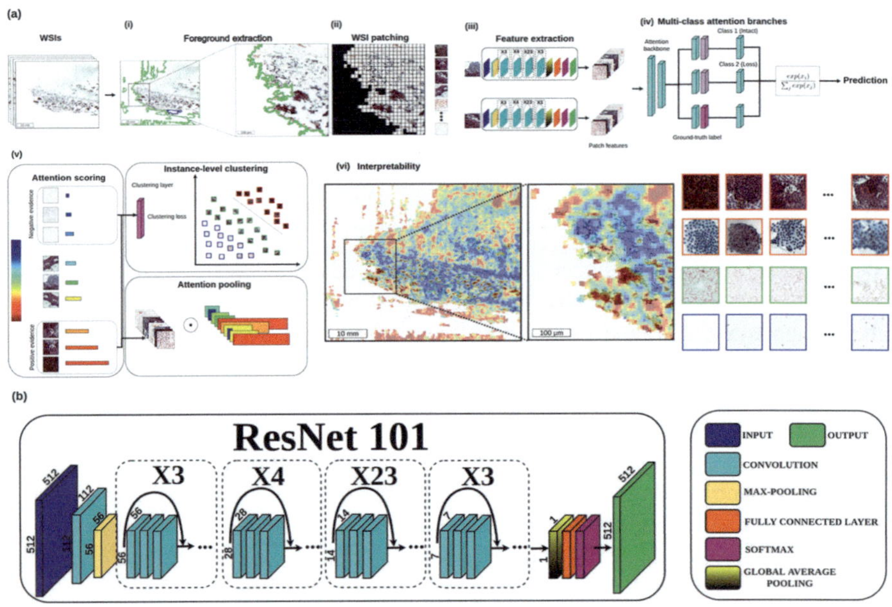

Fig. 34 An overview of the annotation-free deep learning architecture for determining the BRAFV600E status from thyroid FNA and TP WSIs stained with Papanicolaou is shown in (**a**) (i) Foreground segmentation. WSI patching (ii). (iii) Extracting features. (iv) Creating slide-level representations by combining patch-based characteristics. (v) Regions on the slide are ranked by the attention network, which assigns attention ratings according to how relevant they are to the diagnosis (left). To condense features into slide-level representations, attention pooling weights patches based on these scores (bottom right). Positive and negative occurrences are distinguished by regions that are patched strongly (red) and weakly (blue) (top right). (vi) To emphasize important ROIs for interpretation, attention scores are displayed as heatmaps. (**b**) The suggested deep learning model's architecture. (Wang et al., 2023)

according to their individual attention scores. Consequently, the clustering process that distinguishes positive from negative instances is supervised by representative samples that are heavily patched (shown by red regions) and lightly patched (represented by blue regions). (vi) To find ROIs and decipher the critical morphology utilized for diagnosis, a heatmap visualization may be created using the attention ratings. The detailed architecture of the suggested deep learning model utilized as a feature extractor is shown in Fig. 34b (Wang et al., 2023).

On the other hand, LSTM networks excel in analyzing sequential data from ultrasound imaging, offering a unique capability to process temporal patterns that static CNN approaches might overlook. In LSTM-based analysis, (Rathinakumar & Priya, 2023) US images are preprocessed similarly, with techniques like resizing, normalization, and augmentation enhancing data quality. Extracted features such as texture, size, and shape are fed into the LSTM network, which is specifically designed to handle temporal dependencies and long-term contextual data. This ability enables LSTM to analyze sequences of ultrasound frames, capturing subtle changes in nodule characteristics over time, such as evolving textures or motions indicative of malignancy. The training process uses labeled datasets like the Thyroid Cancer Dataset, where the LSTM optimizes its weights through backpropagation and gradient descent, minimizing prediction errors. With a precision rate of 98.7% and an accuracy of 90%, LSTM networks are highly effective in distinguishing benign from malignant nodules, significantly reducing false negatives—a critical requirement for early-stage cancer detection. The SR-LSTM algorithm further enhances diagnostic performance, achieving high recall and F1-scores, making it ideal for improving clinical decision-making and reducing unnecessary biopsies in resource-limited settings (Fig. 35).

The iterative procedure of the GTO algorithm used to optimize the SR-LSTM model's hyperparameters is described in the pseudocode. To find the optimal solution, a random population must be created, the objective function must be assessed, and the population must be sorted. The program then updates the coordinates and computes LS, JS, and VD by carrying out a number of actions, including Forage Moving, Choosing a Suitable Area, and Attacking Target. The optimal hyperparameters are returned if the predetermined criteria are satisfied after the model accuracy and MSE are evaluated. If not, the procedure is repeated until either the maximum number of iterations or convergence is achieved.

Both methods complement each other by addressing distinct aspects of thyroid cancer detection. While CNN-based CAD systems excel in extracting spatial features and associating them with genetic markers, LSTM networks bring the advantage of sequential analysis, capturing temporal variations in nodule characteristics. Together, these technologies represent a transformative step in thyroid cancer diagnostics, offering high precision, reduced diagnostic variability, and enhanced clinical utility. Their integration into clinical workflows enables radiologists and oncologists to make more accurate, timely, and informed decisions, ultimately improving patient outcomes (Sharma et al., 2023).

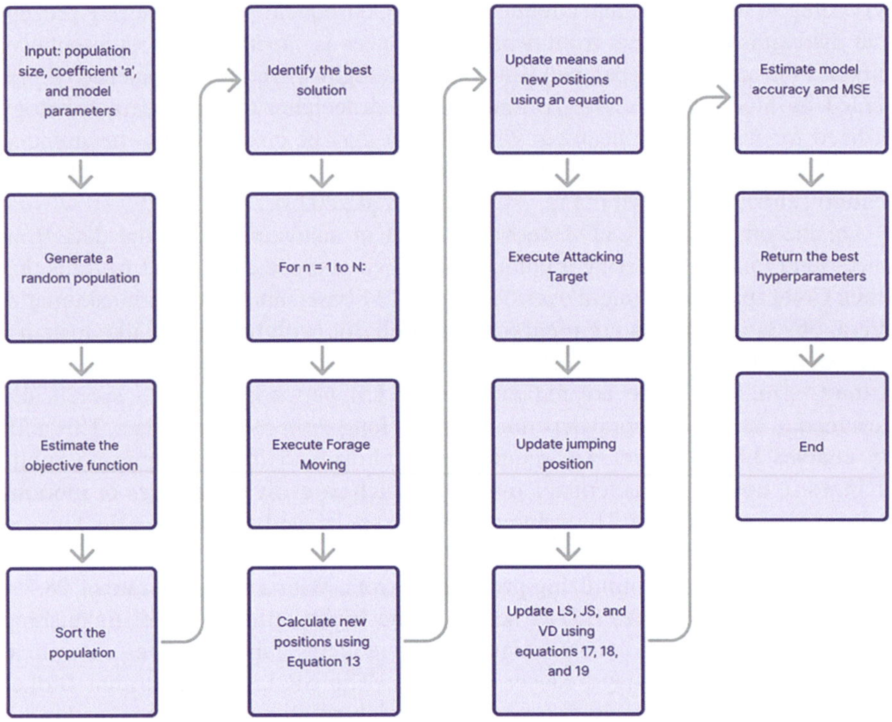

Fig. 35 Pseudocode describing iterative process of the GTO algorithm

4.8 Brain Cancer

Brain cancer, specifically glioblastoma, is one of the most aggressive and challenging forms of cancer to diagnose and treat. The use of 3D Convolutional Neural Networks (3D-CNNs) for MRI analysis has significantly advanced the precision of glioblastoma detection, leveraging the spatial and volumetric nature of MRI scans. Unlike traditional CNNs, which process 2D image slices, 3D-CNNs analyze volumetric data, capturing inter-slice dependencies and spatial context across all three dimensions. This capability makes 3D-CNNs particularly effective in differentiating glioblastomas from low-grade gliomas by identifying subtle and complex features within MRI scans (Mzoughi et al., 2020).

The methodology begins with data preprocessing, where MRI images are normalized, segmented, and augmented to ensure consistency and variability in the training data. Segmentation is critical, as it isolates the brain region from surrounding tissues and removes irrelevant features. The MRI volumes are then fed into a 3D-CNN architecture, which consists of convolutional layers that apply 3D filters, extracting low- to high-level features from the input data. Pooling layers reduce the spatial dimensions while preserving critical features, optimizing the computational load (Mzoughi et al., 2020; Nagendram et al., 2023).

The training phase involves labeled datasets, such as those from public sources like The Cancer Imaging Archive (TCIA) or specialized medical repositories. The 3D-CNN model uses backpropagation and optimization algorithms like Adam or SGD (Stochastic Gradient Descent) to minimize prediction errors and learn to categorize MRI volumes into three categories: glioblastoma, low-grade glioma, or healthy tissue. By preventing overfitting, sophisticated regularization strategies like dropout guarantee that the model performs effectively when applied to new data (Nagendram et al., 2023). Unlike conventional techniques that rely on 2D slice-by-slice analysis, which might lose important spatial information, 3D-CNNs have the benefit of processing whole MRI volumes in a single run. The model can detect characteristic glioblastoma traits including necrotic cores, increased tumor areas, and surrounding edema thanks to this comprehensive analysis, which is crucial for a precise diagnosis (Mzoughi et al., 2020).

Metrics like as sensitivity, specificity, precision, recall, and total accuracy are used to assess the model's performance. The resilience of 3D-CNNs in detecting genuine positive cases is demonstrated by their reported sensitivity of around 92% in glioblastoma detection. Grad-CAM (Gradient-weighted Class Activation Mapping) or comparable visualization approaches are used to highlight MRI areas that impacted the model's conclusions in order to improve clinical interpretability and ensure alignment with radiological results (Song et al., 2020; Nagendram et al., 2023) (Fig. 36).

Implementation in real-world scenarios involves integrating the 3D-CNN model into clinical workflows, allowing radiologists to use AI-assisted tools for faster and more accurate diagnosis. This system not only reduces diagnostic variability but also aids in planning personalized treatment strategies. For example, accurate delineation of glioblastomas can guide surgical interventions, radiotherapy planning, and monitoring of treatment responses (Figs. 37, 38, 39 and 40).

In conclusion, the use of 3D-CNNs for glioblastoma detection represents a transformative approach in medical imaging, offering a robust, accurate, and interpretable method for analyzing MRI data. By leveraging volumetric analysis and advanced neural network architectures, this technique enhances diagnostic precision, ultimately improving patient outcomes and facilitating timely interventions in a critical clinical domain.

4.9 Pancreatic Cancer

Detecting pancreatic cancer through deep learning applied to Endoscopic Ultrasound Fine-Needle Aspiration (EUS-FNA) cytology images represents a significant advancement in early diagnostic capabilities. This approach utilizes convolutional neural networks (CNNs) to analyze cellular morphology and complex cytological patterns indicative of malignancy. The process begins with data preprocessing, where cytology images are enhanced by denoising, normalization, and augmentation techniques, ensuring a robust and diverse dataset. Often, datasets such as those generated in clinical settings or publicly available medical image repositories are

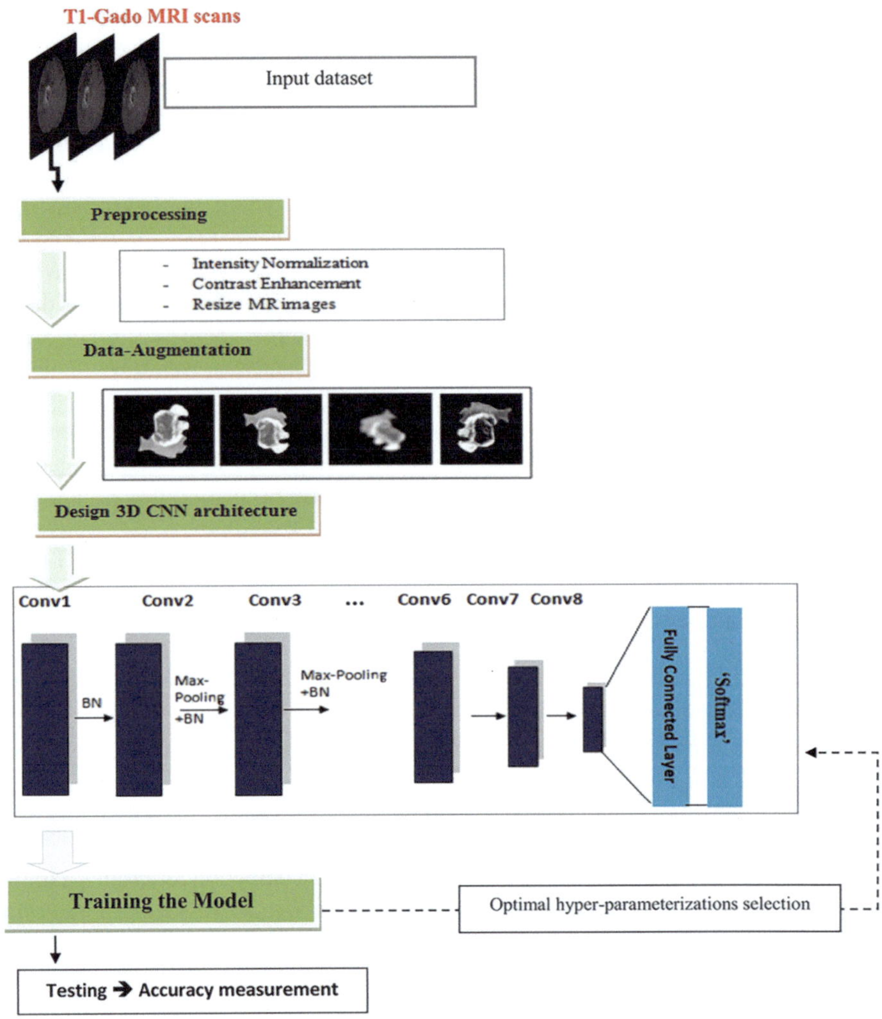

Fig. 36 CNN-based classification approach for glioma brain tumor classification. (Rehman et al., 2021)

employed. Augmentation may involve rotating, scaling, or adjusting brightness to simulate real-world variability and enhance model generalization. CNN architectures, sometimes incorporating transfer learning, are implemented to extract hierarchical features, focusing on identifying anomalies like irregular nuclei and chromatin clumping characteristic of malignant cells (Qin, Ran, et al., 2023a; Qin, Zhang, et al., 2023b) (Fig. 41).

The methodology involves training the CNN with labeled datasets, typically annotated by experienced pathologists. Features are progressively extracted through convolutional layers, which apply filters to detect patterns of cellular irregularities,

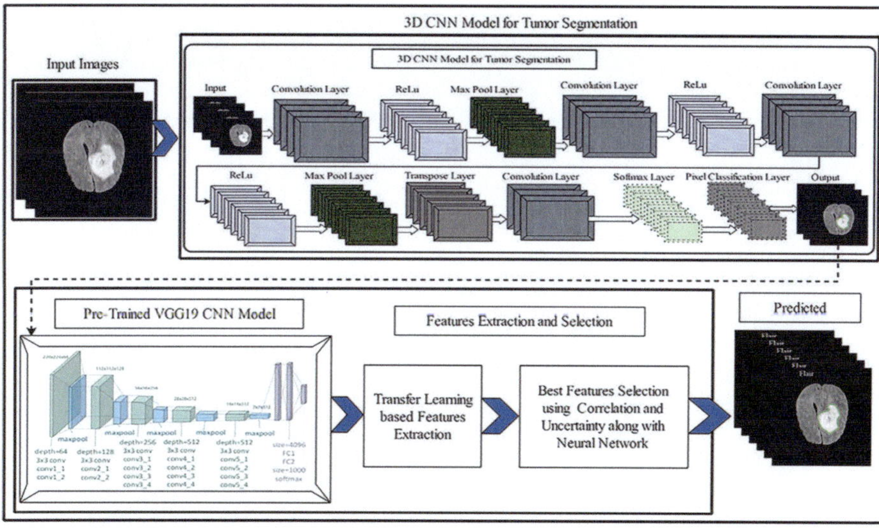

Fig. 37 3D CNN architecture for classifying and detecting tumors. (Rehman et al., 2021)

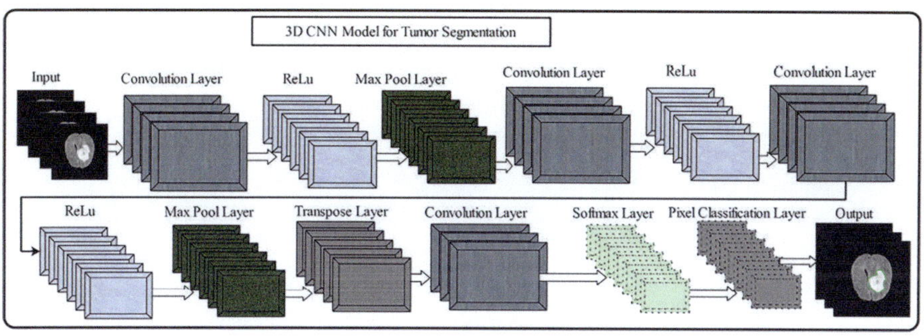

Fig. 38 3D CNN architecture for the extraction of brain tumors. (Rehman et al., 2021)

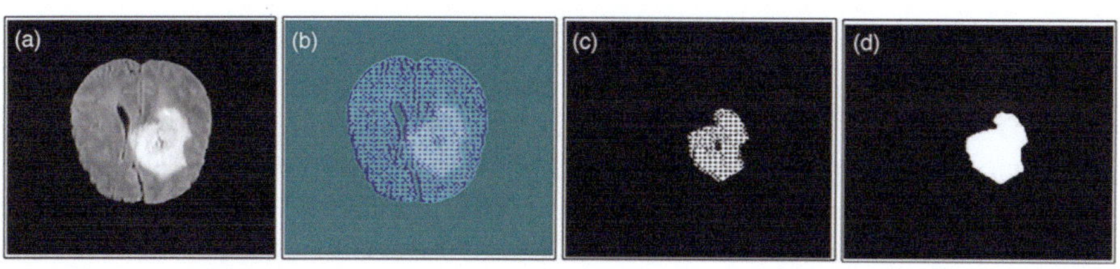

Fig. 39 MRI scan results in 3D CNN. (**a**) shows the original scan, while (**b**) shows the 3D CNN initial findings, (**c**) shows the 3D CNN tumor extracted results, and (**d**) shows the results following morphological procedures for refinement. (Rehman et al., 2021)

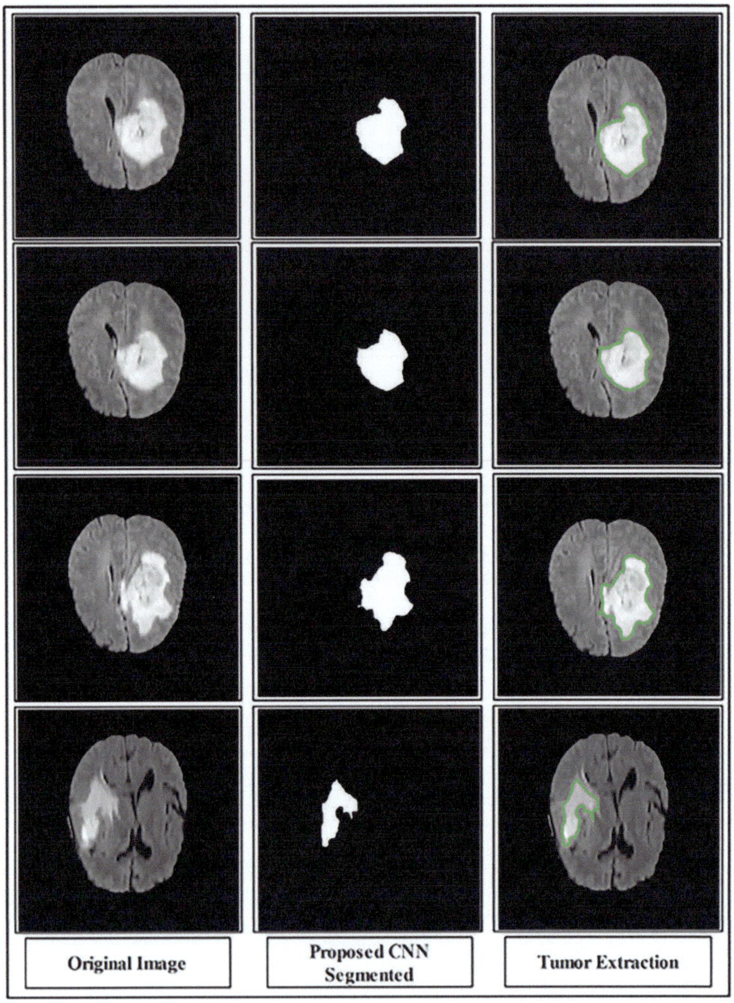

Fig. 40 Brain tumor derived results using 3D CNN. (Rehman et al., 2021)

followed by pooling layers to reduce dimensionality while preserving critical information. The network employs rectified linear unit (ReLU) activation functions for non-linearity, and dropout layers prevent overfitting by randomizing node activation during training. Optimization techniques, such as the Adam optimizer, fine-tune the network's parameters to minimize prediction errors. Loss functions like categorical cross-entropy are used to compare predictions with ground truth labels and iteratively refine model accuracy through backpropagation (Dinesh et al., 2023).

To enhance diagnostic reliability, the pipeline incorporates validation and testing phases, where unseen data are used to evaluate the model's performance. Metrics such as sensitivity, specificity, precision, and F1-score are calculated to assess its

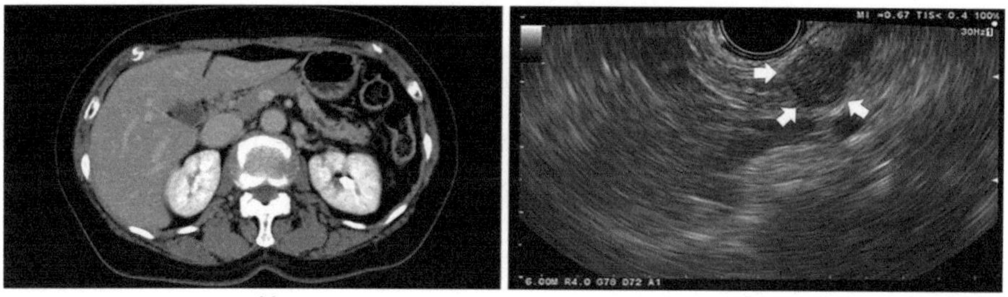

Fig. 41 Pancreatic images on endoscopic ultrasound. (Kitano et al., 2019)

clinical applicability. Sensitivity, in particular, is crucial in minimizing missed malignancies. Heatmaps generated through Class Activation Mapping (CAM) or Grad-CAM provide visual explanations for the network's focus areas, highlighting regions in the cytology images that influenced classification decisions. This interpretability builds trust among clinicians and supports integration into diagnostic workflows (Ahmadvand, 2023).

Once deployed, the model processes new EUS-FNA cytology images by segmenting regions of interest and classifying them as benign, pre-cancerous, or malignant. The ability to analyze cellular structures with an accuracy of approximately 85% aids in identifying early-stage pancreatic cancer, which is critical given the disease's typically late presentation and poor prognosis. Moreover, automation reduces inter-observer variability and diagnostic workload for pathologists, enabling more consistent and faster results. The system's integration with existing clinical tools, such as computer-aided diagnosis platforms, facilitates streamlined workflows and better patient management (Qin, Ran, et al., 2023a).

In conclusion, deep learning on EUS-FNA cytology images represents a transformative step in pancreatic cancer diagnostics. By leveraging CNNs and advanced preprocessing techniques, this method provides reliable, reproducible, and interpretable results, significantly improving early detection rates and guiding timely treatment decisions. The potential for broader adoption depends on continued refinement of algorithms, expansion of annotated datasets, and real-world validation in diverse clinical settings (Table 1).

5 Challenges and Limitations of AI ML in Current Scenario

Even while AI and ML hold great potential for cancer diagnosis, there are still several obstacles to overcome. The requirement for sizable, superior datasets is one of the biggest obstacles. Large volumes of labeled data are necessary for training many AI models, and acquiring such datasets for uncommon tumors might be challenging. Furthermore, there can be significant variations in the quality of medical imaging between various healthcare facilities, and models developed on one dataset

Table 1 Comparison of AI-based techniques for cancer detection: sensitivity, accuracy, and dataset utilization

Technique	Algorithm	Sensitivity	Accuracy	Dataset utilization	Current stage of development
CNN	Convolutional Neural Networks	High in imaging tasks (e.g., lung, breast)	Very high, especially with large datasets	Requires large, annotated datasets for best performance	Well-developed, widely used in clinical settings for imaging-based tasks; performs best with extensive labeled datasets.
GAN	Generative Adversarial Networks	Can enhance sensitivity by generating synthetic data	Dependent on quality of synthetic data	Useful for augmenting small datasets by generating realistic images	Still in development, mainly for augmenting training data, can be paired with CNNs to improve performance, ongoing research in synthetic data quality.
LSTM	Long Short-Term Memory Networks	Best for sequential or time-series data (e.g., progression tracking)	Lower than CNNs in imaging tasks	Works well with longitudinal datasets (e.g., clinical data, genomics)	Primarily used in time-series analysis, still being explored in cancer prediction and tracking over time, especially in patient data.
XGBoost	eXtreme Gradient Boosting	High in structured clinical data (e.g., genomics)	High accuracy with structured data	Effective with small to moderate datasets, requires feature engineering	Mature, widely used in structured data analysis, especially genomics; requires careful feature selection.
ResNet	Residual Networks	Very high, especially for co mplex images	High for image classification tasks	Requires large datasets for optimal performance	Developed and well-established, particularly in medical image analysis, effective for deep feature extraction from complex datasets like CT scans and MRIs.
SVM + Radiomics	Support Vector Machine + Radiomics	High for tumor classification	High for distinguishing malignant from benign tumors	Works well with smaller datasets; radiomics helps in feature extraction	Mature, widely applied for tumor classification, especially in imaging modalities like CT and MRI

3D CNN	3D Convolutional Neural Networks	High in volumetric data analysis (e.g., brain cancer)	Very high for 3D imaging tasks	Computationally intensive; requires large 3D datasets	Emerging, gaining traction in brain cancer and other volumetric imaging tasks; requires significant computational resources for training.
EUS-FNA Cytology	Endoscopic Ultrasound Fine Needle Aspiration (EUS-FNA)	High for pancreatic cancer diagnosis	High in distinguishing malignant from benign masses	Relies on expert cytologists and limited dataset size for training	Still evolving; AI is being integrated into cytology to improve diagnostic speed and accuracy, particularly for early pancreatic cancer detection.

might not function well on data from another because of variations in imaging methods or apparatus.

The explainability of AI models is another difficulty. Due to their opaque decision-making processes, many of the most potent AI models—like deep neural networks—are sometimes referred to as "black boxes." Their acceptance in clinical practice, where physicians must have faith in the AI system's suggestions, may be severely hampered by this lack of interpretability. The widespread use of AI models in actual healthcare settings remains a difficulty despite continuous efforts to make them more interpretable and explicable, such as the use of Grad-CAM to visualize regions of emphasis in medical pictures (Iqbal, Javed, Sadia, Qureshi, et al., 2021b).

Moreover, the integration of AI and ML into clinical workflows requires significant changes to healthcare infrastructure and physician training. The seamless incorporation of AI tools into everyday clinical practice requires not only robust technical systems but also healthcare professionals who are trained to work alongside AI systems and use their outputs effectively. This process of integrating AI into clinical workflows is still in its early stages, and its widespread adoption will require collaboration between technology developers, healthcare providers, and policymakers.

The ongoing research into AI and ML applications in cancer detection is moving the medical field toward a future where early, accurate diagnosis is possible for a wide range of cancers. From leveraging deep learning in medical imaging to integrating genomic data, AI is helping to uncover patterns and insights that were previously hidden from human clinicians. AI is expected to become a vital weapon in the battle against cancer as data becomes more accessible and technology advances, allowing for more individualized treatment regimens and faster, more accurate diagnoses. To fully utilize AI in cancer treatment, however, issues with data quality, model interpretability, and clinical integration must be resolved (Sebastian & Peter, 2022).

6 Conclusion

In conclusion, the integration of advanced computational methodologies in cancer detection has marked a transformative phase in medical diagnostics. Traditional diagnostic methods, although effective, often suffer from limitations such as inter-observer variability, delayed detection, and dependence on invasive techniques. The adoption of AI and machine learning (ML) has significantly bridged these gaps by offering precision, scalability, and speed in interpreting complex medical data. Techniques like CNNs, GANs, LSTM networks, XGBoost, and 3D-CNNs have demonstrated remarkable efficacy in identifying and classifying various cancer types, leveraging vast datasets and intricate algorithms to uncover subtle patterns that often escape human observation.

For instance, CNNs excel in processing image data, making them indispensable in diagnosing skin, breast, and lung cancers through high-resolution imaging modalities. GANs address the challenge of limited data, especially in liver cancer detection, by generating realistic synthetic datasets to train models. Radiomics combined with SVMs enables a nuanced understanding of tumor characteristics, aiding in personalized treatment planning for prostate cancer. Similarly, hybrid approaches such as ultrasound-based ML with XGBoost for ovarian cancer and LSTMs for thyroid cancer exemplify how tailored methodologies can cater to specific diagnostic challenges.

Despite these advancements, challenges such as data heterogeneity, model interpretability, and real-world deployment persist. Addressing these through explainable AI, robust validation techniques, and interdisciplinary collaboration is vital to ensure widespread clinical adoption. Furthermore, the emphasis on early-stage detection underscores the importance of refining these technologies to reduce mortality and improve patient outcomes.

The future of cancer diagnostics lies in synergizing computational tools with clinical expertise. By embracing innovations in AI and ML, we can pave the way for proactive, accurate, and accessible cancer care, ultimately transforming the landscape of oncology. These strides underline the need for continued research, robust policy frameworks, and global cooperation to maximize the potential of these groundbreaking technologies.

References

Abdelrahman, L., Ghamdi, M. A., Collado-Mesa, F., & Abdel-Mottaleb, M. (2021). Convolutional neural networks for breast cancer detection in mammography: A survey. *Computers in Biology and Medicine, 131*, 104248.

Adamu, S., Alhussian, H., Aziz, N., Abdulkadir, S. J., Alwadin, A., Abubakar Imam, A., Abdullahi, M., Garba, A., & Saidu, Y. (2024). The future of skin cancer diagnosis: A comprehensive systematic literature review of machine learning and deep learning models. *Cogent Engineering, 11*(1).

Ahmadvand, P. (2023). *A deep learning approach for classification of pancreatic adenocarcinoma whole-slide pathology images*. University of British Columbia.

Akbari, H., Halig, L. V., Schuster, D. M., Osunkoya, A., Master, V., Nieh, P. T., Chen, G. Z., & Fei, B. (2012). Hyperspectral imaging and quantitative analysis for prostate cancer detection. *Journal of Biomedical Optics, 17*(7), 0760051.

Alabdulkreem, E., Elmannai, H., Saad, A., Kamil, I. S., & Elaraby, A. (2024). Deep learning-based classification of melanoma and non-melanoma skin cancer. *Traitement Du Signal, 41*(1), 213–223.

Alqahtani, H., Kavakli, M., & Kumar, G. (2019). Applications of generative adversarial networks (GANs): An updated review. *Archives of Computational Methods in Engineering., 28*.

Al-Yasriy, H. F., Al-Husieny, M. S., Mohsen, F. Y., Khalil, E. A., & Hassan, Z. S. (2020). Diagnosis of lung cancer based on CT scans using CNN. *IOP Conference Series Materials Science and Engineering, 928*(2), 022035.

Alzubaidi, L., Zhang, J., Humaidi, A. J., Al-Dujaili, A., Duan, Y., Al-Shamma, O., Santamaría, J., Fadhel, M. A., Al-Amidie, M., & Farhan, L. (2021). Review of deep learning: Concepts, CNN architectures, challenges, applications, future directions. *Journal of Big Data, 8*(1), 53.

Anwar, S. M., Majid, M., Qayyum, A., et al. (2018). Medical image analysis using convolutional neural networks: A review. *Journal of Medical Systems, 42*, 226.

Bhattacharjee, S., Singh, Y. J., & Ray, D. (2017). Comparative performance analysis of machine learning classifiers on ovarian cancer dataset. *Third International Conference on Research in Computational Intelligence and Communication Networks (ICRCICN), 2017*, 213–218.

Bi, W. L., Hosny, A., Schabath, M. B., Giger, M. L., Birkbak, N. J., Mehrtash, A., Allison, T., Arnaout, O., Abbosh, C., Dunn, I. F., Mak, R. H., Tamimi, R. M., Tempany, C. M., Swanton, C., Hoffmann, U., Schwartz, L. H., Gillies, R. J., Huang, R. Y., & Aerts, H. J. W. L. (2019 Mar). Artificial intelligence in cancer imaging: Clinical challenges and applications. *CA: a Cancer Journal for Clinicians, 69*(2), 127–157.

Boeken, T., Feydy, J., Lecler, A., Soyer, P., Feydy, A., Barat, M., & Duron, L. (2023). Artificial intelligence in diagnostic and interventional radiology: Where are we now? *Diagnostic and Interventional Imaging, 104*(1), 1.

Cabral, B. P., Braga, L. A. M., Syed-Abdul, S., & Mota, F. B. (2023). Future of artificial intelligence applications in cancer care: A global cross-sectional survey of researchers. *Current Oncology, 30*, 3432.

Chen, J., Zhao, F., Sun, Y., & Yin, Y. (2020). Improved XGBoost model based on genetic algorithm. *International Journal of Computer Applications in Technology, 62*(3), 240. (2020).

Chen, Z.-H., Lin, L., Wu, C.-F., Li, C.-F., Xu, R.-H., & Sun, Y. (2021). Artificial intelligence for assisting cancer diagnosis and treatment in the era of precision medicine. *Cancer Communications, 41*, 1100.

Cui, Y., Li, Y., Xing, D., Bai, T., Dong, J., & Zhu, J. (2021). Improving the prediction of benign or malignant breast masses using a combination of image biomarkers and clinical parameters. *Frontiers in Oncology, 11*.

Dabeer, S., Khan, M. M., & Islam, S. (2019). Cancer diagnosis in histopathological image: CNN based approach. *Informatics in Medicine Unlocked, 16*, 100231.

Dimitrakopoulos, C. M., & Beerenwinkel, N. (2017). Computational approaches for the identification of cancer genes and pathways. *WIREs Systems Biology and Medicine, 9*, e1364. https://doi.org/10.1002/wsbm.1364

Dimitriadis, A., Trivizakis, E., Papanikolaou, N., et al. (2022). Enhancing cancer differentiation with synthetic MRI examinations via generative models: A systematic review. *Insights Into Imaging, 13*, 188.

Dinesh, M. G., Bacanin, N., Askar, S. S., & Abouhawwash, M. (2023). Diagnostic ability of deep learning in detection of pancreatic tumour. *Scientific Reports, 13*(1), 9725.

Forghani, R., Savadjiev, P., Chatterjee, A., Muthukrishnan, N., Reinhold, C., & Forghani, B. (2019 Jul). Radiomics and artificial intelligence for biomarker and prediction model development in oncology. *Computational and Structural Biotechnology Journal, 12*(17), 995–1008.

Frid-Adar, M., Klang, E., Amitai, M., Goldberger, J., & Greenspan, H. (2018). *Synthetic data augmentation using GAN for improved liver lesion classification* (pp. 289–293).

Yunendah Nur Fu'adah et al 2020 IOP conference series: Materials science and engineering 982 012005.

Gholipour, M., Khajouei, R., Amiri, P., et al. (2023). Extracting cancer concepts from clinical notes using natural language processing: A systematic review. *BMC Bioinformatics, 24*, 405.

Gong, X., Zheng, B., Xu, G., Chen, H., & Chen, C. (2021). Application of machine learning approaches to predict the 5-year survival status of patients with esophageal cancer. *Journal of Thoracic Disease, 13*(11), 6240–6251.

Gouda, W., Sama, N. U., Al-Waakid, G., Humayun, M., & Jhanjhi, N. Z. (2022). Detection of skin cancer based on skin lesion images using deep learning. *Healthcare (Basel, Switzerland), 10*(7), 1183.

Guo, R., Zhao, Z., Wang, T., Liu, G., Zhao, J., & Gao, D. (2020). Degradation state recognition of piston pump based on ICEEMDAN and XGBoost. *Applied Sciences (Basel, Switzerland), 10*(18), 6593.

Habchi, Y., Himeur, Y., Kheddar, H., Boukabou, A., Atalla, S., Chouchane, A., Ouamane, A., & Mansoor, W. (2023). AI in thyroid cancer diagnosis: Techniques, trends, and future directions. *Systems., 11*.

Hsiao, Y.-W., Tao, C.-L., Chuang, E. Y., & Lu, T.-P. (2021). A risk prediction model of gene signatures in ovarian cancer through bagging of GA-XGBoost models. *Journal of Advanced Research, 30*, 113–122.

Huang, Gao Liu, Zhuang van der Maaten, Laurens Weinberger, Kilian. (2017). Densely connected convolutional networks.

Huang, S., Cai, N., Pacheco, P. P., Narrandes, S., Wang, Y., & Xu, W. (2018). Applications of support vector machine (SVM) learning in cancer genomics. *Cancer Genomics & Proteomics, 15*, 41. https://doi.org/10.21873/cgp.20063

Hunter, B., Hindocha, S., & Lee, R. W. (2022). The role of artificial intelligence in early cancer diagnosis. *Cancers, 14*(6), 1524. https://doi.org/10.21873/cgp.20063

Ikeda, Y., Doman, K., Mekada, Y., & Nawano, S. (2021). Lesion image generation using conditional GAN for metastatic liver cancer detection. *Journal of Image and Graphics, 9*(1), 27–30.

Iqbal, M. J., Javed, Z., Sadia, H., et al. (2021a). Clinical applications of artificial intelligence and machine learning in cancer diagnosis: Looking into the future. *Cancer Cell International, 21*, 270.

Iqbal, M. J., Javed, Z., Sadia, H., Qureshi, I. A., Irshad, A., Ahmed, R., Malik, K., Raza, S., Abbas, A., Pezzani, R., & Sharifi-Rad, J. (2021b). Clinical applications of artificial intelligence and machine learning in cancer diagnosis: looking into the future. *Cancer Cell International, 21*(1), 270.

Irmak, E. (2021). Multi-classification of brain tumor MRI images using deep convolutional neural network with fully optimized framework. *Iranian Journal of Science and Technology, Transactions of Electrical Engineering, 45*.

Jain, A. (2016). Deep learning for computer vision–introduction to convolution neural networks. *Analytics Vidhya*, April, 4.

Joseph, S. S., & Dennisan, A. (2022). Optimised CNN based brain tumour detection and 3D reconstruction. *Computer Methods in Biomechanics and Biomedical Engineering Imaging & Visualization, 11*(3), 796–811.

Joshua, E. S. N., Bhattacharyya, D., Chakkravarthy, M., & Byun, Y. (2021). 3D CNN with visual insights for early detection of lung cancer using gradient-weighted class activation. *Journal of Healthcare Engineering, 2021*, 1–11.

Kazem, M. A. (2017). Predictive models in cancer management: A guide for clinicians. *The Surgeon: Journal of the Royal Colleges of Surgeons of Edinburgh and Ireland, 15*(2), 93–97.

Khanna, S., Srivastava, S., Khanna, I., & Pandey, V. (2020). Current challenges and opportunities in implementing AI/ML in cancer imaging: Integration, development, and adoption perspectives. *Journal of Advanced Analytics in Healthcare Management, 4*(10).

Kitano, M., Yoshida, T., Itonaga, M., Tamura, T., Hatamaru, K., & Yamashita, Y. (2019). Impact of endoscopic ultrasonography on diagnosis of pancreatic cancer. *Journal of Gastroenterology, 54*(1), 19–32.

Krizhevsky, A., Sutskever, I., & Hinton, G. E. (2017). ImageNet classification with deep convolutional neural networks. *Communications of the ACM, 60*(6), 84–90.

Kwak, M. S., Eun, Y.-G., Lee, J.-W., & Lee, Y. C. (2021). Development of a machine learning model for the prediction of nodal metastasis in early T classification oral squamous cell carcinoma: SEER-based population study. *Head & Neck, 43*(8), 2316–2324.

Li, C., Zhang, Y., Weng, Y., Wang, B., & Li, Z. (2023). Natural language processing applications for computer-aided diagnosis in oncology. *Diagnostics, 13*.

Liu, D., Zhang, X., Zheng, T., Shi, Q., Cui, Y., Wang, Y., & Liu, L. (2021). Optimisation and evaluation of the random forest model in the efficacy prediction of chemoradiotherapy for advanced cervical cancer based on radiomics signature from high-resolution T2 weighted images. *Archives of Gynecology and Obstetrics, 303*(3), 811–820.

Mohapatra, S., Abhishek, N. V. S., Bardhan, D., Ghosh, A. A., & Mohanty, S. (2021). Comparison of MobileNet and ResNet CNN architectures in the CNN-based skin cancer classifier model. In S. N. Mohanty, G. Nalinipriya, O. P. Jena, & A. Sarkar (Eds.), *Machine learning for healthcare applications*.

Mzoughi, H., Njeh, I., Wali, A., Slima, M. B., BenHamida, A., Mhiri, C., & Mahfoudhe, K. B. (2020). Deep multi-scale 3D convolutional neural network (CNN) for MRI gliomas brain tumor classification. *Journal of Digital Imaging, 33*(4), 903–915.

Nagalakshmi, R., & Priya, R. (2023). Predicting thyroid disease using optimized multi-layered long short-term memory (LSTM) neural network with residual attention mechanism. *International Journal of Engineering Trends and Technology, 71*(7), 75–84.

Nagendram, S., Singh, A., Harish Babu, G., Joshi, R., Pande, S. D., Ahammad, S. K. H., Dhabliya, D., & Bisht, A. (2023). Stochastic gradient descent optimisation for convolutional neural network for medical image segmentation. *Open Life Sciences, 18*(1), 20220665.

Patil, S. D., Deore, P. J., & Patil, V. B. (2023). Automated classification of ovarian tumors using GLCM and Tamura features with Light GBM classifier. In *Recent advances in material, manufacturing, and machine learning* (pp. 755–760).

Perdonà, S., Cavadas, V., Di Lorenzo, G., Damiano, R., Chiappetta, G., Del Prete, P., Franco, R., Azzarito, G., Scala, S., Arra, C., De Sio, M., & Autorino, R. (2010). Prostate cancer detection in the "grey area" of prostate-specific antigen below 10 ng/ml: Head-to-head comparison of the updated PCPT calculator and Chun's nomogram, two risk estimators incorporating prostate cancer antigen 3. *European Urology, 59*(1), 81–87.

Qin, Z., Liu, Z., Zhu, P., & Xue, Y. (2020). A GAN-based image synthesis method for skin lesion classification. *Computer Methods and Programs in Biomedicine, 195*, 105568.

Qin, X., Ran, T., Chen, Y., Zhang, Y., Wang, D., Zhou, C., & Zou, D. (2023a). Artificial intelligence in endoscopic ultrasonography-guided fine-needle aspiration/biopsy (EUS-FNA/B) for solid pancreatic lesions: Opportunities and challenges. *Diagnostics (Basel, Switzerland), 13*(19).

Qin, X., Zhang, M., Zhou, C., Ran, T., Pan, Y., Deng, Y., Xie, X., Zhang, Y., Gong, T., Zhang, B., Zhang, L., Wang, Y., Li, Q., Wang, D., Gao, L., & Zou, D. (2023b). A deep learning model using hyperspectral image for EUS-FNA cytology diagnosis in pancreatic ductal adenocarcinoma. *Cancer Medicine, 12*(16), 17005–17017.

Quazi, S. (2022). Artificial intelligence and machine learning in precision and genomic medicine. *Medical Oncology, 39*, 120.

Ramzan, F., Khan, M. U. G., Iqbal, S., Saba, T., & Rehman, A. (2020). Volumetric segmentation of brain regions from MRI scans using 3D convolutional neural networks. *IEEE Access, 8*, 103697–103709.

Rao, C., & Liu, Y. (2020). Three-dimensional convolutional neural network (3D-CNN) for heterogeneous material homogenization. *Computational Materials Science, 184*(109850), 109850.

Rasool, S., Ali, M., Shahroz, H. M., Hussain, H. K., & Gill, A. Y. (2024). Innovations in AI-powered healthcare: Transforming cancer treatment with innovative methods. *BULLET : Jurnal Multidisiplin Ilmu, 3*(1), 118–128.

Rastogi, A., Nigam, N., Bhatia, V., Sureka, B., Jain, P., & Bihari, C. (2019). EUS-guided FNA in diagnosing pancreatic lesions: Strength and cytological spectrum. *Journal of Cytology, 36*(4), 189.

Rathinakumar, N., & Priya, R. (2023). Predicting thyroid disease using optimized multi-layered long short-term memory (LSTM) neural network with residual attention mechanism. *International Journal of Engineering Trends and Technology, 71*(7), 75–84.

Rehman, A., Khan, M. A., Saba, T., Mehmood, Z., Tariq, U., & Ayesha, N. (2021). Microscopic brain tumor detection and classification using 3D CNN and feature selection architecture. *Microscopy Research and Technique, 84*(1), 133–149.

Rennie, J. (1996). From the editors: Reasons for hope. *Scientific American, 275*(3), 6. https://doi.org/10.2307/24993329. Accessed 2 Dec 2024.

Sangariyavanich, E., Ponthongmak, W., Tansawet, A., Theera-Ampornpunt, N., Numthavaj, P., McKay, G. J., Attia, J., & Thakkinstian, A. (2023). Systematic review of natural language processing for recurrent cancer detection from electronic medical records. *Informatics in Medicine Unlocked, 41*, 101326.

Sebastian, A. M., & Peter, D. (2022). Artificial intelligence in cancer research: Trends, challenges and future directions. *Life (Basel, Switzerland), 12*(12), 1991.

Shah, A., Shah, M., Pandya, A., Sushra, R., Sushra, R., Mehta, M., Patel, K., & Patel, K. (2023). A comprehensive study on skin cancer detection using artificial neural network (ANN) and convolutional neural network (CNN). *Clinical eHealth, 6*, 76.

Sharma, R., Mahanti, G. K., Panda, G., Rath, A., Dash, S., Mallik, S., & Hu, R. (2023). A framework for detecting thyroid cancer from ultrasound and histopathological images using deep learning, meta-heuristics, and MCDM algorithms. *Journal of Imaging, 9*(9).

Singh, S., Kumar, V., Verma, H. K., & Singh, D. (2006). SVM based system for classification of microcalcifications in digital mammograms. In *2006 international conference of the IEEE engineering in medicine and biology society*.

Song, J., Chen, C., He, S., Chen, W., Su, J., Yuan, D., Sun, F., & Zhu, J. (2020). Is there a nonlinear relationship between dietary protein intake and prostate-specific antigen: proof from the national health and nutrition examination survey (2003–2010). *Lipids in Health and Disease, 19*(1), 82.

Sun, P., Wang, D., Mok, V. C., & Shi, L. (2019). Comparison of feature selection methods and machine learning classifiers for radiomics analysis in glioma grading. *IEEE Access, 7*, 102010–102020.

Tapani, K. T., Vanhatalo, S., & Stevenson, N. J. (2019). Time-varying EEG correlations improve automated neonatal seizure detection. *International Journal of Neural Systems, 29*(4), 1850030.

Tapani, K. T., Nevalainen, P., Vanhatalo, S., & Stevenson, N. J. (2022). Validating an SVM-based neonatal seizure detection algorithm for generalizability, non-inferiority and clinical efficacy. *Computers in Biology and Medicine, 145*, 105399.

Targ, S., Almeida, D., & Lyman, K. (2016). Resnet in resnet: Generalizing residual architectures. arXiv (Cornell University).

Varan, M., Azimjonov, J., & Maçal, B. (2023). Enhancing prostate cancer classification by leveraging key radiomics features and using the fine-tuned linear SVM algorithm. *IEEE Access, 11*, 88025–88039.

Wang, C.-W., Muzakky, H., Lee, Y.-C., Lin, Y.-J., & Chao, T.-K. (2023). Annotation-free deep learning-based prediction of thyroid molecular cancer biomarker BRAF (V600E) from cytological slides. *International Journal of Molecular Sciences, 24*(3), 2521.

Wei, L., Huang, Y., Chen, Z., Lei, H., Qin, X., Cui, L., & Zhuo, Y. (2021). Artificial intelligence combined with big data to predict lymph node involvement in prostate cancer: A population-based study. *Frontiers in Oncology, 11*, 763381.

Wu, C., Wang, Q., Li, H., Wang, X., Liu, X., Shi, H., & Zhang, Y. (2015). DWI-associated entire-tumor histogram analysis for the differentiation of low-grade prostate cancer from intermediate–high-grade prostate cancer. *Abdominal Radiology, 40*(8), 3214–3221.

Wu, Z., Shen, C., & Van Den Hengel, A. (2019). Wider or deeper: Revisiting the RESNET model for visual recognition. *Pattern Recognition, 90*, 119–133.

Yoon, J., Lee, E., Koo, J. S., Yoon, J. H., Nam, K.-H., Lee, J., Jo, Y. S., Moon, H. J., Park, V. Y., & Kwak, J. Y. (2020). Artificial intelligence to predict the BRAFV600E mutation in patients with thyroid cancer. *PLoS One, 15*(11), e0242806.

Yuan, K. C., Tsai, L. W., Lee, K. H., Cheng, Y. W., Hsu, S. C., Lo, Y. S., & Chen, R. J. (2020). The development an artificial intelligence algorithm for early sepsis diagnosis in the intensive care unit. *International Journal of Medical Informatics, 141*, 104176.

Yuan, Y., Wang, R., Luo, M., Zhang, Y., Guo, F., Bai, G., Yang, Y., & Zhao, J. (2022). A machine learning approach using XGBoost predicts lung metastasis in patients with ovarian cancer. *BioMed Research International*.

Zhao, J., Li, D., Kassam, Z., Howey, J., Chong, J., Chen, B., & Li, S. (2020). Tripartite-GAN: Synthesizing liver contrast-enhanced MRI to improve tumor detection. *Medical Image Analysis, 63*, 101667.

Artificial Intelligence and Machine Learning in Improving Diagnostic Accuracy

Srikruthi Neriyanuri, Srivatsa Venkata Sai Palepu,
Prerna Vats, Bhavika Baweja, Rajeev Nema,
Monisha Banerjee, and Atar Singh Kushwah

Abstract

Cancer is a global concern that has far-reaching consequences for health, society, and the economy. In 2022, there were an anticipated 20 million new cancer cases and 9.7 million cancer deaths worldwide. Cancer affects 1 out of 5 people in their lifespan, and 1 out of 9 men and 1 in 12 women die due to of it. Ninety percent of death occurring due to advanced or metastasize cancer, this emphasizes the importance of early diagnosis and treatment. In reality, early detection, diagnosis, and treatment improve quality of life and overall survival for almost cancers.

Concurrently, scientists are working hard in identifying and developing basic to complex cancer biomarkers to tackle and breakdown carcinogenesis and

Prerna Vats and Bhavika Baweja have been equally contributed to this chapter.

S. Neriyanuri
School of Computers and Intelligent Systems, Manipal University, Jaipur, Rajasthan, India

S. V. S. Palepu
Department of Computer Science and Engineering, Manipal University, Jaipur, Rajasthan, India

P. Vats · B. Baweja · R. Nema
Department of Biosciences, Manipal University, Jaipur, Rajasthan, India

M. Banerjee
Department of Zoology, University of Lucknow, Lucknow, India

A. S. Kushwah (✉)
Women's Biomedical Research Institute, Icahn School of Medicine at Mount Sinai, New York, NY, USA
e-mail: atarskushwah@gmail.com; as_kushwah@hotmail.com

R. Nema et al. (eds.), *Advances in Cancer Detection, Prediction, and Prognosis Using Artificial Intelligence and Machine Learning*,
https://doi.org/10.1007/978-981-96-9346-7_4

develop early diagnostic and potential treatment regimen. Recent advancement in Artificial inelegancy (AI) and Machine Learning (ML) helps to improve the accuracy, speed, and personalization of cancer detection and prediction. These approaches are currently being implemented in cancer diagnosis process, ranging from imaging analysis to genetic profiling, providing unparalleled accuracy.

AI helps to identify the tumor specific genetic variants and molecular markers and allowing to tailored more accurate diagnosis and treatment regimen. ML algorithms help to predict which genetic biomarker is most likely to cause cancer progression, directing targeted therapy decisions. Furthermore, AI-powered liquid biopsy methods advance non-invasive cancer screening by detecting circulating tumor DNA and other blood-based biomarkers. These advances enable early diagnosis and real-time monitoring of cancer progression and treatment response.

Nevertheless, these advantages, problems persist, such as the requirement for big, diverse datasets to train AI/ML models, concerns about algorithm openness, and the integration of AI/ML approaches into clinical practices. Regulatory and ethical concerns must also be addressed, notably regarding data privacy and the potential of algorithmic prejudice. In this book chapter, we emphasized the importance of AI and ML in improving cancer detection by providing more accurate, faster, and tailored diagnostic choices.

Keywords

Artificial Intelligence (AI) · Cancer diagnosis · Machine learning (ML) · Early detection and accuracy · Interpretability and transparency of AI andML

1 Introduction

The implementation of artificial intelligence (AI) and machine learning (ML) in health and medical has become revolutionary advancement, especially when it comes to improving cancer diagnostic precision (Sharma et al., 2023). The way diseases are identified and treated might be completely changed by incorporating these approaches into clinical practice, which would enhance patient outcomes and make healthcare delivery systems more effective (Alowais et al., 2023). The application of AI and ML in diagnostics, the difficulties in putting them into reality, and the prospects for further study and application will all be covered in this thorough examination (Sufyan et al., 2023). Advance information technology makes it possible to create computational models that can learn from enormous volumes of data from electronic medical records (EMRs), which makes it possible to identify intricate patterns linked to a number of illnesses (Xiao et al., 2018). This feature is especially important since it improves the interpretability of data, enabling medical practitioners to base their judgments on the knowledge these models produce.

Furthermore, computer-aided diagnostic (CAD) systems have become more popular in recent years. CAD systems help doctor to discover lesions and provide supplemental diagnostic information by combining imaging, medical image processing, and sophisticated computational analysis (Du et al., 2022). By minimizing human error, AI integration into imaging improves diagnostic accuracy while simultaneously lightening the strain for medical practitioners. AI's potential to enhance diagnostic procedures is demonstrated by its capacity to evaluate big CT image datasets and identify questionable lesions.

When it comes to some illnesses, AI has shown useful in identifying ailments like pancreatic cancer, which is infamously hard to find in its early stages. Improvements in optical technology and artificial intelligence have resulted in the creation of new diagnostic instruments that increase the precision and effectiveness of pancreatic cancer diagnosis (Ahmed Taha et al., 2024). These developments are essential because individuals with this aggressive type of cancer have far higher survival rates when detected early.

Another field where AI and ML are making major advances is the Internet of Medical (IoM). By examining data gathered from numerous medical devices and wearable technology, Albakri and Alqahtani clarify how the combination of deep learning models with the IoM improves the accuracy of diagnostic judgments (Albakri & Alqahtani, 2023). By identifying trends and patterns in patient data, this skill makes it possible to provide more individualized healthcare. In addition to increasing diagnostic precision, AI and IoM enable medical professionals to make data-driven choices that may enhance patient outcomes.

The creation of AI-based headache diagnostic models that increase diagnostic precision for non-specialists is highlighted by Katsuki et al. (Katsuki et al., 2023). These models guarantee that patients receive correct diagnoses while streamlining the diagnostic procedure and saving time that would otherwise be spent on medical interviews (Zhai et al., 2023). AI's potential to improve clinical practice in a variety of medical specialties is highlighted by its capacity to help diagnose complicated ailments like headaches (Cerda et al., 2024).

Additionally, the use of AI in dermatology has yielded encouraging outcomes, especially in the detection of skin malignancies. According to Haggenmüller et al., AI systems are now able to diagnose skin lesions with diagnostic accuracy on par with that of skilled dermatologists (Haggenmüller et al., 2021). Increasing survival rates requires early melanoma identification, and using AI approaches can greatly increase the efficacy of screening initiatives.

There are still obstacles to overcome despite the encouraging developments in AI and ML applications in diagnostics (Kaushik et al., 2021). Data privacy, ethical ramifications, and the possibility of bias in AI algorithms must all be carefully considered when integrating these approaches into healthcare workflows (Harishbhai Tilala et al., 2024.). In order to prevent escalating already-existing health inequities, Rajkomar et al. stress the significance of guaranteeing fairness in machine learning models. AI systems run the risk of reinforcing biases in historical data as they learn from it, which could result in unequal healthcare results (Rajkomar et al., 2018).

Furthermore, improving diagnosis accuracy requires constant advancements in machine learning techniques. In their discussion of the advantages and disadvantages of machine learning in the healthcare sector, An et al. emphasize the necessity of continued research to improve these approaches (An et al., 2023). Because healthcare data is dynamic, machine learning models must adapt to continue being effective at accurately diagnosing illnesses. Notable is also the part machine learning plays in predictive analytics. Weng describes how clinical predictive analytics using machine learning can improve the accuracy and dependability of diagnostic procedures (Weng, 2020). Large datasets can be used to train machine learning models to find patterns that guide clinical judgment and, eventually, enhance patient care.

AI and ML have the ability to improve operational efficiency in healthcare systems in addition to boosting diagnosis accuracy. In order to prioritize investments in diagnostic systems according to factors like cost, productivity, and diagnostic quality, Cho and Hong suggest that machine learning can be applied to hospital operations management (Cho and Hong, 2023). Better healthcare delivery and more efficient use of resources are possible outcomes of this strategic approach.

With further research and development opening the door for creative solutions, the use of AI and ML in diagnostics has a bright future. According to Sree, machine learning techniques allow computers to evaluate enormous volumes of patient data, producing insights that have the potential to revolutionize diagnostic procedures and raise the standard of treatment generally (Sree, 2023). The future of medical diagnostics will be significantly shaped by the integration of AI and ML as the healthcare landscape changes (Fig. 1).

The application of machine learning and artificial intelligence to healthcare diagnostics marks a substantial breakthrough in the area. These approaches simplify the delivery of individualized healthcare, increase operational efficiency, and improve diagnostic accuracy. However, the effective application of AI and ML in clinical practice depends on resolving the issues of bias, data privacy, and ethical

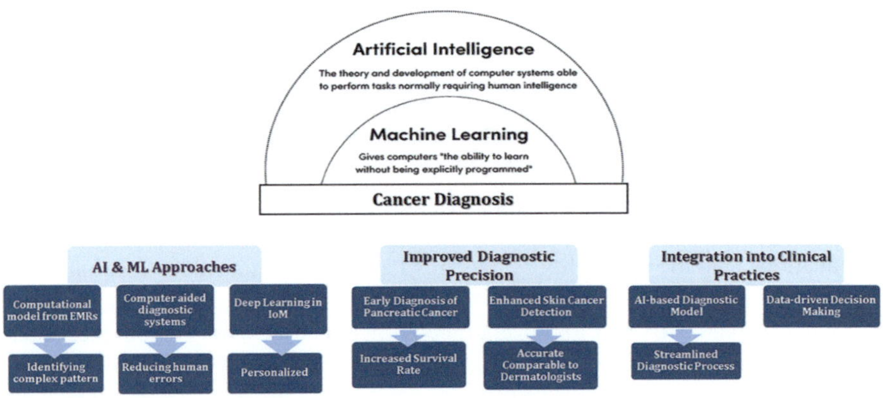

Fig. 1 Overview of AI and ML in cancer diagnosis and accuracy

considerations. With further research, AI and ML have the potential to completely transform healthcare diagnostics, paving the way for a time when prompt and accurate diagnoses are the rule rather than the exception.

2 Recent Developments and Challenges in Cancer Diagnostics

In recent years, the booming field of cancer diagnostics has seen substantial upheaval, with ground-breaking technology developments and creative techniques that are changing cancer detection, treatment, and monitoring (Fig. 2) (Kaushik et al., 2021). The introduction of liquid biopsy technology, which represents a paradigm shift in cancer detection, has enabled non-invasive detection of circulating tumor DNA (ctDNA), circulating tumor cells (CTCs), cell-free DNA (cfDNA), and extracellular vesicles from blood samples, allowing for early cancer detection, disease progression tracking, and even minimal residual disease (MRD) assessment without the need for invasive tissue biopsies (Magro et al., 2024). Liquid biopsies, along with NGS and PCR techniques, allow for full genetic profiling, identifying significant mutations and molecular changes in a single experiment (Li et al., 2024b).

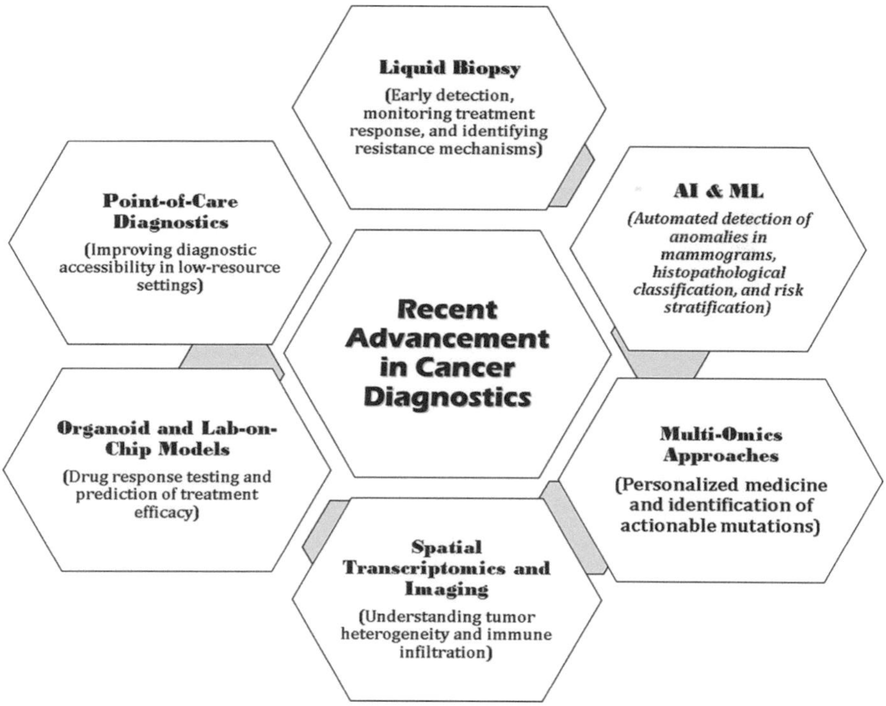

Fig. 2 Recent technological advancement in cancer diagnosis

To increase diagnostic accuracy, researchers are currently using multi-omics techniques that include genomes, transcriptomics, proteomics, and metabolomics, as well as major advanced imaging approaches such as PET-CT scanning (Strati et al., 2023). The ability to generate such comprehensive genetic profiles of tumors has transformed personalized medicine, allowing oncologists to choose targeted therapies based on each patient's unique genetic characteristics, resulting in more effective treatment strategies and better patient outcomes (Parisi et al., 2023).

Despite breakthrough cancer detection technology, several problems remain, with tumor heterogeneity serving as a substantial impediment. Cancer is not a homogeneous illness; it varies widely between people and reveals unique genetic and molecular variations even within various parts of the same tumor (Kushwah et al., 2024). This heterogeneity complicates the development of sensitive and precise diagnostic techniques capable of detecting and identifying cancer in all of its forms and stages (Di Sario et al., 2023). False positives are an important concern as they may trigger are concerned and lead to harmful unnecessary treatment.

Furthermore, the cost of such tests makes them inaccessible, especially in low- and middle-income communities. Furthermore, regulatory barriers impede clinical adoption since strict validation procedures are necessary to ensure safety, accuracy, and dependability, which frequently results in delayed availability (Gomis-Pastor et al., 2024). Integrating multi-omics with advanced imaging approaches generates a large amount of data that necessitates advanced computational analysis and bioinformatics, which are not widely available, and the lack of a standardized protocol across institutions exacerbates results variability.

In order to solve current diagnostic challenges, AI and ML provide transformative solutions by analyzing large datasets, adapting to patient-specific data, and improving predictive power in complex diseases (Murchan et al., 2021) (Fig. 3). Their integration streamlines workflows, reduces clinician workload, and supports evidence-based, data-driven decision-making in real time (Dlamini et al., 2020). AI algorithms trained on large datasets may detect minute morphological changes in tissues and cells, sometimes outperforming human specialists in diagnosing early-stage malignancies or modest recurrences (Mavropoulos et al., 2023). Machine learning is also used to evaluate large genetic and molecular datasets, revealing patterns that traditional analytical methods may miss (Unger and Kather, 2024).

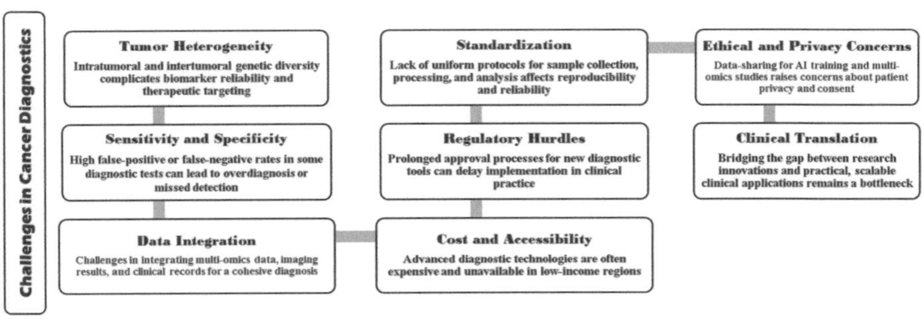

Fig. 3 Current challenges in cancer diagnosis

3 AI and ML Approaches in Cancer Diagnostics

3.1 AI in Imaging Diagnostics

Technological developments are transforming medical imaging diagnostics by increasing the accuracy and efficiency of analyzing complex radiological pictures. X-rays, CT scans, and MRIs generate massive amounts of data, making the detection of tiny abnormalities more difficult. Early detection tools, such as computer-aided diagnosis (CAD) systems, are increasing, particularly in illnesses like cancer, where earlier discovery can have a significant impact on treatment outcomes. In mammography, these approaches serve as a key second layer of review in high-volume settings, reducing false positives and negatives and assisting radiologists in breast cancer identification, as noted by McKinney et al., 2020.

Applications of technology go beyond mammography to include the detection of diseases such neurological illnesses, lung diseases, and skin cancer. For example, Esteva et al. (2019) showed that sophisticated imaging approaches can categorize skin lesions with surprising accuracy, extending expert diagnostic capabilities to underserved communities (Esteva et al., 2019). Similar to how imaging advances in neurology help with the early detection of diseases like multiple sclerosis and Alzheimer's, they also make it possible to identify subtle pulmonary abnormalities in lung diagnostics. These developments are expanding access to expert-level care globally in addition to improving diagnostic accuracy.

Yet issues like guaranteeing these instruments' dependability and resolving biases in their use continue to be crucial. The quality and representativeness of training data are critical to the accuracy of diagnostic tools, and the absence of different datasets may jeopardize results for specific groups. To guarantee fair and efficient application in clinical practice, these systems must undergo thorough evaluation and standardization. Technology will be able to raise the bar for healthcare delivery worldwide if these obstacles are removed.

3.2 Genomic and Precision Medicine

Personalized and genetic medicine represent a revolutionary shift away from conventional "one-size-fits-all" treatment modalities. Precision medicine uses genetic insights from genome sequencing and biomarkers to personalize interventions to a person's unique genetic profile, in contrast to conventional therapy methods (Kaushik et al., 2021). The effectiveness of therapy is greatly increased by this individualized approach, especially for complicated illnesses like cancer. By addressing genetic abnormalities, tailored medicines, for example, have shown impressive success in treating particular subtypes of breast cancer, opening the door to more efficient and patient-centred care.

The field of precision medicine is advancing thanks in large part to AI. Researchers can effectively organize and analyze large genomic datasets to find important biomarkers and genetic variants that affect therapy responses by utilizing machine learning. AI could create customized cancer treatments, improving results for tumors with substantial genetic diversity, according to a study by Luo et al., (2020). The development of radio-genomics, which combines genomic and imaging data to improve diagnostic accuracy, is also aided by these AI breakthroughs (Luo et al., 2020).

By facilitating the quick screening of compounds against certain genetic markers, AI speeds up drug discovery in addition to improving diagnostics and treatment strategies. The time and expense involved in drug development are greatly decreased by this invention. The importance of AI-driven approaches in locating drugs with significant therapeutic promise and accelerating their progression from research to clinical use was highlighted by Gazola et al., (2024). Notwithstanding these advances, issues like guaranteeing a range of genetic representation in datasets and converting genomic discoveries into useful medical treatments continue to be crucial topics of attention to guarantee the fair and efficient use of precision medicine around the world.

3.3 Natural Language Processing for Electronic Health Records

Electronic health records, or EHRs, have greatly enhanced healthcare data management by providing a single repository for patient data (Hossain et al., 2023). However, a significant portion of this data—such as patient histories, clinical notes, and diagnostic reports—is stored in unstructured formats. Even though unstructured text often provides valuable medical insights, traditional data analysis approaches struggle because of the absence of language and format norms. These conventional approaches struggle with language variability and the complexity of free-text medical data (Khurshid et al., 2022).

Natural Language Processing (NLP) has emerged as a revolutionary tool to address these problems. By converting unstructured EHR data into structured, analyzable representations, NLP assists healthcare professionals in identifying trends, deriving significant insights, and streamlining processes (Yang et al., 2022). NLP algorithms, for instance, are being used in oncology to track the progression of the disease, identify patients who are eligible for clinical trials, and evaluate the effectiveness of treatments (Houssein et al., 2023). NLP can be used to analyze follow-up clinical notes and forecast health outcomes by spotting subtle signs of relapse or improvement that could otherwise go unnoticed. This feature demonstrates how NLP has the potential to transform the accuracy of diagnoses and the delivery of healthcare (Jagannatha et al., 2019).

Considering its immense potential, NLP faces numerous challenges in the healthcare sector. Domain-specific training is essential due to the complexity of medical terminology, which calls for specialized models that understand the nuances of clinical language (Kaswan et al., 2021). Additionally, there are privacy and ethical concerns that must be appropriately addressed due to the sensitive nature of healthcare data (Koleck et al., 2019). As NLP approaches advance, it is expected that they will be able to process and comprehend increasingly complex medical data, which will increase their integration into clinical research and patient treatment (Kulshrestha et al., 2021). This breakthrough could lead to previously unimaginable opportunities for tailored therapy and medical care improvement.

4 How Machine Learning Improves Diagnostic Accuracy?

By employing vast datasets to find intricate patterns that increase clinical understanding, boost accuracy, and reduce human error, machine learning (ML) algorithms are revolutionizing diagnostic accuracy. ML can help doctors diagnose a variety of illnesses by examining minute details in imaging and lab data. For example, in radiology, ML models such as convolutional neural networks (CNNs) have shown remarkable success in improving the detection of breast cancer using multiparametric MRI, which combines diffusion-weighted imaging (DWI) and dynamic contrast-enhanced imaging (DCE) to detect malignancies more accurately than traditional evaluations (Daimiel Naranjo et al., 2021). CNNs provide real-time clinical decision-making with greater precision than manual procedures, and they are also useful in distinguishing dental lesions and other anomalies (Chai et al., 2022). The identification of obstructive coronary artery disease has been greatly enhanced in cardiovascular diagnostics by the application of machine learning algorithms to cardiac computed tomographic angiogram. These machine learning models have shown more accuracy in identifying coronary artery blockages and forecast fractional flow reserve, an important metric in assessing arterial function (Coenen et al., 2018). In addition to reducing diagnostic errors, this improved predictive capability makes it possible to implement customized interventions that improve patient outcomes. By adjusting to new data and improving diagnostic parameters, continuously developing machine learning models offer dynamic support in a variety of medical domains, thereby simplifying processes and lowering the possibility of error in complicated instances (Richens et al., 2020). Alongside further development, ML has the potential to be a crucial tool for improving diagnostic processes and attaining better, patient-focused healthcare.

4.1 Predictive Models for Risk Stratification

ML algorithms have revolutionized risk stratification, allowing clinicians to assess patient risk with unprecedented accuracy by analyzing diverse, high-dimensional healthcare data. Predictive models using deep learning, particularly time-series models, have shown effectiveness in predicting patient outcomes by identifying trends in health data that might be indicative of future complications or disease progression. For example, Maweu et al. (2021) demonstrated how deep learning models analyze healthcare data over time to inform early interventions and create individualized treatment plans, thus helping to prevent adverse health events before they occur (Maweu et al., 2021). These predictive capabilities in risk stratification support proactive healthcare approaches, shifting the focus from reactive to preventive care.

In oncology, ML-driven predictive algorithms utilizing image-based CT scan features stratify patients according to their cancer risk, which is essential in planning treatment priorities and allocating healthcare resources effectively (Litjens et al., 2016). This capability enables oncologists to identify high-risk patients who may benefit from more aggressive treatment while managing lower-risk cases conservatively, optimizing both patient outcomes and resource utilization. In psychiatry, ML models have also proven valuable in risk stratification for conditions like major depressive disorder. Here, algorithms combining clinical and neuroimaging data help in diagnosing and stratifying patients, particularly high-risk groups, ensuring targeted interventions that address their specific needs (Ho et al., 2022). By learning from multiple data sources, ML models are setting new standards in risk prediction, facilitating personalized, evidence-based care that supports long-term health management across various specialties.

4.2 AI in Pathology: Automation and Consistency

AI is transforming pathology, particularly in histopathological image analysis, by providing consistent, accurate interpretations that reduce variability inherent in human diagnosis. Through deep learning models, AI can analyze vast amounts of image data in prostate, lung, and breast cancer cases, identifying complex tissue patterns and disease markers with precision. For instance, AI has been shown to improve accuracy in diagnosing breast cancer biopsies by standardizing interpretations and efficiently analyzing cellular structures, which helps reduce human error (Litjens et al., 2016). Additionally, AI can quantify biomarkers, like HER2 in breast cancer, essential for guiding targeted therapies, making diagnostics both faster and more reliable (Campanella et al., 2019).

AI's impact on pathology extends to laboratory efficiency, enabling high-volume sample processing with quicker turnaround times and streamlined workflows. Automated systems in pathology labs can manage large data sets and prioritize urgent cases, ensuring timely diagnosis and resource optimization (Soto et al., 2022). This automation enhances clinical workflows and supports pathologists by freeing them from repetitive tasks, allowing them to focus on complex cases. Overall, AI-driven pathology improves diagnostic accuracy, consistency, and accessibility, marking a significant advancement in healthcare.

4.3 AI-Assisted Clinical Decision Support Systems

AI-assisted clinical decision support systems (CDSS) are revolutionizing healthcare by providing practitioners with advanced tools to make informed, data-driven clinical decisions. By integrating vast amounts of patient data—from lab results and imaging data to genetic information and patient history—AI-powered CDSS deliver

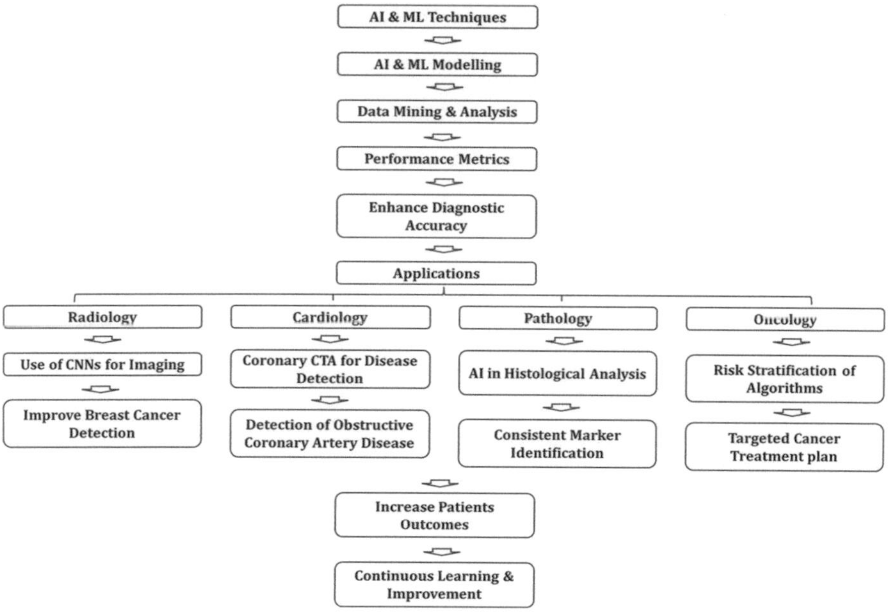

Fig. 4 Overview about how machine learning improves diagnostic accuracy

tailored, evidence-based recommendations, improving diagnostic accuracy and treatment outcomes. These systems can analyze complex inputs to suggest optimal treatment strategies or highlight potential risks, particularly beneficial in cancer care, where they help oncologists predict patient responses to therapies. For example, AI-driven CDSS in oncology have shown promise in determining patient-specific responses to chemotherapy, enabling oncologists to tailor regimens more accurately to each patient's needs (Hanis et al., 2022). Such precision in treatment planning helps maximize therapeutic benefits while minimizing adverse effects, which is crucial for complex cancer cases (Fig. 4).

Beyond oncology, AI-based CDSS also play a significant role in cardiovascular and neurological disease management by analyzing complex, multidimensional data sets to predict disease progression or evaluate treatment risks. For example, in cardiovascular care, predictive models can anticipate complications such as heart failure based on real-time patient data, guiding timely interventions and resource allocation (Richens et al., 2020). In neurology, CDSS can assist in the diagnosis of conditions like Alzheimer's by interpreting imaging data, genetic markers, and cognitive test results, helping clinicians detect early-stage disease that may benefit from early intervention. Moreover, AI-CDSS enhances clinical efficiency by automating routine analyses, which reduces the cognitive burden on healthcare providers and allows them to focus on high-priority decision-making, ultimately improving both patient management and healthcare delivery (Aggarwal et al., 2021).

5 Case Studies and Success Stories

5.1 AI/ML in Breast Cancer Detection

Through advancing imaging analysis's sensitivity and specificity, AI and ML approaches are revolutionizing the diagnosis of breast cancer. AI is becoming a very useful tool in population-based screening programs for breast cancer detection, with AI-augmented mammography systems showing sensitivity rates of over 90% and specificity comparable to that of professional radiologists. According to the 2024 study by Díaz et al., automated systems may diagnose cancer with a diagnostic accuracy that is comparable to that of skilled radiologists. This helps radiologists find subtle cancers and reduces inaccurate diagnoses (Díaz et al., 2024).

AI models in breast cancer screening largely rely on convolutional neural networks (CNNs), which excel at analyzing 2D and 3D imaging data due to their pattern recognition capabilities. In South Korea, the prospective AI-STREAM cohort study found that AI-based computer-aided detection (CADe/x) systems significantly reduced the rate of missed cancers by identifying subtle patterns often overlooked by human observers (Chang et al., 2022). AI interpretability techniques, such as heatmaps, highlight specific areas in mammograms that led to model predictions, providing radiologists with visual support and facilitating greater trust in AI recommendations (Rodríguez-Ruiz et al., 2019). These interpretable visualizations make it easier for clinicians to verify AI-detected abnormalities, enhancing diagnostic confidence and fostering AI adoption in clinical workflows.

Beyond clinical accuracy, AI's role in breast cancer detection includes cost reduction and workflow efficiency. This efficiency is particularly impactful in high-volume screening centers, where quick, reliable assessments are essential. However, the deployment of AI in breast cancer diagnostics must address potential biases in training data to ensure equitable outcomes across diverse demographic groups. Ethical concerns regarding data privacy, along with the need for regulatory oversight to standardize AI in healthcare, are central to achieving responsible AI integration (Díaz-Rodríguez et al., 2023).

5.2 AI/ML in Lung Cancer

AI and ML are equally revolutionary in lung cancer diagnostics, with a significant focus on analyzing CT scans to detect and assess lung nodules. Deep learning models trained on large datasets are enhancing early-stage lung cancer detection, which is crucial for improving survival rates (Doppalapudi et al., 2021). A 2023 study demonstrated that AI-assisted CT interpretation could increase early diagnosis rates by over 30% compared to traditional approaches, underscoring the potential for more timely and effective lung cancer interventions (Najjar, 2023). In order to increase diagnostic robustness, lung cancer detection algorithms frequently use ensemble learning approaches to aggregate several model predictions. Furthermore, models make use of interpretive tools such as SHAP (Shapley Additive Explanations) values, which illustrate how each variable—such as nodule size and density—influences the final diagnosis and aid in explaining the model's choice (Ahmed et al., 2024a). This interpretability is essential since CT scans are intricate 3D images, and in order to build clinician trust, AI's "black-box" predictions need to be somewhat transparent (Ennab and Mcheick, 2024).

AI also helps physicians prioritize critical situations and cut down on pointless treatments by helping to stratify patients according to risk. An AI-driven model successfully distinguished between benign and malignant nodules in a 2021 trial conducted in the Netherlands, reducing unnecessary biopsies and accelerating treatment for high-risk patients (Peled et al., 2021). However, concerns persist: low CT image quality and disparities in imaging technology often led to issues for lung cancer diagnostic models. Future research will likely focus on enhancing AI's real-time application capabilities for regular screenings and refining the technique to maintain accuracy despite these challenges.

5.3 AI/ML in Gynecological Cancer

Advances in AI and ML have shown promise for early diagnosis and more accurate treatment of gynecological cancers, especially ovarian and cervical tumors. Deep learning models applied to colposcopy images have outperformed conventional Pap smear sensitivity in the diagnosis of cervical cancer, making them a useful tool in screening programs (Mathivanan et al., 2024). The promise of AI in cervical cancer screening was highlighted by the 2021 study by Ali et al., which shown that AI models detected early-stage cervical lesions with sensitivity levels of up to 95%. This is particularly true in low-resource locations where access to conventional diagnostic methods is restricted (Ali et al., 2021). These advancements have been largely attributed to CNN architectures, which are skilled at processing colposcopic pictures and have improved screening systems' accessibility and accuracy.

ML models that analyze transvaginal ultrasound images are now able to accurately distinguish between benign and malignant cysts in relation to ovarian cancer. By correctly classifying cysts according to complexity and form, the AI model considerably decreased the need for needless surgical procedures. These deep learning-based image identification methods are anticipated to lower ovarian cancer morbidity rates and promote early therapies (Whitney et al., 2024). Additionally, AI's capacity to produce interpretable results—by displaying which areas the model prioritized using techniques like Grad-CAM (Gradient-weighted Class Activation Mapping)—ensures transparency, promotes physician adoption, and improves patient care outcomes.

5.4 Gene Expression, Multi-Cancer Prediction, and Diagnosis

AI-assisted gene expression analysis has created new opportunities for concurrent multi-cancer detection, enabling early identification of different cancer types. Single tests can screen for numerous cancers thanks to AI models based on gene expression data that can identify biomarkers for multiple tumors (Elemento et al., 2021). According to a 2023 study by Marinovich et al., machine learning algorithms might detect lung, breast, and colorectal cancers all at once with over 85% sensitivity and specificity (Marinovich et al., 2023). This would be a revolutionary approach to multi-cancer screening.

By classifying patients according to their expected prognosis, AI-based gene expression analysis also facilitates individualized therapy planning. Santeramo et al. (2024), for instance, used an AI model to evaluate transcriptome data for ovarian and breast malignancies, enabling medical professionals to identify individuals at high risk of recurrence and adjust treatment accordingly (Rawat & Nayak, 2021). A possible avenue for developing a thorough diagnostic profile that supports precision oncology and improves early intervention outcomes is the integration of multimodal data, such as fusing gene expression profiles with imaging data (Santeramo et al., 2024).

6 Challenges Specific to AI and Machine Learning in Cancer Diagnosis

6.1 Data Quality and Quantity

AI and ML algorithms applied to cancer diagnosis pose significant challenges related to quality and quantity of data, affecting the efficacy, reliability, and generalisability of diagnostic models (Abbas et al., 2024). Accessibility of data is a significant concern, as HIPAA and GDPR impose limitations to the access and disclosure of the patient's data, even for research purposes, further complicated by data confinement within the proprietary systems of healthcare institutions (Thapa and Camtepe, 2021). Another critical challenge is quality and standardization of medical imaging data which is crucial for developing accurate diagnostic tools. Uniform imaging and annotation protocols across institutions can mitigate variability caused by differences in equipment, resolution, and formatting. Initiatives such as developing universal standards for MRI and CT imaging parameters and metadata annotations are critical to reducing heterogeneity (McKinney et al., 2020). Additionally, the variability in parameters, resolution, and formatting across different sources makes it challenging to develop consistent algorithms across different healthcare settings. Poor-quality annotations greatly impact the accuracy of diagnostic model predictions, as the malignant and benign cases labelled inaccurately requires inputs from experienced medical professionals for creating high-quality labels, which is an expensive and time-consuming process (Hosny et al., 2018). Additionally, the predictive ability of machine learning algorithms is frequently hindered by the presence of noise in clinical data, such as inconsistent measurements or insufficient information, which may result in inaccurate diagnosis (Esteva et al., 2019).

Furthermore, the need for diverse patient demographics, cancer stages, and clinical presentations complicates data quantity. To perform well across diverse populations, AI models need many examples of cancers in different age groups, ethnicities, and genetic backgrounds. Following patients through their cancer journey requires longitudinal data collection, which presents another quantitative challenge. Temporal data collection is essential for understanding disease progression and treatment responses, but long-term datasets are scarce (Hart et al., 2024). To generalize across populations, ML models need large and diverse datasets. However, data scarcity hinders cancer diagnosis, especially for rare cancers or specific demographic groups (Elmore et al., 2015). In small datasets, overfitting can occur, causing the model to focus on training data patterns rather than generalisable features, reducing diagnostic reliability in clinical settings. Federated learning allows models to train on distributed data without direct access, but it has technical challenges like ensuring model performance across diverse data sources (Yang et al., 2019). Overall, the dual challenges of data quality and quantity in AI/ML for cancer diagnosis underscore the need for collaborative, standardized data management efforts to fully harness AI's diagnostic potential while ensuring patient safety and data integrity (Topol, 2019). Establishing centralized, secure, and ethically governed data repositories can address issues related to data accessibility and fragmentation. Collaborative frameworks like federated learning allow institutions to train AI models on distributed data without directly sharing sensitive information, ensuring privacy compliance under regulations such as GDPR and HIPAA (Williamson and Prybutok, 2024). To improve annotation accuracy, hybrid annotation methods combining expert-driven insights with AI-based pre-labelling can be implemented. AI tools can perform initial labelling, which human experts refine, significantly reducing costs and time while maintaining high-quality data. Advanced preprocessing algorithms can identify and correct inconsistencies in clinical data, improving model training. For example, outlier detection systems can flag incomplete or noisy records for exclusion or refinement.

Collaborations between academia, healthcare institutions, and industry can provide the resources and expertise required for large-scale data collection and analysis. Funding programs that incentivize creating open-access datasets for rare cancers can bridge critical gaps.

6.2 Interpretability and Transparency of AI/ML Models

The interpretability and transparency of AI and ML models in cancer diagnosis remain critical challenges that must be addressed to enhance clinical trust and adoption. Current AI systems often function as "black boxes," making it difficult for clinicians to understand how decisions are made. This lack of clarity hinders accountability and raises concerns about safety, particularly when AI models are deployed in high-stakes medical scenarios (Caruana et al., 2015). Their complexity makes it difficult to understand how they generate diagnostic conclusions, hindering trust and transparency. This lack of explainability raises concerns among medical professionals who must rely on these systems while being accountable for treatment decisions (Nazir et al., 2024). The intricacy of medical imaging and cancer diagnosis complicates interpretability issues. In contrast to simpler tasks, it encompasses intricate patterns and complex feature interactions, frequently surpassing the capabilities of even expert radiologists. The inability of AI to clearly explain its conclusions, key features, and confidence levels poses significant challenges in identifying patterns that may elude human perception (Ghassemi et al., 2020). The necessity for explainability is crucial, particularly for unanticipated or possibly inaccurate diagnoses, to ensure patient safety and enhance system reliability.

By assigning specific features to outcomes, interpretability frameworks like SHAP (SHapley Additive exPlanations) and LIME (Local Interpretable Model-agnostic Explanations) have been developed to shed light on model predictions (Ahmed et al., 2024b). However, these approaches still have drawbacks, especially when dealing with extremely complex data like imaging or genomics, which are crucial for the diagnosis of cancer. Addressing these interpretabilities and transparency issues, multidisciplinary efforts are needed to improve model explainability. For example, hybrid models that combine machine reasoning and oncology domain knowledge can be developed, ensuring that AI-driven cancer diagnostics are both clinically applicable and scientifically reliable (Abbas et al., 2024). Detailed documentation outlining AI model training data, decision pathways, and known limitations must be provided to end-users. This transparency fosters trust and ensures clinicians understand the contexts in which the model performs optimally. Structured programs to train clinicians in understanding and leveraging AI tools are essential. These initiatives should focus on enhancing digital literacy and explaining the scope and limitations of AI outputs.

6.3 Regulatory and Ethical Issues

The adoption of AI in cancer diagnostics brings critical regulatory and ethical challenges that must be addressed to ensure the safe and equitable application of these approaches. A key issue lies in the dynamic nature of AI systems, which evolve through continuous learning from new data. This creates challenges for regulatory frameworks traditionally designed for static medical devices. To address this, adaptive regulatory approaches, such as "Software as a Medical Device" (SaMD) guidelines, must evolve to include mechanisms for ongoing monitoring, validation, and performance tracking of AI systems across updates (Fraser et al., 2023). Establishing a centralized database for model version control and mandatory re-validation for significant updates could enhance reliability and trust in these systems.

Bias and fairness in AI models are critical ethical concerns, as unrepresentative datasets can lead to disparities in diagnostic accuracy across demographic groups. Proactive strategies, such as mandating diverse and representative datasets during development, implementing bias-detection algorithms, and requiring equity audits as part of the regulatory approval process, can mitigate this risk (Boppiniti, 2023).

Collaboration between regulators, AI developers, and healthcare providers is crucial to enforce such standards.

Data privacy and security are additional priorities in AI deployment, as healthcare datasets are highly sensitive. Techniques such as federated learning, which enables training AI models on distributed data without centralizing it, and advanced anonymization methods like differential privacy, can balance the need for robust datasets with compliance to regulations like GDPR and HIPAA (Williamson and Prybutok, 2024). Further, ensuring transparency in data usage and obtaining informed consent from patients strengthens ethical safeguards.

To enhance interpretability and transparency, regulators should encourage the use of explainable AI frameworks, such as SHAP or LIME, alongside clinical validation processes. Training healthcare professionals to understand AI outputs and integrating these tools into existing workflows will foster trust and adoption. A unified effort involving international stakeholders is needed to create global, adaptive regulatory standards that balance innovation with safety and ethical compliance.

6.4 Validation and Clinical Trials

The validation and clinical implementation challenges of AI and ML systems in cancer diagnosis offer numerous obstacles that require careful consideration to guarantee safe and effective integration into healthcare environments (Elemento et al., 2021). A primary challenge resides in the validation process, wherein AI systems must exhibit not only statistical accuracy but also clinical significance and dependability across varied patient demographics (Aminizadeh et al., 2024). The validation of these systems necessitates comprehensive clinical trials and real-world testing, which can be long-lasting, costly, and logistically challenging. In contrast to conventional medical devices or treatments, AI systems persistently learn and adapt, raising concerns regarding the validation of systems that may change over time and the assurance of their consistent and reliable performance during clinical application (Li et al., 2024a).

Another major drawback is the lack of standardised benchmarks and robust evaluation protocols for AI models in cancer diagnostics. Traditional diagnostic tools undergo validation through highly regulated clinical trials, whereas numerous AI-based solutions are evaluated using retrospective datasets that may not accurately represent real-world complexities (Kann et al., 2021). Moreover, dependence on metrics such as accuracy or AUC-ROC for validation fails to comprehensively reflect the clinical significance of a model's predictions, including their influence on treatment decisions or patient outcomes (Liao et al., 2022). The clinical implementation is hindered by regulatory challenges, as AI models must adhere to strict healthcare regulations that require accuracy, interpretability, reproducibility, and ethical compliance (Esmaeilzadeh, 2024).

The practical incorporation of AI models into the clinical workflows poses further challenges. Numerous AI systems necessitate specialised computational infrastructure, which may be unfeasible in resource-limited environments. Moreover, healthcare professionals frequently lack the requisite training to accurately interpret AI outputs, potentially resulting in resistance or improper utilization of these approaches (Ahmad et al., 2021). In order to address these challenges, iterative validation through prospective studies in actual clinical environments is crucial, enabling models to adjust to variability and achieve acceptance among healthcare practitioners. Cooperative initiatives among AI developers, clinicians, and regulatory agencies are essential to create thorough validation frameworks that emphasise patient safety and outcomes (Mennella et al., 2024).

7 Future Directions

7.1 Personalized Diagnostics

There are exciting opportunities to transform healthcare delivery and improve patient outcomes with the potential developments of AI/ML in personalized diagnostics for cancer diagnosis. Precision oncology is expected to progress through the identification of biomarkers that can predict cancer type, progression, and individual responses to targeted therapies (Kaushik et al., 2021). This is made possible by the integration of multiple data sources, including genomic data, medical imaging, clinical records, and real-time patient monitoring.

In oncology, where knowledge of the unique characteristics of each patient's illness can have a substantial impact on treatment choices and results, the move towards precision medicine is especially encouraging (Mennella et al., 2024). Developing explainable AI (XAI) frameworks to enhance the interpretability and dependability of AI systems in clinical settings is a promising approach. In order to help physicians better grasp the reasoning behind diagnostic recommendations, future AI models are anticipated to use hybrid approaches that combine deep learning with rule-based techniques and domain expertise (Sirocchi et al., 2024). When paired with machine learning models for pattern recognition and risk assessment, the development of AI-driven liquid biopsy approaches—which analyze circulating tumor DNA (ctDNA) or other biomarkers from non-invasive specimens like blood—offers significant promise for the early detection and monitoring of cancer (Zakari et al., 2024).

As AI systems are being developed to include lifestyle characteristics, environmental exposures, and patient preferences in addition to clinical data in their diagnostic assessments, the shift towards more patient-centered diagnostic approaches is anticipated to accelerate. This thorough diagnostic process could lead to more effective and individualized treatment plans. Wearable technology and continuous monitoring systems working together may provide real-time health data for AI systems to analyze, making it easier to identify early warning signs of cancer recurrence or treatment side effects and encouraging more proactive and preventive care approaches. AI will also be able to continuously improve its diagnostic accuracy through input from clinical results thanks to advancements in real-time monitoring and adaptive learning systems, guaranteeing that models remain in line with growing medical knowledge (Esteva et al., 2019). AI-driven diagnostics combined with augmented and virtual reality (AR/VR) may enhance the visualization and interpretation of complex imaging and molecular data, offering new tools for a previously unheard-of level of cancer comprehension (Shen et al., 2017). In order to achieve these milestones, oncologists, data scientists, and regulatory agencies must work together to create moral and uniform procedures that guarantee the safe and fair application of AI systems for individualized cancer diagnosis (Topol, 2019).

7.2 AI in Remote and Global Diagnostics

Advancements in artificial intelligence (AI) are on the rise for remote and global cancer diagnostics, which could help bridge the healthcare access gap and improve early detection in underserved areas. AI-driven diagnostic tools, including automated imaging systems and decision-support algorithms, possess significant potential to address disparities in healthcare access by delivering precise and efficient diagnostics in regions with restricted specialist availability (Khalifa and Albadawy, 2024). AI-enabled mobile apps and portable imaging devices can analyze medical pictures in real time, enabling primary care physicians in remote or resource-constrained places to identify cancer early and refer patients for prompt treatment. Without the need for physical infrastructure or specialized knowledge, these systems can remotely share and analyze data via cloud-based platforms, giving users access to sophisticated diagnostics (Sachdeva et al., 2024).

One crucial strategy is the incorporation of federated learning, which addresses the problems of data availability and heterogeneity by allowing AI systems to be trained on a variety of geographically dispersed datasets while maintaining patient privacy (Yang et al., 2022). By minimizing biases that may arise from datasets biased towards high-resource settings, this methodology ensures that AI models are adaptable and representative of global demographics. Additionally, by enabling AI systems to understand and integrate medical records, imaging reports, and genetic data from a variety of languages and formats, developments in natural language processing (NLP) can help overcome language barriers (Shen et al., 2017).

Collaborations between governments, healthcare providers, and technology companies are essential for the widespread adoption of AI tools in order to enhance global diagnosis, particularly in low- and middle-income countries (LMICs). There will likely be an increase in efforts to integrate AI into telemedicine platforms, which will improve access to high-quality cancer diagnoses by enabling remote consultations and expert second views (Razzak et al., 2018). Global cancer diagnostics, especially for early-stage cancers, will also depend heavily on affordable AI-powered point-of-care devices that may detect cancer biomarkers through blood, saliva, or other non-invasive methods (Heitzer et al., 2019).

In order to ensure that AI approaches are available, culturally sensitive, and in line with the unique requirements of many global populations, significant expenditures in infrastructure, training, and ethical governance are required in order to achieve these improvements (Ghassemi et al., 2020). These developments all point to the expansion of access to high-quality cancer diagnostics in the future, which could improve cancer outcomes worldwide and lessen healthcare inequities.

7.3 Collaborative AI Systems

The future of collaborative AI systems in cancer diagnosis highlights the incorporation of multi-disciplinary expertise, utilizing advancements in AI and improving human-AI collaboration. These systems seek to integrate the diagnostic expertise of clinicians with the computational efficacy of AI to enhance accuracy and decision-making. A primary focus is the implementation of explainable AI (XAI) frameworks to connect clinicians with AI systems, clarifying and making actionable the rationale behind diagnostic predictions (Zaidan, 2024). Collaborative platforms are progressively engineered to facilitate real-time data sharing among researchers and institutions globally, promoting the creation of universally trained models that integrate diverse patient data while addressing regional disparities in cancer diagnosis (Saroğlu et al., 2024). Furthermore, human-computer multi-disciplinary teams (MDTs) are being investigated for intricate diagnostic situations, such as pulmonary nodule assessment, where AI enhances human decision-making rather than substituting it (Yang et al., 2023). Future research will concentrate on creating hybrid AI systems that integrate domain-specific knowledge with machine learning abilities to tackle context-specific issues, such as diagnosing rare cancers or analysing intricate imaging datasets (Patra et al., 2024). Collaborative initiatives will also encompass the integration of IoT and 5G approaches, which can improve remote diagnostics and optimise workflows in cancer care (Saroğlu et al., 2024). These advancements indicate a transformative period for collaborative AI, wherein collective expertise and state-of-the-art technology incorporate to enhance the precision, inclusivity, and efficiency of cancer diagnostics.

8 Conclusion

AI and ML are rapidly reshaping the diagnostic landscape, offering tools that improve the accuracy, speed, and personalization of medical diagnoses. While challenges such as data quality, interpretability, and ethical considerations remain, the potential benefits of AI in diagnostics are undeniable. As research and innovation continue, AI will become an integral part of modern healthcare, enhancing clinicians' ability to diagnose diseases more accurately and provide tailored treatments to patients worldwide.

Acknowledgement We would like to express our deepest gratitude to all those who contributed to the successful completion of this book chapter. We sincerely thank Icahn School of Medicine at Mount Sinai, New York, for their valuable insights and support during the research and writing process.

Finally, we extend our thanks to our families and friends for their unwavering encouragement and understanding throughout these endeavours.

This chapter is a testament to the collaborative spirit of everyone involved, and we are deeply appreciative of their contributions.

References

Abbas, S., Asif, M., Rehman, A., Alharbi, M., Khan, M. A., & Elmitwally, N. (2024). Emerging research trends in artificial intelligence for cancer diagnostic systems: A comprehensive review. *Heliyon, 10*, e36743. https://doi.org/10.1016/j.heliyon.2024.e36743

Aggarwal, R., Sounderajah, V., Martin, G., Ting, D. S. W., Karthikesalingam, A., King, D., Ashrafian, H., & Darzi, A. (2021). Diagnostic accuracy of deep learning in medical imaging: A systematic review and meta-analysis. *npj Digital Medicine, 4*, 1–23. https://doi.org/10.1038/s41746-021-00438-z

Ahmad, Z., Rahim, S., Zubair, M., & Abdul-Ghafar, J. (2021). Artificial intelligence (AI) in medicine, current applications and future role with special emphasis on its potential and promise in pathology: Present and future impact, obstacles including costs and acceptance among pathologists, practical and philosophical considerations. A comprehensive review. *Diagnostic Pathology, 16*, 24. https://doi.org/10.1186/s13000-021-01085-4

Ahmed Taha, B., Kadhim, A. C., Addie, A. J., Haider, A. J., Azzahrani, A. S., Raizada, P., Rustagi, S., Chaudhary, V., & Arsad, N. (2024). Advancing cancer diagnostics through multifaceted optical biosensors supported by nanomaterials and artificial intelligence: A panoramic outlook. *Microchemical Journal, 205*, 111307. https://doi.org/10.1016/j.microc.2024.111307

Ahmed, M., Bibi, T., Khan, R.A., & Nasir, S. (2024a). Enhancing breast cancer diagnosis in mammography: Evaluation and integration of convolutional neural networks and explainable AI. https://doi.org/10.48550/arXiv.2404.03892

Ahmed, S., Shamim Kaiser, M., Hossain, M. S., & Andersson, K. (2024b). A comparative analysis of LIME and SHAP interpreters with explainable ML-based diabetes predictions. *IEEE Access, 13*, 37370. https://doi.org/10.1109/ACCESS.2024.3422319

Albakri, A., & Alqahtani, Y. M. (2023). Internet of medical things with a Blockchain-assisted smart healthcare system using metaheuristics with a deep learning model. *Applied Sciences, 13*, 6108. https://doi.org/10.3390/app13106108

Ali, M. M., Ahmed, K., Bui, F. M., Paul, B. K., Ibrahim, S. M., Quinn, J. M. W., & Moni, M. A. (2021). Machine learning-based statistical analysis for early stage detection of cervical cancer. *Computers in Biology and Medicine, 139*, 104985. https://doi.org/10.1016/j.compbiomed.2021.104985

Alowais, S. A., Alghamdi, S. S., Alsuhebany, N., Alqahtani, T., Alshaya, A. I., Almohareb, S. N., Aldairem, A., Alrashed, M., Bin Saleh, K., Badreldin, H. A., Al Yami, M. S., Al Harbi, S., & Albekairy, A. M. (2023). Revolutionizing healthcare: The role of artificial intelligence in clinical practice. *BMC Medical Education, 23*, 689. https://doi.org/10.1186/s12909-023-04698-z

Aminizadeh, S., Heidari, A., Dehghan, M., Toumaj, S., Rezaei, M., Jafari Navimipour, N., Stroppa, F., & Unal, M. (2024). Opportunities and challenges of artificial intelligence and distributed systems to improve the quality of healthcare service. *Artificial Intelligence in Medicine, 149*, 102779. https://doi.org/10.1016/j.artmed.2024.102779

An, Q., Rahman, S., Zhou, J., & Kang, J. J. (2023). A comprehensive review on machine learning in healthcare industry: Classification, restrictions, opportunities and challenges. *Sensors, 23*, 4178. https://doi.org/10.3390/s23094178

Boppiniti, S. T. (2023). Data ethics in AI: Addressing challenges in machine learning and data governance for responsible data science. *International Scientific Journal for Research, 5.*

Campanella, G., Hanna, M. G., Geneslaw, L., Miraflor, A., Werneck Krauss Silva, V., Busam, K. J., Brogi, E., Reuter, V. E., Klimstra, D. S., & Fuchs, T. J. (2019). Clinical-grade computational pathology using weakly supervised deep learning on whole slide images. *Nature Medicine, 25*, 1301–1309. https://doi.org/10.1038/s41591-019-0508-1

Caruana, R., Lou, Y., Gehrke, J., Koch, P., Sturm, M., & Elhadad, N. (2015). *Intelligible models for HealthCare: Predicting pneumonia risk and hospital 30-day readmission, in: Proceedings of the 21th ACM SIGKDD international conference on knowledge discovery and data mining, KDD '15* (pp. 1721–1730). Association for Computing Machinery. https://doi.org/10.1145/2783258.2788613

Cerda, I. H., Zhang, E., Dominguez, M., Ahmed, M., Lang, M., Ashina, S., Schatman, M. E., Yong, R. J., & Fonseca, A. C. G. (2024). Artificial intelligence and virtual reality in headache disorder diagnosis, classification, and management. *Current Pain and Headache Reports, 28*, 869–880. https://doi.org/10.1007/s11916-024-01279-7

Chai, Z.-K., Mao, L., Chen, H., Sun, T.-G., Shen, X.-M., Liu, J., & Sun, Z.-J. (2022). Improved diagnostic accuracy of ameloblastoma and odontogenic keratocyst on cone-beam CT by artificial intelligence. *Frontiers in Oncology, 11*. https://doi.org/10.3389/fonc.2021.793417

Chang, Y.-W., An, J. K., Choi, N., Ko, K. H., Kim, K. H., Han, K., & Ryu, J. K. (2022). Artificial intelligence for breast cancer screening in mammography (AI-STREAM): A prospective multicenter study design in Korea using AI-based CADe/x. *Journal of Breast Cancer, 25*, 57–68. https://doi.org/10.4048/jbc.2022.25.e4

Cho, Y. S., & Hong, P. C. (2023). Applying machine learning to healthcare operations management: CNN-based model for malaria diagnosis. *Healthcare, 11*, 1779. https://doi.org/10.3390/healthcare11121779

Coenen, A., Kim, Y.-H., Kruk, M., Tesche, C., De Geer, J., Kurata, A., Lubbers, M. L., Daemen, J., Itu, L., Rapaka, S., Sharma, P., Schwemmer, C., Persson, A., Schoepf, U. J., Kepka, C., Hyun Yang, D., & Nieman, K. (2018). Diagnostic accuracy of a machine-learning approach to coronary computed tomographic angiography–based fractional flow reserve. *Circulation. Cardiovascular Imaging, 11*, e007217. https://doi.org/10.1161/CIRCIMAGING.117.007217

Daimiel Naranjo, I., Gibbs, P., Reiner, J. S., Lo Gullo, R., Sooknanan, C., Thakur, S. B., Jochelson, M. S., Sevilimedu, V., Morris, E. A., Baltzer, P. A. T., Helbich, T. H., & Pinker, K. (2021). Radiomics and machine learning with multiparametric breast MRI for improved diagnostic accuracy in breast cancer diagnosis. *Diagnostics, 11*, 919. https://doi.org/10.3390/diagnostics11060919

Di Sario, G., Rossella, V., Famulari, E. S., Maurizio, A., Lazarevic, D., Giannese, F., & Felici, C. (2023). Enhancing clinical potential of liquid biopsy through a multi-omic approach: A systematic review. *Frontiers in Genetics, 14*. https://doi.org/10.3389/fgene.2023.1152470

Díaz, O., Rodríguez-Ruíz, A., & Sechopoulos, I. (2024). Artificial intelligence for breast cancer detection: Technology, challenges, and prospects. *European Journal of Radiology, 175*, 111457. https://doi.org/10.1016/j.ejrad.2024.111457

Díaz-Rodríguez, N., Del Ser, J., Coeckelbergh, M., López de Prado, M., Herrera-Viedma, E., & Herrera, F. (2023). Connecting the dots in trustworthy artificial intelligence: From AI principles, ethics, and key requirements to responsible AI systems and regulation. *Information Fusion, 99*, 101896. https://doi.org/10.1016/j.inffus.2023.101896

Dlamini, Z., Francies, F. Z., Hull, R., & Marima, R. (2020). Artificial intelligence (AI) and big data in cancer and precision oncology. *Computational and Structural Biotechnology Journal, 18*, 2300–2311. https://doi.org/10.1016/j.csbj.2020.08.019

Doppalapudi, S., Qiu, R. G., & Badr, Y. (2021). Lung cancer survival period prediction and understanding: Deep learning approaches. *International Journal of Medical Informatics, 148*, 104371. https://doi.org/10.1016/j.ijmedinf.2020.104371

Du, W., He, B., Luo, X., & Chen, M. (2022). Diagnostic value of artificial intelligence based on CT image in benign and malignant pulmonary nodules. *Journal of Oncology, 2022*, 5818423. https://doi.org/10.1155/2022/5818423

Elemento, O., Leslie, C., Lundin, J., & Tourassi, G. (2021). Artificial intelligence in cancer research, diagnosis and therapy. *Nature Reviews. Cancer, 21*, 747–752. https://doi.org/10.1038/s41568-021-00399-1

Elmore, J. G., Longton, G. M., Carney, P. A., Geller, B. M., Onega, T., Tosteson, A. N. A., Nelson, H. D., Pepe, M. S., Allison, K. H., Schnitt, S. J., O'Malley, F. P., & Weaver, D. L. (2015).

Diagnostic concordance among pathologists interpreting breast biopsy specimens. *JAMA, 313*, 1122–1132. https://doi.org/10.1001/jama.2015.1405

Ennab, M., & Mcheick, H. (2024). Enhancing interpretability and accuracy of AI models in healthcare: A comprehensive review on challenges and future directions. *Frontiers in Robotics and AI, 11*, 1444763. https://doi.org/10.3389/frobt.2024.1444763

Esmaeilzadeh, P. (2024). Challenges and strategies for wide-scale artificial intelligence (AI) deployment in healthcare practices: A perspective for healthcare organizations. *Artificial Intelligence in Medicine, 151*, 102861. https://doi.org/10.1016/j.artmed.2024.102861

Esteva, A., Robicquet, A., Ramsundar, B., Kuleshov, V., DePristo, M., Chou, K., Cui, C., Corrado, G., Thrun, S., & Dean, J. (2019). A guide to deep learning in healthcare. *Nature Medicine, 25*, 24–29. https://doi.org/10.1038/s41591-018-0316-z

Fraser, A. G., Biasin, E., Bijnens, B., Bruining, N., Caiani, E. G., Cobbaert, K., Davies, R. H., Gilbert, S. H., Hovestadt, L., Kamenjasevic, E., Kwade, Z., McGauran, G., O'Connor, G., Vasey, B., & Rademakers, F. E. (2023). Artificial intelligence in medical device software and high-risk medical devices—A review of definitions, expert recommendations and regulatory initiatives. *Expert Review of Medical Devices, 20*, 467–491. https://doi.org/10.1080/17434440.2023.2184685

Gazola, A. A., Lautert-Dutra, W., Archangelo, L. F., dos Reis, R. B., & Squire, J. A. (2024). Precision oncology platforms: Practical strategies for genomic database utilization in cancer treatment. *Molecular Cytogenetics, 17*, 28. https://doi.org/10.1186/s13039-024-00698-w

Ghassemi, M., Naumann, T., Schulam, P., Beam, A. L., Chen, I. Y., & Ranganath, R. (2020). A review of challenges and opportunities in machine learning for health. In *AMIA Summits on Translational Science proceedings 2020* (pp. 191–200).

Gomis-Pastor, M., Berdún, J., Borrás-Santos, A., De Dios López, A., Fernández-Montells Rama, B., García-Esquirol, Ó., Gratacòs, M., Ontiveros Rodríguez, G. D., Pelegrín Cruz, R., Real, J., Bachs I Ferrer, J., & Comella, A. (2024). Clinical validation of digital healthcare solutions: State of the art, challenges and opportunities. *Healthcare, 12*, 1057. https://doi.org/10.3390/healthcare12111057

Haggenmüller, S., Krieghoff-Henning, E., Jutzi, T., Trapp, N., Kiehl, L., Utikal, J. S., Fabian, S., & Brinker, T. J. (2021). Digital natives' preferences on Mobile artificial intelligence apps for skin cancer diagnostics: Survey study. *JMIR mHealth and uHealth, 9*, e22909. https://doi.org/10.2196/22909

Hanis, T. M., Islam, M. A., & Musa, K. I. (2022). Diagnostic accuracy of machine learning models on mammography in breast cancer classification: A meta-analysis. *Diagnostics, 12*, 1643. https://doi.org/10.3390/diagnostics12071643

Harishbhai Tilala, M., Kumar Chenchala, P., Choppadandi, A., Kaur, J., Naguri, S., Saoji, R., & Devaguptapu, B. (2024). Ethical considerations in the use of artificial intelligence and machine learning in health care: A comprehensive review. *Cureus, 16*, e62443. https://doi.org/10.7759/cureus.62443

Hart, M. L., Kato, R., & Rolauffs, B. (2024). Editorial: Artificial intelligence in predicting, determining and controlling cell phenotype or tissue function in inflammatory diseases. *Frontiers in Immunology, 15*. https://doi.org/10.3389/fimmu.2024.1443534

Heitzer, E., Haque, I. S., Roberts, C. E. S., & Speicher, M. R. (2019). Current and future perspectives of liquid biopsies in genomics-driven oncology. *Nature Reviews. Genetics, 20*, 71–88. https://doi.org/10.1038/s41576-018-0071-5

Ho, C. S. H., Chan, Y. L., Tan, T. W. K., Tay, G. W. N., & Tang, T. B. (2022). Improving the diagnostic accuracy for major depressive disorder using machine learning algorithms integrating clinical and near-infrared spectroscopy data. *Journal of Psychiatric Research, 147*, 194–202. https://doi.org/10.1016/j.jpsychires.2022.01.026

Hosny, A., Parmar, C., Quackenbush, J., Schwartz, L. H., & Aerts, H. J. W. L. (2018). Artificial intelligence in radiology. *Nature Reviews. Cancer, 18*, 500–510. https://doi.org/10.1038/s41568-018-0016-5

Hossain, E., Rana, R., Higgins, N., Soar, J., Barua, P. D., Pisani, A. R., & Turner, K. (2023). Natural language processing in electronic health records in relation to healthcare decision-making: A systematic review. *Computers in Biology and Medicine, 155*, 106649. https://doi.org/10.1016/j.compbiomed.2023.106649

Houssein, E. H., Mohamed, R. E., & Ali, A. A. (2023). Heart disease risk factors detection from electronic health records using advanced NLP and deep learning techniques. *Scientific Reports, 13*, 7173. https://doi.org/10.1038/s41598-023-34294-6

Jagannatha, A., Liu, F., Liu, W., & Yu, H. (2019). Overview of the first natural language processing challenge for extracting medication, indication, and adverse drug events from electronic health record notes (MADE 1.0). *Drug Safety, 42*, 99–111. https://doi.org/10.1007/s40264-018-0762-z

Kann, B. H., Hosny, A., & Aerts, H. J. W. L. (2021). Artificial intelligence for clinical oncology. *Cancer Cell, 39*, 916–927. https://doi.org/10.1016/j.ccell.2021.04.002

Kaswan, K. S., Gaur, L., Dhatterwal, J. S., & Kumar, R. (2021). AI-based natural language processing for the generation of meaningful information electronic health record (EHR). In *Data, in: Advanced AI techniques and applications in bioinformatics*. CRC Press.

Katsuki, M., Shimazu, T., Kikui, S., Danno, D., Miyahara, J., Takeshima, R., Takeshima, E., Shimazu, Y., Nakashima, T., Matsuo, M., & Takeshima, T. (2023). Developing an artificial intelligence-based headache diagnostic model and its utility for non-specialists' diagnostic accuracy. *Cephalalgia, 43*, 03331024231156925. https://doi.org/10.1177/03331024231156925

Kaushik, M., Chandra Joshi, R., Kushwah, A. S., Gupta, M. K., Banerjee, M., Burget, R., & Dutta, M. K. (2021). Cytokine gene variants and socio-demographic characteristics as predictors of cervical cancer: A machine learning approach. *Computers in Biology and Medicine, 134*, 104559. https://doi.org/10.1016/j.compbiomed.2021.104559

Khalifa, M., & Albadawy, M. (2024). AI in diagnostic imaging: Revolutionising accuracy and efficiency. *Computer Methods and Programs in Biomedicine Update, 5*, 100146. https://doi.org/10.1016/j.cmpbup.2024.100146

Khurshid, S., Reeder, C., Harrington, L. X., Singh, P., Sarma, G., Friedman, S. F., Di Achille, P., Diamant, N., Cunningham, J. W., Turner, A. C., Lau, E. S., Haimovich, J. S., Al-Alusi, M. A., Wang, X., Klarqvist, M. D. R., Ashburner, J. M., Diedrich, C., Ghadessi, M., Mielke, J., Eilken, H. M., McElhinney, A., Derix, A., Atlas, S. J., Ellinor, P. T., Philippakis, A. A., Anderson, C. D., Ho, J. E., Batra, P., & Lubitz, S. A. (2022). Cohort design and natural language processing to reduce bias in electronic health records research. *npj Digital Medicine, 5*, 1–14. https://doi.org/10.1038/s41746-022-00590-0

Koleck, T. A., Dreisbach, C., Bourne, P. E., & Bakken, S. (2019). Natural language processing of symptoms documented in free-text narratives of electronic health records: A systematic review. *Journal of the American Medical Informatics Association, 26*, 364–379. https://doi.org/10.1093/jamia/ocy173

Kulshrestha, S., Dligach, D., Joyce, C., Baker, M. S., Gonzalez, R., O'Rourke, A. P., Glazer, J. M., Stey, A., Kruser, J. M., Churpek, M. M., & Afshar, M. (2021). Prediction of severe chest injury using natural language processing from the electronic health record. *Injury, 52*, 205–212. https://doi.org/10.1016/j.injury.2020.10.094

Kushwah, A. S., Masood, S., Mishra, R., & Banerjee, M. (2024). Genetic and epigenetic alterations in DNA repair genes and treatment outcome of chemoradiotherapy in cervical cancer. *Critical Reviews in Oncology/Hematology, 194*, 104240. https://doi.org/10.1016/j.critrevonc.2023.104240

Li, Y.-H., Li, Y.-L., Wei, M.-Y., & Li, G.-Y. (2024a). Innovation and challenges of artificial intelligence technology in personalized healthcare. *Scientific Reports, 14*, 18994. https://doi.org/10.1038/s41598-024-70073-7

Li, J.-Y., Zuo, L.-P., Xu, J., & Sun, C.-Y. (2024b). Clinical applications of circulating tumor DNA in hematological malignancies: From past to the future. *Blood Reviews, 68*, 101237. https://doi.org/10.1016/j.blre.2024.101237

Liao, W.-W., Hsieh, Y.-W., Lee, T.-H., Chen, C., & Wu, C. (2022). Machine learning predicts clinically significant health related quality of life improvement after sensorimotor rehabilitation interventions in chronic stroke. *Scientific Reports, 12*, 11235. https://doi.org/10.1038/s41598-022-14986-1

Litjens, G., Sánchez, C. I., Timofeeva, N., Hermsen, M., Nagtegaal, I., Kovacs, I., Hulsbergen-van de Kaa, C., Bult, P., van Ginneken, B., & van der Laak, J. (2016). Deep learning as a tool for increased accuracy and efficiency of histopathological diagnosis. *Scientific Reports, 6*, 26286. https://doi.org/10.1038/srep26286

Luo, S., Xu, J., Jiang, Z., Liu, L., Wu, Q., Leung, E. L.-H., & Leung, A. P. (2020). Artificial intelligence-based collaborative filtering method with ensemble learning for personalized lung cancer medicine without genetic sequencing. *Pharmacological Research, 160*, 105037. https://doi.org/10.1016/j.phrs.2020.105037

Magro, D., Venezia, M., & Rita Balistreri, C. (2024). The omics approaches and liquid biopsies: Advantages, limitations, applications. *Medicine in Omics, 11*, 100039. https://doi.org/10.1016/j.meomic.2024.100039

Marinovich, M. L., Wylie, E., Lotter, W., Lund, H., Waddell, A., Madeley, C., Pereira, G., & Houssami, N. (2023). Artificial intelligence (AI) for breast cancer screening: BreastScreen population-based cohort study of cancer detection. *eBioMedicine, 90*, 104498. https://doi.org/10.1016/j.ebiom.2023.104498

Mathivanan, S. K., Francis, D., Srinivasan, S., Khatavkar, V., P K, & Shah, M. A. (2024). Enhancing cervical cancer detection and robust classification through a fusion of deep learning models. *Scientific Reports, 14*, 10812. https://doi.org/10.1038/s41598-024-61063-w

Mavropoulos, A., Johnson, C., Lu, V., Nieto, J., Schneider, E. C., Saini, K., Phelan, M. L., Hsie, L. X., Wang, M. J., Cruz, J., Mei, J., Kim, J. J., Lian, Z., Li, N., Boutet, S. C., Wong-Thai, A. Y., Yu, W., Lu, Q.-Y., Kim, T., Geng, Y., Masaeli, M. (Mahdokht), Lee, T. D., & Rao, J. (2023). Artificial intelligence-driven morphology-based enrichment of malignant cells from body fluid. *Modern Pathology, 36*, 100195. https://doi.org/10.1016/j.modpat.2023.100195

Maweu, B. M., Shamsuddin, R., Dakshit, S., & Prabhakaran, B. (2021). Generating healthcare time series data for improving diagnostic accuracy of deep neural networks. *IEEE Transactions on Instrumentation and Measurement, 70*, 1–15. https://doi.org/10.1109/TIM.2021.3077049

McKinney, S. M., Sieniek, M., Godbole, V., Godwin, J., Antropova, N., Ashrafian, H., Back, T., Chesus, M., Corrado, G. S., Darzi, A., Etemadi, M., Garcia-Vicente, F., Gilbert, F. J., Halling-Brown, M., Hassabis, D., Jansen, S., Karthikesalingam, A., Kelly, C. J., King, D., Ledsam, J. R., Melnick, D., Mostofi, H., Peng, L., Reicher, J. J., Romera-Paredes, B., Sidebottom, R., Suleyman, M., Tse, D., Young, K. C., De Fauw, J., & Shetty, S. (2020). International evaluation of an AI system for breast cancer screening. *Nature, 577*, 89–94. https://doi.org/10.1038/s41586-019-1799-6

Mennella, C., Maniscalco, U., Pietro, G. D., & Esposito, M. (2024). Ethical and regulatory challenges of AI approaches in healthcare: A narrative review. *Heliyon, 10*, e26297. https://doi.org/10.1016/j.heliyon.2024.e26297

Murchan, P., Ó'Brien, C., O'Connell, S., McNevin, C. S., Baird, A.-M., Sheils, O., Ó Broin P, & Finn, S. P. (2021). Deep learning of histopathological features for the prediction of tumour molecular genetics. *Diagnostics, 11*, 1406. https://doi.org/10.3390/diagnostics11081406

Najjar, R. (2023). Redefining radiology: A review of artificial intelligence integration in medical imaging. *Diagnostics, 13*, 2760. https://doi.org/10.3390/diagnostics13172760

Nazir, A., Hussain, A., Singh, M., & Assad, A. (2024). *Deep learning in medicine: Advancing healthcare with intelligent solutions and the future of holography imaging in early diagnosis* (Multimedia tools and applications). https://doi.org/10.1007/s11042-024-19694-8

Parisi, C., Tagliamento, M., Belcaid, L., Aldea, M., Bayle, A., Remon-Masip, J., Italiano, A., Planchard, D., Besse, B., & Barlesi, F. (2023). Circulating tumor DNA in clinical trials for solid tumors: Challenges and current applications. *Journal of Liquid Biopsy, 1*, 100007. https://doi.org/10.1016/j.jlb.2023.100007

Patra, A., Biswas, P., Behera, S. K., Barpanda, N. K., Sethy, P. K., & Nanthaamornphong, A. (2024). Transformative insights: Image-based breast cancer detection and severity assessment through advanced AI techniques. *Journal of Intelligent Systems, 33*. https://doi.org/10.1515/jisys-2024-0172

Peled, N., Fuchs, V., Kestenbaum, E. H., Oscar, E., & Bitran, R. (2021). An update on the use of exhaled breath analysis for the early detection of lung cancer. *Lung Cancer: Targets and Therapy, 12*, 81–92. https://doi.org/10.2147/LCTT.S320493

Rajkomar, A., Hardt, M., Howell, M. D., Corrado, G., & Chin, M. H. (2018). Ensuring fairness in machine learning to advance health equity. *Annals of Internal Medicine, 169*, 866–872. https://doi.org/10.7326/M18-1990

Rawat, L., & Nayak, V. (2021). Ursolic acid disturbs ROS homeostasis and regulates survival-associated gene expression to induce apoptosis in intestinal cancer cells. *Toxicology Research, 10*(3), 369–375.

Razzak, M. I., Naz, S., & Zaib, A. (2018). Deep learning for medical image processing: Overview, challenges and the future. In N. Dey, A. S. Ashour, & S. Borra (Eds.), *Classification in BioApps: Automation of decision making* (pp. 323–350). Springer. https://doi.org/10.1007/978-3-319-65981-7_12

Richens, J. G., Lee, C. M., & Johri, S. (2020). Improving the accuracy of medical diagnosis with causal machine learning. *Nature Communications, 11*, 3923. https://doi.org/10.1038/s41467-020-17419-7

Rodríguez-Ruiz, A., Krupinski, E., Mordang, J.-J., Schilling, K., Heywang-Köbrunner, S. H., Sechopoulos, I., & Mann, R. M. (2019). Detection of breast cancer with mammography: Effect of an artificial intelligence support system. *Radiology, 290*, 305–314. https://doi.org/10.1148/radiol.2018181371

Sachdeva, S., Bhatia, S., Harrasi, A. A., Shah, Y. A., Anwer, M. K., Philip, A. K., Shah, S. F. A., Khan, A., & Halim, S. A. (2024). Unraveling the role of cloud computing in health care system and biomedical sciences. *Heliyon, 10*, e29044. https://doi.org/10.1016/j.heliyon.2024.e29044

Santeramo, R., Damiani, C., Wei, J., Montana, G., & Brentnall, A. R. (2024). Are better AI algorithms for breast cancer detection also better at predicting risk? A paired case-control study. *Breast Cancer Research, 26*, 25. https://doi.org/10.1186/s13058-024-01775-z

Saroğlu, H. E., Shayea, I., Saoud, B., Azmi, M. H., El-Saleh, A. A., Saad, S. A., & Alnakhli, M. (2024). Machine learning, IoT and 5G approaches for breast cancer studies: A review. *Alexandria Engineering Journal, 89*, 210–223. https://doi.org/10.1016/j.aej.2024.01.043

Sharma, R., Gulati, A., & Chopra, K. (2023). Artificial intelligence (AI) and machine learning (ML): An innovative cross-talk perspective and their role in the healthcare industry. In D. K. Yadav & A. Gulati (Eds.), *Artificial intelligence and machine learning in healthcare* (pp. 9–38). Springer. https://doi.org/10.1007/978-981-99-6472-7_2

Shen, D., Wu, G., & Suk, H.-I. (2017). Deep learning in medical image analysis. *Annual Review of Biomedical Engineering, 19*, 221–248. https://doi.org/10.1146/annurev-bioeng-071516-044442

Sirocchi, C., Bogliolo, A., & Montagna, S. (2024). Medical-informed machine learning: Integrating prior knowledge into medical decision systems. *BMC Medical Informatics and Decision Making, 24*, 186. https://doi.org/10.1186/s12911-024-02582-4

Soto, J. T., Weston Hughes, J., Sanchez, P. A., Perez, M., Ouyang, D., & Ashley, E. A. (2022). Multimodal deep learning enhances diagnostic precision in left ventricular hypertrophy. *European Heart Journal Digital Health, 3*, 380–389. https://doi.org/10.1093/ehjdh/ztac033

Sree, P. K. (2023). Machine learning for quality in health care – A comprehensive review. *Biomedical Journal of Scientific & Technical Research, 51*, 42925–42929.

Strati, A., Markou, A., Kyriakopoulou, E., & Lianidou, E. (2023). Detection and molecular characterization of circulating tumour cells: Challenges for the clinical setting. *Cancers, 15*, 2185. https://doi.org/10.3390/cancers15072185

Sufyan, M., Shokat, Z., & Ashfaq, U. A. (2023). Artificial intelligence in cancer diagnosis and therapy: Current status and future perspective. *Computers in Biology and Medicine, 165*, 107356. https://doi.org/10.1016/j.compbiomed.2023.107356

Thapa, C., & Camtepe, S. (2021). Precision health data: Requirements, challenges and existing techniques for data security and privacy. *Computers in Biology and Medicine, 129*, 104130. https://doi.org/10.1016/j.compbiomed.2020.104130

Topol, E. (2019). *Deep medicine: How artificial intelligence can make healthcare human again.* Hachette UK.

Unger, M., & Kather, J. N. (2024). Deep learning in cancer genomics and histopathology. *Genome Medicine, 16*, 44. https://doi.org/10.1186/s13073-024-01315-6

Weng, W.-H. (2020). Machine learning for clinical predictive analytics. In L. A. Celi, M. S. Majumder, P. Ordóñez, J. S. Osorio, K. E. Paik, & M. Somai (Eds.), *Leveraging data science for Global Health* (pp. 199–217). Springer. https://doi.org/10.1007/978-3-030-47994-7_12

Whitney, H. M., Yoeli-Bik, R., Abramowicz, J. S., Lan, L., Li, H., Longman, R. E., Lengyel, E., & Giger, M. L. (2024). AI-based automated segmentation for ovarian/adnexal masses and their internal components on ultrasound imaging. *Journal of Medical Imaging, 11*, 044505. https://doi.org/10.1117/1.JMI.11.4.044505

Williamson, S. M., & Prybutok, V. (2024). Balancing privacy and Progress: A review of privacy challenges, systemic oversight, and patient perceptions in AI-driven healthcare. *Applied Sciences, 14*, 675. https://doi.org/10.3390/app14020675

Xiao, C., Choi, E., & Sun, J. (2018). Opportunities and challenges in developing deep learning models using electronic health records data: A systematic review. *Journal of the American Medical Informatics Association, 25*, 1419–1428. https://doi.org/10.1093/jamia/ocy068

Yang, Q., Liu, Y., Chen, T., & Tong, Y. (2019). Federated machine learning: Concept and applications. *ACM Transactions on Intelligent Systems and Technology, 10*(2), 1–19. https://doi.org/10.1145/3298981

Yang, X., Chen, A., PourNejatian, N., Shin, H. C., Smith, K. E., Parisien, C., Compas, C., Martin, C., Costa, A. B., Flores, M. G., Zhang, Y., Magoc, T., Harle, C. A., Lipori, G., Mitchell, D. A., Hogan, W. R., Shenkman, E. A., Bian, J., & Wu, Y. (2022). A large language model for electronic health records. *Npj Digital Medicine, 5*, 1–9. https://doi.org/10.1038/s41746-022-00742-2

Yang, L., Yang, D., Yao, M., & Bai, C. (2023). Concept and prospect of the Human-Computer Multi-Disciplinary team (MDT) in pulmonary nodule evaluation. *Clinical eHealth, 6*, 172–181. https://doi.org/10.1016/j.ceh.2023.11.003

Zaidan, D. F. Z. (2024). AI-driven decision support systems for clinical diagnosis: Develops AI-driven decision support systems to aid clinicians in making accurate diagnoses. *Journal of Bioinformatics and Artificial Intelligence, 4*, 84–95.

Zakari, S., Niels, N. K., Olagunju, G. V., Nnaji, P. C., Ogunniyi, O., Tebamifor, M., Israel, E. N., Atawodi, S. E., & Ogunlana, O. O. (2024). Emerging biomarkers for non-invasive diagnosis and treatment of cancer: A systematic review. *Frontiers in Oncology, 14*, 1405267. https://doi.org/10.3389/fonc.2024.1405267

Zhai, K., Yousef, M. S., Mohammed, S., Al-Dewik, N. I., & Qoronfleh, M. W. (2023). Optimizing clinical workflow using precision medicine and advanced data analytics. *PRO, 11*, 939. https://doi.org/10.3390/pr11030939

Advances in Image Processing and Pattern Recognition

Manish Mathuria and Dinesh Kumar Saini

Abstract

Image processing and pattern recognition are pivotal components of modern computer vision, enabling machines to interpret and understand visual data. This chapter delves into the latest advances in these fields, exploring cutting-edge techniques, algorithms, and applications. The proliferation of digital images and videos has created an overwhelming need for efficient and accurate image processing and pattern recognition methods. Traditional approaches have limitations, prompting researchers to develop innovative solutions. Recent advances in image processing and pattern recognition have transformed the field, enabling innovative applications. Continued research will drive further breakthroughs, fostering a future where machines seamlessly understand and interact with visual data.

1 Introduction

Image processing is the field of study that deals with the manipulation and analysis of digital images. It involves using algorithms and mathematical models to enhance, restore, compress, or extract information from images. **Pattern recognition** is a field of artificial intelligence that focuses on identifying and classifying patterns within data. It involves developing algorithms and techniques to recognize patterns in various forms, such as images, audio signals, text, and numerical data. Image

M. Mathuria
Manipal University Jaipur, Jaipur, India

D. K. Saini (✉)
Department of IoT and Intelligent Systems, Manipal University Jaipur,
Jaipur, Rajasthan, India
e-mail: Dineshkumar.saini@jaipur.manipal.edu

R. Nema et al. (eds.), *Advances in Cancer Detection, Prediction, and Prognosis Using Artificial Intelligence and Machine Learning*,
https://doi.org/10.1007/978-981-96-9346-7_5

processing and pattern recognition are two interconnected fields that have become increasingly crucial in various aspects of modern life. Their importance stems from their ability to extract meaningful information from visual data and make intelligent decisions based on that information. **Computer vision** is a field of artificial intelligence that deals with enabling computers to understand and interpret visual information. It involves developing algorithms and techniques to process, analyze, and extract meaningful information from images and videos (Litjens et al., 2017).

Vision is our most advanced sense, making images crucial to human perception. Unlike humans, who are restricted to the visual band of the electromagnetic spectrum, imaging technologies have expanded our visual capabilities across the entire spectrum, from gamma to radio waves. This enables machines to capture and process images from unconventional sources, including ultrasound, electron microscopy, and computer-generated imagery. As a result, digital image processing has evolved into a diverse and multifaceted field, encompassing a broad range of applications that transcend traditional visual boundaries (Kumar & Kumar, 2018).

The boundaries between image processing, image analysis, and computer vision are blurry and debated among experts. Some define image processing narrowly as transforming images to images, but this distinction is limiting. For instance, calculating an image's average intensity, yielding a numerical output, wouldn't qualify. Computer vision, a subset of artificial intelligence, aims to replicate human vision, enabling computers to learn, infer, and act upon visual inputs. Image analysis, or image understanding, occupies the middle ground, bridging image processing and computer vision. Artificial intelligence, still in its infancy, faces slower-than-expected progress, while image analysis integrates techniques from both image processing and computer vision to extract meaningful information (Kumar & Kumar, 2018).

The spectrum from image processing to computer vision is seamless, but a useful framework categorizes processes into low-, mid-, and high-level operations. Low-level processes, such as noise reduction, contrast enhancement, and image sharpening, take images as input and output, focusing on primitive operations. Mid-level processing involves segmentation, object description, and classification, extracting attributes like edges, contours, and object identities from images. High-level processing interprets recognized objects, enabling image analysis and, ultimately, cognitive functions akin to human vision, such as understanding and inference (Litjens et al., 2017).

The intersection of image processing and image analysis lies in object recognition, where digital image processing encompasses not only image-to-image transformations but also attribute extraction and object recognition. This book defines digital image processing broadly, covering processes from image acquisition and preprocessing to character recognition and description. For instance, automated text analysis involves image processing techniques like character segmentation, description, and recognition. In contrast, interpreting the text's meaning falls under image analysis or computer vision, depending on the complexity. Digital image processing, as defined here, has far-reaching applications in valuable fields, and the concepts explored in subsequent chapters provide the foundation for methods used in these areas (Bengio, 2009).

Humans possess unparalleled pattern recognition abilities, surpassing those found in the biological world. In contrast, current machine recognition capabilities, although impressive, fall short of human capabilities, particularly in interpreting complex images and generalizing knowledge. Nevertheless, recognition machines play a vital role in daily life, automating essential tasks such as barcode scanning, check processing, quality inspection, fingerprint analysis, mail sorting, and speech recognition, making modern life inconceivable without them.

1.1 Digital Image Processing

An image can be mathematically represented as a two-dimensional function, f(x,y), where x and y are spatial coordinates, and the amplitude at each point (x,y) represents the image's intensity or gray level. When these coordinates and intensity values are discrete and finite, the image is classified as digital. Digital image processing involves manipulating these digital images using computers. A digital image consists of a finite number of picture elements, also known as image elements, pels, or pixels—with pixel being the most commonly used term. Each pixel has a specific location and value, forming the fundamental building blocks of digital images (Gonzalez & Woods, 2018).

1.2 Recognition of Classify Patterns

In image pattern recognition, a pattern refers to the spatial arrangement of distinct features. A pattern class comprises patterns sharing common characteristics. Machine-based pattern recognition involves automated techniques that categorize patterns into their corresponding classes. Specifically, a pattern recognition system receives unknown patterns or pattern sets as input and assigns a class label to each, effectively identifying and grouping similar patterns.

The recognition process involves four stages: sensing, preprocessing, feature extraction, and classification. In image processing, sensing generates 2D or higher-dimensional signals. Preprocessing enhances and restores images, while classification assigns class labels to unknown patterns based on extracted features. There are three primary image pattern classification approaches: (1) prototype matching, (2) optimum statistical classifiers, and (3) neural networks. These methods differ in the level of engineering required to transform raw data into processable formats. Recognition performance hinges on feature discriminative power. Prototype-based classification seeks unique, detectable features for simple classification, as seen in bank-check processors using stylized fonts. Statistical classifiers optimize performance through parameter selection, emphasizing both feature quality and classifier design. Neural networks can operate on engineered features or generate their own representations from raw data, eliminating the need for feature engineering (Gonzalez & Woods, 2018).

The three recognition approaches share a common trait: they rely on parameters that must be specified or learned from representative patterns. These patterns can be labeled, where each pattern's class is known, or unlabeled, where the data's pattern nature is clear, but class assignments are unknown. Labeled data examples include character recognition, where character samples are collected and labeled with their corresponding identities (0–9, a-z). In contrast, unlabeled data is often used for clustering, where the goal is to identify cluster centers that serve as prototypes for the underlying pattern classes within the data (Gonzalez & Woods, 2018; Pratt, 2014).

When working with labeled data, a dataset is typically divided into training (50%), validation (25%), and test sets (25%). The training set is used to generate classifier parameters through an iterative process, where the classifier adjusts parameters based on error corrections. After training, the validation set evaluates and compares different designs against performance objectives. Once a design is selected, the test set assesses its real-world performance on unseen patterns. If training/validation results diverge from test results, it indicates overfitting, requiring revisions to the system architecture. This process is known as supervised learning, where the system learns from labeled data. In contrast, unsupervised learning involves systems learning pattern classes from unlabeled data. This chapter focuses on supervised learning, encompassing approaches from feature-based learning to deep learning, where systems automatically learn features from raw data without human specification (Pratt, 2014).

1.3 Applications of Image Processing (IP) and Pattern Recognition (PR) (Table 1)

1.4 The World's Dependence on Image Processing Applications and Tools

Image processing has become an integral part of our modern world, impacting various industries and aspects of daily life. Its applications are vast and diverse, ranging from healthcare to entertainment and security. Here's a breakdown of how the world depends on image processing (Jain, 1989):

- Healthcare
  - Medical Imaging: X-rays, MRI scans, CT scans, and ultrasounds rely heavily on image processing to enhance clarity, detect abnormalities, and aid in diagnosis and treatment planning.
  - Surgical Assistance: Image-guided surgery uses real-time imaging to assist surgeons in precise procedures, reducing risks and improving outcomes.
- Entertainment
  - Film and Television: Special effects, editing, and restoration of old films heavily depend on image processing techniques.

Table 1 Applications of Image Processing (IP) and Pattern Recognition (PR) Application IP Techniques PR Techniques Industry Description

Application	IP techniques	PR techniques	Industry	Description
Facial Recognition	Edge detection, feature extraction	Neural networks, machine learning	Security	Identifying individuals based on facial features.
Medical Image Analysis	Image segmentation, feature extraction	Deep learning, convolutional neural networks	Healthcare	Detecting abnormalities in medical images (e.g., tumors, fractures).
Autonomous Vehicle Navigation	Image segmentation, object detection	Computer vision, machine learning	Automotive	Enabling vehicles to perceive and navigate their surroundings.
Product Quality Inspection	Image segmentation, defect detection	Machine learning, statistical analysis	Manufacturing	Inspecting products for defects and ensuring quality.
Remote Sensing	Image enhancement, feature extraction	Machine learning, deep learning	Environmental Science	Analyzing satellite images for land use classification, disaster monitoring, and climate change studies.
Document Analysis	Optical character recognition (OCR), image segmentation	Machine learning, statistical analysis	Information Technology	Extracting text from documents and images.
Fingerprint Recognition	Feature extraction, matching	Machine learning, statistical analysis	Security	Identifying individuals based on their fingerprints.
Augmented Reality	Image tracking, object detection	Computer vision, machine learning	Entertainment, Gaming	Overlaying digital information on real-world scenes.
Sports Analysis	Object tracking, motion analysis	Computer vision, machine learning	Sports	Analyzing sports footage for player performance, strategy, and statistics.
Art and Design	Image editing, manipulation	Creative algorithms, generative models	Art, Design	Creating digital art and designs.

Pratt (2014)

  - Video Games: Image processing is used to create realistic graphics, animations, and virtual environments.
- Security
  - Surveillance: Cameras and facial recognition systems use image processing to monitor public spaces, identify individuals, and detect suspicious activities.
  - Biometrics: Fingerprint, iris, and facial recognition systems rely on image processing for accurate identification and authentication.
- Manufacturing
  - Quality Control: Image processing is used to inspect products for defects, ensuring quality and efficiency.
  - Automated Assembly: Visual guidance systems use image processing to guide robots in assembly tasks.
- Retail
  - Product Identification: Barcode scanners and QR code readers use image processing to identify products and track inventory.
  - Virtual Try-On: Online retailers use image processing to allow customers to virtually try on clothes.
- Autonomous Systems
  - Self-Driving Cars: Image processing is essential for cars to perceive their surroundings, identify objects, and navigate safely.
  - Drones: Image processing is used for aerial photography, mapping, and inspection tasks.
- Environmental Monitoring
  - Satellite Imagery: Image processing of satellite data helps monitor deforestation, climate change, and natural disasters.
  - Wildlife Conservation: Image processing is used to track wildlife populations and monitor their habitats.

1.5 Most Popular Image Processing Techniques (Table 2)

1.6 Image Processing Tools (Table 3)

2 Highly Demanding Applications

The demand for image processing has skyrocketed in recent years, driven by the exponential growth of digital data and the increasing need for automated visual analysis. Advances in computer vision, machine learning, and artificial intelligence have opened up new applications across industries, from healthcare and surveillance to autonomous vehicles and space exploration. As a result, image processing

Table 2 Most popular image processing techniques

Sr. No.	Technique	Description	Applications
1	Filtering	Applying mathematical operations to modify pixel values.	Noise reduction, edge detection, sharpening, blurring.
2	Segmentation	Dividing images into meaningful regions.	Object detection, image analysis, medical imaging.
3	Transformation	Converting images from one representation to another.	Fourier transform, wavelet transform, Hough transform.
4	Morphological Operations	Applying shape-based operations to extract features.	Erosion, dilation, opening, closing.
5	Machine Learning and Deep Learning	Using algorithms to learn from large datasets.	Object recognition, image classification, image generation.
6	Feature Extraction	Extracting relevant features from images.	Edge detection, corner detection, texture analysis.
7	Image Enhancement	Improving image quality.	Contrast enhancement, brightness adjustment, histogram equalization.
8	Image Restoration	Removing noise or artifacts from images.	Denoising, deblurring, inpainting.
9	Image Compression	Reducing image file size.	JPEG, PNG, GIF.
10	Image Registration	Aligning multiple images.	Medical image analysis, remote sensing.

Pratt (2014) and Jain (1989)

specialists are in high demand, with expertise sought after in areas such as object detection, segmentation, and recognition. The rise of deep learning has further accelerated this trend, enabling unprecedented accuracy and efficiency in image analysis. Industries are eager to harness the power of image processing to enhance decision-making, improve safety, and drive innovation. Here are some highly demanding applications of image processing (Gonzalez & Woods, 2018):

Medical imaging 1. Tumor detection and segmentation 2. Cancer diagnosis (e.g., mammography, MRI) 3. Brain image analysis (e.g., stroke detection) 4. Cardiovascular disease diagnosis 5. Medical image reconstruction (e.g., CT, MRI)	Aerial imaging 1. Land use classification 2. Crop monitoring 3. Disaster response 4. Environmental monitoring 5. Geospatial mapping
Surveillance and security 1. Facial recognition and authentication 2. Object detection and tracking 3. Intrusion detection 4. License plate recognition 5. Biometric identification (e.g., fingerprint, iris)	Forensic analysis 1. Image enhancement 2. Facial recognition 3. Fingerprint analysis 4. Footwear analysis 5. Digital evidence processing

Table 3 Image processing tools

Sr. No.	Tool	Description	URL
1	Adobe Photoshop	A professional image editing software with a wide range of features.	https://www.adobe.com/products/photoshop.html
2	GIMP	A free and open-source image editing software.	https://www.gimp.org/
3	ImageJ	An image processing and analysis platform for scientific research.	https://imagej.net/nih-image/
4	MATLAB	A high-performance computing environment for technical computing.	https://www.mathworks.com/
5	OpenCV	An open-source computer vision library.	https://opencv.org/
6	scikit-image	A Python library for image processing.	https://scikit-image.org/
7	PIL (Python Imaging Library)	A Python library for image processing.	https://pillow.readthedocs.io/
8	TensorFlow	An open-source machine learning framework.	https://www.tensorflow.org/
9	PyTorch	An open-source machine learning framework.	https://pytorch.org/
10	Keras	A high-level API for deep learning.	https://keras.io/

Gonzalez and Woods (2018), Pratt (2014) and Jain (1989)

Autonomous vehicles 1. Object detection and classification 2. Lane detection and tracking 3. Pedestrian detection 4. Traffic signal recognition 5. Scene understanding	Space exploration 1. Planetary image analysis 2. Object detection (e.g., asteroids) 3. Terrain mapping 4. Atmospheric analysis 5. Image compression
Quality inspection 1. Defect detection (e.g., manufacturing) 2. Product classification 3. Texture analysis 4. Color inspection 5. Barcode recognition	Robotics and computer vision 1. Object recognition 2. Scene understanding 3. Motion tracking 4. Gesture recognition 5. Human-robot interaction

2.1 Challenges

The high demand applications of image processing pose significant challenges, testing the limits of current technologies and expertise. One major hurdle is the sheer volume and complexity of data, with applications such as surveillance and autonomous vehicles generating terabytes of images per second. Additionally, image quality issues, noise, and variability in lighting conditions can severely impact accuracy. Scalability and real-time processing requirements also pose significant challenges, particularly in applications like object detection and tracking. Furthermore, ensuring robustness, security, and privacy in image processing systems is crucial, especially in sensitive domains like healthcare and law enforcement. The need for large annotated datasets, computational resources, and energy efficiency also creates bottlenecks. Moreover, addressing domain-specific challenges, such as variability in medical images or adverse weather conditions in autonomous vehicles, demands specialized expertise. Finally, the interpretability and explainability of deep learning-based image processing models remain significant concerns, necessitating ongoing research and innovation to overcome these challenges (Litjens et al., 2017).

Key challenges:

- Data volume and complexity
- Image quality and noise
- Scalability and real-time processing
- Robustness, security, and privacy
- Large annotated datasets
- Computational resources and energy efficiency
- Domain-specific challenges
- Model interpretability and explainability

2.2 New Improvement

Recent breakthroughs in image processing have significantly enhanced the performance and efficiency of highly demanding applications. Advances in deep learning architectures, such as convolutional neural networks (CNNs) and generative adversarial networks (GANs), have improved image recognition accuracy and robustness. New techniques like transfer learning, attention mechanisms, and ensemble methods have further boosted performance. Improved optimization algorithms and specialized hardware like graphics processing units (GPUs) and tensor processing units (TPUs) have accelerated processing speeds. Additionally, the integration of image processing with other technologies like computer vision, natural language processing, and Internet of Things (IoT) has opened up new applications (Kumar & Kumar, 2018).

Notable improvements include:

- Enhanced object detection and tracking
- Improved image segmentation and segmentation
- Advanced image restoration and enhancement
- Increased robustness to noise and variability
- Real-time processing capabilities
- Improved edge AI and IoT applications

2.3 Latest Research

Current research on highly demanding applications of image processing focuses on advancing deep learning architectures, exploring new techniques, and addressing domain-specific challenges. Some notable research directions include (Chen et al., 2020):

- Explainable AI (XAI) for image processing to improve model interpretability
- Edge AI and distributed computing for real-time processing
- Quantum computing for accelerated image processing
- 3D and 4D image processing for enhanced visualization
- Multimodal image analysis for fusion of visual and non-visual data

3 Why Advancement Is Required?

The ever-increasing demand for accurate and efficient image processing has created a pressing need for advancements in image processing techniques. The requirement for improvement stems from various factors, including:

- Increased complexity of applications (e.g., autonomous vehicles, medical diagnostics)
- Growing volume and diversity of image data
- Demand for real-time processing and analysis
- Need for improved accuracy, robustness, and reliability
- Emerging applications in fields like extended reality, robotics, and healthcare

3.1 Recent Failures and Security Breaks

Recent years have witnessed several high-profile failures and several alarming security breaks of image processing systems.

3.1.1 Limitations and Vulnerabilities of These Technologies

- Misidentification of people of color by facial recognition systems (e.g., Amazon Rekognition, Google Cloud Vision)
- Self-driving car accidents due to faulty object detection (e.g., Uber, Tesla)
- Misdiagnosis by medical imaging AI systems (e.g., Google Health, IBM Watson)
- Inaccurate surveillance footage analysis (e.g., London Metropolitan Police)
- Biased image classification by social media platforms (e.g., Facebook, Twitter)

3.1.2 Integrity and Confidentiality of Visual Data Related Incidents Include

- Deepfake attacks on facial recognition systems (e.g., Facebook, Google)
- Adversarial attacks on self-driving car vision systems (e.g., Tesla, Uber)

- Data breaches in medical imaging repositories (e.g., healthcare organizations)
- Unauthorized access to surveillance footage (e.g., city surveillance systems)
- Manipulation of images on social media platforms (e.g., Twitter, Instagram)

3.2 Recent Research Publications

2023–2024 (Table 4)

3.3 International Organization with Research Projects in Image Processing

Sr. No.	Organization	Project name	Focus area
1	European Union	Horizon Europe	Various projects related to AI, computer vision, and image analysis.
2	National Science Foundation (NSF), USA	Computer Vision and Image Understanding (CVUI) Program	Supports fundamental research in computer vision and image understanding.
3	National Institutes of Health (NIH), USA	Biomedical Informatics and Computational Biology (BIC) Program	Funds research on image analysis for biomedical applications.
4	Deutsche Forschungsgemeinschaft (DFG), Germany	Collaborative Research Centers (CRC)	Supports interdisciplinary research projects, including those in image processing and pattern recognition.
5	Japan Society for the Promotion of Science (JSPS)	KAKENHI Grants	Provides funding for research projects in various fields, including computer science and engineering.
6	Chinese Academy of Sciences (CAS)	Strategic Priority Research Program (SPRP)	Funds large-scale research projects in areas such as information technology and artificial intelligence.
7	Indian Council of Scientific and Industrial Research (CSIR)	National Science and Technology Plan	Supports research projects in various scientific and technological areas.
8	Massachusetts Institute of Technology (MIT), USA	Computer Science and Artificial Intelligence Laboratory (CSAIL)	Conducts research on a wide range of topics, including computer vision and machine learning.
9	Stanford University, USA	Stanford Artificial Intelligence Laboratory (SAIL)	Focuses on research in artificial intelligence, including computer vision and natural language processing.
10	University of Oxford, UK	Department of Engineering Science	Engages in research projects related to computer vision, machine learning, and robotics.

Table 4 International organization with research projects in image processing

Sr. No.	Keyword	No. of result	Publisher	Type
1	Image Processing	**39,890**	IEEE	Conferences
2	Image Processing	**16,666**	IEEE	Journals
3	Image Processing	**1,569**	IEEE	Early Access Articles
4	Image Processing	**183**	IEEE	Books
5	Image Processing	**26**	IEEE	Standards
6	Image Processing	**301**	IEEE	Magazines
7	Pattern Recognition	**11,676**	IEEE	Conferences
8	Pattern Recognition	**2005**	IEEE	Journals
9	Pattern Recognition	**211**	IEEE	Early access articles
10	Pattern Recognition	**28**	IEEE	Books
11	Pattern Recognition	**1**	IEEE	Standards
12	Pattern Recognition	**42**	IEEE	Magazines

3.4 Advance Image Processing in Top Companies/ Research Organizations

Sr. No.	Organization	Project area	Focus
1	NVIDIA	Deep Learning and Neural Networks	Generative Adversarial Networks (GANs), Convolutional Neural Networks (CNNs), Transformer architectures
2	Google AI	Deep Learning and Neural Networks, Medical Image Analysis, Augmented Reality (AR) and Virtual Reality (VR)	GANs, CNNs, Computer-aided diagnosis (CAD), Object tracking and recognition, Image-based rendering
3	OpenAI	Deep Learning and Neural Networks	GANs, Transformer architectures, Explainable AI (XAI)
4	Stanford University	Deep Learning and Neural Networks, Medical Image Analysis, Augmented Reality (AR) and Virtual Reality (VR)	CNNs, Medical image segmentation, Human-computer interaction
5	Massachusetts Institute of Technology (MIT)	Deep Learning and Neural Networks, Medical Image Analysis	CNNs, Computer-aided diagnosis (CAD), Explainable AI (XAI)
6	National Institutes of Health (NIH)	Medical Image Analysis	Computer-aided diagnosis (CAD), Medical image segmentation
7	European Commission	Medical Image Analysis, Remote Sensing	Computer-aided diagnosis (CAD), Land cover classification, Change detection
8	NASA	Remote Sensing	Land cover classification, Object detection in satellite imagery, Change detection

Sr. No.	Organization	Project area	Focus
9	European Space Agency (ESA)	Remote Sensing	Land cover classification, Object detection in satellite imagery, Change detection
10	Max Planck Institute for Intelligent Systems	Remote Sensing, Biometrics	Object detection in satellite imagery, Facial recognition
11	University of Oxford	Remote Sensing, Biometrics	Object detection in satellite imagery, Fingerprint recognition
12	Johns Hopkins University	Medical Image Analysis, Remote Sensing	Medical image segmentation, Change detection
13	University of Maryland	Remote Sensing	Change detection
14	University of California, Los Angeles (UCLA)	Medical Image Analysis	Image-guided surgery
15	Imperial College London	Medical Image Analysis	Image-guided surgery
16	University of Notre Dame	Biometrics	Iris recognition
17	University of New South Wales	Biometrics	Iris recognition
18	Apple	Augmented Reality (AR) and Virtual Reality (VR)	Object tracking and recognition
19	Microsoft	Augmented Reality (AR) and Virtual Reality (VR), Biometrics	Object tracking and recognition, Facial recognition
20	Meta Platforms	Augmented Reality (AR) and Virtual Reality (VR)	Object tracking and recognition
21	University of Southern California	Augmented Reality (AR) and Virtual Reality (VR)	Image-based rendering
22	University of Toronto	Augmented Reality (AR) and Virtual Reality (VR)	Image-based rendering
23	Carnegie Mellon University	Augmented Reality (AR) and Virtual Reality (VR)	Human-computer interaction
24	Google	Federated Learning	Training machine learning models on decentralized data
25	Samsung	Federated Learning	Training machine learning models on decentralized data
26	Harvard University	AI ethics and bias	Addressing issues of fairness, bias, and privacy in AI systems

4 The Role of AI in Advancement in Image Processing

Artificial Intelligence (AI) plays a transformative role in advancing image processing and pattern recognition, revolutionizing the field with unprecedented accuracy, efficiency, and scalability. AI-powered algorithms, particularly deep learning techniques such as Convolutional Neural Networks (CNNs) and Generative Adversarial Networks (GANs), enable significant improvements in image enhancement, restoration, segmentation, and object detection. Moreover, AI-driven pattern recognition capabilities facilitate accurate classification, identification, and prediction in various applications, including facial recognition, biometrics, medical imaging, and autonomous systems. AI's ability to learn from vast datasets and adapt to new scenarios allows for real-time processing, improved decision-making, and enhanced customer experiences. Furthermore, AI-driven image processing and pattern recognition find applications in diverse industries, including healthcare, surveillance, robotics, and aerospace, driving innovation and growth. As AI technology continues to evolve, we can expect even more sophisticated image processing and pattern recognition capabilities, fueling breakthroughs in fields like computer vision, natural language processing, and predictive analytics (Gonzalez & Woods, 2018; Pratt, 2014; Jain, 1989).

4.1 Benefits of AI in Advancement in Image Processing

- Improved accuracy and efficiency
- Enhanced decision-making
- Increased productivity
- Reduced costs
- Improved customer experience

4.2 Challenges with AI in Image Processing

- Data quality and availability
- Computational complexity
- Overfitting and bias
- Explainability and interpretability
- Security and privacy concerns

5 Deep Learning Based Image Processing

Deep learning-based image processing has revolutionized the field of computer vision, enabling unprecedented accuracy and efficiency in image analysis and manipulation. Leveraging convolutional neural networks (CNNs), recurrent neural networks (RNNs), and generative adversarial networks (GANs), deep learning algorithms can learn complex patterns and features from large datasets, allowing for remarkable advancements in image processing tasks such as denoising, deblurring, segmentation, and object detection. Deep learning-based techniques also facilitate image enhancement, restoration, and reconstruction, with applications in medical imaging, surveillance, and multimedia processing (Bengio, 2009; He et al., 2016).

Furthermore, deep learning-based image processing enables:

- Image classification and recognition
- Object detection and tracking
- Facial recognition and biometrics
- Image-to-image translation and generation
- Image super-resolution and enhancement

5.1 Key Architectures Include

- U-Net
- ResNet
- Inception
- YOLO (You Only Look Once)
- GANs (Generative Adversarial Networks)

6 Advances in Cancer Detection Techniques

Advances in cancer detection techniques based on image processing and pattern recognition have been remarkable, transforming the medical landscape. Deep Learning-Based Approaches have taken center stage, enabling computers to automatically analyze medical images and detect abnormalities. These approaches have improved cancer diagnosis and treatment by:

- Enhancing Image Quality: Techniques like denoising and deblurring improve image clarity, allowing for more accurate diagnoses.
- Automated Detection: Algorithms can detect tumors, calcifications, and other abnormalities, reducing human error.
- Pattern Recognition: Computers can identify patterns in medical images, facilitating early cancer detection (Kaur & Garg, 2023).

6.1 Recent Developments Include

- Digital Single-Operator Cholangioscopy (D-SOC): This technique uses artificial intelligence to analyze images of the bile ducts, detecting cancerous cells.
- Convolutional Neural Networks (CNNs): CNNs have improved cancer diagnosis accuracy by analyzing medical images and identifying patterns (He et al., 2016).

6.2 Key Applications of Image Processing and Pattern Recognition in Cancer Detection Include

- Breast Cancer Detection: Automated analysis of mammography images

Ya-nan Zhang, Ke-rui XIA, Chang-yi LI, Ben-li WEI, Bing Zhang	The paper discusses a breast cancer recognition method based on image processing, covering aspects such as detection, segmentation, registration, and fusion. It explores the application of machine learning techniques, including supervised learning, unsupervised learning, deep learning, and reinforcement learning, in breast cancer examination. The study highlights the potential of reinforcement learning in breast cancer image processing and emphasizes the need for addressing privacy protection concerns for breast cancer patients (Zhang et al., 2021)
Hasnae Zerouaoui and Ali Idri	A comprehensive review of machine learning and image processing techniques for breast cancer imaging was conducted through a structured literature review of 530 papers. The results showed that diagnosis was the most common medical task, with deep learning techniques widely used for classification. Mammograms were the most used imaging modality, and public and private datasets were utilized. Image processing techniques included pre-processing, segmentation, feature extraction, and feature selection. The review highlighted trends and gaps in current research, providing insights for future studies to improve breast cancer diagnosis and treatment (Zerouaoui & Idri, 2021)
David Murcia-Gómez, Ignacio Rojas-Valenzuela, and Olga Valenzuela	Early diagnosis of cancer is very important as it significantly increases the chances of appropriate treatment and survival. To this end, Deep Learning models are increasingly used in the classification and segmentation of histopathological images, as they obtain high accuracy index and can help specialists. In most cases, images need to be preprocessed for these models to work correctly. In this paper, a comparative study of different preprocessing methods and deep learning models for a set of breast cancer images is presented. For this purpose, the statistical test ANOVA with data obtained from the performance of five different deep learning models is analyzed. An important conclusion from this test can be obtained; from the point of view of the accuracy of the system, the main repercussion is the deep learning models used, however, the filter used for the preprocessing of the image, has no statistical significance for the behavior of the system (Murcia-Gomez et al., 2022)
Rafika Harrabi and Ezzedine Ben Braiek	This article presents a new color image segmentation method that combines multilevel thresholding and data fusion techniques. The approach involves merging multiple realizations of the same image to increase information quality and achieve more accurate segmentation. The method consists of two steps: First, Optimal multilevel thresholding using the two-stage Otsu optimization approach to identify significant histogram peaks. Second, merging images represented in different color spaces using Dempster-Shafer evidence theory to achieve a final reliable and accurate segmentation result. The method is demonstrated through medical color image segmentation, and its classification accuracy is evaluated through a comparative study with existing techniques. The results show satisfactory segmentation performance, highlighting the effectiveness and superiority of the proposed method (Harrabi & Ben Braiek, 2012)

- Lung Cancer Detection: Analysis of CT scans for tumor detection

Yongbai Zhang, Wenpeng Huang, Hao Jiao and Lei Kang	Radiomics is a growing field in medical imaging that analyzes high-dimensional image features to improve diagnosis, prognosis, and treatment planning. In nuclear medicine, radiomics uses imaging methods like PET and SPECT to evaluate biomarkers. This review focuses on radiomics in lung cancer, discussing its potential benefits and limitations. The radiomics workflow involves image acquisition, segmentation, feature extraction, and model building. The review covers radiomics-based prediction models for lung cancer diagnosis, differentiation, prognosis, and efficacy evaluation, and addresses current challenges and future directions for integrating radiomics into clinical practice (Zhang et al., 2024)
Nidhi S. Nadkarni, Sangam Borkar	Lung cancer is a prevalent and deadly disease, and early detection is crucial. Computed tomography (CT) scans are used for identification, but visual interpretation can be error-prone. This paper proposes an automated approach for lung cancer detection in CT scan images using image processing techniques, including median filtering, segmentation, and feature extraction, followed by classification using Support Vector Machine (SVM) (Nadkarni & Borkar, 2019)
Pablo Borrelli, John Ly, Reza Kaboteh, Johannes Ulén, Olof Enqvist, Elin Trägårdh, and Lars Edenbrandt	Researchers developed an AI-based method to automatically detect lung lesions and calculate total lesion glycolysis (TLG) on FDG PET-CT scans. The study, which included 112 patients with suspected or confirmed lung cancer, showed promising results, with the AI-tool demonstrating 90% sensitivity, 88% positive predictive value, and 100% negative predictive value. The AI-based method also strongly correlated with manual TLG measurements, particularly in small- to medium-sized tumors. This breakthrough has the potential to prioritize patient care and focus on those with potentially malignant lesions, revolutionizing the detection and treatment of lung cancer (Borrelli et al., 2021)

- Skin Cancer Classification: AI-powered classification of skin lesions

S. P. Angelin Claret, Jose Prakash Dharmian and A. Muthu Manokar	Researchers explored the use of artificial intelligence (AI) to improve skin cancer diagnosis by combining convolutional neural networks (CNNs) and discrete wavelet transformation (DWT). The proposed approach uses DWT to extract features from skin lesion images, which are then used to train a classification model. The results showed a sensitivity of 94% and specificity of 91%, outperforming existing machine learning algorithms. This study demonstrates the effectiveness of the DWT-based approach in accurately classifying skin cancer lesions, aiding in early detection and diagnosis (Claret et al., 2024)
Apeksha Pandey, Manepalli Sai Teja, Parul Sahare, Vipin Kamble, Mayur Parate, and Mohammad Farukh Hashmi	A skin disease detection methodology is presented, combining image processing techniques, non-local means denoising, and convolutional neural networks (CNNs) with sparse dictionary learning. This multi-stage approach enhances input data quality, extracts meaningful features, and improves classification performance. The method addresses noise robustness, feature extraction, and classification accuracy challenges. Results show an average Peak Signal to Noise ratio (PSNR) of 33.59 dB for the HAM-10000 dataset and 34.37 dB for the ISIC-2019 dataset. The CNN model achieves an accuracy of 85.61% for the HAM-10000 dataset and 81.23% for the ISIC-2019 dataset, comparable to existing approaches (Pandey et al., 2024)

Ulzii-Orshikh Dorj, Keun-Kwang Lee, Jae-Young Choi & Malrey Lee	This paper presents an intelligent skin cancer classification system using deep convolutional neural networks (CNNs) and Error-Correcting Output Codes Support Vector Machines (ECOC SVM). The system utilizes a pre-trained AlexNet CNN model to extract features from RGB images of skin cancers, which are then classified into four types: basal cell carcinoma, squamous cell carcinoma, actinic keratosis, and melanoma. The results show high accuracy, sensitivity, and specificity rates, with maximum average accuracy of 95.1%, sensitivity of 98.9%, and specificity of 94.17%. The system demonstrates high performance in classifying skin cancer, making it a valuable tool for dermatologists and healthcare professionals (Dorj et al., 2018)

6.3 Role of AI in Cancer Detection

Artificial intelligence (AI) plays a significant role in cancer detection, improving accuracy, speed, and patient outcomes. Here are some ways AI contributes to cancer detection (Zhang et al., 2023):

- Image Analysis
  - Computer vision: AI algorithms analyze medical images (e.g., X-rays, CT scans, MRI) to detect abnormalities, such as tumors.
  - Image segmentation: AI helps separate tumors from surrounding tissue, making it easier to analyze.
- Data Analysis
  - Pattern recognition: AI identifies patterns in large datasets, including medical histories, genetic profiles, and imaging data.
  - Predictive modeling: AI builds models to predict cancer risk, diagnosis, and treatment outcomes.
- Diagnostic Assistance
  - Decision support systems: AI-powered systems provide doctors with diagnostic suggestions and recommendations.
  - Second opinion: AI offers a virtual second opinion, reducing errors and improving diagnosis accuracy.
- Liquid Biopsy Analysis
  - Circulating tumor DNA analysis: AI helps analyze circulating tumor DNA in blood samples to detect cancer.
- Patient Stratification
  - Risk assessment: AI identifies high-risk patients and predicts treatment outcomes.
  - Personalized medicine: AI helps tailor treatment plans to individual patients based on their genetic profiles and medical histories.

Benefits:

1. Improved accuracy: AI reduces false positives and negatives.
2. Increased efficiency: AI automates repetitive tasks, freeing up doctors' time.
3. Enhanced patient outcomes: AI facilitates early detection, leading to better treatment outcomes.
4. Reduced costs: AI minimizes unnecessary procedures and tests.

Challenges and Limitations:

1. Data quality and availability: AI requires high-quality, diverse data to learn and improve.
2. Regulatory frameworks: Clear guidelines are needed to ensure AI safety and efficacy.
3. Clinical validation: AI-driven diagnostic tools must be clinically validated before widespread adoption.
4. Explainability and transparency: AI decisions must be interpretable and transparent to build trust.

7 Conclusion

This research has explored the advancements in image processing and pattern recognition, with a focus on the role of artificial intelligence (AI) and deep learning in these fields. The study has demonstrated the significant impact of image processing on various applications, including cancer detection, and highlighted the need for continued advancements in these areas. The results of this research have shown that AI-powered image processing techniques, such as deep learning-based architectures, can significantly improve the accuracy and efficiency of image analysis tasks. Furthermore, the study has identified key challenges and limitations in current image processing techniques and highlighted areas for future research.

In the context of cancer detection, this research has demonstrated the potential of image processing and AI to improve diagnosis accuracy and reduce false positives. The study's findings suggest that deep learning-based image processing techniques can be effective in detecting cancerous patterns in medical images. Overall, this research contributes to the growing body of knowledge on image processing and AI, and highlights the importance of continued innovation in these fields. The study's findings have significant implications for various applications, including healthcare, and demonstrate the potential of image processing and AI to improve outcomes and save lives.

Recommendations for Future Research

1. Further exploration of deep learning-based architectures for image processing and pattern recognition.

2. Investigation of the application of image processing and AI in other medical imaging modalities, such as MRI and CT scans.
3. Development of more robust and efficient image processing techniques for real-time applications.
4. Examination of the potential of image processing and AI in other fields, such as robotics and autonomous vehicles.

References

Bengio, Y. (2009). Deep learning for pattern recognition. In *Proceedings of the international conference on Neural Information Processing Systems* (pp. 1–9).

Borrelli, P., Ly, J., Kaboteh, R., et al. (2021). AI-based detection of lung lesions in [18F]FDG PET-CT from lung cancer patients. *EJNMMI Physics, 8*, 32.

Chen, L., Zhang, J., & Dai, J. (2020). Image processing and pattern recognition for autonomous vehicles. In *Proceedings of the IEEE international conference on Intelligent Transportation Systems* (pp. 1–8).

Claret, S. P. A., Dharmian, J. P., & Manokar, A. M. (2024). Artificial intelligence-driven enhanced skin cancer diagnosis: Leveraging convolutional neural networks with discrete wavelet transformation. *Egyptian Journal of Medical Human Genetics, 25*, 50.

Dorj, U. O., Lee, K. K., Choi, J. Y., et al. (2018). The skin cancer classification using deep convolutional neural network. *Multimedia Tools and Applications, 77*, 9909–9924.

Gonzalez, R. C., & Woods, R. E. (2018). *Digital image processing* (4th ed.). Pearson.

Harrabi, R., & Ben Braiek, E. (2012). Color image segmentation using multi-level thresholding approach and data fusion techniques: Application in the breast cancer cells images. *Journal on Image and Video Processing, 2012*, 11.

He, K., Zhang, X., Ren, S., & Sun, J. (2016). Deep residual learning for image recognition. In *Proceedings of the IEEE conference on computer vision and pattern Recognition* (pp. 770–778).

Jain, A. K. (1989). *Fundamentals of digital image processing*. Prentice Hall.

Kaur, C., & Garg, U. (2023). Artificial intelligence techniques for cancer detection in medical image processing: A review. In *Proceedings of Elsevier on material today* (Vol. 81, Part 2, pp. 806–809).

Kumar, P., & Kumar, R. (2018). Image processing and pattern recognition for surveillance systems. In *Proceedings of the IEEE international conference on Advanced Video and Signal-Based Surveillance* (pp. 1–6).

Litjens, G., Sanchez, C. I., Timofeeva, N., Hermsen, M., Nagtegaal, I., & van Ginneken, B. (2017). Image processing and pattern recognition for medical Image analysis. *Medical Image Analysis, 35*, 485–495.

Murcia-Gomez, D., Rojas-Valenzuela, I., & Valenzuela, O. (2022). Impact of image preprocessing methods and deep learning models for classifying histopathological breast cancer images. *Applied Sciences, 12*, 11375.

Nadkarni, N. S., & Borkar, S. (2019). Detection of lung cancer in CT images using Image processing. In *2019 3rd international conference on trends in electronics and informatics (ICOEI), Tirunelveli, India* (pp. 863–866).

Pandey, A., Teja, M. S., Sahare, P., et al. (2024). Skin cancer classification using non-local means denoising and sparse dictionary learning based CNN. *Journal of Electrical Systems and Inf Technol, 11*, 36.

Pratt, W. K. (2014). *Digital image processing* (4th ed.). Wiley.

Zerouaoui, H., & Idri, A. (2021). Reviewing machine learning and image processing based decision-making systems for breast cancer imaging. *Journal of Medical Systems, 45*, 8.

Zhang, Y. N., Xia, K. R., Li, C. Y., Wei, B. L., & Zhang, B. (2021). Review of breast cancer pathologigcal image processing. *BioMed Research International, 2021*, 1994764., 7 pages.

Zhang, B., Shi, H., & Wang, H. (2023). Machine learning and AI in cancer prognosis, prediction, and treatment selection: A critical approach. *Journal of Multidisciplinary Healthcare, 26*(16), 1779–1791. https://doi.org/10.2147/JMDH.S410301. PMID: 37398894; PMCID: PMC10312208.

Zhang, Y., Huang, W., Jiao, H., & Kang, L. (2024). PET radiomics in lung cancer: Advances and translational challenges. *EJNMMI Physics, 11*(1), 81. https://doi.org/10.1186/s40658-024-00685-5. PMID: 39361110; PMCID: PMC11450131.

Genomics and Bioinformatics in the Discovery and Validation of Diagnostic Biomarkers

Saumya Dubey, Jai Singh, Grijesh Jaiswal, Abhishek Kumar Verma, Sandesh P. Telang, Ananya Srivastava, Jaikee Kumar Singh, and Sandeep Kumar Srivastava

Abstract

Genomics and bioinformatics are transforming diagnostic biomarker discovery and validation, offering unprecedented insights into disease molecular mechanisms. The rapid development of high-throughput technologies, specifically next-generation sequencing (NGS), has transformed cancer research by facilitating the exploration of genetic variations such as single nucleotide polymorphisms (SNPs), copy number variations (CNVs), gene expression, and protein levels. When paired with advanced computational tools, these technologies allow for the comprehensive analysis of genomic, transcriptomic, and epigenomic data at a large scale, ultimately enabling the identification of distinctive molecular signatures linked to disease conditions. Bioinformatics plays a crucial role in the management, analysis, and interpretation of this data, employing methods such as machine learning, network analysis, and integrative genomics to identify potential biomarkers with marked sensitivity and specificity. These biomarkers have the capacity to aid in prompt diagnosis, oversee the advancement of a disease, and forecast responses to treatment, ultimately enhancing patient outcomes. Moreover, bioinformatics tools assist in validating these biomarkers in various cohorts, ensuring reproducibility and reliability. These technologies have heightened the importance of precision oncology, resulting in an increased need for the identification and validation of biomarkers. This review delves into the integration of genomics and bioinformatics in biomarker discovery, emphasizing essential methodologies, challenges, and prospective paths for personalized

S. Dubey · J. Singh · G. Jaiswal · A. K. Verma · S. P. Telang · A. Srivastava · J. K. Singh
S. K. Srivastava (✉)
Structural Biology & Bioinformatics Laboratory, Department of Biosciences, Manipal University Jaipur, Jaipur, Rajasthan, India
e-mail: sandeepkumar.srivastava@jaipur.manipal.edu

R. Nema et al. (eds.), *Advances in Cancer Detection, Prediction, and Prognosis Using Artificial Intelligence and Machine Learning*,
https://doi.org/10.1007/978-981-96-9346-7_6

diagnostics. Bioinformatics had a major impact on cancer research by revealing new biomarkers, driver mutations, carcinogenic pathway, and therapeutic targets. Incorporating multi-omics data, network analysis, and advanced machine learning (ML) techniques will be crucial in future biomarker discovery and prediction of patient prognosis.

Keywords

Oncology · Bioinformatics · Biomarker discovery · Machine learning (ML) · Personalized diagnostics

1 Introduction

Biomarkers are evidentiary indicators of biological or pathogenic processes, or physiological responses to an exposure or intervention, including therapeutic interventions. They can be in the form of molecular, histologic, radiographic, or physiologic (Cagney et al., 2018). Genomics and bioinformatics both are revolutionized the research in the field of cancer by the identification and validation of prognostic markers with remarkable precision. High-throughput technologies such as next-generation sequencing (NGS) have been the backbone of the field of precision oncology by enabling the comprehensive assessment of genetic, transcriptional, and protein variation (Clark & Lillard Jr, 2024). This chapter investigates the methodologies and tools which are redefining the research in cancer biomarkers, highlighting the vital role of bioinformatics in organizing and interpreting the large sets of data. By utilizing sophisticated multi-omics approaches and machine learning algorithms, researchers can identify the complicated biomarker signatures that promote the early detection, prognosis, and customized treatment therapies (Perera-Bel et al., 2019). This chapter underlines the role of interdisciplinary approaches in improving cancer diagnostics *via* discussing the difficulties related to the collection of data, maintaining the quality and the implementation of biomarkers into clinical procedures (Dhillon et al., 2023).

1.1 Overview of Bioinformatics and Biostatistical Methods and Tools Used in Biomarker Research and Discovery

Bioinformatics with biostatistics will play an important role in managing and analyzing copious amount of datasets generated by high-throughput technologies, such as genomics, proteomics, and metabolomics. This section focuses on the techniques and resources of biostatistics and bioinformatics which were employed in the research of biomarkers, emphasizing their relevance and application in the cancer studies. Bioinformatics tools are playing an important role in the processing of large datasets, in the integration of diverse data types and conducting comprehensive analysis of them. Resources along with sequence alignment software, gene expression analysis platforms, and comprehensive databases for storing and retrieving biological information (Perera-Bel et al., 2019). Bioinformatics is complemented by biostatistical methods, including hypothesis testing, regression analysis, and survival analysis, which provide a robust technique for data analysis. Advanced statistical approaches, such as machine learning and Bayesian inference, are increasingly employed to enhance the predictive power and accuracy of cancer biomarker models (Clark & Lillard Jr, 2024).

1.2 Public Resources and Open-Source Tools

In the field of cancer research, public resources and open-source methods are indispensable and they offer accessible platforms for analysis, sharing, and collaboration with data. They democratize access to cutting-edge technologies for enabling researchers worldwide to contribute and benefit from collective scientific advancements. Through these tools, it is possible to expedite the discovery, validation, and clinical application of biomarkers, thereby directly translating into better patient outcomes and understanding of cancer. Public resources with prime examples like The Cancer Genome Atlas (TCGA) and the Gene Expression Omnibus (GEO) are playing a vital role in cancer research. TCGA is useful in the findings of biomarkers for cancer since it offers the comprehensive genomic information across various cancer types (Informatics Tools for Cancer Research, n.d.). Similarly, GEO facilitates several studies involving biomarkers using their enormous collection of gene expression information (Klemm et al., 2024). Open-source tools are also playing a very crucial role in providing powerful analytical capabilities and fostering innovation for cancer research. Digital Biomarker Discovery Pipeline (DBDP) is a platform that offers end-to-end solutions for developing digital biomarkers by using data collected from mobile health (mHealth) and wearable devices (Rius et al., 2024).

1.3 Genomics and Biomarker Discovery

Genomics offers a profoundly important understanding of the underlying genetics of cancer by elucidating the intricate details of genetic materials of an organism. These findings further drives discoveries in biomarkers, such as molecular markers that would be of use in diagnosis, prognosis, and monitoring the response of treatment in cancer patients. Genomics coupled with biomarker research drive a paradigm shift toward precision medicine which allows targeted therapies to be developed according to the specific genetic profile of each patient. In the field of cancer research, NGS enables the rapid and complete analysis of genetic variations which is ranging from SNPs to copy number variations and gene-expression profiles. These advancements in technologies have significantly accelerated the pace of genomic research and biomarker discovery, that enabling a deeper understanding of the molecular basis of cancer (Dienstmann et al., 2015). The Cancer Genome Atlas (TCGA) is one of large-scale genomic initiatives, which have generated a large amount of data on the genetic alterations underlying various cancer types. Such invaluable resources have become helpful in identifying sudden changes (key mutations) and genomic aberrations, that drive tumorigenesis and influence therapeutic responses. As an example, comprehensive genomic profiling of TCGA samples has indicated that most samples contain alterations that are potentially targetable, which indicates the complexity and therapeutic potential of the cancer genome (Dienstmann et al., 2015). Discovery of biomarkers has thus emerged as the critical objective within cancer research, identifying at the molecular level detecting, diagnosing and following both progressiveness and treatment of a certain type of cancer. Molecular markers or more commonly referred to as biomarkers are vital components in studies of cancer. Biomarkers may be made of DNA, RNA, proteins, or metabolites. Biomarkers serve as useful markers for the measurement of the disease state, prognosis, and response to therapy. Most importantly, the recent technological boom has experienced a furious acceleration of genetic biomarkers through the use of next-generation sequencing (Clark & Lillard Jr, 2024). In addition, the holistic view of multi-omics approaches, including proteomics, lipidomics, and metabolomics, now allows determination of a greater number of biomarkers. Among these biomarkers, significant protein biomarkers with good diagnostic and prognostic potential have been identified through that powerful tool which has appeared within proteomics for high throughput analyses by mass spectrometry (Iftikhar et al., 2022).

1.4 Genomic Technologies

This marked a new era for biomedical research ushered by the genomic technologies, in turn transforming the scope of knowledge related to the disease and therapy. Thus, through the introduction of next-generation sequencing, unprecedented precision is bestowed on genomic information analyses and manipulations. Since this next-generation sequencing makes possible rapid genome-wide sequencing, this technique is instrumental in finding out several genetic markers responsible for cancer along with several other diseases. This technological breakthrough has made possible the development of targeted therapies tailored to individual patient needs, a major step toward personalized medicine (Lau et al., 2024). One of the landmark developments in genetic engineering-CRISPR-Cas9 gene-editing-empowers scientists to define and modify the DNA sequence directly. This tool has phenomenal potential for correcting genetic disease, defining gene function for man and beast, or developing innovative therapies for genetic-based diseases. Recent advances add to the list the powers of CRISPR tools in editing the epigenome and the mitochondrial compartment, which will change medical science forever (Smith, 2021). By using, complicated computational tools and machine learning algorithms, scientists can now effectively analyze and interpret the large-scale datasets in genomic data.

This is a real revolution and transformation in the field of life sciences with the advent and application of genomic technologies. With clear meanings in the field of genetics, genomics offers insights into molecular variation in health and disease. This enables researchers to build better instruments for diagnostics and to perform more targeted therapy, and new single cell sequencing technologies unleashed revolution in understanding heterogeneity, at the level of one cell and hence clinical outcomes based on those phenomena. These technologies allow analysis at the single-cell level, providing unprecedented insight into the complex genetic landscape of tumors. Such a granular level of analysis can identify rare cell populations, characterize clonal evolution, and even discover novel therapeutic targets-all of which will further the progress of precision oncology (Cunha, 2017).

1.5 Different Types of Genomic Biomarkers

Genomic biomarkers represent the frontier in cancer research, transforming that particular field into an area of study of the mechanisms of disease and opening avenues for the development of more targeted therapies. This chapter considers the multiple categories of genomic biomarkers, as well as their importance and use in cancer.

1.5.1 Single Nucleotide Polymorphisms

SNPs are variations in a single DNA nucleotide at specified positions in the genome and associated with specific diseases, including cancer. Analysis of certain SNPs can help to identify those at increased risk for certain types of cancer (Hapmap, 2003).

1.5.2 Copy Number Variations

CNVs are the number of copies of DNA segments, and associated with changes in gene dosage, and several diseases, including cancer. Analysis of CNVs will allow determining those at higher risk for cancer and assist in making decisions regarding treatment (Redon et al., 2006).

1.5.3 Gene Expression Biomarkers

The differentiation of mRNA transcripts that occur in cells or tissues marks the gene expression biomarkers. They help in identification of genes that are overexpressed or under expressed in the cancer cells and probably give leads into disease mechanism or possible therapeutic strategies (Hanahan & Weinberg, 2011).

1.5.4 MicroRNA Biomarkers

Small and non-coding RNAs that are also known as miRNAs are the regulator of the gene expression. Inhibition or aberrant expression of miRNAs has been tied to cancer pathogenesis as well as its progression (Calin & Croce, 2006).

1.5.5 Epigenetic Biomarkers

Gene expression changes can occur through epigenetic events that do not involve change in the DNA sequence of the gene itself. An application of epigenetic markers provides crucial insights about how cancers develop, and progress, as well as how they respond to a given therapy (Baylin & Jones, 2011).

2 Role of Bioinformatics in Biomarker Discovery

New breakthroughs in next-generation sequencing (NGS) have made it easier to use computer tools to study gene expression, microRNA expression, protein expression, single nucleotide polymorphisms (SNPs), and other changes in genes. Cancer treatment based on genomics uses screening technology and bioinformatics tools to customize cancer drugs for each patient based on their unique genetic makeup. This method boosts the ability to uncover and validate biomarkers that play a key role in diagnosing cancer, figuring out how it might progress, and coming up with personalized treatment plans (Clark & Lillard Jr, 2024).

2.1 Tools for Data Analysis and Visualization

Effective data visualization is crucial in bioinformatics, especially in biomarker discovery, where large scale datasets from genomic and proteomic studies need to be presented in interpretable formats. Many computational tools have been implemented for this, which support its use in preprocessing as well as statistical analysis and visualization of results (Perera-Bel et al., 2019). Tools such as R, combined with Bioconductor, provide an integrated toolbox for statistical analysis and visualization in bioinformatics (Huber et al., 2015). Packages like the DESeq2 and limma are dedicated to differential gene expression analysis, which is very important for the detection of potential biomarkers. The flexibility of Python is largely due to its libraries, which include Pandas for comprehensive data processing and Matplotlib/Seaborn for generating diverse charts. These are essential in the preparation of large datasets and the production of comprehensive biomarker visualizations. It also enhances predictive biomarker analysis with interoperability with machine learning frameworks. Clustergrammer is an online application designed for interactive heatmaps generation, enabling academics to inspect hierarchical clustering that occurs on high-dimensional datasets (Fernandez et al., 2017). Cytoscape allows the visualization and exploration of molecular interaction networks. Protein-protein and gene interaction mapping helps in the identification of biomarkers that describe biological processes. Its extensibility through the use of plugins enables functional enrichment analysis, making it indispensable in systems biology research (Shannon et al., 2003). Galaxy streamlines bioinformatics operations by consolidating data processing and visualization under an accessible, web-based interface. This way, the researchers carry out omics data analysis without needing to have programming expertise while also sharing workflow collaborations, making it an indispensable instrument for biomarker discovery within a team environment. Shiny enables researchers to create online apps that are interactive for visualizing biomarker data. Tableau would convert raw data into interactive visual dashboards, enabling researchers to intuitively analyze complex biomarker datasets. Not specifically designed for bioinformatics, it is flexible enough to enable the integration of omics data that helps identify patterns and sets up the visualization of results. Spotfire offers sophisticated visual analytics for multi-omics datasets. Its capacity to perform comprehensive analysis over extensive datasets and provide interaction through visualizations facilitates the identification of patterns, outliers, and possible biomarkers, thus aiding both exploratory and confirmatory research. Qlucore specializes in all visualization of multi-omics data. They make it easy by giving real-time feedback to help across the whole research process. Techniques such as PCA, volcano plots, and hierarchical clustering enable fast identification of relevant biomarkers for genomic and proteomic research.

IGV stands for Integrative Genomics Viewer: This is a tool specially designed to analyze high throughput sequencing data, including the alignments and variations. The interactive interface allows scientists to quickly scan genetic areas, hence obtaining a candidate biomarker. DeepTools is a Python-based toolbox that has been developed to analyze high-throughput sequencing data. Its heatmap and profile-plotting capabilities are particularly useful for analyzing epigenomic data that will aid in the discovery of biomarkers. KEGG Mapper allows pathway-oriented visualization of genes or proteins in biological pathways. This tool enables researchers to elucidate the functional functions of biomarkers in pathways of disease mechanisms and signal transduction. Such platforms as MetaboAnalyst are based on machine learning, statistical analysis, and visualization capabilities for identifying biomarkers. They provide pathway enrichment and metabolite set analysis, letting researchers find crucial biomarkers in omics datasets.

2.2 Data Management and Integration

2.2.1 Integration of Prior Biological Knowledge into Multi-Omics Analysis

Biomarker discovery relies on effective data management and integration, integrating bioinformatical data with clinical and prior biological knowledge. Screening studies focus on integrating data from different molecular levels, such as DNA, gene expression, and protein expression. Horizontal data integration involves bioinformatical techniques and screening studies, while vertical data integration collects data from different molecular levels. Vertical data integration aids in identifying more intricate targets, as most genetic variations in genome-wide association studies reside in noncoding areas. The eQTL studies are the example of data integration, and these studies integrated with gene expression and sequence data, along with miRNA-mRNA pairs based on miRNA target predictions. The incorporation of external information is beneficial in several cases.

1. The signal of individual biomolecules can be weak, making it difficult to predict their expression over noise levels and corrections for multiple tests. In such instances, the combined signal from multiple molecules which are grouped into modules of interconnected molecules like pathways, might be sufficiently strong. For example, the "Kyoto Encyclopaedia of Genes and Genomes" (KEGG) pathway database contains approximately five hundred pathways compared to thousands of annotated genes (Kanehisa & Goto, 2000).

2. In an alternative scenario, there may be too many molecules and the molecules in question are differentially abundant between conditions. Prior knowledge can help in producing large number of molecules, which interpreted as single pathway, might be represented disproportionately. The discovery may be stabilized by directing the process toward molecules of known importance or by targeting the most interesting targets, such as actionable molecules with known agents.
3. External knowledge is often used to highlight biologically meaningful combinations, such as protein and protein interactions (PPI), miRNA with mRNA pairs, gene networks of co-expression, or regulation networks. In recent years, methods have been proposed using prior knowledge during classification, shifting the focus from gene level to pathway level (Szklarczyk et al., 2019).
4. The question shifts towards the predictability of the gene expression in a pathway when prior knowledge of the pathway is incorporated. More sophisticated models can attempt to represent the topology of the network and consider the connections between the genes along with the mere membership information.

   Internal validation methods are used to reduce false-positive findings in biomarker screening studies. These methods split data into training and test sets, allowing predictive models to be built and evaluated on different data sets. However, it is crucial not to use validation data before performance estimation. A common mistake is featuring selection on the entire dataset, building a predictive model using selected features, and evaluating the prediction performance in a cross-validation scheme. All stages in model building must be validated, and feature selection must be performed in each fold for realistic performance estimates. Clinical evaluation is crucial after analytical validation for successful clinical application. Clinical trials are essential for testing the clinical validity and utility of biomarkers, as they are conducted on low-dimensional data with many samples for a few variables. These trials are costly and require strong hypotheses from earlier stages in the biomarker discovery process. The clinical utility of a biomarker determines whether testing leads to better outcomes than standard care.

2.3 Data Analysis and Interpretation

Bioinformatics tools play a critical role in managing and interpreting the vast datasets derived from genomic data repositories. One of the most used platforms is cBioPortal that provides access to pre-analyzed data, enabling researchers to analyze and interpret somatic mutations, race/ethnicity, gene expression patterns, copy number variations, methylation, and phosphorylation (Gao et al., 2013). This tool also provides features like heatmaps and gene network visualizations. Example of interpretation and reporting of data: When a tumor is sequenced by using NGS technique, the bioinformatic pipelines result standard variant in the form of files (vcf, maf, or bed formats10), which contain a list of tens to hundreds of somatically altered genes that have point mutations, amplifications, deletions, insertions, and fusions. The clinicians in boards of molecular tumor those are responsible for analysis of NGS results, the main challenge lies in determining the clinical meaning of genomic alterations, which is referred as variant interpretation. Clinicians will base their therapeutic decisions solely on the evidence that demonstrates the variant's ability to predict the effectiveness of a drug. Numerous databases are available to aid this process, each varying in detail, curation and comprehensiveness. These include clinical trial databases (like "ClinicalTrials.gov," EU-CTR), predictive biomarker sites ("mycancergenome.org," GKDB, TARGET, CIViC, and CGI), and Guidelines for treatment (NCCN, FDA, and ACMG). For interpretation of somatic variant, clinicians often use academic resources like PubMed or Google Scholar to find case report and preclinical studies. Another approach for somatic alterations identification, especially for new variants, is by using public genomic resources or computational prediction tools. It plays key role to identify key mutations, predict their functional impacts, understand their effects on signalling pathways, and explore their potential to activate immune responses, such as through neoantigen production by mutated fragments of DNA.

However, several challenges remain. Automation is critical for streamlining the reporting of genomic findings, as clinicians depend on these reports to make informed treatment decisions, yet consolidating biomarker knowledge into a unified system has proven difficult, particularly since many biomarkers are rare and their responses to treatment can vary widely. For instance, mutations in EGFR exons 19 and 21 may predict a favorable response to certain drugs, while exon 20 mutations in the same gene might indicate the opposite. Similarly, identical mutations can have contrasting effects depending on the type of cancer. Drug efficacy is another area of concern.

3 Validation of Diagnostic Biomarkers

Biomarker validation is an important stage in determining the clinical effectiveness of proposed biomarker for diagnosis (Davis et al., 2020; Rifai et al., 2006). It includes careful evaluation to verify that biomarkers are precisely and consistently produce signal for certain biological states or diseases (Borrebaeck, 2017). This step fills the gap between the identification of bio markers and clinical implementation, ensuring that the findings are not only of statistical importance but also physiologically and therapeutically relevant (Hayes, 2015). The validation step of diagnostic biomarkers can be divided into two category experimental validation and statistical and computational validation. Each step has its specific role in increasing confidence in the biomarker's potential to provide accurate and useful diagnostic insights (Fig. 1).

3.1 Experimental Validation of Diagnostic Biomarkers

Experimental validation of diagnostic biomarkers includes laboratory-based procedures for confirming the biological and diagnostic efficacy of a biomarker (Davis

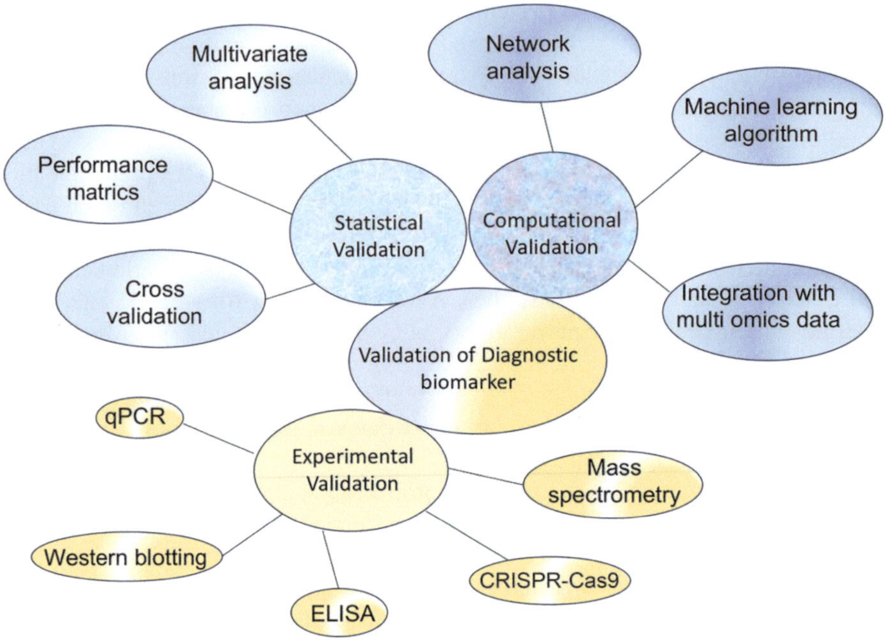

Fig. 1 Different techniques involve in the validation of diagnostic biomarkers

et al., 2020; Ahmad et al., 2023). The fundamental purpose of experimental validation of diagnostic biomarkers is to demonstrate that the biomarker can accurately detect or monitor linked diseases; therefore this step involves testing biomarkers in cell lines, animal models, and clinical samples (Rifai et al., 2006; Chau et al., 2008). To evaluate the functional sensitivity and specificity of biomarkers by experimental validation, different techniques such as qPCR (Quantitative Polymerase Chain Reaction), Western blotting, ELISA (Enzyme-Linked Immunosorbent Assay), Mass spectrometry, and CRISPR-Cas9 genome editing are used as key techniques.

3.1.1 qPCR

Quantitative PCR (qPCR) is a highly accurate and precise approach for validating diagnostic biomarkers by quantification of nucleic acids like DNA, RNA, and miRNAs. It is especially useful in identification of gene expression changes, detecting copy number variations, and verifying pathogen-specific biomarkers (Piccoli et al., 2019; Ruijter et al., 2013). High sensitivity and repeatability natures of qPCR make it ideal for clinical applications for different samples such as tissue, blood, and bodily fluids (Engstrom-Melnyk et al., 2015). qPCR is widely used for validation of diagnostic biomarkers related to diseases such as cancer, infectious diseases, and genetic disorders. For example, it can easily assess gene expression alterations in tumor suppressors or oncogenes, detect CNVs or fusion genes associated with genomic instability, or can find low-abundance miRNAs in plasma (Gossing et al., 2020). Critical factors such as primer design, sample quality, and the incorporation of normalization controls are essential to ensure the precision of outcomes (de Ronde et al., 2018).

3.1.2 Western Blotting

Western blotting has progressed from a subjective and manual method to a highly accurate and quantitative assay suitable for biomarker identification, because of recent advances in antibody production, automation, gel and membrane manufacture, and very sensitive detective machines (Owen et al., 2024). Western blotting is used to separate proteins based on their molecular weights through SDS-PAGE, subsequently transferred onto an absorbent membrane and then specific proteins present on the membrane can be detected through the use of antibody probes (Gershoni & Palade, 1983). Gelsolin is one such biomarker candidate in HIV-associated neurocognitive disorders and can be detected using Western blot assay and a variety of anti-gelsolin antibodies (Haverland et al., 2010). Different types of cancer diagnostic biomarkers can be validated through western blot for example gastric cancer (Humphries et al., 2014), lung cancer (Shen et al., 2006), and colorectal cancer (Hamelin et al., 2011).

3.1.3 ELISA

The Enzyme-Linked Immunosorbent Assay (ELISA) is commonly employed as a highly sensitive and efficient technology for confirming soluble biomarkers in biological fluids like serum, plasma, and urine. It utilizes the interaction between antigens and antibodies to accurately measure proteins, cytokines, and antigens with great specificity (Porstmann & Kiessig, 1992). This method holds promise for conducting high-throughput analysis on numerous samples, enabling the evaluation of multiple proteins in less sample. Moreover, the assays utilized for validation tests can seamlessly transition to clinical applications, as ELISA analysis is commonly employed in clinical settings for the evaluation of specific proteins. ELISA can be used for validation of different diagnostic biomarkers for infectious diseases, inflammation, and cancers (Zangar et al., 2006).

3.1.4 Mass spectrometry

A very effective analytical method for confirming diagnostic biomarkers at the proteome and metabolomic levels is mass spectrometry (MS). It offers unmatched sensitivity, specificity, and precision for identifying and measuring proteins, metabolites, and peptides in intricate biological materials (Crutchfield et al., 2016). In order to determine molecular weight, structure, and composition, mass spectrometry works by ionizing molecules to produce charged species (ions), accelerating these ions through an electric field, separating them based on their mass-to-charge ratio (m/z) in a mass analyzer (such as a time-of-flight, quadrupole, or ion trap), and precisely detecting them to measure their m/z. For cancer, heart disease, and metabolic disorders, mass spectrometry is very helpful in verifying biomarkers when a comprehensive molecular investigation is required (Vasan, 2006).

3.1.5 CRISPR-Cas9

A cutting-edge genome-editing technique called CRISPR-Cas9 is transforming the validation of genetic biomarkers. It allows for the accurate modification of DNA sequences to investigate the potential biomarkers' functional significance (Dixit et al., 2024). CRISPR-Cas9 helps test biomarkers by guiding genetic or regulating accurate genome changes in the target genome. This method allows researchers to knock, activate or change individual genes to test their effects on cellular schemes, phenotypes of the disease, or treatment reactions. CRISPR-Cas9 provides direct functional evidence of biomarker involvement by analyzing changes in expression, cell behaviour, or disease progression caused by gene editing, ensuring the reliability of its diagnosis, prognosis, and treatment monitoring (Bhattacharjee et al., 2022).

In order to rigorously validate diagnostic biomarkers and make sure they are both dependable and clinically relevant, statistical, and computational techniques are essential. By establishing a framework for evaluating a biomarker's performance, consistency, and applicability across various patient populations, these techniques ultimately validate its appropriateness for therapeutic application (Zhang & Drabier, 2013).

3.1.6 Statistical Validation of Diagnostic Biomarkers

Statistical validation involves applying quantitative analysis to assess a biomarker's ability to differentiate between disease and non-disease states, as well as its diagnostic accuracy and predictability. Key statistical techniques include:

(i) **Cross-Validation and Resampling**: Cross-validation and resampling are crucial for assessing diagnostic biomarkers. Cross-validation divides data into training and testing sets (e.g., K-Fold, LOOCV, stratification) in order to assess performance and guarantee unbiased estimation. Resampling is used to create a number of datasets (bootstrap, jackknife, etc.) in order to evaluate variability and resilience. These methods aid in avoiding overfitting, estimating crucial parameters (sensitivity, specificity, AUC, etc.) and optimizing biomarkers for clinical usage (Dakna et al., 2010).

(ii) **Performance metrics:** Performance metrics for statistical validation of diagnostic biomarkers are critical for assessing accuracy, reliability, and clinical values like sensitivity (true positive rate), specificity (true negative rate), positive predictive value (ppv), negative predictive value (npv), and accuracy (Levinson, 2009).

(iii) **Multivariate Statistical Analysis**: Techniques like Principal Component Analysis (PCA), correlation analysis, multiple linear and logistic regression, Multivariate Analysis of Variance (MANOVA), and Receiver Operating Characteristic (ROC) curve analysis are all part of multivariate statistical analysis for diagnostic biomarker validation. An indicator of the biomarker's capacity to differentiate between distinct diagnostic categories is the area under the curve (AUC), which is produced by graphing the Receiver Operating Characteristic (ROC) curve's sensitivity versus 1-specificity at different threshold levels (Søreide, 2009). These techniques help in assessing the relationships between biomarkers, predicting diagnostic outcomes, and evaluating the performance and reliability of biomarker panels.

3.1.7 Computational Validation of Diagnostic Biomarkers

Computational methods, including bioinformatics and machine learning, are increasingly used to validate and refine diagnostic biomarkers by analysing large-scale biological data. These techniques allow for the integration and interpretation of complex, high-dimensional data from genomics, proteomics, transcriptomics, and clinical studies.

(i) **Integration with multi-omics data**: Combining multi-omics data (including genomics, transcriptomics, proteomics, and metabolomics) offers a comprehensive perspective for validation of biomarkers associated with particular diseases. Methods such as principal component analysis (PCA) and canonical correlation analysis (CCA) facilitate the merging of various data types to validate diagnostic biomarkers. This integration emphasizes essential biological pathways, allowing for the identification of possible diagnostic markers by linking molecular profiles to clinical outcomes (Srivastava & Creek, 2019).

(ii) **Machine learning algorithm**: Machine learning (ML) models such as Random Forests, Support Vector Machines (SVM), and logistic regression are used to predict illness outcomes from large datasets to validate diagnostic biomarker. Unsupervised learning techniques, such as k-means clustering, can reveal new disease subgroups, whereas supervised learning is effective for predicting diagnostic outcomes. Feature selection strategies like LASSO increase model accuracy by focusing on the most significant biomarkers (Chang et al., 2021).

(iii) **Network analysis**: In order to find biomarkers that are essential to disease processes, network analysis looks at molecular relationships and pathways. Protein-protein interaction (PPI) networks are constructed using programs like Cytoscape and STRING, which show how possible biomarkers interact with biological systems. Critical biomarkers for disease detection are highlighted by systems biology models that integrate multi-omics data to simulate biological processes, while pathway enrichment analysis finds pathways that are over-represented with illness-related biomarkers (Stevens et al., 2014).

4 Drugs and Biomarkers Co-development

Co-development of medications and biomarkers is a term that describes the process of simultaneously discovering and developing therapeutic agents in addition to the diagnostic tests that correspond to those therapeutic agents. This strategic approach is gaining importance in the field of personalised medicine, which is a sector in which the incorporation of biomarker information into the process of drug development improves the effectiveness of therapy and maximises the outcomes for patients. Biomarkers, which are defined as quantitative indications of biological processes or responses to therapeutic treatments, have become extremely important in the process of enhancing patient stratification, monitoring therapeutic responses, and guiding clinical decision-making throughout the whole lifecycle of drug development (Hendrikse et al., 2022; Diaz-Uriarte et al., 2022).

This technique of co-development not only speeds up the regulatory approval procedure, but it also greatly cuts down on the costs that are connected with bringing novel medications to market. When compared to coordinated development efforts, which can boost success rates and reduce timeframes, traditional drug development can take more than a decade and entail costs that exceed one billion dollars. In a Phase III study of non-small cell lung cancer (NSCLC), for example, the success rate increased from 28% to 62% as a result of biomarker-guided development (Shimazawa & Ikeda, 2016; Bakhtiar, 2008).

A further benefit of this technique is that it enhances patient safety by ensuring that individuals receive medications that are tailored to their particular biomarker profiles. The possibility that patients would receive therapies that are either ineffective or dangerous is considerably reduced as a result of this. Although there are many advantages to be gained from the co-development of biomarkers and drugs, there are also a number of challenges that arise from this approach. Among these issues include the intricacies of test selection, the design of clinical trials, and the demand for successful collaboration between enterprises in the diagnostic and biopharmaceutical industries. Another aspect that contributes to the difficulty of navigating the process is the presence of regulatory impediments. In order to successfully navigate the various requirements for diagnostics and drugs, it is necessary to engage in careful preparation and to align one's approach (Day & Siu, 2016; Kuhlmann & Wensing, 2006).

An example that stands out in the field of medication and biomarker co-development is the application of pembrolizumab, which is an IgG4 monoclonal antibody that targets PD-1, for the treatment of non-small cell lung cancer (NSCLC). When the medication was initially studied across a number of trials that were collectively referred to as KEYNOTE, stratified analysis found that patients who had a tumor PD-L1 expression of at least 50% saw greatly increased benefits from the treatment. The ensuing trials were designed with this understanding in mind, which ultimately resulted in pembrolizumab being designated as a first-line therapy for patients with advanced non-small cell lung cancer and high PD-L1 expression in the year 2016. Additionally, it was the first anti-PD-1 drug that was licensed for use as a first-line treatment in this particular patient cohort at that time. On the other hand, nivolumab, which is another PD-1 inhibitor, was examined concurrently but did not go through the same stratification process. It was eventually approved as a first-line treatment in May 2020 for patients who had a tumor PD-L1 expression of at least 1% in conjunction with ipilimumab (Biomarkers at FDA | FDA – U.S. Food and Drug Administration, n.d.).

One of the earliest and most influential examples of effective drug and biomarker co-development is trastuzumab, which is produced and sold under the brand name Herceptin. The FDA granted approval for trastuzumab in 1998 for the treatment of HER2-positive metastatic breast cancer. This drug played a crucial role in the development of targeted cancer therapy. In addition to being used as a combination therapy, it was also used as a monotherapy in the second line of treatment. Following its debut, a number of tyrosine kinase inhibitors (TKIs) that target the HER2 protein have been developed and authorized. These TKIs include neratinib, lapatinib, and tucatinib, among others. The therapeutic applications of these agents have been further expanded as a result of the discovery of HER2 overexpression and gene amplification in a variety of malignancies (Biomarkers at FDA I FDA – U.S. Food and Drug Administration, n.d.).

5 Challenges in Biomarker Discovery and Validation

5.1 Data Complexity and Heterogeneity

Data Complexity It is a major challenge in the biomarker discovery, which involves the interaction between various factors such as environmental, lifestyle, and genetic. The advancements in the multi-omics and machine learning have helped in identification of the robust biomarkers. The utilization of the multi-omics data helps in the understanding of disease mechanism, as seen in projects like The Cancer Genome Atlas (TCGA) and Alzheimer's Disease Neuroimaging Initiative (ADNI) (Huang et al., 2023). Other techniques such as canonical correlation analysis (CCA) aid in identification of the variable across different levels, and improve biomarker discovery process in condition such as inflammatory bowel disease (Pusa & Rousu, 2024). The hybrid ensemble feature selection (HEFS) method enriches the biomarker identification by combining various strategies such as resampling and feature reduction (Claude et al., 2024). Despite advancements, it is still a challenge to translating metabolomics data into clinically useful biomarkers, often due to inadequate study designs and data analysis methods (Considine, 2019). Data complexity in biomarker discovery occurs by integration of the multi-omics data which requires advanced analytical tools, methods such as Stability CCA, a method which enhances variable selection stability, facilitates in biomarker development in complex conditions such as inflammatory bowel disease (Pusa & Rousu, 2024). Data complexity occurs in the biomarker discovery process from high-dimensional datasets, which have more features than the samples lead to complicated analysis. The conventional tool like differential expression fails to account for feature interactions shows necessity for more stable approach like molecular feature selection tool (MFeaST); it identifies immune-related biomarkers in prostate adenocarcinoma (Gerolami et al., 2022).

Data Heterogeneity Heterogeneity in biomarker discovery shows significant challenge in the understanding of the disease. Diseases have got multiple subtypes and for the biomarker identification tailored approach is required, as a single biomarker may not be universally acceptable. There is a significant impact of disease heterogeneity on biomarker discovery; each disease subtype has unique or distinct biomarker that requires specific discovery strategies. Heterogeneous diseases require more than twice the sample sizes compared to the homogeneous disease to achieve adequate statistical power (Wallstrom et al., 2013). Data heterogeneity in the discovery of the biomarker is important. One study revealed 22 upregulated and 4 downregulated genes across diverse etiological factors, indicating shared molecular mechanism in breast cancer, which can lead to innovative biomarker-targeted therapies (Hossain et al., 2024). Data heterogeneity arises from the variable clinical presentation and response to therapy in immunological diseases, leading to unequal dispersion of biomarker expression (Glehr et al., 2024).

5.2 Reproducibility and Standardization

5.2.1 Data Reproducibility

It is a critical issue in biomedical research, particularly in context of machine learning and genomic studies. The reproducibility of the biomarkers is altered by various factors, which include data integration, methodological consistency, and the inherent variability of the biological data. Factors which affect the data reproducibility are dataset integration, methodological consistency, and technical variability. Combination of the multiple datasets may enhance data reproducibility. For example, integrating five SNP datasets reduced the failure rate of biomarkers from 93% to 62% in Parkinson's disease as reported by (Ameli et al., 2024). In microbiome studies inconsistent processing and small sizes hinder reproducibility. A new method DADA2 pipeline and recursive Ensemble Feature Selection improved reproducibility across multiple datasets (Rojas-Velazquez et al., 2024). The variation in the genotyping platforms and selection criteria leads to discrepancies in biomarker identification, as observed in Genome-Wide Association Studies (GWAS) (Ameli et al., 2024). Data reproducibility in biomarker discovery is often low, with studies showing minimum overlap in identified biomarkers; here in this study reproducibility score is introduced to quantify the reproducibility for a dataset using univariate hypothesis testing (Forouzandeh et al., 2022).

5.2.2 Data Standardization

There is a distinguishable belief in technology regarding cancer bio markers, especially in policy reports, where they are anticipated to authorize higher-quality, safer, and more efficient cancer therapy while lowering health-care costs. Bio markers can provide highly specific and individualized insights into an individual's health, such as predicting susceptibility to cancer, heart disease, and neurological problems, as well as tracking disease development and guiding therapy decisions (Mishra & Verma, 2010). One of the most significant obstacles in standardizing diagnostic bio markers is the inherent diversity of biomarker discovery and validation methods. However, the journey between discovery and therapeutic application is filled with challenges. Sample variability presents a significant challenge in the field of biomarker discovery. Various laboratories may employ different methods, instruments, or reagents for biomarker analysis, resulting in result discrepancies. For instance, differences in assay methodologies, such as enzyme-linked immunosorbent assays

(ELISA) or mass spectrometry, can influence the sensitivity and specificity of biomarker identification, leading to inconsistencies in outcomes.

An important element of biomarker standardization is defining clinical endpoints—specific and reproducible measures of clinical outcomes for validating biomarkers. Nonetheless, there is often a lack of agreement on these endpoints within the scientific and clinical communities. In the absence of standardized clinical endpoints, evaluating the actual value of a biomarker in clinical practice becomes challenging, resulting in inconsistent implementation. Major consequence of the absence of standardized biomarker methodologies is the potential for incorrect diagnoses, which can have a direct effect on patient safety. Non-standardized biomarker tests may result in false positives or false negatives, posing serious risks for patient care. A false positive refers to a situation where a biomarker test mistakenly signals the existence of a disease or condition, resulting in unwarranted tests, treatments, or interventions. For instance, receiving a false positive result in cancer screening might result in unnecessary biopsies, radiation, or chemotherapy, all of which carry their own risks and side effects. Early identification is crucial for improving patient outcomes in diseases such as cancer and cardiovascular diseases. False negatives can delay treatment and allow the diseases to grow untreated.

5.3 Ethical and Privacy Concerns

Ethical concern raised in the lab, is bioethical, is how informed consent forms are submitted from patients of the cancer, and how the digital datasets generated from their samples are managed and protected. As bioinformatics technology advances and new biomarkers emerge, the ethical issues surrounding consent become more difficult. Traditional consent models may no longer be enough, as patients may not completely understand how their data will be utilized in future research or therapeutic applications (Blanchard, 2016).

5.3.1 The Current Frameworks for Genetic Data Privacy

Several rules and regulations have been implemented around the world to address privacy concerns with genetic data. For example, the Health Insurance Portability and Accountability Act (HIPAA) in the United States establishes privacy safeguards for health data, whereas the General Data Protection Regulation (GDPR) in the European Union specifies the use and protection of personal data, including genetic information.

5.3.2 Genetic Privacy Concern

One of the most important privacy concerns of diagnostic biomarkers is the sensitivity of genetic data. Genomic data, which can be used to discover biomarkers, is highly personalized and unique to each individual. The potential for this information to be exploited, shared, or accessed without consent creates serious ethical concerns. Bioinformatics frequently requires enormous datasets to detect important patterns in genetic sequences. As genomic data becomes more important in personalized medicine, questions regarding how it is stored, shared, and accessed have grown. Centralized repositories containing genetic data may become targets for hackers or data breaches. The sharing of this data between institutions, whether for research or therapeutic cooperation, raises concerns regarding data ownership and security.

5.3.3 Data Breach and Cybersecurity

With an increasing amount of health-related data being collected and evaluated, cyberattacks and data breaches constitute a major worry. A compromise in genetic data or clinical information could lead to a variety of negative effects, including identity theft, inappropriate treatment, or unlawful disclosure of medical history. When diagnostic biomarker data are compromised, the repercussions can be far-reaching. It jeopardizes not only individuals' privacy, but also public trust in healthcare facilities and bioinformatics research. These issues may impede the advancement of medical research and the creation of novel medicines.

6 Future Approach

6.1 Multi-omics Approaches

Multi-omic technologies include the genome, transcriptome, proteome, and metabolome. This chapter describes some of the advanced technologies that can change the way diagnostic biomarker discovery and validation are performed.

Development of more complex computational tools for data integration is a significant advancement of multi-omics research. Future approaches will be powerful artificial intelligence and machine learning for the analysis of the mass and complexity of datasets generated by the multi-omics studies (Advances in Multi-Omics Biomarkers for Cancer Diagnosis and Prognosis, 2024). Dynamic biomarkers could be created by combining multi-omics with real-time data collection technologies such as wearables and continuous monitoring systems. They would reflect changes in health status for the patient, therefore allowing timely interventions with potentially improved disease management (Milner Jr & Lennerz, 2024). Multi-omics approaches will drive the future of personalized and precision medicine. Data derived from genomics, transcriptomics, proteomics, and metabolomics will empower investigators to develop very highly individualized diagnosis and therapy strategies. From this comprehensive view, new biomarkers that can predict responses in individuals to treatments can be identified for improved healthcare solutions (Milner Jr & Lennerz, 2024). Future multi-omics approaches will revolutionize early detection of cancer by integrating data from the multiple omics platforms so that the biomarkers become indicative of the earliest possible stages of cancer. This could significantly improve patient outcomes because it is possible to diagnose earlier and more precisely (Dhillon et al., 2023). Tumor heterogeneity, being the existence of diverse populations of cells in a tumor, is the most significant hurdle encountered in the diagnosis and treatment of cancer. A multi-omics approach is the best solution which not only offers a full-fledged understanding of genetic and molecular heterogeneity but also helps to identify a good set of biomarkers capable of capturing the complete tumor heterogeneity and hence making possible the targeted therapies and their successful execution (Raufaste-Cazavieille et al., 2022).

6.2 Artificial Intelligence and Machine Learning

The integration of Artificial Intelligence with Machine Learning is going to revolutionize the biomarker discovery and validation. With this section, we analyze some of the breakthrough methods and impacts of AI and ML in such a domain.

One of the most promising applications of AI and ML is the integration and analysis of massive amounts of multi-omics data. The future AI systems will likely use more complex algorithms for integrating genomics, proteomics, metabolomics, and other omics data. This integrated approach will reveal complex patterns and interactions that cannot be perceived through the naked eye by using the traditional methods and thus facilitate more discoveries of novel biomarkers (Vazquez-Levin et al., 2023). Predictive modelling is quite promising with the integration of AI with ML. These predictive capacities would go a long way in forming approaches to personal medicine wherein treatments are devised to cater to the specifics of the patient. For example, the treatment plan may be advised on the basis of the genetic and molecular profile of the patient by using AI-driven models (Carini & Seyhan, 2024). The wearable devices and continuous monitoring systems are a combination of AI and real-time data-collecting technologies. For instance, continuous glucose monitors with AI algorithms can potentially detect early complications of diabetes. It will thus receive early medical care attention (Pham et al., 2024). Future AI will enhance patient selection, stratification, and monitoring, ensuring more efficient and cost-effective clinical trials. With the availability of vast amounts of electronic health record data and other data, AI can help identify who will be a good candidate for the clinical trial and what his or her response might be to the experimental treatment. To accelerate the new biomarkers identification and develop the novel therapies, these AI-driven clinical trials are promising (Kim et al., 2020). Then, ethics and regulations around AI and ML advances will be scorching discussions. Privacy of patients' data, mitigated biases in AI algorithms, and strong validation frameworks will prove vital in integrating the technologies into clinical practice (Kasula, 2021).

6.3 Personalized Medicine

This phenomenon treatment personalization, where treatments are personalized based on the biological signature of the patient. Use diagnostic biomarkers to predict disease risk, early disease detection, and appropriate plans for treatment with respect to the genetic, environmental, and lifestyle conditions of an individual. Advances in genomic technologies, such as whole-genome sequencing (WGS) and

CRISPR-based gene editing, will revolutionize the means of biomarker discovery and validation. These technologies will identify the rare genetic variants and the mutations that have associations with diseases, hence diagnosing early and implementing targeted and tailored treatment approaches (Vicente et al., 2020). Wearable devices and live monitoring technologies are poised to revolutionize personalized medicine scenarios. Wearable devices collecting health data in real-time will provide insight into the current physiological state of an individual, allowing dynamically identified biomarkers, with their integration into multi-omics/AI analysis for timely interventions/personalized treatment adjustment (Sadee et al., 2023). Personalized medicine will bring revolutionary changes in cancer treatment and find out biomarkers, indicating which response pattern the person is likely to develop in future treatments. Tumors are heterogenic and different, so treatments should be devised individually. Therefore, it would better target therapies and outcomes for patients (Pavez Loriè et al., 2021).

7 Conclusion

The rapid advances in genetics and bioinformatics have transformed the development and validation of diagnostic biomarkers, offering unprecedented prospects to improve precision medicine. This review emphasizes the importance of merging cutting-edge genomic technologies with strong computational tools for the discovery, validation, and co-development of biomarkers with therapeutic medicines. Despite these advancements, challenges such as data complexity, heterogeneity, and ethical concerns remain stressing the importance of standardized protocols and collaborative efforts. The validation of biomarkers, through both experimental and computational methods, guarantees their reliability and therapeutic value, thus bridging the gap between research and practical applications. Furthermore, the advent of machine learning, artificial intelligence, and multi-omics techniques offers promising avenues/innovative channels for resolving these issues and facilitating the advancement of personalized medicine strategies. In conclusion, the collaborative applications of bioinformatics, genomics, and advanced biostatistical approaches are driving groundbreaking advancements in diagnostic solutions. Ongoing collaboration across disciplines, technological advancements, and commitment to ethical norms are critical to take full advantage of biomarkers potential to revolutionize the healthcare sector.

References

Ahmed, S., Kim, J. and Raza, M.H. (eds.) (2024). Advances in Multi-Omics Biomarkers for Cancer Diagnosis and Prognosis. Singapore: Springer. https://link.springer.com/collections/bjidecgadd

Ahmad, A., Imran, M., & Ahsan, H. (2023). Biomarkers as biomedical bioindicators: Approaches and techniques for the detection, analysis, and validation of novel biomarkers of diseases. *Pharmaceutics, 15*(6), 1630.

Ameli, A., Peña-Castillo, L., & Usefi, H. (2024). Assessing the reproducibility of machine-learning-based biomarker discovery in Parkinson's disease. *Computers in Biology and Medicine, 174*, 108407. https://doi.org/10.1016/j.compbiomed.2024.108407

Bakhtiar, R. (2008). Biomarkers in drug discovery and development. *Journal of Pharmacological and Toxicological Methods, 57*(2), 85–91. https://doi.org/10.1016/j.vascn.2007.10.002

Baylin, S. B., & Jones, P. A. (2011). A decade of exploring the cancer epigenome—Biological and translational implications. *Nature Reviews Cancer, 11*(10), 726–734. https://doi.org/10.1038/nrc3130

Bhattacharjee, G., Gohil, N., Khambhati, K., Mani, I., Maurya, R., Karapurkar, J. K., Gohil, J., Chu, D. T., Vu-Thi, H., Alzahrani, K. J., & Show, P. L. (2022). Current approaches in CRISPR-Cas9 mediated gene editing for biomedical and therapeutic applications. *Journal of Controlled Release, 343*, 703–723.

U.S. Food and Drug Administration (FDA). Biomarkers at the FDA. https://www.fda.gov/science-research/about-science-research-fda/biomarkers-fda

Blanchard, A. (2016). Mapping ethical and social aspects of cancer biomarkers. *New Biotechnology, 33*(6), 763–772.

Borrebaeck, C. A. (2017). Precision diagnostics: Moving towards protein biomarker signatures of clinical utility in cancer. *Nature Reviews Cancer, 17*(3), 199–204.

Cagney, D. N., Sul, J., Huang, R. Y., Ligon, K. L., Wen, P. Y., & Alexander, B. M. (2018). The FDA NIH Biomarkers, EndpointS, and other Tools (BEST) resource in neuro-oncology. *Neuro-Oncology, 20*(9), 1162–1172.

Calin, G. A., & Croce, C. M. (2006). MicroRNA signatures in human cancers. *Nature Reviews Cancer, 6*(11), 857–866. https://doi.org/10.1038/nrc1997

Carini, C., & Seyhan, A. A. (2024). Tribulations and future opportunities for artificial intelligence in precision medicine. *Journal of Translational Medicine, 22*(1), 411. https://doi.org/10.1186/s12967-024-05067-0

Chang, C. H., Lin, C. H., & Lane, H. Y. (2021). Machine learning and novel biomarkers for the diagnosis of Alzheimer's disease. *International Journal of Molecular Sciences, 22*(5), 2761.

Chau, C. H., Rixe, O., McLeod, H., & Figg, W. D. (2008). Validation of analytic methods for biomarkers used in drug development. *Clinical Cancer Research, 14*(19), 5967–5976.

Clark, A. J., & Lillard, J. W., Jr. (2024). A comprehensive review of bioinformatics tools for genomic biomarker discovery driving precision oncology. *Genes, 15*(8), 1036. https://doi.org/10.3390/genes15081036

Claude, E., Leclercq, M., Thébault, P., Droit, A., & Uricaru, R. (2024). Optimizing hybrid ensemble feature selection strategies for transcriptomic biomarker discovery in complex diseases. *NAR Genomics and Bioinformatics, 6*(3), lqae079. https://doi.org/10.1093/nargab/lqae079

Considine, E. C. (2019). The search for clinically useful biomarkers of complex disease: A data analysis perspective. *Metabolites, 9*(7), 126. https://doi.org/10.3390/metabo9070126

Crutchfield, C. A., Thomas, S. N., Sokoll, L. J., & Chan, D. W. (2016). Advances in mass spectrometry-based clinical biomarker discovery. *Clinical Proteomics, 13*, 1–12.

Cunha, A. (2017). Genomic technologies—From tools to therapies. *Genome Medicine, 9*, 1–2. https://doi.org/10.1186/s13073-017-0462-9

Dakna, M., Harris, K., Kalousis, A., Carpentier, S., Kolch, W., Schanstra, J. P., Haubitz, M., Vlahou, A., Mischak, H., & Girolami, M. (2010). Addressing the challenge of defining valid proteomic biomarkers and classifiers. *BMC Bioinformatics, 11*, 1–16.

Davis, K. D., Aghaeepour, N., Ahn, A. H., Angst, M. S., Borsook, D., Brenton, A., Burczynski, M. E., Crean, C., Edwards, R., Gaudilliere, B., & Hergenroeder, G. W. (2020). Discovery and validation of biomarkers to aid the development of safe and effective pain therapeutics: Challenges and opportunities. *Nature Reviews Neurology, 16*(7), 381–400.

Day, D., & Siu, L. L. (2016). Approaches to modernize the combination drug development paradigm. *Genome Medicine, 8*(1), 115. https://doi.org/10.1186/s13073-016-0369-x

de Ronde, M. W., Ruijter, J. M., Moerland, P. D., Creemers, E. E., & Pinto-Sietsma, S. J. (2018). Study design and qPCR data analysis guidelines for reliable circulating miRNA biomarker experiments: A review. *Clinical Chemistry, 64*(9), 1308–1318.

Dhillon, A., Singh, A., & Bhalla, V. K. (2023). A systematic review on biomarker identification for cancer diagnosis and prognosis in multi-omics: From computational needs to machine learning and deep learning. *Archives of Computational Methods in Engineering, 30*(2), 917–949. https://doi.org/10.1007/s11831-022-09821-9

Diaz-Uriarte, R., Gómez de Lope, E., Giugno, R., Fröhlich, H., Nazarov, P. V., Nepomuceno-Chamorro, I. A., Rauschenberger, A., & Glaab, E. (2022). Ten quick tips for biomarker discovery and validation analyses using machine learning. *PLoS Computational Biology, 18*(8), e1010357. https://doi.org/10.1371/journal.pcbi.1010357

Dienstmann, R., Jang, I. S., Bot, B., Friend, S., & Guinney, J. (2015). Database of genomic biomarkers for cancer drugs and clinical targetability in solid tumors. *Cancer Discovery, 5*(2), 118–123. https://doi.org/10.1158/2159-8290.CD-14-1118

Dixit, S., Kumar, A., Srinivasan, K., Vincent, P. D. R., & Ramu Krishnan, N. (2024). Advancing genome editing with artificial intelligence: Opportunities, challenges, and future directions. *Frontiers in Bioengineering and Biotechnology, 11*, 1335901.

Engstrom-Melnyk, J., Rodriguez, P. L., Peraud, O., & Hein, R. C. (2015). Clinical applications of quantitative real-time PCR in virology. In *Methods in microbiology* (Vol. 42, pp. 161–197). Academic.

Fernandez, N. F., Gundersen, G. W., Rahman, A., Grimes, M. L., Rikova, K., Hornbeck, P., & Ma'ayan, A. (2017). Clustergrammer, a web-based heatmap visualization and analysis tool for high-dimensional biological data. *Scientific Data, 4*(1), 1–12. https://doi.org/10.1038/sdata.2017.151

Forouzandeh, A., Rutar, A., Kalmady, S. V., & Greiner, R. (2022). Analyzing biomarker discovery: Estimating the reproducibility of biomarker sets. *PLoS One, 17*(7), e0252697. https://doi.org/10.1371/journal.pone.0252697

Gao, J., Aksoy, B. A., Dogrusoz, U., Dresdner, G., Gross, B., Sumer, S. O., Sun, Y., Jacobsen, A., Sinha, R., Larsson, E., & Cerami, E. (2013). Integrative analysis of complex cancer genomics and clinical profiles using the cBioPortal. *Science Signaling, 6*(269), pl1-pl1. https://doi.org/10.1126/scisignal.2004088

Gerolami, J., Wong, J. J. M., Zhang, R., Chen, T., Imtiaz, T., Smith, M., Jamaspishvili, T., Koti, M., Glasgow, J. I., Mousavi, P., & Renwick, N. (2022). A computational approach to identification of candidate biomarkers in high-dimensional molecular data. *Diagnostics, 12*(8), 1997. https://doi.org/10.3390/diagnostics12081997

Gershoni, J. M., & Palade, G. E. (1983). Protein blotting: Principles and applications. *Analytical Biochemistry, 131*(1), 1–15.

Glehr, G., Riquelme, P., Kronenberg, K., Lohmayer, R., López-Madrona, V. J., Kapinsky, M., Schlitt, H. J., Geissler, E. K., Spang, R., Haferkamp, S., & Hutchinson, J. A. (2024). Restricting datasets to classifiable samples augments discovery of immune disease biomarkers. *Nature Communications, 15*(1), 5417. https://doi.org/10.1038/s41467-024-49094-3

Gossing, W., Frohme, M., & Radke, L. (2020). Biomarkers for liquid biopsies of pituitary neuroendocrine tumors. *Biomedicine, 8*(6), 148.

Hamelin, C., Cornut, E., Poirier, F., Pons, S., Beaulieu, C., Charrier, J. P., Haidous, H., Cotte, E., Lambert, C., Piard, F., & Ataman-Önal, Y. (2011). Identification and verification of heat shock protein 60 as a potential serum marker for colorectal cancer. *The FEBS Journal, 278*(24), 4845–4859.

Hanahan, D., & Weinberg, R. A. (2011). Hallmarks of cancer: The next generation. *Cell, 144*(5), 646–674. https://doi.org/10.1016/j.cell.2011.02.013

Hapmap, C. A. (2003). The international HapMap project: The international HapMap consortium. *Nature, 426*, 789–796. https://doi.org/10.1038/nature02168

Haverland, N., Pottiez, G., Wiederin, J., & Ciborowski, P. (2010). Immunoreactivity of anti-gelsolin antibodies: Implications for biomarker validation. *Journal of Translational Medicine, 8*, 1–10.

Hayes, D. F. (2015). Biomarker validation and testing. *Molecular Oncology, 9*(5), 960–966.

Hendrikse, N. M., Llinares Garcia, J., Vetter, T., Humphreys, A. J., & Ehmann, F. (2022). Biomarkers in medicines development—From discovery to regulatory qualification and beyond. *Frontiers in Medicine, 9*, 878942. https://doi.org/10.3389/fmed.2022.878942

Hossain, M. U., Mehak, M. A., Hasan, S. S., Sakib, M. N., Rahman, A. N., Bhattacharjee, A., Chowdhury, Z. M., Ahammad, I., Sohag, M. M. H., Das, K. C., & Salimullah, M. (2024). Dissecting breast cancer heterogeneity through transcriptomics insights of diverse etiological factors for common biomarker discovery. *bioRxiv*, pp. 2024-10. https://doi.org/10.1101/2024.10.20.619262

Huang, H. H., Li, J., & Cho, W. C. (2023). Integrative analysis for complex disease biomarker discovery. *Frontiers in Bioengineering and Biotechnology, 11*, 1273084. https://doi.org/10.3389/fbioe.2023.1273084

Huber, W., Carey, V. J., Gentleman, R., Anders, S., Carlson, M., Carvalho, B. S., Bravo, H. C., Davis, S., Gatto, L., Girke, T., & Gottardo, R. (2015). Orchestrating high-throughput genomic analysis with Bioconductor. *Nature Methods, 12*(2), 115–121. https://doi.org/10.1038/nmeth.3252

Humphries, J. M., Penno, M. A., Weiland, F., Klingler-Hoffmann, M., Zuber, A., Boussioutas, A., Ernst, M., & Hoffmann, P. (2014). Identification and validation of novel candidate protein biomarkers for the detection of human gastric cancer. *Biochimica et Biophysica Acta (BBA)-Proteins and Proteomics, 1844*(5), 1051–1058.

Iftikhar, M., Younis, A., Lee, Y. S., & Shehzad, A. (2022). Introduction to cancer biomarkers. In *Cancer biomarkers in diagnosis and therapeutics* (pp. 1–26). Springer. https://doi.org/10.1007/978-981-16-5759-7_1

Informatics Tools for Cancer Research. https://www.cancer.gov/about-nci/organization/cssi/resources/informatics-tools

Kanehisa, M., & Goto, S. (2000). KEGG: Kyoto encyclopedia of genes and genomes. *Nucleic Acids Research, 28*(1), 27–30. https://doi.org/10.1093/nar/28.1.27

Kasula, B. Y. (2021). Ethical and regulatory considerations in AI-driven healthcare solutions. *International Meridian Journal, 3*(3), 1–8. https://meridianjournal.in/index.php/IMJ/article/view/23

Kim, H., Kim, E., Lee, I., Bae, B., Park, M., & Nam, H. (2020). Artificial intelligence in drug discovery: A comprehensive review of data-driven and machine learning approaches. *Biotechnology and Bioprocess Engineering, 25*, 895–930. https://doi.org/10.1007/s12257-020-0049-y

Klemm, J. D., Singer, D. S., & Mesirov, J. P. (2024). Transforming cancer research through informatics. *Cancer Discovery, 14*(10), 1779–1782. https://doi.org/10.1158/2159-8290.CD-24-0604

Kuhlmann, J., & Wensing, G. (2006). The applications of biomarkers in early clinical drug development to improve decision-making processes. *Current Clinical Pharmacology, 1*(2), 185–191. https://doi.org/10.1007/978-3-540-49529-1_3

Lau, C. H., Liang, Q. L., & Zhu, H. (2024). Next-generation CRISPR technology for genome, epigenome and mitochondrial editing. *Transgenic Research, 33*, 1–35. https://doi.org/10.1007/s11248-024-00404-x

Levinson, S. S. (2009). Clinical validation of biomarkers for predicting risk. *Advances in Clinical Chemistry, 48*, 1–25.

Milner, D. A., Jr., & Lennerz, J. K. (2024). Technology and future of multi-cancer early detection. *Life, 14*(7), 833. https://doi.org/10.3390/life14070833

Mishra, A., & Verma, M. (2010). Cancer biomarkers: Are we ready for the prime time? *Cancers, 2*(1), 190–208.

Owen, C., Fader, K. A., & Hassanein, M. (2024). Western blotting: Evolution of an old analytical method to a new quantitative tool for biomarker measurements. *Bioanalysis, 16*(5), 319–328.

Pavez Loriè, E., Baatout, S., Choukér, A., Buchheim, J. I., Baselet, B., Dello Russo, C., Wotring, V., Monici, M., Morbidelli, L., Gagliardi, D., & Stingl, J. C. (2021). The future of personalized medicine in space: From observations to countermeasures. *Frontiers in Bioengineering and Biotechnology, 9*, 739747. https://doi.org/10.3389/fbioe.2021.739747

Perera-Bel, J., Leha, A., & Beissbarth, T. (2019). Bioinformatic methods and resources for biomarker discovery, validation, development, and integration. In *Predictive*

Biomarkers in Oncology: Applications in Precision Medicine (pp. 149–164). https://doi.org/10.1007/978-3-319-95228-4_11

Pham, T. D., Teh, M. T., Chatzopoulou, D., Holmes, S., & Coulthard, P. (2024). Artificial intelligence in head and neck cancer: Innovations, applications, and future directions. *Current Oncology, 31*(9), 5255–5290. https://doi.org/10.3390/curroncol31090389

Piccoli, S., Mehta, D., Vitaliti, A., Allinson, J., Amur, S., Eck, S., Green, C., Hedrick, M., Hopper, S., Ji, A., & Joyce, A. (2019). 2019 white paper on recent issues in bioanalysis: FDA immunogenicity guidance, gene therapy, critical reagents, biomarkers and flow cytometry validation (part 3–recommendations on 2019 FDA immunogenicity guidance, gene therapy bioanalytical challenges, strategies for critical reagent management, biomarker assay validation, flow cytometry validation & CLSI H62). *Bioanalysis, 11*(24), 2207–2244.

Porstmann, T., & Kiessig, S. T. (1992). Enzyme immunoassay techniques an overview. *Journal of Immunological Methods, 150*(1–2), 5–21.

Pusa, T., & Rousu, J. (2024). Stable biomarker discovery in multi-omics data via canonical correlation analysis. *PLoS One, 19*(9), e0309921. https://doi.org/10.1371/journal.pone.0309921

Raufaste-Cazavieille, V., Santiago, R., & Droit, A. (2022). Multi-omics analysis: Paving the path toward achieving precision medicine in cancer treatment and immuno-oncology. *Frontiers in Molecular Biosciences, 9*, 962743. https://doi.org/10.3389/fmolb.2022.962743

Redon, R., Ishikawa, S., Fitch, K. R., Feuk, L., Perry, G. H., Andrews, T. D., Fiegler, H., Shapero, M. H., Carson, A. R., Chen, W., & Cho, E. K. (2006). Global variation in copy number in the human genome. *Nature, 444*(7118), 444–454. https://doi.org/10.1038/nature05329

Rifai, N., Gillette, M. A., & Carr, S. A. (2006). Protein biomarker discovery and validation: The long and uncertain path to clinical utility. *Nature Biotechnology, 24*(8), 971–983.

Rius, C., Liu, Y., Sixto-Costoya, A., Valderrama-Zurián, J. C., & Lucas-Dominguez, R. (2024). State of open science in cancer research. *Clinical and Translational Oncology, 26*, 1–9. https://doi.org/10.1007/s12094-024-03468-7

Rojas-Velazquez, D., Kidwai, S., Kraneveld, A. D., Tonda, A., Oberski, D., Garssen, J., & Lopez-Rincon, A. (2024). Methodology for biomarker discovery with reproducibility in microbiome data using machine learning. *BMC Bioinformatics, 25*(1), 26. https://doi.org/10.1186/s12859-024-05639-3

Ruijter, J. M., Pfaffl, M. W., Zhao, S., Spiess, A. N., Boggy, G., Blom, J., Rutledge, R. G., Sisti, D., Lievens, A., De Preter, K., & Derveaux, S. (2013). Evaluation of qPCR curve analysis methods for reliable biomarker discovery: Bias, resolution, precision, and implications. *Methods, 59*(1), 32–46.

Sadee, W., Wang, D., Hartmann, K., & Toland, A. E. (2023). Pharmacogenomics: Driving personalized medicine. *Pharmacological Reviews, 75*(4), 789–814. https://doi.org/10.1124/pharmrev.122.000810

Shannon, P., Markiel, A., Ozier, O., Baliga, N. S., Wang, J. T., Ramage, D., Amin, N., Schwikowski, B., & Ideker, T. (2003). Cytoscape: A software environment for integrated models of biomolecular interaction networks. *Genome Research, 13*(11), 2498–2504. https://doi.org/10.1101/gr.1239303

Shen, J., Behrens, C., Wistuba, I. I., Feng, L., Lee, J. J., Hong, W. K., & Lotan, R. (2006). Identification and validation of differences in protein levels in normal, premalignant, and malignant lung cells and tissues using high-throughput Western Array and immunohistochemistry. *Cancer Research, 66*(23), 11194–11206.

Shimazawa, R., & Ikeda, M. (2016). Drug–diagnostic co-development: Challenges and issues. *Expert Review of Molecular Diagnostics, 16*(2), 187–204. https://doi.org/10.1586/14737159.2016.1132163

Smith, J. (2021). The next 20 years of human genomics must be more equitable and more open. *Nature, 590*(7845), 183–184. https://doi.org/10.1038/d41586-021-00328-0

Søreide, K. (2009). Receiver-operating characteristic curve analysis in diagnostic, prognostic and predictive biomarker research. *Journal of Clinical Pathology, 62*(1), 1–5.

Srivastava, A., & Creek, D. J. (2019). Discovery and validation of clinical biomarkers of cancer: A review combining metabolomics and proteomics. *Proteomics, 19*(10), 1700448.

Stevens, A., De Leonibus, C., Hanson, D., Dowsey, A. W., Whatmore, A., Meyer, S., Donn, R. P., Chatelain, P., Banerjee, I., Cosgrove, K. E., & Clayton, P. E. (2014). Network analysis: A new approach to study endocrine disorders. *Journal of Molecular Endocrinology, 52*(1), R79–R93.

Szklarczyk, D., Gable, A. L., Lyon, D., Junge, A., Wyder, S., Huerta-Cepas, J., Simonovic, M., Doncheva, N. T., Morris, J. H., Bork, P., & Jensen, L. J. (2019). STRING v11: Protein–protein association networks with increased coverage, supporting functional discovery in genome-wide experimental datasets. *Nucleic Acids Research, 47*(D1), D607–D613. https://doi.org/10.1093/nar/gky1131

Vasan, R. S. (2006). Biomarkers of cardiovascular disease: Molecular basis and practical considerations. *Circulation, 113*(19), 2335–2362.

Vazquez-Levin, M. H., Reventos, J., & Zaki, G. (2023). Artificial intelligence: A step forward in biomarker discovery and integration towards improved cancer diagnosis and treatment. *Frontiers in Oncology, 13*, 1161118. https://doi.org/10.3389/fonc.2023.1161118

Vicente, A. M., Ballensiefen, W., & Jönsson, J. I. (2020). How personalised medicine will transform healthcare by 2030: The ICPerMed vision. *Journal of Translational Medicine, 18*, 1–4. https://doi.org/10.1186/s12967-020-02316-w

Wallstrom, G., Anderson, K. S., & LaBaer, J. (2013). Biomarker discovery for heterogeneous diseases. *Cancer Epidemiology, Biomarkers & Prevention, 22*(5), 747–755. https://doi.org/10.1158/1055-9965.EPI-12-1236

Zangar, R. C., Daly, D. S., & White, A. M. (2006). ELISA microarray technology as a high-throughput system for cancer biomarker validation. *Expert Review of Proteomics, 3*(1), 37–44.

Zhang, F. and Drabier, R., 2013. *Bioinformatics and statistics: Computational discovery, verification, and validation of functional biomarkers.*

Computer-Based Wearable Devices for Remote Patient Monitoring

Anji Anura and Debashis Maji

1 Introduction to Wearable Devices

Advancements in flexible electronics and miniaturized diagnostic devices together with the need for personalized healthcare monitoring have led to the tremendous boost in wearable devices market. Wearable sensors are projected to develop from 327.6 million in 2021 to 1487 million 2030, at a compound annual growth rate (CAGR) of 18.3% (Vaghasiya et al., 2023). These wearable devices are capable of monitoring a plethora of vital biochemical signatures like concentrations of various metabolites, salts, amino acids, hormonal levels, etc. as well as biophysical signatures like measurement of heart rate (HR), pulse rate, temperature, record electrical activities like electrocardiogram (ECG), electroencephalogram (EEG), electromyogram (EMG), hydration levels, etc. of our body for a holistic understanding of a person's physiological state. In addition, various environmental factors like pollution levels, various contaminants, etc. may also be detected which may affect our body directly or indirectly (Dias & Cunha, 2018). Point-of-care (POC) diagnostic devices are among the most popular wearable devices meant to gather these biochemical and biophysical signatures (Ongaro et al., 2022). It is reported that ~75–80% of deaths in developing nations is due to lack of proper early diagnosis and delay in timely treatment of the subjects (Valera et al., 2021). Advancements in microfluidic technology coupled with novel printing based fabrication strategies as well as use of low cost eco-sustainable materials like paper and fabrics as device

A. Anura
Department of Biochemistry, Christian Medical College Vellore, Vellore, India

D. Maji (✉)
Department of Sensor and Biomedical Technology, School of Electronics Engineering (SENSE), Vellore Institute of Technology, Vellore, India
e-mail: debashis.maji@vit.ac.in

R. Nema et al. (eds.), *Advances in Cancer Detection, Prediction, and Prognosis Using Artificial Intelligence and Machine Learning*,
https://doi.org/10.1007/978-981-96-9346-7_7

substrates have led to tremendous research work in developing POC devices meeting the World Health Organization's ASSURED criteria which aims at making diagnostic devices more affordable, sensitive, specific, and user friendly with rapid and robust detection capabilities which may be minimal or completely equipment free, and deliverable to end-users (Naseri et al., 2022).

Wearable sensors have formed the major modality of modern day human-machine interface (HMI) (Vaghasiya et al., 2023). Capabilities of seamless body conformality of these wearable POC sensors, real time continuous data monitoring and remote transmission of clinically relevant data have led to wide acceptance of these products felicitating proper and timely diagnosis by medical experts. This tremendous power of continuous data transmission, real time patient monitoring, establishing secured data connectivity, and tele health based diagnostic solutions using small processing devices have led to the advent of body-based computers or wearable computers. Ranging from simple mobile phones or fitness trackers to sophisticated research over epidermal antennas for data transmission or processing, today's wearable healthcare sector has witnessed tremendous revolution (Vaghasiya et al., 2023).

Wearable computing was coined somewhere in the mid 1990s which referred to a processing device which always remained on with a heads-on-display and an interfacing unit (Cliff, 2005). Computer based wearables possess the capabilities of not only sensing physiological/biochemical changes, or simple data transmissions, but even computational capabilities to process and interpret diagnostically relevant information. Quite often this processing capability remains totally integrated within the sensor system which may even be conformally worn over the body as integrated processors or carried along with one like mobile phones, wrist watches, smart glasses, etc. According to Steve Mann, in order to successfully address the above purpose, wearable computers should possess few fundamental properties like it should be affixed or worn over the body, secondly it should be able to transfer data without/minimal user intervention and, finally it should always operate, i.e., function in real time with intermittent sleep periods (Mann, 1997). Further attributed of wearables include comfort in wearing for long periods, i.e., the device should not be big and bulky and not generate much heat or too many cables.

Initial applications of wearable computing were related to industrial usage where availability of resource data to the workers in the form of mounted displays, etc. for guidance and updation was essential since most often industrial workers need to use both hands as well as maintain eye contact over materials (Caudell & Mizell, 1992). 1n 1990s Boeing came up with the idea of a head-mounted augmented reality display that could automate the assembly line process. Use of technical drawings, datasheets, and video clips at the point of application became a very popular choice. Researchers at the Carnegie Mellon University developed wearable devices to inspect imperfections in aircraft skin panels reducing 50% time taken otherwise. Further, potential applications of wearables were also observed in military (Caudell & Mizell, 1992). Use of wearable handheld devices became helpful in controlling fellow military personals, provide appropriate commands during need, identify between friendly and hostile forces as well as help is developing strategies for fighting against them. These initial advancements paved the way for its use even in the healthcare domain (Cliff, 2005). Although implantable devices like pacemakers had been developed long back, user control over them is not possible. Today's modern wearable devices have the potential to continuously monitor health vitals in real time thereby providing immediate diagnosis, analyze physiological performance of athletes in sports, as well as accelerate healing rates of critically admitted patients like those suffering from burns or ulcer wounds, etc. (Gnanasambanthan & Maji, 2023).

2 Concept of Remote Patient Monitoring

Development of novel flexible and wearable sensors coupled with the ease of data transmission from the target point, increasing data connectivity, and seamless integration of different devices with cloud services has led to the new concept of physiological data collection and monitoring of the same at a remote location with automated analysis and feedback. This concept, commonly known as remote patient monitoring, significantly changes the traditional outlook pertaining to timely monitoring of body vitals by nurses with manual recording followed by its monitoring by a medical practitioner (Majumder et al., 2017). With increasing aging population (above 65 years) and complex medical ailments coupled with costly medical drugs and hospital expenses, traditional monitoring at hospitals is getting replaced by remote monitoring process and finds more favourable by patients as they can be in their homes doing their regular work fitted with wireless flexible on-body sensors continuously monitoring their vitals in a non-invasive and unobtrusive manner (Majumder et al., 2017). These not only cut medical costs, but also boost the patients mentally by being surrounded by their near and dear ones during the onset of any ailment. These wearable sensors may be designed over different flexible and body conformal substrates like fabrics, papers, polymers, skin patches, tattoos, etc. with wireless transmission of the acquired data using flexible antennas, compact Bluetooth module or even tiny NFC chips as highlighted in details in the subsequent sections. Figure 1 demonstrates the schematic of a typical remote patient monitoring system encompassing need for body wearables, signal transmission through various computing modalities, incorporation of AI/ML for data processing followed by data transmission towards home gateways and IoT based web services.

3 Computing Modalities for Wearable Devices

Use of wearable devices have paved the way for decentralized approach towards healthcare boosting personalized medical care than being traditional hospital-centric. Cutting-edge material research and design developments in past few years have led to development of myriad of novel wearable healthcare products. Vivid computing modalities like use of smartphones, near field communication devices, even flexible antennas integrated with the device coupled with tremendous growth

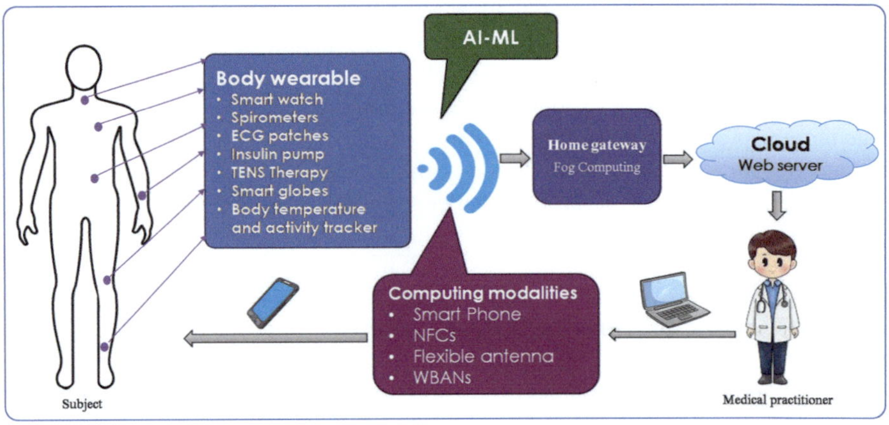

Fig. 1 Schematic representation of a remote patient monitoring system connected to various body wearables for physilogical parameter monitoring and connected to a cloundserver for data transmision using wireless modalities like smartphones, antennas, NFCs, etc. Raw signals are processed remotely and interpreted by an expert medicalpractitioner and the clinically relevant data is send back to the patient by the medical practitioner using various computing modalities like smartphones, etc

and implication of artificial intelligence (AI) /machine learning (ML) and internet-of-things (IoT) applications have boosted digital healthcare. Coupling these computing modalities with wearable healthcare devices aims at developing personalized medicine, auto-regulated medical/drug doses, bio-signal monitoring in real time, communication with healthcare experts, education about digital healthcare, etc. Below we discuss some of the few most popular wearable computers that are becoming most popular for realization of the above.

3.1 Flexible Antennas

Flexible and wearable antennas form an important source of wireless power delivery as well as data transmission in the wearable domain and have witnessed tremendous applicability in healthcare domain. Various materials like polymers (neoprene rubber, wool, cellulose and silk), textiles, nanomaterials like graphene, inorganic materials like metal inks, MXenes, ceramics, etc. have been explored to realize wearable antennas (El Gharbi et al., 2020).

Requirements of these wearable antennas include a working range from 900 MHz to 38 GHz with typical size less than ~25 cm^2 (El Gharbi et al., 2020)· The major challenge involved in realization of these wearable antennas involves proper design considerations to ensure effective signal coupling between the skin surface and the antenna geometry when exposed to external electromagnetic source. Further, performance management during high body deformations like bending, twisting associated with stability over time, signal fading due to body shadowing effect as well as interference with external environmental factors like washability, humidity, temperature, proximity to other antennas or devices, etc. are some of the major issues that needs to be considered for its practical usability (Alam & Ben Hamida, 2015).

Skin mounted epidermal antennas are one of the latest developments in this domain. Rogers group from Northwestern University have developed several epidermal sensors with highly skin conformal and flexible antennas for wireless transmission of physiological signals (Kim et al., 2011) as well as neonatal care (Chung et al., 2019). In 2022, Lin et al. developed flexible antennas through digital embroidery using liquid metal fibers which were created by injecting galinstan (Ga68.5/In21.5/Sn10) into a tubing containing perfluoro alkoxy alkane (PFA) (Lin et al., 2022).

3.2 Smartphones

World Health Organization (WHO) in 2011 defined implication of mobile usage for health monitoring as mHealth—a component of eHealth (Doupis et al., 2020). Ubiquitous nature of smartphones has made them an indispensable yet powerful tool for connection with wearable devices in this digital era. Seamless integration, fast processing capabilities, and affordable costs driven with IoT enabled connectivity have made them convenient as well as a reliable source of real time data monitoring unit, with healthcare feedbacks prompting a new paradigm of preliminary self-assessment of the disease by the subjects themselves with potential life-saving implications as well (Wall et al., 2023). Monitoring of real time physiological parameters may even be so sensitive and accurate that they may be able to detect any discernible signs even prior to that of by trained medical practitioners. Within almost one decade since its inception, mHealth technology significantly affects techniques in which healthcare affects both patients as well as healthcare practitioners by providing sensory data acquisition with feedback, remote patient monitoring, tele-health services, etc.

3.2.1 Applications of Smartphone-Based Diagnosis

Evolution of smartphone industry has led to the rapid development of various real-time health monitoring and management applications (apps) for users to keep track upon their daily biophysical (like ECG, EMG, EEG, fall detection, gait analysis, temperature, SpO_2, etc) (Turanli et al., 2024; Leu et al., 2018; Jeong et al., 2022) and biochemical (like metabolites, salts, etc) (Sempionatto et al., 2022; Banik et al., 2021; Zhang et al., 2022a) activities.

Jeong et al. in 2022 developed a signal acquisition device for real-time monitoring of biophysical signals like ECG/EMG signal using a smartphone followed by real-time data processing and monitor using cloud servers to efficiently predict before the onset of a heart attack for cardiac patients (Jeong et al., 2022). Further, smartphones have also been used for assessment of mental, emotional, and physical stress (Mohino-Herranz et al., 2015), obesity management (Choi et al., 2015a), etc. Built-in microphones within smartphones have also been used to perform lung function test (LFT) by using a high-resolution time-frequency spectrum analysis of the received sound signals. LFTs may adversely affect lung function due to presence of asthma, bronchitis, emphysema, chronic obstructive pulmonary disease (COPD), cystic fibrosis, etc. (Thap et al., 2016). Apart from basic real time monitoring of physiological parameters through smartphone, its use in case of rehabilitation for Parkinson's disease (PD) management is also noteworthy. Since smartphones are fitted with different movement sensors like accelerometers, gyroscopes, etc. irregular gait or walking pattern can be easily monitored for patients suffering from PD preventing fatal risk of falling. Use of auditory cues like sonification may also be applied through smartphones, where sound signals are used as biofeedbacks to the PD patients helps in rectifying their gait and improving motor balance (Nown et al., 2023).

Biochemical analytes on the other hand are measured by several platforms like electrochemical, colorimetric, fluorescent, chemiluminescent, etc. (Zhang et al., 2022a). Among these, electrochemical readouts involve quantitative and highly accurate results, whereas colorimetric analysis depends upon color change due to presence of any analytes and is regarded as qualitative or semi-quantitative in nature. However, modern smartphone based image analysis technique overrides these shortcomings by promoting quantitative results making colorimetric technique as one of the most promising and reliable source of preliminary disease screening in the modern era (Lingadharini et al., 2024). Most enzyme based colorimetric POC devices for determination of metabolic biomarkers involve oxidation of the target analyte using the oxidase enzyme producing H_2O_2 which is reduced to H_2O by another enzyme namely peroxidase and in-turn oxidises a colorimetric agent that produces color intensities proportional to the concentration of H_2O_2. This is in turn proportional to the target analyte which can be easily interpreted using smartphone based image processing techniques from different analytes like blood, sweat, saliva, urine, tear, etc. (Ates et al., 2022). Recently, several researchers have also explored the use of various nanoparticles (NPs), metal oxides, metal oxide frameworks (MOFs), etc. with above catalytic properties to produce color changes (Zhang et al., 2022b; Ortiz-Gómez et al., 2017). In 2021, Feng et al. reported the use of zinc nanorods which were hydrothermally grown over Whatman filter papers and used for simultaneous detection of glucose and uric acid. They obtained a low detection limit of 3 μmol L^{-1} and 4 μmol L^{-1} for glucose and uric acid, respectively using post processing of smartphone captured images for quantitative interpretation (Feng et al., 2021). In 2022, Zheng et al. used platinum nanoparticles as peroxidase mimic and successfully detected glucose and uric acid levels using a smartphone and the image processing interpreted limit of detection was observed to be 0.004 mM and 0.003 mM for glucose and uric acid respectively (Zheng et al., 2022). Recently, Taeha et al. developed a smartphone based glucose detection using colorimetric assay by immobilizing glucose and ferrocene over on paper functionalized with polyaniline (PANI) nanoparticles demonstrating high sensitivity, quick response time and a low detection limit of 1.25 μM along with high stability of the developed colors (Lee et al., 2024). In this regard, recently, our group has also developed a fabric based glucose sensor with beeswax as microchannel walls and PVC clear sheets as potential stencils for the microchannel/test well design. Glucose levels from 10 mM to 50 mM were easily detected using image processing technique through smartphones with 1 mM being the minimum level by using potassium iodide as the coloring agent (Lingadharini et al., 2024). Similar technologies have been used for detection of metabolites other than body metabolites like body salts, bacterial infections, water contaminants, and other biomolecules (Al-Hawary et al., 2023; Jalal et al., 2017; Liu et al., 2022; Ramalingam et al., 2021). Lee et al. in 2023 fabricated a colorimetric based lateral flow immunoassay (LFIA) which incorporated specific protein (CBP31-BC) linker to attach to antibodies of SARS-CoV-2 within as quickly as 15 min and demonstrated a detection limit of 5×10^4 copies/ml (Lee et al., 2023). Recently, Tao Wu et al. in 2024 modified a 3D paper-based analytical device (μPAD) into a loop-mediated isothermal amplification (LAMP) device with a closed lateral flow (LF) strip to enable quick, economical and visual mode of detection of mycobacteria capable of detecting it as low as 10 fg reaction^{-1} (Wu et al., 2024).

Another notable applications of smartphones are as imaging device (SID) and in optical microscopy for disease diagnosis, parasite detection, etc. (Banik et al., 2021). Recent advancements in mobile phone based microscopy have found useful applications in cytological studies. A smartphone based blood analyzer for determining the concentration of RBCs, WBCs, and hemoglobin was developed by Zhu et al. from just ~10 μL of blood sample (Zhu et al., 2013). Cesaretti et al. developed a SID using a specialized lens called BLIPS lens attached over a smartphone camera using sticky tape for identification of macro-vesicular steatosis (MS) responsible for liver graft failures (Cesaretti et al., 2018). Chang et al. developed a smartphone-based fluorescence microscopic system for screening of tuberculosis (TB). Obtained images were used to classify TB objects from others using support vector machines (SVM) with an accuracy of 89.2% ± 2.1% from sputum smear samples (Chang et al., 2012). Smartphone based imaging has also been used in cancer diagnosis with improved imaging resolution using Deep Learning techniques. Uthoff et al. in 2018 used the above technique for screening oral cancer (Uthoff et al., 2018). Images obtained were uploaded to the cloud and classified using pre-trained convoluted neural network (CNN) technique demonstrating similar specificity and sensitivity as that by a remote specialist with provision of even downloading the entire report from the cloud. Thus, smartphones have gone way past its traditional means of a communicating device to an important diagnostic tool in the hands of healthcare professionals particularly in resource limited settings.

3.2.2 Attributes of Smartphone-Based Diagnosis

Smartphones act as a decentralized mode of communication between a patient and the medical practitioner, which may pose several challenges. The market potential of any digital data generated from the patients is of immense value and needs to be protected with robust data standards and security protocols. These continuous streams of data may be mined for generating valuable information and thousands of such mined datasets may in turn be used for effective training of AI/ML based wearables for further improvements of future generations of the devices (Wall et al.,

2023; Galan et al., 2020). However, if these security protocols are breached or hacked, high loss of medical information may result in its misuse by the darker community by leveraging high cost of selling medical data, posing significant doubts in their reliability as well as chances of potential data manipulation with an attempt to introduce massive viral attacks over reputed medical companies. Another aspect pertains to the equitable access of healthcare data to all irrespective of age, gender, caste, or social status so that implications of trained data for a better device feedback (Gajarawala & Pelkowski, 2021).

3.3 Near Field Communication Devices

Near field communication based devices (NFCs) work on the principle of electrical energy exchange between two coils when place within close proximity (less than 5 cm). Signal frequency used here is 13.56 MHz and works based on ISO/IEC14443 standards allowing both power as well as data transfer rates of up to 424 kbit s^{-1} (Olenik et al., 2021). Further, these NFCs generate sufficient power (~10 mW) to run most low-powered electronics and sensors wirelessly. All of the above parameters make NFCs an important candidate as a future wearable computing modality filling the gap between other wireless communication devices like Bluetooth and Wi-Fi. Recent developments include use of NFC based wearable sensors, NFC based POC devices and even NFC based implantable devices (Olenik et al., 2021).

3.3.1 Applications of NFC-Based Diagnosis

Modern healthcare applications have seen the use of NFCs in data transmission in various forms starting from its integration over flexible substrates like PDMS or polyimide to more sophisticated on-skin or epidermal POC systems to even "tattoo" based sensors for determination of various biophysical measurements like ECG, skin temperature, strain, etc. (Kim et al., 2017; Lee et al., 2018) as well as biochemical sensing of analytes through sweat (Xu et al., 2019). Rogers group in 2019 use skin as communication medium for demonstration of virtual reality (VR) and augmented reality (AR) through epidermal sensors and actuators and demonstrated wireless data transmission using NFC antennas conformally mounted over the akin surface (Yu et al., 2019). Kang et al. in 2021 developed a low-cost system for monitor heart rate and temperature using a wireless NHC system over different areas like fingertip, wrist, temple, and neck regions to transmit the acquired data and analyzed it using a smartphone based application for real-time data acquisition and processing (Kang et al., 2021). The device used dual layer of coiled antennas for improved signal strength and data transmission sandwiched between two black encapsulating medical grade polymer layers with a lower adhesive layer ensuring skin compatibility. Recently, Conti et al. in 2023 reported a flexible NFC transponder over a paper substrate which consisted of a NFC transceiver, along with printed graphene resistors as pH and strain sensors with a PEDOT:PSS based humidity sensor. The assembly was integrated with a printed solar cell and a silver antenna (Conti et al., 2023).

3.3.2 Attributes of NFC-Based Diagnosis

Stringent requirement of these NFCs is their flexibility particularly when associated with any wearable or POC device. Even a slight crease in the metal thin film based coils would result in signal and data loss, demanding excellent stability and high flexibility of these devices. Researchers have explored the use of liquid metals like GalnSn and EGaln alloys, in small microchannels to serve as antennal coils (Teng et al., 2019). Further, cost of these NFCs need to be as minimal as possible so as to seamlessly integrate with the sensors. Low cost materials like paper and cellulose have been explored in this regard using printable inks (Conti et al., 2023). NFC based sensors are designed as planar 2D structures which restricts them to minimum of few cm in length to work in the 13.56 MHz range. Stacking multiple layers of the coil may further assist in reducing the overall dimensions as well as boost power efficiency as reported by Kim J et al. (Kim et al., 2017). NFCs for implants further need to be highly biocompatible to reduce risks of infection. Additionally, they may also be developed using bioresorbable materials like molybdenum and magnesium which typically gets resorbed in to the body without forming any toxic by-product (Kang et al., 2016; Choi et al., 2020). However, untimely degradation of the device and limited lifetime are some of the drawbacks that still need to be worked out before practical implementation of these devices. Nevertheless, due to its high security reasons, uses of NFCs are bound to see tremendous applications in the near future for wireless communication with wearable devices.

3.4 Wearable Body Area Network

Ever increasing growth of wearable market and their integration with wearable computers have led to the development of a much wider network of connectivity known as the wearable body area network (WBAN or BAN). A survey at the United Kingdom in 206 estimated that almost 80% of the people wanted to have a WBAN network with them in near future (Lambert, 2016).

3.4.1 WBAN Architecture

Pramanik et al. proposed a general architecture of WBAN systems comprising of 3 basic levels (Pramanik et al., 2019). Level 1 attempts to integrate all body worn devices (implantable, patchable, or simply affixed over body) to transmit all the data to a common local processing unit, instead of each individual wearable device having their own. Level 2 deals with protocols of delivering these multiplexed information to personal handheld devices like smartphones, laptops, etc. near the patients or their relatives as raw data for continuous monitoring using local networks like Bluetooth, NFC, WiFi, etc. Level 3 finally related to transmission of these data to healthcare professionals for proper diagnosis using internet services.

3.4.2 Integration Levels

These various levels of integration call for seamless co-ordination between the various sensors and the local processing unit. Thus, sensors may now be made even further smaller and flexible connected to a flexible antenna for transmitting data to the local controller. Data communication is again a very important factor which should be free from any local interference within the body worn devices or even from others within its vicinity and maybe controlled through proper gateway protocols. Security of the data transmission also plays important part as highlighted in Sect. 3.2.1. Low powered communication technologies such as Bluetooth or WiFi may be used to connect the central server to the local smartphones as a single sinking device. Hence sufficient bandwidth of these should be maintained to avoid any delay or signal overlapping. Further secured internet link (use of VPN, IPSec, SSL, TLS, etc.) is very critical to ensure data transmitted from the local server is not breached as it goes to the medical practitioner. Appropriate alert mechanisms should also be triggered during identification of any abnormal medical data so as to intervene the patiently at the earliest. Last, but not the least, these a central medical servers must be maintained at hospitals to store and process these real-time data with high security for future usage like training of medical devices using AI/ML, improving personalized care, educational purposes, etc. (Bhatti et al., 2022).

3.4.3 Wireless Technologies

It may be worth noting here that the wireless technologies driving the above WBANs are different from the wireless sensor networks (WSNs). Notable differences include limited number of nodes included in WBANs as compared to much higher number in WSNs. Further, there are no redundant nodes present in WBANs unlike in WSNs and each sensor node may be of different sizes as well as have varying power level requirements. Additionally, heterogeneity in the functioning of each sensor adds to network complexities. Thus, there may not be a single wireless technique that may be applicable for all the integrated wearable sensors. These techniques may be divided into three basic categories viz. inter-BAN, intra-BAN and beyond-BAN (Bhatti et al., 2022; Khan & Pathan, 2018). Inter-BAN networks include use of Ultra-Wideband (UWB), Bluetooth (IEEE 802.15.1), BLE (Bluetooth Low Energy), Zigbee (IEEE 802.15.4), WiFi (IEEE 802.11), etc. which are mainly used for communicating with wearables devices and the controller, i.e., level 1 only. Intra-BAN network involves communication between the local controller with the personal devices like mobiles, laptops, etc. and may use the above technologies like Bluetooth, BLE, Zigbee, WiFi, or even Universal Mobile Telecommunications System (UMTS/GPRS), radio-frequency identification tags(RFID), etc. Finally, the level 3 systems involving connections beyond-BAN, i.e., with the doctors or medical practitioners may include internet services, General Packet Radio Service (GPRS), Global System for Mobile Communications (GSM), WiFi, cellular networks, etc.

Processing of the data collected from the wearable devices forms a major challenge which involves elucidation of some meaningful information about the body vitals of the patient. This data processing requires computational capabilities which can be addressed through some local hubs (known as edge/fog computing) or even a global server (known as cloud computing) using Internet-of-things (IoT) gateways (Mukhopadhyay et al., 2022). Cloud computing refers to utilization of processing and storage resources of global servers (acting as clouds) by these sensors individually. Thus, with billions of sensors demanding the IoT gateway, issue of bottlenecking and data latency is quite high. Additionally, access to the client and security issues would be highly challenging. On the other hand, edge or fog computing refers to utilization of storage and computing capabilities of a local hub for meaningful data interpretation by each wearable sensor. They help in reducing unwanted bandwidth utilization and data bottlenecking by individual sensors as well as improves the security of the data since it stays close to the user near a local server. These local fog centers may in turn move the processed data for cloud service for further data processing/mining, etc.

Use of IoT has exemplified the user's access to the external environment seamlessly becoming a global communicating platform. IoT enabled wearable technology primarily includes three components viz. a publisher, broker, and subscriber (Saminathan & Geetha, 2018). The publisher consists of end users with connected sensors and smart devices to record biophysical parameters of an individual like ECG, EEG, temperature, strain, etc. or biochemical parameters like body glucose levels, salt concentrations, etc. These physiological data are sent to a broker who is responsible for processing it and forwarding it to the cloud server. The subscriber then gets access to these data continuously in real-time and is responsible for its tracking using a smartphone, watch, etc. The publisher may also communicate to the patient about his physiological health status post processing.

4 Powering of Wearable Devices

One of the major issues faced by wearable devices is the need for continuous power supply to ensure continuous monitoring of any physiological parameter. Conventional methods utilize incorporation of batteries to power up sensor functionalities and associated signal conditioning circuits, wireless connectivity, etc. However, limited energy storage capabilities and need for continuous replacements have let to exploration of alternative techniques for continuous power supply. Energy harvesting technique is one of the most popular techniques of exploring this possibility since it exploits naturally occurring phenomenon from within the vicinity of the wearable device for power generation (Mukhopadhyay et al., 2022), i.e., either from the environment or from the wearer's body itself. Most common energy sources for healthcare based energy harvesting devices involve exploiting mechanical movements of heartbeat and body falls or changes in body temperature, blood pressure, etc. (Gljušćić et al., 2019). Ever since its conception, energy harvesting methods have played critical roles in leveraging potential energy solutions to wearable POC devices by not only delivering an unobstructive power supply but also allowing the sensor to naturally perform its work without any interference from additional power circuits (Mukhopadhyay et al., 2022).

Energy harvesting solutions can be realized through various techniques like harnessing of piezoelectric, triboelectric, or thermoelectric property of materials, or even through photovoltaic method, electromagnetic method, chemical reaction based (using body metabolites) (Ates et al., 2022), or combinations of these in many hybrid systems (Yin et al., 2021). By varying the materials and principle of its usage, triboelectric nanogenerators (TENG) (Pan & Zhang, 2019) or piezoelectric nanogenerators (PENG) (Covaci & Gontean, 2020) are capable of generating current or voltage levels of sufficient magnitudes to drive wearable sensors. PENG devices rely on physical movements of body parts by the users like walking, running, exercising, etc. whereby either stress over a piezoelectric material like zinc oxide nanorods, lead zirconate titanate, barium titanate, etc. produce an electrical output whereas TENG rely on repeated contact and separation of materials with varying electronegativities triggering current flow or voltage production and may be used in several modes (Nitha & Chandrasekhar, 2023). Although these devices are capable of generating high power densities of even upto 810 mW m^{-2} and stable towards high mechanical deformations, generation of alternating current at the output needs conversion to direct mode for sensor integration. Further material durability and practical realization of a robust, highly flexible and fatigue-free electrodes in miniaturized form integrated into an a wearable device still remains challenging (Liu et al., 2019). Thermoelectric technique uses body heat to power wearable devices. Here, uses of conductive polymers, hybrid organic–inorganic materials, liquid metals, etc. have been explored for use in thermoelectric based power generators (Ren et al., 2021). Further, compared to the above, these devices are bigger in size with less power generation capabilities. Photovoltaic materials involve use of solar-powered wearable sensors converting daylight energy to power wearable devices. However trade-off between miniaturization and power efficiency coupled with skin conformality, flexibility, and washability are still some inherent challenges required to be overcome for using it as a reliable power source (Zhao et al., 2022). Electromagnetic method of wireless powering of wearables refers to use of antennas integrated with wearable systems for both power supply as well as data transmission. Advancements on flexible technology have made possible realization of highly flexible and stretchable antennas for numerous healthcare applications (Nie et al., 2021). Although this technique looks promising source of powering as well data retrieving from wearables, robustness of the flexible antennas towards various environmental factors like temperature, humidity, washability as well as presence of other similar devices may pose significant problems. Another modality of powering wearable devices is through use of chemical energy of bio-fuel cells which uses enzymes as catalysts for converting chemical reaction to electrical energy. Notable examples include use of lactate oxidase, glucose oxidase, etc. which are available in body sweat and can be used as sources of fuel in bio-fuel cells to detect changes in pH levels, body salts, metabolites like glucose, lactate and uric acid, etc. (Song et al., 2021). Limitations include continuous availability, stability, as well as device miniaturization and system integration for unobstructed and continuous use (Manjakkal et al., 2021).

Thus, to overcome these above shortcomings, use of supercapacitors in wearable devices may be explored using materials like CNT based yarns, textile based electrodes with 2D heterostructures, etc. (Choi et al., 2015b; Wu et al., 2020). However, device flexibility, toxicity, deformability as well as raggedness towards continuous usage need to be explored rigorously to ascertain its usage in wearable applications.

5 Applications of Wearable Sensing Devices in Healthcare

Wearable sensors are capable of monitoring a plethora of vital biochemical as well as biophysical signatures of our body for a holistic understanding of a person's physiological state. Figure 2 summarizes the various types of wearable sensors along with various platforms for its development with sensing modalities as well as disease detection with detailed discussion highlighted below.

5.1 Measurement of Biochemical Signals

One of the major applications of wearable sensors includes determination of physiological well being of any individual. Most physiological signals can be broadly categorized as biochemical and biophysical signals and has become increasingly popular due to limitations in wearability, breathability, and comfort of conventional sensors (Lingadharini & Maji, 2025). Among the plethora of materials used for making POC devices like acrylic, silicon, glass, etc., use of flexible materials like polymers and recently biodegradable materials like paper and fabric has become very popular in addressing green point-of-care devices (GPOCDs) (Lingadharini & Maji, 2025). Various low cost patterning techniques like screen printing, stencil printing, wax printing, ink stamping, etc. have led to the realization of various novel and economical POC devices.

Analysis of biochemical signals includes determination of body metabolites, salts, amino-acids, proteins, hormonal balance, antibiotic and viral loads, etc. within our body (Lingadharini & Maji, 2025). As mentioned earlier these can be measured by several platforms like electrochemical, colorimetric, fluorescent, chemiluminescent, etc. with electrochemical mode being the most accurate, reliable and direct

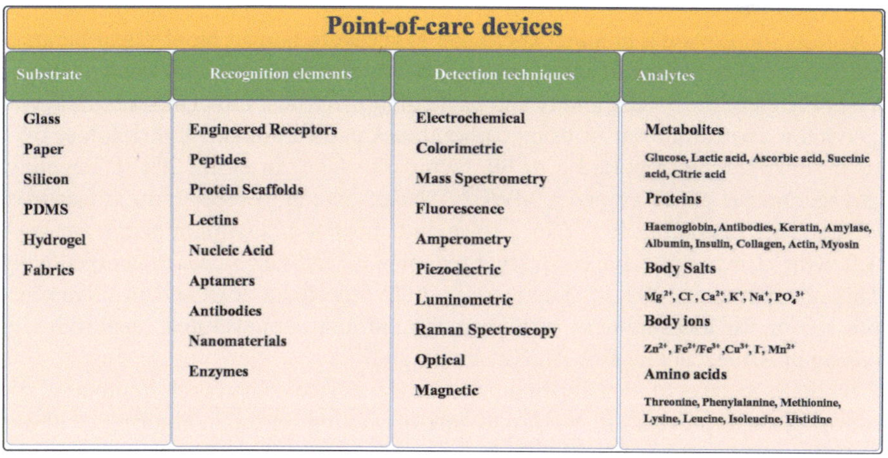

Fig. 2 Schematic illustration of POC devices highlighting various substrates used, recognition elements employed along with detection techniques and analytes detected

quantitative modality present. Dungchai et al. in 2009 reported the development of first paper-based electrochemical device (μPED) for detection of glucose, lactate, and uric acid levels (Dungchai et al., 2009). After several years of development, Li et al. in 2015, developed an improved μPED involving zinc oxide nanowires (ZnO-NWs) for glucose detection through electrochemical technique (Li et al., 2015). The increased surface area resulted in a highly efficient biosensor with limit of detection (LOD) as low as 59.9 μM. In 2016, a novel pop-up based POC device was fabricated by Wang et al. for early detection of diabetic ketoacidosis (DKA) in pediatric patients through measurement of beta-hydroxybutyrate (BHB) levels (Wang et al., 2016). In 2020, Valentine et al. developed a porous CNT-based electrodes for electrochemical detection of glucose over paper and obtained an LOD of almost 2.93 mM (Valentine et al., 2020). Nongyao et al. reported development of a three-dimensional electrochemical paper-based analytical device for determining uric acid levels and dopamine (Nontawong et al., 2022). In 2021, an impedance biosensor for the detection of the SARS-CoV-2 spike protein was developed by Muhammad Ali Ehsan et al. using the IgG anti-SARS-CoV-2 spike antibody and the sensor demonstrated a very low limit of quantification of ~0.25 fg mL^{-1} accurately (Ehsan et al., 2021).

Similarly, fabric based biochemical sensors have also been developed by several researchers to address device washability and improved ruggedness towards prolonged usage. In 2019, Zhao et al. developed an enzymatic glucose biosensor over an elastic gold fiber and showed a LOD of 6 μM (Zhao et al., 2019). Wang et al. reported a simple method of realizing fabric based electrochemical sensor using active coatings of CNT for detection of various analytes like skin pH, glucose, as well as salts like Na^+, K^+, and Ca^{2+} in real time (Wang et al., 2018).

5.2 Measurement of Biophysical Signals

POC sensors provide a unique experience to measure human biophysical parameters over the skin surface in a continuous and nonintrusive technique using advances in flexible electronics technology and packaging materials. Liu et al. in 2017 developed a low cost strain sensor using carbon black and carboxymethyl cellulose over paper substrate and reported a gauge factor of ~4.3 (Liu et al., 2017). Recently, Thiyagarajan et al. developed a MWCNT based screen printed strain sensor over polyethylene terephthalate (PET) surface and showed a maximum gauge factor of 31.9 with a response and recovery time of ~13 ms and ~ 17 ms respectively (Thiyagarajan et al., 2022a). Xue et al. in 2022 reported a flexible capacitive pressure sensor for measurement of human motion and demonstrated capabilities of sensing pressures as low as 2 Pa (Xue et al., 2022).

In 2018, Sadri et al. developed a paper based device coated with fluoroalkyl trichlorosilanes to prevent effects of moisture or wetting during monitoring of physiological signals like ECG, EMG, etc. over the skin surface (Sadri et al., 2018). Carnerio et al. in 2020 developed a patient-specific EEG headband using screen printed silver flake electrodes which were almost equivalent to standard Ag/AgCl electrodes (Carneiro et al., 2020). Li et al. in 2022 developed hydrogel-paper patch for measurement of both glucose (biochemical) as well as ECG (biophysical) signals in real time (Li et al., 2022). In the same year, Masihi et al. developed a MWCNT/PDMS composite over a polyurethane substrate along with a screen printed layer of silver to create flexible dry electrodes for ECG monitoring and demonstrated better performance than traditional Ag/AgCl electrodes (Masihi et al., 2022). Thiyagarajan et al. in 2021 developed a MWCNT/PDMS composite based breath sensor fixed over a face mask to monitor variations in breathing pattern (Thiyagarajan et al., 2021). The same group in 2022 developed a MWCNT based skin temperature sensor on paper with a sensitivity of $-0.0685\%\ C^{-1}$ which may be used for non-contact based body temperature measurement (Thiyagarajan et al., 2022b). Fabric based temperature sensor was also developed by Kuzubasoglu et al. in 2020 using CNT ink over soft taffeta fabric and showed highest sensitivity between room temperature and 50 °C (Kuzubasoglu et al., 2021).

The associations between various biochemical analytes like metabolites (glucose, lactate, uric acid, etc.), salts (Na^+, K^+, Mg^{2+}, etc.), etc. and biophysical signals like heart rate (HR), pulse rate, temperature, ECG, EMG, EEG, etc. in the human body are intricate and interconnected. Glucose and lactate are important metabolic sources. Excess glucose can cause hyperglycemia which affects HR and BP due to increased blood viscosity and vascular resistance. Lactate created during anaerobic metabolism frequently rises with physical activity impacting HR and BP as the body adjusts to metabolic demands with higher levels also causing organ failure due to lactic acidosis. Uric acid a result of purine metabolism is linked to cardiovascular health. Elevated uric acid levels can cause hypertension and increases HR and BP as a result of endothelial dysfunction and oxidative stress as well as cause gout, renal diseases, etc. Electrolyte imbalances causes anomaly in normal body functioning with high sodium levels causing restlessness (hypernatremia). Temperature regulation is essential for metabolic homeostasis since it influences enzyme activity and metabolic rates. The high temperature can raise heart rate and blood pressure as the body responds to higher metabolic demands affecting the ECG signal in turn. EMG monitors muscle activity and can reflect metabolic states, with greater EMG activity during exercise being associated with higher glucose and lactate levels. EEG which measures brain activity can be influenced by metabolic changes for example hypoglycemia can modify EEG patterns indicating altered neuronal function. Together these indicators portray a representation of the body's physiological and metabolic condition emphasizing the interconnectivity of the biochemical and biophysical changes occurring.

5.3 Detection of Bacterial and Viral Biomarkers

Advancements in wearable POC systems are now capable of screening bacterial infections like E. coli, Staphylococcus aureus, Helicobacter pylori, etc. (Mazur et al., 2023) and viral disease attacks like dengue, hepatitis, EBOLA, AIDS, influenza, coronavirus, etc. (Ozer & Henry, 2021) with devices made over paper as well as fabric surfaces (Lingadharini & Maji, 2025). Among the most common types of analytical devices, paper based devices (PADs) are most popular employing lateral flow assays (LFAs) for disease identification. LFAs are 2D PADs that consist of a sample entry pad, conjugate pad, a nitrocellulose membrane, and an absorbent pad. Most LFAs generate two test signals namely a test line (T-line) and a control line (C-line). The T-line mostly consists of target analyte specific antibodies, nanoparticles, enzymes, etc. for its easy identification often through visual marking followed by C-line which has antibodies not specific to the analytes under consideration and is as indication that the analyte solution has passed the T-line marking for proper test validation.

Bacterial infections may be detected through direct or indirect methods. Direct method involves affixing the bacteria with specific enzyme tagged antibodies while the indirect method involves detection of by-products formed or presence of specific biomarkers like, toxins (Mazur et al., 2020), proteases (Alhogail et al., 2016), volatile organic compounds (VOCs) (Chen et al., 2014), etc. released by them. PADs for viral disease detection may be classified based upon the biorecognition element that need to be immobilized over the target substrate, as antibody/antigen based, nucleic acid based or affinity based. Antibody/antigen type PADs can be used for development of immunosensors for virus detection. Since antibodies are produced few days after a viral attack, these are used as biorecognition elements here which are often loaded with various nanoparticles for enhanced detection. Chen et al. developed a fluorescence based lateral flow immunoassay (LFIA) for detection of anti-SARS-CoV-2 IgG using specific antibodies loaded with lanthanide-doped polystyrene nanoparticles (LPN) (Chen et al., 2020). Similar other detection techniques like chemiluminescence, colorimetric, electrochemical, etc. have also been used for detection of SARS-CoV2 virus (Ozer & Henry, 2021). Nucleic acid based PADs explore use of genomes that code these viruses as specific biomarkers for its detection unlike antibodies. Here DNA/RNA probes are used as the recognition element and specific primers are designed to identify the target nucleic acid type and hence the virus. Recently, Li et al. in 2021 reported an entropy-driven amplification strategy based LFA for nucleic acid based detection of influenza virus H1N1-RNA (Li et al., 2021). The third category for virus detection includes use of other materials like glycoproteins, aptamers, etc. and has been used for the detection of red-spotted grouper nervous necrosis virus (RGNNV-CP) (Liu et al., 2020), p24-HIV (Gogola et al., 2021), etc.

5.4 Sleep Monitoring Sensors

Sleep is a vital process by our body which helps us in maintaining proper health and well-being. Poor sleep quality adversely affects our health and mental conditions leading to obesity, heart diseases, elevated mood swings, hampered immune system, etc. Thus, it is very vital to properly ascertain sleep patterns particularly for patients suffering from the sleep disorders (Kwon et al., 2021). Present method involves use of polysomnography (PSG), which monitors physiological changes during the sleep process. However, complicated setup, high cost, and uncomfortable usage with long running wires pose severe patient discomfort and may not show actual sleep data. Modern wearable systems with flexible electronic sensing patches provide a much reliable and comfortable readings for sleep monitoring as listed below (Kwon et al., 2021).

Advances in wearable and flexible technology have developed devices for monitoring of EEG (electroencephalogram) signals form brain. Changes during the brain signal may occur during any of the five stages of sleeping (W, N1, N2, N3, and REM) and may be easily monitored via these wearable sensors (Berry et al., 2017). Difficulties in attaching electrodes over the brain have even led researches to place them over the forehead for a reliable contact (Levendowski et al., 2017), or even a headband with dry electrodes affixed near the forehead and back of head (Arnal et al., 2020), or even measurement near the ear (known as cEEGrid) containing multiple electrodes arranged in a structure that attaches behind the ear in an arc (Sterr et al., 2018) or even over earplugs (Mikkelsen et al., 2019).

Monitoring of heart-activity is another important parameter during the sleep process as the heart rate decreases as we go from the awake state to N3 state and increases again from REM (rapid eye movement) state to normal as we awaken. ECG measurement is regarded as the gold standard for this measurement. Yoon et al. in 2018 developed a compact wireless ECG patch with integrated 3-axis accelerometer to monitor heart rate as well as body movements during sleep (Yoon et al., 2018). Ilen et al. in 2019 developed a pull-up-pant type wearable device for infants with flexible fabric based ECG sensors to be worn over the abdomen region along with integrated accelerometer, gyroscope, and a magnetic sensor and clearly distinguished between REM, non-REM, and awake states (Ilen et al., 2019).

Pulse measurement has also recently become very popular due to small and light weight devices developed to be fitted at the fingertip (Hartmann et al., 2019) or as ring (Kinnunen et al., 2020) and can also be used to study heart rate (HR) and heart rate variability (HRV). Manoni et al. in 2019 developed a nose mounted PPG device with accelerometer for detection of sleep apnea along with different postures while sleeping (Manoni et al., 2020).

Respiration monitoring is another important parameter for measurement during sleep monitoring to detect any abnormality caused due to sleep apnea and hypopnea. Respiratory inductance plethysmograph (RIP) belts are most common in this sense and much comfortable than nose mounted sensors for monitoring air flow through nostrils. In this regard, Ding et al. developed a TENG based chest belts for recording RIP with high accuracy (Ding et al., 2018). Ramı'rez et al., 2019 further miniaturized the same through a stretchable tattoo-like strain sensor having a composite of graphene, ultrathin granular palladium, and PEDOT:PSS with enhanced piezoresistivity and enclosed within a PDMS coating for mounting over the skin surface for measuring heartbeat and breath rate (Ramírez et al., 2019).

Although the above examples highlight the advancements made in monitoring of sleep patterns due to modern technologies, still clinical implementation will require time with further advancements in device reliability, on-skin packaging, data retrieval and monitoring, etc. with reduced form factor (Kwon et al., 2021).

5.5 Detection of Cancer Biomarkers

Cancer is a universal disease with a high mortality rate and its instances are expected to double within the next 50 years globally (Sanko & Kuralay, 2023). Each cancer has its unique molecular change such as up/downregulations of biological molecules in cancer cells. These biological molecules have been identified as biomarkers and used as a target analyte for cancer diagnosis which may be specifically, DNA, RNA, hormones, enzymes, or even certain cells present in fluids such as urine, serum, blood, or plasma (Sanko & Kuralay, 2023). Since the level of cancer biomarkers is very insufficient at an early stage cancer, there is a need for technological advances that can be more accurate, ultra-sensitive, trustworthy, and selective to biomarkers. Past decade has witnessed development of innovative, economically feasible strategies for detecting cancer biomarkers over electrochemical, optical, etc. platforms demonstrating high sensitivity, low limit of detection, excellent specificity, etc.

Electrochemical detection technique forms the most preferred sensing modality due to its high accuracy and in particular label-free detection techniques are further preferred due to simplicity of the process and reducing label-based chances of contamination. Various biorecognition elements are used in this regard like nucleic-acids, aptamers, antibodies, whole cells, etc. (Sanko & Kuralay, 2023). Pothipr et al. in 2022 developed a label-free electrochemical sensor for detection of breast cancer using two screen printed carbon electrodes decorated with AuNP-dye/ poly(3-aminobenzylamine)/$MoSe_2$ for simultaneous detection of its antigen 15–3 and microRNA-21 for enhanced accuracy. The sensor demonstrated a detection limit of 1.2 fM for microRNA-21 detection (Pothipor et al., 2022). Zhang F et al. designed a label-free aptasensor based electrochemical device for detection of cancer antigen 125, by immobilizing the aptamers over nickel hexacyanoferrate nanocubes/polydopamine functionalized graphene electrodes which showed a lower detectable limit of 0.076 pg/mL (Zhang et al., 2022c). An antibody based electrochemical sensor for detection of lung cancer was developed by Zhang Y et al. through detection of cytokeratin fragment antigen 21–1 (CYFRA 21-1). An antibody-antigen-antibody sandwich structure was formed over a gold electrode coated with poly(ε-caprolactone)-b-poly(ethylene oxide) block copolymer and demonstrated a lower limit of detection of 0.25 pg/mL (Zhang et al., 2021). Above advancements highlight the growing need for development of rapid POC based early detection of cancer biomarkers. Further incorporation of AI/ML in its early prediction may assists similar devices further and along with integration over the internet advancement of such diseases may be tracked even remotely highlighting further insights into its progression mechanisms.

6 Use of AI/ML in Wearable Sensing

POC biosensors have several advantages, including lower sample volume, on-site screening, and cost-effectiveness. It also eliminates the requirement for skilled workers in traditional lab-based testing, making it a very valuable tool in a variety of healthcare settings. The integration of biosensors with Artificial Intelligence (AI) and Machine Learning (ML) has also been critical to the healthcare revolution (Bhaiyya et al., 2024). There are various applications and transformational possibilities of AI/ML in enhancing performances of POC biosensors and wearables. AI/ML powered algorithms can analyze and understand complex biological data as well as improve the accuracy, sensitivity, and speed of diagnostic processes in a wide range of healthcare settings. First, as these algorithms can manage enormous amounts of complex biological data, the corresponding biosensors can detect and understand even minor changes in biomarkers which in turn reduce the time spent by medical practitioners in spotting problems at the optimal moment for intervention, critical in the early detection of various diseases. Second, the rapid data processing capabilities of ML-based biosensors are of significant advantage particularly during medical crisis, outbreaks, infectious disease treatment, etc. Biosensors powered by machine learning produce results in near real time, allowing for faster clinical decisions and more effective patient care. Third, ML technology is versatile and adaptive, providing solutions to a wide range of diagnostic difficulties (Bhaiyya et al., 2024).

Machine learning is usually divided into three categories: supervised learning, unsupervised learning, and reinforcement learning. Each type serves a specific purpose and is appropriate for a variety of applications (Morales & Escalante, 2022). Supervised ML is a type of ML where algorithm is trained on a labelled dataset, indicating that the input data is associated with appropriate output labels. The objective of supervised learning is to develop a function that correlates input variables with corresponding output variables. Unsupervised learning is the process of training a model using a data set without labelled responses. This model tried to identify patterns or structures (cluster or association) in the input data such. Reinforcement learning involves training a model through interactions with an environment. The model learns a policy that maximizes cumulative rewards by taking actions and receiving feedback in the form of rewards or penalties. Among three, supervised learning models provide good sensitivity, specificity, and accuracy by carefully training on labelled data sets, where each input (features) is paired with the correct output (label) (Morales & Escalante, 2022). This makes it possible for the model to precisely understand the relationship between inputs and outputs. Sensitivity (true positive rate) is improved by training the model with a diverse set of positive cases, allowing it to identify patterns linked to these cases effectively. Specificity (true negative rate) is increased by ensuring that the model is trained with sufficient negative cases, helping it distinguish between positive and negative outcomes accurately. Finally, learning both positive and negative situations improves accuracy, the model's overall correctness, and improves overall prediction ability of the supervised ML models (Saboor et al., 2022). In addition, supervised ML models provide better performance as compared to others because of quality and diversity of training data, feature selection and engineering, the choice of algorithm and its optimization, and regular evaluation and validation of the model through techniques like cross-validation, threshold adjustment, and hyper parameter tuning. Integration of AI-ML driven diagnostic solutions in POC devices promises great future in disease identification and POC medical treatment. However, here it may be worth mentioning that these technologies need to go through approvals from comprehensive regulatory bodies like FDA, CE markings, etc. and clinical validation to ascertain its safety, accuracy, and reliability before it can be deployed into the market (Bhaiyya et al., 2024).

Among the plethora of POC biosensors, electrochemical sensors being the most commonly used one finds numerous applications for processing quickly large, complicated data sets and identifying minute patterns that may indicate a disease with remarkable sensitivity. For instance, Narayana et al. in 2023 developed lidocaine detection (Kadian et al., 2024) sensor using a wireless microneedle array and demonstrated a detection limit of 0.13 μM with high sensitivity and selectivity toward lidocaine under optimal condition. ML algorithms accurately predicted the concentration and used digital visualization of lidocaine concentration by developing a web application. Similarly, ML-assisted POC sensors have been used in several other electrochemical sensors for prediction of glucose, bacterial detection, portable nitrate biosensors, SARS-CoV-2 (with over 99% accuracy), etc. (Bhaiyya et al., 2024). Similarly ML has been also used for the development of various Lab-on-chip based colorimetric sensors, electrochemiluminescence sensors, e-nose, and other POC sensors (Bhaiyya et al., 2024).

7 Overview and Future Scope of Remote Patient Monitoring Systems

Present chapter provides an overview of the potential application of modern wearable sensors in remote monitoring of body ailments from detection of biochemical and biophysical parameters to even POC detection of several bacterial or viral pathogens as well as identification of cancer biomarkers. Although the later doesn't pertain directly towards applicability of remote monitoring, still requirement of disease diagnosis may demand acquired data transmission to remote medical experts or even complex processing for elucidation of vital information which in turn requires data transmission and connectivity as any other remote monitoring systems. Various computing modalities have also been briefed here highlighting the importance and utility of technological advancements in recording body vitals as well as disease screening along with safe and secured data transmission and processing for further validation and follow-up by medical experts. Thus both computing modalities as well as wearable sensors assist each other in fulfilling the goals of remote health monitoring for an effective care from infants to elderly persons within the ambiance of a home aiming towards a non-intrusive and non-invasive, continuous monitoring of vital health parameters at a relatively low-cost.

However, many challenges still persist that needs to be effectively implemented for its wider acceptance by common masses as highlighted below (Majumder et al., 2017). Firstly, bio-potential signals like ECG, etc. acquired through flexible electrodes, suffer from large motion artifacts resulting in low signal-to-noise ratio and need to be improved to obtain reliable and continuous signals while at the same time enabling the subjects to do their routine work simultaneously. Further, use of novel signal processing techniques may even be explored to improve the SNR. Secondly, need for high computational resources by WBANs connected to multiple sensors over the body may be another limitation on its acceptance. Thirdly, all connected devices should be low powered or preferably utilize energy harvesting modes to account for un-obtrusive power requirement for continuous monitoring applications. Fourthly, efficient privacy and security protocols must be ascertained to ensure protection of sensitive medical data. Finally, the entire remote monitoring system should be cost effective in the sense that it doesn't exceed traditional hospital admission expenses and becomes affordable by the general public which otherwise would defy the sole purpose of personalized healthcare.

References

Alam, M. M., & Ben Hamida, E. (2015). Strategies for optimal MAC parameters tuning in IEEE 802.15.6 wearable wireless sensor networks. *Journal of Medical Systems, 39*, 106.

Al-Hawary, S. I. S., Althomali, R. H., Elov, B. B., Hussn, M., Sapaev, I. B., Obaid, R. F., et al. (2023). Biomedical applications of smartphone-based lateral flow detection systems as a diagnosis tool. *Microchemical Journal, 193*, 109159.

Alhogail, S., Suaifan, G. A. R. Y., & Zourob, M. (2016). Rapid colorimetric sensing platform for the detection of Listeria monocytogenes foodborne pathogen. *Biosensors and Bioelectronics, 86*, 1061–1066.

Arnal, P. J., Thorey, V., Debellemaniere, E., Ballard, M. E., Bou Hernandez, A., Guillot, A., et al. (2020). The Dreem Headband compared to polysomnography for electroencephalographic signal acquisition and sleep staging. *Sleep, 43*.

Ates, H. C., Nguyen, P. Q., Gonzalez-Macia, L., Morales-Narváez, E., Güder, F., Collins, J. J., et al. (2022). End-to-end design of wearable sensors. *Nature Reviews Materials, 7*, 887–907.

Banik, S., Melanthota, S. K., Arbaaz, J. M., Vaz, V. M., Kadambalithaya, I. H., et al. (2021). Recent trends in smartphone-based detection for biomedical applications: a review. *Analytical and Bioanalytical Chemistry, 413*, 2389–2406.

Berry, R. B., Brooks, R., Gamaldo, C., Harding, S. M., Lloyd, R. M., Quan, S. F., et al. (2017). AASM scoring manual updates for 2017 (Version 2.4). *Journal of Clinical Sleep Medicine, 13*, 665–666.

Bhaiyya, M., Panigrahi, D., Rewatkar, P., & Haick, H. (2024). Role of machine learning assisted biosensors in point-of-care-testing for clinical decisions. *ACS Sensors, 9*, 4495–4519.

Bhatti, D. S., Saleem, S., Imran, A., Iqbal, Z., Alzahrani, A., Kim, H., et al. (2022). A survey on wireless wearable body area networks: A perspective of technology and economy. *Sensors, 22*, 7722.

Carneiro, M. R., Almeida, A. T. d., & Tavakoli, M. (2020). Wearable and comfortable e-textile headband for long-term acquisition of forehead EEG signals. *IEEE Sensors Journal, 20*, 15107–15116.

Caudell, T. P., & Mizell, D. W. (1992). Augmented reality: an application of heads-up display technology to manual manufacturing processes. In *Proceedings of the twenty-fifth Hawaii international conference on system sciences* (Vol. 2, pp. 659–669).

Cesaretti, M., Poté, N., Dondero, F., Cauchy, F., Schneck, A. S., Soubrane, O., et al. (2018). Testing feasibility of an accurate microscopic assessment of macrovesicular steatosis in liver allograft biopsies by smartphone add-on lenses. *Microscopy Research and Technique, 81*, 58–63.

Chang, J., Arbeláez, P., Switz, N., Reber, C., Tapley, A., Davis, J. L., et al. (2012). Automated tuberculosis diagnosis using fluorescence images from a mobile microscope. *Medical Image Computing and Computer-Assisted Intervention, 15*, 345–352.

Chen, Q., Li, H., Ouyang, Q., & Zhao, J. (2014). Identification of spoilage bacteria using a simple colorimetric sensor array. *Sensors and Actuators B: Chemical, 205*, 1–8.

Chen, Z., Zhang, Z., Zhai, X., Li, Y., Lin, L., Zhao, H., et al. (2020). Rapid and sensitive detection of anti-SARS-CoV-2 IgG, using lanthanide-doped nanoparticles-based lateral flow immunoassay. *Analytical Chemistry, 92*, 7226–7231.

Choi, A., Kim, J. Y., Jo, S., Jee, J. H., Heymsfield, S. B., Bhagat, Y. A., et al. (2015a). Smartphone-based bioelectrical impedance analysis devices for daily obesity management. *Sensors, 15*, 22151–22166.

Choi, C., Kim, S. H., Sim, H. J., Lee, J. A., Choi, A. Y., Kim, Y. T., et al. (2015b). Stretchable, weavable coiled carbon nanotube/MnO2/polymer fiber solid-state supercapacitors. *Scientific Reports, 5*, 9387.

Choi, Y. S., Hsueh, Y.-Y., Koo, J., Yang, Q., Avila, R., Hu, B., et al. (2020). Stretchable, dynamic covalent polymers for soft, long-lived bioresorbable electronic stimulators designed to facilitate neuromuscular regeneration. *Nature Communications, 11*, 5990.

Chung, H. U., Kim, B. H., Lee, J. Y., Lee, J., Xie, Z., Ibler, E. M., et al. (2019). Binodal, wireless epidermal electronic systems with in-sensor analytics for neonatal intensive care. *Science, 363*, eaau0780.

Cliff, R. (2005). *Wearable computing: A review*. University of Bristol.

Conti, S., Nepa, F., Pascoli, S. D., Brunetti, I., Pimpolari, L., Song, X., et al. (2023). Hybrid flexible NFC sensor on paper. *IEEE Journal on Flexible Electronics, 2*, 4–10.

Covaci, C., & Gontean, A. (2020). Piezoelectric energy harvesting solutions: A review. *Sensors, 20*, 3512.

Dias, D., & Cunha, J. P. S. (2018). Wearable health devices—Vital sign monitoring, systems and technologies. *Sensors, 18*, 2414.

Ding, X., Cao, H., Zhang, X., Li, M., & Liu, Y. (2018). Large scale triboelectric nanogenerator and self-powered flexible sensor for human sleep monitoring. *Sensors, 18*, 1713.

Doupis, J., Festas, G., Tsilivigos, C., Efthymiou, V., & Kokkinos, A. (2020). Smartphone-based technology in diabetes management. *Diabetes Therapy, 11*, 607–619.

Dungchai, W., Chailapakul, O., & Henry, C. S. (2009). Electrochemical detection for paper-based microfluidics. *Analytical Chemistry, 81*, 5821–5826.

Ehsan, M. A., Khan, S. A., & Rehman, A. (2021). Screen-printed graphene/carbon electrodes on paper substrates as impedance sensors for detection of coronavirus in nasopharyngeal fluid samples. *Diagnostics, 11*, 1030.

El Gharbi, M., Fernández-García, R., Ahyoud, S., & Gil, I. (2020). A review of flexible wearable antenna sensors: design, fabrication methods, and applications. *Materials, 13*, 3781.

Feng, L.-X., Tang, C., Han, X.-X., Zhang, H.-C., Guo, F.-N., Yang, T., et al. (2021). Simultaneous and sensitive detection of multiple small biological molecules by microfluidic paper-based analytical device integrated with zinc oxide nanorods. *Talanta, 232*, 122499.

Gajarawala, S. N., & Pelkowski, J. N. (2021). Telehealth benefits and barriers. *The Journal for Nurse Practitioners, 17*, 218–221.

Galan, E. A., Zhao, H., Wang, X., Dai, Q., Huck, W. T. S., & Ma, S. (2020). Intelligent microfluidics: The convergence of machine learning and microfluidics in materials science and biomedicine. *Matter, 3*, 1893–1922.

Gljušćić, P., Zelenika, S., Blažević, D., & Kamenar, E. (2019). Kinetic energy harvesting for wearable medical sensors. *Sensors, 19*, 4922.

Gnanasambanthan, H. G., & Maji, D. (2023). Development of a flexible and wearable microelectrode array patch using a screen-printed masking technique for accelerated wound healing. *ACS Applied Electronic Materials, 5*, 4426–4436.

Gogola, J. L., Martins, G., Gevaerd, A., Blanes, L., Cardoso, J., Marchini, F. K., et al. (2021). Label-free aptasensor for p24-HIV protein detection based on graphene quantum dots as an electrochemical signal amplifier. *Analytica Chimica Acta, 1166*, 338548.

Hartmann, V., Liu, H., Chen, F., Qiu, Q., Hughes, S., & Zheng, D. (2019). Quantitative comparison of photoplethysmographic waveform characteristics: effect of measurement site. *Frontiers in Physiology, 10*, 198.

Ilen, E., Leinonen, N. A., Ranta, J., Airaksinen, M., Haataja, L., & Vanhatalo, S. (2019). User experience of wearable infant sleep monitoring system for medical research and diagnostics. In *Proceedings of the 19th World Textile Conference* (p. 6). Ghent University, Ghent.

Jalal, U. M., Jin, G. J., & Shim, J. S. (2017). Paper–plastic hybrid microfluidic device for smartphone-based colorimetric analysis of urine. *Analytical Chemistry, 89*, 13160–13166.

Jeong, J.-W., Lee, W., & Kim, Y.-J. (2022). A real-time wearable physiological monitoring system for home based healthcare applications. *Sensors, 22*, 104.

Kadian, S., Sahoo, S. S., Kumari, P., & Narayan, R. J. (2024). Machine learning enabled onsite electrochemical detection of lidocaine using a microneedle array integrated screen printed electrode. *Electrochimica Acta, 475*, 143664.

Kang, S.-K., Murphy, R. K. J., Hwang, S.-W., Lee, S. M., Harburg, D. V., Krueger, N. A., et al. (2016). Bioresorbable silicon electronic sensors for the brain. *Nature, 530*, 71–76.

Kang, M. H., Lee, G. J., Yun, J. H., & Song, Y. M. (2021). NFC-based wearable optoelectronics working with smartphone application for untact healthcare. *Sensors, 21*, 878.

Khan, R. A., & Pathan, A.-S. K. (2018). The state-of-the-art wireless body area sensor networks: A survey. *International Journal of Distributed Sensor Networks, 14*, 1550147718768994.

Kim, D.-H., Lu, N., Ma, R., Kim, Y.-S., Kim, R.-H., Wang, S., et al. (2011). Epidermal electronics. *Science, 333*, 838–843.

Kim, J., Gutruf, P., Chiarelli, A. M., Heo, S. Y., Cho, K., Xie, Z., et al. (2017). Miniaturized battery-free wireless systems for wearable pulse oximetry. *Advanced Functional Materials, 27*, 1604373.

Kinnunen, H., Rantanen, A., Kenttä, T., & Koskimäki, H. (2020). Feasible assessment of recovery and cardiovascular health: accuracy of nocturnal HR and HRV assessed via ring PPG in comparison to medical grade ECG. *Physiological Measurement, 41*, 04NT01.

Kuzubasoglu, B. A., Sayar, E., Cochrane, C., Koncar, V., & Bahadir, S. K. (2021). Wearable temperature sensor for human body temperature detection. *Journal of Materials Science: Materials in Electronics, 32*, 4784–4797.

Kwon, S., Kim, H., & Yeo, W.-H. (2021). Recent advances in wearable sensors and portable electronics for sleep monitoring. *iScience, 24*, 102461.

Lambert, V. (2016). Empowering patients to track their health through wearable tech. Ph.D. Dissertation. Available online: https://www.raconteur.net/empowering-patients-to-track-their-health-through-wearable-tech/. Accessed 14 Jan 2024.

Lee, S. P., Ha, G., Wright, D. E., Ma, Y., Sen-Gupta, E., Haubrich, N. R., et al. (2018). Highly flexible, wearable, and disposable cardiac biosensors for remote and ambulatory monitoring. *npj Digital Medicine, 1*, 2.

Lee, A. S., Kim, S. M., Kim, K. R., Park, C., Lee, D.-G., Heo, H. R., et al. (2023). A colorimetric lateral flow immunoassay based on oriented antibody immobilization for sensitive detection of SARS-CoV-2. *Sensors and Actuators B: Chemical, 379*, 133245.

Lee, T., Park, J., Oh, S. H., Cheong, D. Y., Roh, S., You, J. H., et al. (2024). Glucose oxidase activity colorimetric assay using redox-sensitive electrochromic nanoparticle-functionalized paper sensors. *ACS Omega, 9*, 15493–15501.

Leu, F. Y., Ko, C. Y., You, I., Choo, K.-K. R., & Ho, C.-L. (2018). A smartphone-based wearable sensors for monitoring real-time physiological data. *Computers & Electrical Engineering, 65*, 376–392.

Levendowski, D. J., Ferini-Strambi, L., Gamaldo, C., Cetel, M., Rosenberg, R., & Westbrook, P. R. (2017). The accuracy, night-to-night variability, and stability of frontopolar sleep electroencephalography biomarkers. *Journal of Clinical Sleep Medicine, 13*, 791–803.

Li, X., Zhao, C., & Liu, X. (2015). A paper-based microfluidic biosensor integrating zinc oxide nanowires for electrochemical glucose detection. *Microsystems & Nanoengineering, 1*, 15014.
Li, S., Meng, H.-M., Zong, H., Chen, J., Li, J., Zhang, L., et al. (2021). Entropy-driven amplification strategy-assisted lateral flow assay biosensor for ultrasensitive and convenient detection of nucleic acids. *Analyst, 146*, 1668–1674.
Li, T., Liang, B., Ye, Z., Zhang, L., Xu, S., Tu, T., et al. (2022). An integrated and conductive hydrogel-paper patch for simultaneous sensing of chemical–electrophysiological signals. *Biosensors and Bioelectronics, 198*, 113855.
Lin, R., Kim, H.-J., Achavananthadith, S., Xiong, Z., Lee, J. K. W., Kong, Y. L., et al. (2022). Digitally-embroidered liquid metal electronic textiles for wearable wireless systems. *Nature Communications, 13*, 2190.
Lingadharini, P., & Maji, D. (2025). Eco-sustainable point-of-care devices: Progress in paper and fabric based electrochemical and colorimetric biosensors. *Talanta, 285*, 127397.
Lingadharini, P., Shirsat, A., Gardi, P., Kore, S., Joshi, V., Patra, R., et al. (2024). A cost-effective and facile technique for realizing fabric based microfluidic channels using beeswax and PVC stencils. *Analytical Methods, 16*, 3372–3384.
Liu, H., Jiang, H., Du, F., Zhang, D., Li, Z., & Zhou, H. (2017). Flexible and degradable paper-based strain sensor with low cost. *ACS Sustainable Chemistry & Engineering, 5*, 10538–10543.
Liu, Z., Li, H., Shi, B., Fan, Y., Wang, Z. L., & Li, Z. (2019). Wearable and implantable triboelectric nanogenerators. *Advanced Functional Materials, 29*, 1808820.
Liu, J., Qin, Q., Zhang, X., Li, C., Yu, Y., Huang, X., et al. (2020). Development of a novel lateral flow biosensor combined with aptamer-based isolation: Application for rapid detection of grouper nervous necrosis virus. *Frontiers in Microbiology, 11*, 886.
Liu, J., Xing, Y., Xue, B., & Zhou, X. (2022). Nanozyme enhanced paper-based biochip with a smartphone readout system for rapid detection of cyanotoxins in water. *Biosensors and Bioelectronics, 205*, 114099.
Majumder, S., Mondal, T., & Deen, M. J. (2017). Wearable sensors for remote health monitoring. *Sensors, 17*, 130.
Manjakkal, L., Yin, L., Nathan, A., Wang, J., & Dahiya, R. (2021). Energy autonomous sweat-based wearable systems. *Advanced Materials, 33*, 2100899.
Mann, S. (1997). An historical account of the 'WearComp' and 'WearCam' inventions developed for applications in 'personal imaging'. In *Digest of Papers. First International Symposium on Wearable Computers* (pp. 66–73).
Manoni, A., Loreti, F., Radicioni, V., Pellegrino, D., Della Torre, L., Gumiero, A., et al. (2020). A new wearable system for home sleep apnea testing, screening, and classification. *Sensors, 20*, 7014.
Masihi, S., Panahi, M., Maddipatla, D., Hanson, A. J., Fenech, S., Bonek, L., et al. (2022). Development of a flexible wireless ecg monitoring device with dry fabric electrodes for wearable applications. *IEEE Sensors Journal, 22*, 11223–11232.
Mazur, F., Tran, H., Kuchel, R. P., & Chandrawati, R. (2020). Rapid detection of listeriolysin o toxin based on a nanoscale liposome–gold nanoparticle platform. *ACS Applied Nano Materials, 3*, 7270–7280.
Mazur, F., Tjandra, A. D., Zhou, Y., Gao, Y., & Chandrawati, R. (2023). Paper-based sensors for bacteria detection. *Nature Reviews Bioengineering, 1*, 180–192.
Mikkelsen, K. B., Tabar, Y. R., Kappel, S. L., Christensen, C. B., Toft, H. O., Hemmsen, M. C., et al. (2019). Accurate whole-night sleep monitoring with dry-contact ear-EEG. *Scientific Reports, 9*, 16824.
Mohino-Herranz, I., Gil-Pita, R., Ferreira, J., Rosa-Zurera, M., & Seoane, F. (2015). Assessment of mental, emotional and physical stress through analysis of physiological signals using smartphones. *Sensors, 15*, 25607–25627.
Morales, E. F., & Escalante, H. J. (2022). Chapter 6 – A brief introduction to supervised, unsupervised, and reinforcement learning. In A. A. Torres-García, C. A. Reyes-García, L. Villaseñor-Pineda, & O. Mendoza-Montoya (Eds.), *Biosignal processing and classification using computational learning and intelligence* (pp. 111–129). Academic.

Mukhopadhyay, S. C., Suryadevara, N. K., & Nag, A. (2022). Wearable sensors for healthcare: Fabrication to application. *Sensors, 22*, 5137.

Naseri, M., Ziora, Z. M., Simon, G. P., & Batchelor, W. (2022). ASSURED-compliant point-of-care diagnostics for the detection of human viral infections. *Reviews in Medical Virology, 32*, e2263.

Nie, H.-K., Xuan, X.-W., & Ren, G.-J. (2021). Wearable antenna pressure sensor with electromagnetic bandgap for elderly fall monitoring. *AEU – International Journal of Electronics and Communications, 138*, 153861.

Nitha, P. K., & Chandrasekhar, A. (2023). Marriage between metal-organic frameworks/covalent-organic frameworks and triboelectric nanogenerator for energy harvesting —A review. *Materials Today Energy, 37*, 101393.

Nontawong, N., Ngaosri, P., Chunta, S., Jarujamrus, P., Nacapricha, D., Lieberzeit, P. A., et al. (2022). Smart sensor for assessment of oxidative/nitrative stress biomarkers using a dual-imprinted electrochemical paper-based analytical device. *Analytica Chimica Acta, 1191*, 339363.

Nown, T. H., Upadhyay, P., Kerr, A., Andonovic, I., Tachtatzis, C., & Grealy, M. A. (2023). A mapping review of real-time movement sonification systems for movement rehabilitation. *IEEE Reviews in Biomedical Engineering, 16*, 672–686.

Olenik, S., Lee, H. S., & Güder, F. (2021). The future of near-field communication-based wireless sensing. *Nature Reviews Materials, 6*, 286–288.

Ongaro, A. E., Ndlovu, Z., Sollier, E., Otieno, C., Ondoa, P., Street, A., et al. (2022). Engineering a sustainable future for point-of-care diagnostics and single-use microfluidic devices. *Lab on a Chip, 22*, 3122–3137.

Ortiz-Gómez, I., Salinas-Castillo, A., García, A. G., Álvarez-Bermejo, J. A., de Orbe-Payá, I., Rodríguez-Diéguez, A., et al. (2017). Microfluidic paper-based device for colorimetric determination of glucose based on a metal-organic framework acting as peroxidase mimetic. *Microchimica Acta, 185*, 47.

Ozer, T., & Henry, C. S. (2021). Paper-based analytical devices for virus detection: Recent strategies for current and future pandemics. *TrAC Trends in Analytical Chemistry, 144*, 116424.

Pan, S., & Zhang, Z. (2019). Fundamental theories and basic principles of triboelectric effect: A review. *Friction, 7*, 2–17.

Pothipor, C., Bamrungsap, S., Jakmunee, J., & Ounnunkad, K. (2022). A gold nanoparticle-dye/poly(3-aminobenzylamine)/two dimensional MoSe2/graphene oxide electrode towards label-free electrochemical biosensor for simultaneous dual-mode detection of cancer antigen 15-3 and microRNA-21. *Colloids and Surfaces B: Biointerfaces, 210*, 112260.

Pramanik, P. K. D., Nayyar, A., & Pareek, G. (2019). Chapter 7 – WBAN: Driving e-healthcare beyond telemedicine to remote health monitoring: Architecture and protocols. In H. D. Jude & V. E. Balas (Eds.), *Telemedicine technologies* (pp. 89–119). Academic.

Ramalingam, S., Collier, C. M., & Singh, A. (2021). A paper-based colorimetric aptasensor for the detection of gentamicin. *Biosensors, 11*, 29.

Ramírez, J., Rodriquez, D., Urbina, A. D., Cardenas, A. M., & Lipomi, D. J. (2019). Combining high sensitivity and dynamic range: wearable thin-film composite strain sensors of graphene, ultrathin palladium, and PEDOT:PSS. *ACS Applied Nano Materials, 2*, 2222–2229.

Ren, W., Sun, Y., Zhao, D., Aili, A., Zhang, S., Shi, C., et al. (2021). High-performance wearable thermoelectric generator with self-healing, recycling, and Lego-like reconfiguring capabilities. *Science Advances, 7*, eabe0586.

Saboor, A., Usman, M., Ali, S., Samad, A., Abrar, M. F., & Ullah, N. (2022). A method for improving prediction of human heart disease using machine learning algorithms. *Mobile Information Systems, 2022*, 1410169.

Sadri, B., Goswami, D., & Martinez, R. V. (2018). Rapid fabrication of epidermal paper-based electronic devices using razor printing. *Micromachines, 9*, 420.

Saminathan, S., & Geetha, K. (2018). Real-time health care monitoring system using IoT. *International Journal of Engineering & Technology, 7*, 484–488.

Sanko, V., & Kuralay, F. (2023). Label-free electrochemical biosensor platforms for cancer diagnosis: Recent achievements and challenges. *Biosensors, 13*, 333.

Sempionatto, J. R., Lasalde-Ramírez, J. A., Mahato, K., Wang, J., & Gao, W. (2022). Wearable chemical sensors for biomarker discovery in the omics era. *Nature Reviews Chemistry, 6*, 899–915.

Song, Y., Mukasa, D., Zhang, H., & Gao, W. (2021). Self-Powered Wearable Biosensors. *Accounts of Materials Research, 2*, 184–197.

Sterr, A., Ebajemito, J. K., Mikkelsen, K. B., Bonmati-Carrion, M. A., Santhi, N., della Monica, C., et al. (2018). Sleep EEG derived from behind-the-ear electrodes (cEEGrid) compared to standard polysomnography: A proof of concept study. *Frontiers in Human Neuroscience, 12*, 452.

Teng, L., Ye, S., Handschuh-Wang, S., Zhou, X., Gan, T., & Zhou, X. (2019). Liquid metal-based transient circuits for flexible and recyclable electronics. *Advanced Functional Materials, 29*, 1808739.

Thap, T., Chung, H., Jeong, C., Hwang, K.-E., Kim, H.-R., Yoon, K.-H., et al. (2016). High-resolution time-frequency spectrum-based lung function test from a smartphone microphone. *Sensors, 16*, 1305.

Thiyagarajan, K., Rajini, G. K., & Maji, D. (2021). Flexible, highly sensitive paper-based screen printed MWCNT/PDMS composite breath sensor for human respiration monitoring. *IEEE Sensors Journal, 21*, 13985–13995.

Thiyagarajan, K., Rajini, G. K., & Maji, D. (2022a). Fully printed MWCNT strain sensor over paper substrate for human motion monitoring. *Flexible and Printed Electronics, 7*, 045003.

Thiyagarajan, K., Rajini, G. K., & Maji, D. (2022b). Cost-effective, disposable, flexible, and printable MWCNT-based wearable sensor for human body temperature monitoring. *IEEE Sensors Journal, 22*, 16756–16763.

Turanli, M., Ilhan, I., & Yavsan, E. (2024). Development of wearable device and synchronized mobile application to monitor vital signs in real time. *Multimedia Tools and Applications*.

Uthoff, R. D., Song, B., Sunny, S., Patrick, S., Suresh, A., Kolur, T., et al. (2018). Point-of-care, smartphone-based, dual-modality, dual-view, oral cancer screening device with neural network classification for low-resource communities. *PLoS One, 13*, e0207493.

Vaghasiya, J. V., Mayorga-Martinez, C. C., & Pumera, M. (2023). Wearable sensors for telehealth based on emerging materials and nanoarchitectonics. *npj Flexible Electronics, 7*, 26.

Valentine, C. J., Takagishi, K., Umezu, S., Daly, R., & De Volder, M. (2020). Paper-based electrochemical sensors using paper as a Scaffold to create porous carbon nanotube electrodes. *ACS Applied Materials & Interfaces, 12*, 30680–30685.

Valera, E., Jankelow, A., Lim, J., Kindratenko, V., Ganguli, A., White, K., et al. (2021). COVID-19 point-of-care diagnostics: Present and future. *ACS Nano, 15*, 7899–7906.

Wall, C., Hetherington, V., & Godfrey, A. (2023). Beyond the clinic: The rise of wearables and smartphones in decentralising healthcare. *npj Digital Medicine, 6*, 219.

Wang, C.-C., Hennek, J. W., Ainla, A., Kumar, A. A., Lan, W.-J., Im, J., et al. (2016). A paper-based "pop-up" electrochemical device for analysis of beta-hydroxybutyrate. *Analytical Chemistry, 88*, 6326–6333.

Wang, L., Wang, L., Zhang, Y., Pan, J., Li, S., Sun, X., et al. (2018). Weaving sensing fibers into electrochemical fabric for real-time health monitoring. *Advanced Functional Materials, 28*, 1804456.

Wu, T., Wu, X., Li, L., Hao, M., Wu, G., Zhang, T., et al. (2020). Anisotropic boron–carbon hetero-nanosheets for ultrahigh energy density supercapacitors. *Angewandte Chemie International Edition, 59*, 23800–23809.

Wu, T., Shen, C., Zhao, Z., Lyu, M., Bai, H., Hu, X., et al. (2024). Integrating paper-based microfluidics and lateral flow strip into nucleic acid amplification device toward rapid, low-cost, and visual diagnosis of multiple mycobacteria. *Small Methods, 8*, 2400095.

Xu, G., Cheng, C., Yuan, W., Liu, Z., Zhu, L., Li, X., et al. (2019). Smartphone-based battery-free and flexible electrochemical patch for calcium and chloride ions detections in biofluids. *Sensors and Actuators B: Chemical, 297*, 126743.

Xue, H., Li, F., Zhao, H., Lin, X., & Zhang, T. (2022). A paper-based iontronic capacitive pressure sensor for human muscle motion monitoring. *IEEE Electron Device Letters, 43*, 2165–2168.
Yin, L., Kim, K. N., Lv, J., Tehrani, F., Lin, M., Lin, Z., et al. (2021). A self-sustainable wearable multi-modular E-textile bioenergy microgrid system. *Nature Communications, 12*, 1542.
Yoon, H., Hwang, S. H., Choi, S. H., Choi, J.-W., Lee, Y. J., Jeong, D.-U., et al. (2018). Wakefulness evaluation during sleep for healthy subjects and OSA patients using a patch-type device. *Computer Methods and Programs in Biomedicine, 155*, 127–138.
Yu, X., Xie, Z., Yu, Y., Lee, J., Vazquez-Guardado, A., Luan, H., et al. (2019). Skin-integrated wireless haptic interfaces for virtual and augmented reality. *Nature, 575*, 473–479.
Zhang, Y., Wang, X., Fang, X., Yuan, X., Yang, H., & Kong, J. (2021). Label-free electrochemical immunoassay for detecting CYFRA 21-1 using poly(ε-caprolactone)-b-poly(ethylene oxide) block copolymer. *Microchemical Journal, 165*, 106119.
Zhang, M., Cui, X., & Li, N. (2022a). Smartphone-based mobile biosensors for the point-of-care testing of human metabolites. *Materials Today Bio, 14*, 100254.
Zhang, W.-Y., Zhang, H., & Yang, F.-Q. (2022b). An economical and portable paper-based colorimetric sensor for the determination of hydrogen peroxide-related biomarkers. *Chem, 10*, 335.
Zhang, F., Fan, L., Liu, Z., Han, Y., & Guo, Y. (2022c). A label-free electrochemical aptasensor for the detection of cancer antigen 125 based on nickel hexacyanoferrate nanocubes/polydopamine functionalized graphene. *Journal of Electroanalytical Chemistry, 918*, 116424.
Zhao, Y., Zhai, Q., Dong, D., An, T., Gong, S., Shi, Q., et al. (2019). Highly stretchable and strain-insensitive fiber-based wearable electrochemical biosensor to monitor glucose in the sweat. *Analytical Chemistry, 91*, 6569–6576.
Zhao, R., Gu, Z., Li, P., Zhang, Y., & Song, Y. (2022). Flexible and wearable optoelectronic devices based on perovskites. *Advanced Materials Technologies, 7*, 2101124.
Zheng, J., Zhu, M., Kong, J., Li, Z., Jiang, J., Xi, Y., et al. (2022). Microfluidic paper-based analytical device by using Pt nanoparticles as highly active peroxidase mimic for simultaneous detection of glucose and uric acid with use of a smartphone. *Talanta, 237*, 122954.
Zhu, H., Sencan, I., Wong, J., Dimitrov, S., Tseng, D., Nagashima, K., et al. (2013). Cost-effective and rapid blood analysis on a cell-phone. *Lab on a Chip, 13*, 1282–1288.

Overview of Computational Approaches for Cancer Prognosis

Piyush Agrawal

Abstract

Cancer is the second leading cause of death globally, after heart disease. As of 2022, the United States alone reported 1.9 million new cases and 609,360 deaths. Cancer is a deadly and complex disease that is very heterogeneous in nature, making it very difficult to treat. To improve patient outcomes, it is crucial to comprehend how cancer behaves at the molecular and clinical levels. Prognosis, the prediction of disease progression and how a patient will respond to treatment, plays a vital role in designing precision treatments and optimizing care. However, the complexjbity of cancer necessitates the use of sophisticated technologies for data integration and analysis. Computational approaches in the past few years have shown enormous potential in characterizing probable signatures that can predict the cancer prognosis. In this chapter, we have provided an overview of these computational approaches and the various modalities they require for prognosis prediction. These can be broadly classified into four categories (i) data types, including multi-omics, imaging, and clinical data, (ii) traditional statistical and advanced machine learning approaches employed to predict disease progression and survival outcomes, (iii) gene signatures, with a particular focus on immune gene signatures, and (iv) various online resources, packages, and tools developed related to cancer prognosis. Finally, we discussed the various limitations associated with the current computational approaches and the potential future perspectives to address these challenges.

P. Agrawal (✉)
Division of Medical Research, SRM Medical College Hospital & Research Centre, SRMIST, Chennai, India
e-mail: piyusha@srmist.edu.in

R. Nema et al. (eds.), *Advances in Cancer Detection, Prediction, and Prognosis Using Artificial Intelligence and Machine Learning*,
https://doi.org/10.1007/978-981-96-9346-7_8

Keywords

Cancer · Prognosis · Survival · Machine learning · Multi-omics

1 Introduction

Cancer is a heterogenous disease which happens due to changes in cell behavior, uncontrolled cell division, genomic aberrations such as mutations, etc. (Demir Karaman & Işık, 2023). Cancer remains one of the most pressing global health challenges, with millions of new cases diagnosed each year. The International Agency for Research on Cancer (IARC) (https://gco.iarc.fr/en) reported around 20 million new cancer cases and 9.7 million deaths in 2022, with lung, breast, and colorectal cancer ranking among the top 3 (Bray et al., 2024).

Accurate prognostic knowledge is crucial for guiding decisions and personalizing better treatment; however, due to the heterogenous nature of the cancer, achieving accurate and actionable prognoses in cancer care is complex. In the past, cancer prognosis has relied majorly on histopathological analyses, tumor staging, and clinical features. However, the limitations of these methods lie in their inability to fully capture the intricate molecular details of individual cancer cases, resulting in variable outcomes and limited predictive power.

Recent advancements in high-throughput omics technologies and computational power have led to a new era of precision oncology (Dai & Shen, 2022). Computational techniques allow the integration of massive multi-omics datasets, patient clinical data, and real-world evidence, providing unprecedented prospects for personalized cancer prognosis. Modern system biology approaches and computational techniques, including artificial intelligence (AI) and machine learning (ML), have allowed us to process, analyze, and interpret the vast amount of data generated in oncology research (Sebastian & Peter, 2022).

These modern approaches have allowed for characterizing accurate and key prognostic markers, the construction of robust predictive models, and insights into the biological mechanisms of cancer progression. While computational approaches hold immense potential for improving cancer prognosis, they also present certain challenges that must be addressed to bring these tools into widespread clinical practice, such as data quality, reproducibility, and the interpretability of complex models.

In this chapter, we have provided an overview of the current computational approaches in cancer prognosis, discussing both established and emerging methodologies. We'll explore how different types of data, gene signatures, ML, deep learning (DL), and network-based analytics are changing the cancer prognosis by looking at concrete examples. Ultimately, the chapter delves into the potential future paths and persistent obstacles that lie ahead in applying these innovations to the field of clinical oncology. General framework of this chapter is provided in Fig. 1.

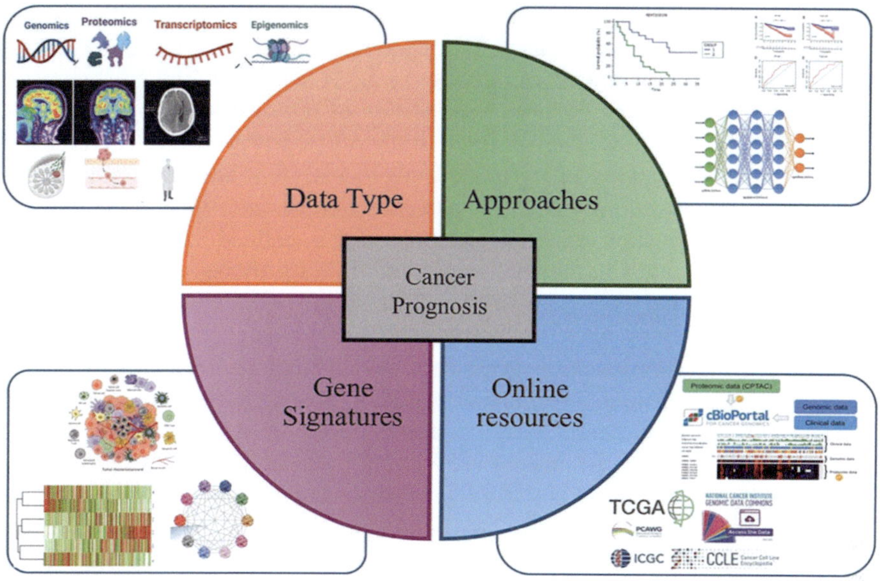

Fig. 1 Overview of framework associated with cancer prognosis prediction. Various data types, approaches, gene signatures, and online resources have been used for predicting prognosis in various cancer types

2 Data Types and Source for Cancer Prognosis

An estimated 29 million new cases of cancer will occur by 2040. For better prognosis and treatment, it is important to categorize tumors into genetically similar subgroups and link these subgroups with clinical outcomes (Weinstein et al., 2013). One or two data types cannot explain the complexity of cancer, as they can only provide limited knowledge about the biological mechanism of the disease, necessitating the utilization of diverse data types to gain comprehensive insight into the disease. Some of these diverse data types include (i) multi-omics data (genome, transcriptome, epigenome, proteome, and metabolome); imaging data; and (iii) clinical data.

Recent advancements in sequencing technologies have created a lot of multi-omics data, mostly related to cancer. For example, the National Cancer Institute-Clinical Proteomic Tumor Analysis Consortium (NCI-CPTAC) has multi-omics data (genomics, transcriptomics, epigenomics, proteomics, and metabolomics) for about 1000 patients with 10 different types of cancer (Li et al., 2023b). Genomics is defined as the study of genomes, which are complete sets of DNAs and all their genes. The data provides information about various mutations, structural variants, copy number variations, etc. Transcriptomics is defined as the study of the transcriptome, i.e., the complete set of RNA transcripts that is expressed in a specific condition or cell. This data provides insights about gene expression patterns in normal vs. diseased conditions, patterns associated with tumor progression and metastasis, splicing patterns, and coding and non-coding RNAs. Epigenomics is the study of chemical modifications that occur on DNA and histone proteins, which are associated with gene regulation, without altering the DNA sequence. The most common form of epigenetic change includes the addition or removal of methyl and acetyl groups. Likewise, proteomics refers to the study of the proteome, i.e., proteins produced or modified in an organism. It allows us to analyze protein expression, post-translational modifications such as phosphorylation and ubiquitination, and protein-protein interactions. Lastly, metabolomics is defined as the study of the metabolites associated with metabolism within the cells, tissues, or organisms and the changes in the metabolic pathways. This data type has allowed us to find more robust biomarkers for prognosis in cancer (Subramanian et al., 2020).

Similar to multi-omics data types, medical imaging technologies in clinical oncology have demonstrated significant potential in cancer diagnosis and prognosis. This data type has recently been a preferred choice over the traditional biopsy-based molecular assays, as cancers are spatially and temporally heterogeneous. Also, in solid cancers, image-based analysis allows capturing intra-tumoral heterogeneity in a non-invasive way. We classify imaging data into two broad categories: (i) radiological data and (ii) pathological data, respectively. Mostly radiological data comes from computed tomography (CT), magnetic resonance imaging (MRI), and positron emission tomography (PET) (Lambin et al., 2012; Vial et al., 2018); on the other hand, pathological data comes from histopathological images taken from tumor tissues.

Next, data types used in cancer prognosis include various clinical data associated with the patient, such as their demography (age, sex, etc.), response to treatment, treatment history, comorbidities, and other lifestyle factors (Zhang et al., 2022b). These clinical data types help researchers and clinicians to understand disease progression, predict patient outcomes, and tailor the intensity of a treatment. Table 1 provides the general information about how these various omics data are associated with cancer prognosis in pan-cancer.

3 Approaches for Prognostic Prediction in Cancer

3.1 Traditional Statistical Approach

Statistical models in the field of medical research and clinical studies have been in use for a long time to understand disease progression and predict patient outcomes. These models outline a structured way for survival data analysis, risk factor identification, and estimating treatment effects. These models use mathematical frameworks to help researchers and clinicians get useful information from censored data. This is common in survival studies where patients drop out or the study's outcome (death or recurrence) hasn't happened by the study's end (Vickers & Cronin, 2010).

Two keystones of conventional statistical survival analysis are the Cox Proportional Hazards Model and the Kaplan-Meier (KM) Estimator.

Table 1 Data types associated with cancer prognosis. Cancer prognosis has been predicted using diverse data types in various cancer types, either as a single data type or by integrating them

Data type	Subtype	Description	Examples
Omics	Genomics	Defined as the study of genomes providing information about mutations, structural variants, copy number variations, etc.	Mutation in *BRCA2, TP53*, in breast, colorectal, ovarian, lung, etc. (Chen et al., 2022). *BRAF, KRAS* mutation in colorectal (Midthun et al., 2019).
	Transcriptomics	Expression of the total set of RNAs in a cell or in a specific condition.	High expression of *DLGAP5, JAG1* in breast (Dickson et al., 2007; Xu et al., 2020). Androgen Receptor (AR) mRNA level in TNBC (Govindan et al., 2020).
	Proteomics	Defined as the study of the total proteome expressed in the cell.	Overexpression of *UPK1B* in bladder cancer; *DEFA5* in colon cancer (Zhou et al., 2020) and *SFTPA2* in lung cancer (Papier et al., 2024).
	Metabolomics	Large-scale study of metabolites and other molecules within cells, biofluids, tissue, and organs.	Altered plasma metabolite is associated with gastric cancer (Cao et al., 2023). Lipid metabolism dysregulation via *HSPC111* in colorectal cancer (Zhang et al., 2022a).
	Posttranslational modifications (PTMs)	Defined as the addition of a chemical group to the proteins after translation.	Phosphorylation role in breast cancer (Theivendran et al., 2020). Glycosylation in liver cancer (Zhang et al., 2024).
	Single-cell	Field of study analyzing individual cells molecular profiles to understand cellular function.	Immune-related state (hc7) showed better outcome in breast cancer (Pang et al., 2024). Novel fibroblast subtype is associated with poor prognosis (Wang et al., 2021b).
	Multi-omics integration	Integrates diverse datasets (genomics, transcriptomics, proteomics, etc.) to gain novel insight about disease biology.	Integration of multi-omics improves the C-index value in multiple cancer types (Chai et al., 2021). Multi-omics integration predicts poor prognosis in low-grade gliomas (Du et al., 2024).

(continued)

Table 1 (continued)

Data type	Subtype	Description	Examples
Medical imaging	Radiological imaging	Medical specialty where imaging techniques are used to create and study pictures inside the body.	Labelled CT image data was used to predict survival outcome in head & neck cancer (Parmar et al., 2015). Labelled CT/FDG PET image data was used to predict survival outcome in lung cancer (Vallières et al., 2015).
	Pathological imaging	Medical specialty where histopathology images are used to analyze pathology	Histopathology images have been used in breast and lung cancer prognosis (Luo et al., 2017; Jiang et al., 2024).
Clinical data		This datatype includes different socio-demographic (age, sex, etc.) and clinical variables (tumor stage, recurrence status, etc.) that affect the disease outcome.	Females show higher survival in adenocarcinoma and squamous cell carcinoma (Islam et al., 2024). Clinical pathological factors along with gene expression are used for pathological complete response (pCR) in breast cancer (Fan et al., 2011).

The KM estimate is widely used and considered one of the best options to measure the fraction of time a patient will live post-treatment. Edward L. Kaplan and Paul Meier developed this non-parametric method in 1958 (Altman, 1990). While computing the KM survival curve, we made three assumptions. We first assume that censored patients have the same survival prospects as those who are still under monitoring and follow-up. Second, we assume that subjects have the same survival probability irrespective of their enrollment time. Lastly, we assume that the event happens at the specified time. The KM estimate is also referred to as the product limit estimate. We compute the KM survival probability using the following Eq. 1.

$$\mathrm{S}t_i = S\left(t_{i-1}\right)\left(1-\frac{d_i}{n_i}\right) \quad (1)$$

where,

- $S(t_i)$ is the survival probability at time t_i.
- $S(t_{i-1})$ is probability of being alive at t_{i-1}.
- n_i is the number of patients alive just before t_i.
- d_i is the number of events at t_i.

S(*t*), estimated probability, is a step function whose value changes only at the time of each event. One may also calculate confidence intervals for the survival probability. KM curve, a plot of the survival probability against time, also helps in providing a data summary that can be used for estimation of median survival time.

Likewise, the Cox proportional hazard model is another regression-based statistical model widely used for survival analysis. It was first introduced in the year 1972 by Sir David Cox (Prentice & Kalbfleisch, 2003). As the KM estimator is useful in survival estimation, it's an example of univariate analysis and does not include other adjustment factors. However, the Cox model works for both quantitative predictor variables as well as categorical variables, assessing the simultaneous effects of several risk factors on the survival rate and time. The Cox model not only models several factors (known as covariates) at the same time to estimate the survival time but also evaluates the effect of individual factors on survival (Abd Elhafeez et al., 2021). In other words, Cox regression explains how an individual factor is influencing the rate of a particular event happening (e.g., recurrence, death) at a particular point in time. This rate is commonly known as the "hazard rate" and is estimated by the following Eq. 2:

$$h(t) = h_0(t) \mathrm{X} \exp\left(\mathrm{b}_{1x1} + \mathrm{b}_{2x2} + \ldots.. + \mathrm{b}_{pxp}\right) \tag{2}$$

where,

- t represents the survival time.
- $h(t)$ represents the determined hazard function by the set of p covariates (x_1, x_2, …, x_p).
- $b_{1,}$ $b_{2,}$ …, b_p are the coefficients which measures the impact of the covariates.
- h_0 represents the baseline hazard and corresponds to the value of the hazard when all the x_i are set to zero. The (*t*) in *h*(*t*) remind us that the hazard may or may not vary with respect to time.

We interpret the hazard ratio (HR) based on its value. If the HR estimate is 1, it means there is no effect; HR <1 means there is a reduction in the hazard, and if the HR >1, it means there is an increase in the hazard.

These methods, which have been in use for prognosis prediction for a long time, have certain limitations that require attention. One of the major limitations of the KM estimate is that it is purely a significance test and cannot offer an estimate of the amount of the variation between the groups and the confidence interval associated with it. Another limitation with this approach is that it only provides unadjusted mortality (and survival) probabilities (Pocock et al., 2002). Similarly, a limitation of the Cox hazard model is that its constant assumption of the hazard ratio over time may not hold in real time. Additionally, the model struggles with time-varying covariates such as body weight, response to treatment, and the mixing of censoring events (Xu et al., 2024).

3.2 Machine Learning Approaches for Cancer Prognosis

Advancements in sequencing technologies have generated vast amounts of data for analysis within the medical research community. However, an accurate prediction of disease outcomes remains an unmet challenge. In recent times, ML methods have gained significant popularity as a solution to this issue, due to their ability to uncover new patterns and relationships within complex datasets. We classify ML into two broad categories: (i) supervised learning and (ii) unsupervised learning. Additionally, we have another technique known as semi-supervised learning, which is a combination of supervised and semi-supervised methods (Sarker, 2021).

Previously, physicians used histological, clinical, and population-based data for prognosis prediction with reasonable accuracy. However, these parameters are insufficient for making robust decisions. Multi-omics data, including imaging data, can obtain new kinds of molecular features, such as gene expression, thanks to ML approaches. We have used these approaches to identify patterns and relationships among complex datasets, as well as for feature extraction. Some of the widely used ML approaches currently used for prognosis include support vector machines (SVM), random forests (RF), decision trees, boosting algorithms, DL packages like convolutional neural networks (CNN), recurrent neural networks (RNN), and multilayer perceptron (MP).

For classification and regression tasks, SVM is a type of supervised machine learning approach. RF is another ensemble learning ML algorithm used for classification and regression tasks that builds multiple decision trees during training and integrates the output from them to yield a unified outcome. These algorithms have been used in many computer studies, such as drug design (Agrawal & Raghava, 2018; Agrawal et al., 2020); peptide therapeutic activity prediction (Agrawal et al., 2018, 2021); and cancer therapy response prediction (Agrawal et al., 2023, 2024). CNN, a type of deep learning, processes structured grid-type data, for example, images. Common applications of this technique include object recognition, audio, and video analysis. Biomedical research has recently gained popularity for this technique, implementing CNN on biopsy images to predict gene expression and potential biomarkers (Chuang et al., 2021). Similarly, RNN is a type of deep learning technique that handles sequential data like time series or natural language. They have been routinely used in applications such as speech recognition because of their memory of previous inputs. Various biomedical applications have also utilized them (Rasmy et al., 2022). The MP technique is similar to traditional neural network techniques, utilizing multiple layers of neurons or nodes. It has the typical three layers: input, hidden, and output. Table 2 enlists the studies implementing various approaches used in prognostic prediction in cancer.

Table 2 Approaches for prognostic prediction in cancer. Here, we have outlined the different approaches (traditional and advanced) developed for predicting prognosis in various cancer types

Sr No	Method	Description	Examples
1	Kaplan-Meier curve	Widely used method and considered the best option to measure the fraction of time a patient will live after treatment.	Li et al. used the approach to predict the survival of breast cancer patients (Li et al., 2023a). Abedi et al. utilized KM analysis to forecast survival rates in lung cancer (Abedi et al., 2019).
2	Cox proportional hazard model	The regression-based statistical model is widely used for survival analysis.	Chen et al. used the Cox model to predict the survival of breast cancer patients (Chen et al., 2016). Su et al. utilized KM analysis to forecast survival rates in lung cancer (Su et al., 2022).
3	Support vector machine (SVM) model	A supervised ML approach, used for classification and regression tasks.	Chen et al. demonstrate the application of the SVM method in predicting breast cancer prognosis (Chen et al., 2011). Xu et al. proposed a 15-gene signature for colon cancer prognosis based on SVM (Xu et al., 2017).
4	Random Forest (RF)	The ML algorithm, based on decision trees integrating the output from various trees to yield a unified outcome.	Jin et al. use this approach in breast cancer prognosis prediction (Jin et al., 2023). Xu et al. used the RF approach to predict the prognosis and survival of gastric cancer patients (Xu et al., 2021a).
5	Boosting algorithm	ML technique that combines multiple weak learners in a sequential method and improves observation upon multiple iterations.	Geeitha et al. used the CAT-gradient boosting approach to predict prognosis in cervical cancer (Geeitha et al., 2024). Song et al. used the XGBoost algorithm to predict prognosis in breast cancer patients (Song et al., 2024).
6	Principal component analysis (PCA)	ML technique that reduces the dimensionality of the big data yet preserves the important features.	Alexe et al. used the PCA approach to predict prognosis in breast cancer (Alexe et al., 2007).

(continued)

Table 2 (continued)

Sr No	Method	Description	Examples
7	Neural network (NN)	This is a form of advanced ML that uses multiple layers of neurons, just like the brain, to process the data.	Ching et al. used an NN approach with no feature extraction to predict prognosis in pan-cancer (Ching et al., 2018). Huang et al. used an NN approach with feature extraction to predict prognosis in breast cancer (Huang et al., 2019).
8	Convolutional neural network (CNN)	A sophisticated type of DL, also known as feedforward neural network, can identify patterns in images, visuals, sequences, and more, enabling it to carry out tasks such as classification.	Bychkov et al. used a CNN approach on H&E images to predict the prognosis of colorectal cancer (Bychkov et al., 2018). Mobadersany et al. implemented a CNN approach on MRI imaging data to predict the prognosis of low-grade gliomas (Mobadersany et al., 2018).

4 Immune Gene Signatures as Prognostic Biomarkers

Within the same genetic basis, tumor cells differ in terms of development, survival, metastases, and responsiveness to treatment. The tumor microenvironment (TME) is a complex, evolving environment composed of three major components: (i) cells such as immune cells, endothelial cells, and fibroblasts; (ii) extracellular components such as cytokines, growth factors, hormones, and extracellular matrix (ECM); and (iii) blood vessels required for the TME's nourishment. The TME not only plays a vital role in tumor initiation, progression, and metastasis but also affects the tumor response to treatment and predicts the survival outcomes (Li et al. 2020a, b).

Over the past 20 years, researchers have developed and validated multiple gene signatures in clinical settings to stratify cancer patients into different risk groups, thereby providing guidance for patient-specific treatment decisions. The first example of such prognostic expression was developed by Bernards et al. in 2002, where they used a panel of 70 gene expression signatures in 295 breast cancer patients and successfully classified them into good and bad prognosis groups (van de Vijver et al., 2002). Likewise, over the years, researchers developed multiple such signatures for different cancer types (Ye et al., 2003; Bullinger et al., 2004; Valk et al., 2004). However, despite the significant number of several prognostic gene signatures, very few have been validated clinically.

We can broadly classify these gene signatures into proliferation, metabolic, hypoxic, epigenetic, and immune gene signatures (ISGs). Immune gene signatures are the most widely used and accurate in prognosis prediction among all these signatures. Recent availability of multi-omics data has allowed researchers to characterize ISGs that correlate with disease progression, response to treatment, and survival outcome. ISGs have the advantage of capturing both innate and adaptive components of the immune system, including macrophage-related, lymphocyte-related, and cell non-specific immune signatures.

Tumor-associated macrophages (TAMs) formed a major component of the innate immune systems in the TME. They are highly heterogeneous cells originating from resident tissue-specific macrophages as well as newly recruited ones. TAMs can polarize into M1 (pro-inflammatory) or M2 (pro-tumor and anti-inflammatory) phenotypes. Multiple cancer types, including liver, breast, colon, and gastric, have used gene signatures derived from macrophages or associated with macrophage infiltration as prognostic signatures. Lymphocyte-based immune signatures play a critical role in predicting cancer prognosis by regulating adaptive immune responses within the TME. Specifically, cytotoxic CD8+ T cells are strongly linked to a good prognosis in cancers such as melanoma, colorectal, and non-small cell lung cancer. This is because they are antitumor in nature and can directly kill tumor cells. Other than macrophage and lymphocyte-related gene signatures, there are other cells that play a pivotal role in cancer prognosis; however, they do not belong to specific immune cell populations, for example, cells associated with antigen processing and presentation, defense response to infection, epithelial cell migration, etc. Table 3 provides a comprehensive list of immune gene signatures derived from macrophages, lymphocytes, and other cells and their role in prognosis in various cancer types.

Table 3 Gene signatures developed for prognostic prediction in cancer. Gene signatures have been developed using diverse types of data, including multi-omics, imaging, and clinical features

Sr. No.	Data type	Signature name	Signature gene list	Description
1	Transcriptomic	Oncotype DX	*Ki67, STK15, BIRC5, CCNB1, MYBL2, MMP11, CTSL2, GRB7, HER2, ER, PGR, BCL2, SCUBE2, GSTM1, BAG1, CD68, ACTB, GAPDH, GUS, RPLPO, TFRC*	The 21-gene signature was used to predict if chemotherapy is beneficial and the chances of breast cancer recurrence. (Paik et al., 2004).
		4-gene signature	*DKC1, FLNA, CSE1L, NSUN5*	A novel gene expression-based signature has been used to predict the diagnosis and prognosis of colorectal cancer (Wang et al., 2022).
2	Epigenetic	8-gene signature	*C13orf18, TMEM150B, SLC22A11, NR0B2, KLC4, LRRC2, ACOX2, AIFM3*	Hypermethylation in the promoter region predicts poor prognosis in colorectal carcinomas (Condelli et al., 2021).
		13-gene signature	*PCDHGA12, HIF3A, EZR, PCDHGA3, TPD52, STAC2, C2orf40, KRT19, NDRG2, KCNH8, CCND2, SIAH2, ITPRIPL1*	Differential methylation was associated with low overall survival in breast cancer patients (Bao et al., 2019).
3	Proteomic	4 protein signatures	*EGFR, IGFBP2, SRC, SRC_pY527*	The Cox regression model, developed using these four proteins, classified colon cancer patients with high accuracy into high- and low-risk groups with overall survival outcomes (Shan et al., 2021).
		7 proteins based signature.	*ATG9A, DCDC2, HPS5, FIL1L, LZTL1, PGTA, SPTN2,*	A predictive model using the seven proteins was associated with immunotherapy responses and overall survival outcomes in NSCLC patients (Mondelo-Macía et al., 2024).
4	Metabolomics	6 metabolic gene signatures	*NAT2, XDH, GPX3, AKR1C4, SPHK1, ADCY5*	6 metabolic genes were strongly correlated with survival and clinical stage in colorectal cancer (Miao et al., 2020).
5	Multi-omics data integration	Cuproptosis-related angiogenesis gene	*SFT2D1*	Cuproptosis-associated angiogenesis and multi-omics analysis characterize the gene associated with prognosis in cervical cancer (Kang et al., 2024).

6	Immune Gene Signature (Macrophage-related)	9-gene signature	*SOCS1, MSR1, TGFB1, CTSC, SOCS3, PTX3, FN1, PSME2, CXCL9*	TAM-based signatures whose higher expression is associated with poor overall survival in glioma patients (Shan et al., 2020).
		9-gene signature	*KIF23, BIN1, LAPTM4A, ERAP2, ATP8B2, FAM118A, RGS16, ELMO1, RAPGEFL1*	Based on these signatures, pancreatic cancer patients in the low-risk group have longer overall survival (Li et al., 2021).
7	Immune Gene Signature (Lymphocyte-related)	8-gene signature	*SERPINE1, CX3CR1, CD200R1, GBP1, IRF1, STAP1, LOX, OR7E47P*	High expression of this gene signature is associated with shorter overall survival in lung cancer patients (Wu et al., 2021).
		5-gene signature	*FPR1, CIITA, KLRC1, TNFRSF6B, WFIKKN1*	High expression of this gene signature is associated with shorter overall survival in bladder cancer patients (Chen et al., 2021).
8	Immune Gene Signature (Mixed)	5-gene signature	*ATG10, IL18RAP, PRKCD, SLC11A1, SPP1*	In hepatocellular carcinoma patients, higher expression of this signature is associated with shorter overall survival (Xu et al., 2021b).
		8-gene signature	*ITGA7, RBM14, DENND4B, LQK1, ZNF709, COL7A1, SP1, NCBP2*	In pancreatic cancer patients, higher expression of this signature is associated with shorter overall survival (Wang et al., 2021a).

5 Bioinformatics Resources for Cancer Prognosis

Advancements in the field of bioinformatics and computational biology techniques have led to the development of several web-based resources, ML-based tools, pipelines, etc., which have revolutionized the field of precision medicine, particularly in cancer diagnosis, prognosis, and improving overall survival outcomes. These advanced techniques have allowed us to analyze big and complex data in a small-time frame with remarkable speed and accuracy, which traditional methods were not capable of.

Projects led by multiple researchers, companies, or consortiums, such as TCGA, PCAWG, ICGC, GTEx, CPTCA, etc., have generated a plethora of data worth hundreds of TBs (terabytes) and have been compiled in the form of databases. A major section of this data is freely accessible to the public, while some parts need permission. These resources provide platforms to leverage vast amounts of data to analyze and come up with more robust biomarkers. They integrate multi-omics data with clinical data to characterize prognostic biomarkers and therapeutic targets. Here, we have compiled all such resources that have been associated with cancer prognosis in a comprehensive manner (Table 4).

6 Current Limitations and Future Perspective

Current computational techniques and approaches have improved our understanding of biological mechanisms associated with cancer initiation, progression, and treatment; however, we are far from achieving a high success rate in its treatment. Here, we have discussed potential challenges associated with current computational approaches in cancer prognosis, as well as potential future perspectives.

The data heterogeneity associated with multi-omics datasets generated via diverse protocols implemented by various platforms poses a major challenge, resulting in inconsistency and biased output. Lack of single and proper standardization techniques in data formatting, processing, and feature selection leads to complicated and biased results (Mohr et al., 2024). Another limitation is the lack of interpretability by advanced machine and DL models. These models function as black boxes, leading to limited interpretability of the data, which does not allow clinicians to design personalized treatment (Hong et al., 2020). The failure of current computational models to perform on diverse datasets stems from their training on specific populations, races, and ethnicities, which does not incorporate information about broader populations. This limits the usage of such models in current clinical settings (Wang et al., 2023). Another limitation is the availability of the data in the form of open-source resources, as many researchers do not have full access to such large-scale data, which could be useful in developing robust ML models. Last but not least, the current computational infrastructure does not meet the need to develop robust AI/DL models incorporating diverse populations, as they require significant computational power.

Table 4 Bioinformatics resources for cancer prognosis. The table provides details of various databases, ML-based algorithms/software, pipelines, and packages developed for cancer prognosis

Sr. No.	Name	Description	Link
1	The Cancer Genome Atlas (TCGA) (Weinstein et al., 2013)	It offers extensive information, including multi-omics and clinical data, for 33 distinct types of cancer.	https://www.cancer.gov/ccg/research/genome-sequencing/tcga
2	Genomic Data Commons (GDC) (Grossman et al., 2016)	Supports several cancer genome programs.	https://gdc.cancer.gov/
3	cBioPortal (Cerami et al., 2012)	This is an open-access resource for interactive exploration of multi-omics cancer genomics data sets.	https://www.cbioportal.org/
4	Catalogue of Somatic Mutations in Cancer (COSMIC) (Tate et al., 2019)	This resource provides information about somatic mutations in cancer.	https://cancer.sanger.ac.uk/cosmic
5	Cancer Cell Line Encyclopedia (CCLE) (Barretina et al., 2012)	The repository contains genomics data from nearly 1000 cancer cell lines.	https://depmap.org/portal/ccle/
6	Human Protein Atlas (HPA) (Uhlén et al., 2015)	This resource provides information about protein expression in cancer tissues.	https://www.proteinatlas.org/
7	ArrayExpress (Athar et al., 2019)	Microarray experiments and gene expression profiles are available in an open-source database.	https://www.ebi.ac.uk/biostudies/arrayexpress
8	Clinical Proteomic Tumor Analysis Consortium (CPTAC) (Rudnick et al., 2016)	The open-source database offers multi-omics data for 10 different types of cancer.	https://proteomics.cancer.gov/programs/cptac
9	SurvExpress (Aguirre-Gamboa et al., 2013)	This online platform provides cancer-wide gene expression data and tools that predict survival analysis.	http://bioinformatica.mty.itesm.mx/SurvExpress
10	Kaplan-Meier Plotter (Győrffy, 2024)	This web-based tool assesses the correlation between gene expression and survival.	https://kmplot.com/analysis/
11	GEPIA (Tang et al., 2017)	The online platform offers multi-dimensional analysis, which includes survival analysis for pan-cancer.	http://gepia.cancer-pku.cn/

(continued)

Table 4 (continued)

Sr. No.	Name	Description	Link
12	TIMER 2.0 (Li, Fu, et al., 2020b)	This comprehensive resource provides a multi-dimensional systematic analysis of immune infiltrate across various cancer types.	http://timer.cistrome.org/
13	DeepSurv (Katzman et al., 2018)	A regression model based on DL is utilized to predict survival.	https://github.com/jaredleekatzman/DeepSurv
14	UCSC Xena Browser (Kuleshov et al., 2016)	An in-silico tool is available for both public and private, multi-omic, and clinical/phenotype data.	https://xena.ucsc.edu/
15	Enrichr (Kuleshov et al., 2016)	The web-based integrative tool examines gene lists and conducts multiple gene enrichment analyses.	https://maayanlab.cloud/Enrichr/#
16	TCGAbiolinks (Colaprico et al., 2016)	The R package is specifically designed for analyzing TCGA data, which includes survival analysis.	https://bioconductor.org/packages/release/bioc/html/TCGAbiolinks.html
17	Seurat (Satija et al., 2015)	The R package was created to analyze single-cell RNAseq data and identify prognostic cellular heterogeneity.	https://satijalab.org/seurat/
18	MOFA (Argelaguet et al., 2018)	This is a python package designed for the unsupervised integration of multimodalities of omics data.	https://github.com/bioFAM/MOFA

In the future, we must address the aforementioned challenges in order to develop computational approaches that can predict a better prognosis for cancer patients and improve their treatment outcomes. We need to develop statistical approaches that not only accurately standardize the multi-omics data generated via multiple platforms, but also incorporate multiple modalities to provide better biological insight into the disease. There is a need to focus on the explainability of the AI/DL models to get better insight about the various features, guiding clinicians in designing personalized treatments. While developing AI/DL models, there is a need to incorporate information from diverse populations, races, and ethnicities, their genetic information, and other clinical features. Lastly, we must make the data available as an open source to the scientific community to facilitate better research. Additionally, we must establish a robust computational infrastructure to support the development of robust DL models.

Acknowledgement This research was supported by SRM Institute of Science & Technology, Kattankulathur, India.

References

Abd Elhafeez, S., D'Arrigo, G., Leonardis, D., Fusaro, M., Tripepi, G., & Roumeliotis, S. (2021). Methods to analyze time-to-event data: The cox regression analysis. *Oxidative Medicine and Cellular Longevity, 2021*, 1302811.

Abedi, S., Janbabaei, G., Afshari, M., Moosazadeh, M., Alashti, M. R., Hedayatizadeh-Omran, A., Alizadeh-Navaei, R., & Abedini, E. (2019). Estimating the survival of patients with lung cancer: What is the best statistical model? *Journal of Preventive Medicine and Public Health = Yebang Uihakhoe chi, 52*(2), 140–144.

Agrawal, P., & Raghava, G. P. S. (2018). Prediction of antimicrobial potential of a chemically modified peptide from its tertiary structure. *Frontiers in Microbiology, 9*(OCT), 2551.

Agrawal, P., Bhalla, S., Chaudhary, K., Kumar, R., Sharma, M., & Raghava, G. P. S. (2018). In silico approach for prediction of antifungal peptides. *Frontiers in Microbiology, 9*(FEB), 323.

Agrawal, P., Mishra, G., & Raghava, G. P. S. (2020). SAMbinder: A web server for predicting S-adenosyl-L-methionine binding residues of a protein from its amino acid sequence. *Frontiers in Pharmacology, 10*, 1690.

Agrawal, P., Bhagat, D., Mahalwal, M., Sharma, N., & Raghava, G. P. S. (2021). AntiCP 2.0: An updated model for predicting anticancer peptides. *Briefings in Bioinformatics, 22*(3), bbaa153.

Agrawal, P., Gopalan, V., & Hannenhalli, S. (2023). Predicting gene expression changes upon epigenomic drug treatment, *bioRxiv: The preprint server for biology*.

Agrawal, P., Jain, N., Gopalan, V., Timon, A., Singh, A., Rajagopal, P. S., & Hannenhalli, S. (2024). Network-based approach elucidates critical genes in BRCA subtypes and chemotherapy response in triple negative breast cancer. *iScience, 27*(5), 109752.

Aguirre-Gamboa, R., Gomez-Rueda, H., Martínez-Ledesma, E., Martínez-Torteya, A., Chacolla-Huaringa, R., Rodriguez-Barrientos, A., Tamez-Peña, J. G., & Treviño, V. (2013). SurvExpress: An online biomarker validation tool and database for cancer gene expression data using survival analysis. *PLoS One, 8*(9), e74250.

Alexe, G., Dalgin, G. S., Ganesan, S., DeLisi, C., & Bhanot, G. (2007). Analysis of breast cancer progression using principal component analysis and clustering. *Journal of Biosciences, 32*(5), 1027–1039.

Altman, D. G. (1990). *Practical statistics for medical research*. Practical Statistics for Medical Research.

Argelaguet, R., Velten, B., Arnol, D., Dietrich, S., Zenz, T., Marioni, J. C., Buettner, F., Huber, W., & Stegle, O. (2018). Multi-omics factor analysis-a framework for unsupervised integration of multi-omics data sets. *Molecular Systems Biology, 14*(6), e8124.

Athar, A., Füllgrabe, A., George, N., Iqbal, H., Huerta, L., Ali, A., Snow, C., Fonseca, N. A., Petryszak, R., Papatheodorou, I., Sarkans, U., & Brazma, A. (2019). ArrayExpress update - from bulk to single-cell expression data. *Nucleic Acids Research, 47*(D1), D711–D715.

Bao, X., Anastasov, N., Wang, Y., & Rosemann, M. (2019). A novel epigenetic signature for overall survival prediction in patients with breast cancer. *Journal of Translational Medicine, 17*(1), 380.

Barretina, J., Caponigro, G., Stransky, N., Venkatesan, K., Margolin, A. A., Kim, S., Wilson, C. J., Lehár, J., Kryukov, G. V., Sonkin, D., Reddy, A., Liu, M., Murray, L., Berger, M. F., Monahan, J. E., Morais, P., Meltzer, J., Korejwa, A., Jané-Valbuena, J., Mapa, F. A., Thibault, J., Bric-Furlong, E., Raman, P., Shipway, A., Engels, I. H., Cheng, J., Yu, G. K., Yu, J., Aspesi, P., De Silva, M., Jagtap, K., Jones, M. D., Wang, L., Hatton, C., Palescandolo, E., Gupta, S., Mahan, S., Sougnez, C., Onofrio, R. C., Liefeld, T., MacConaill, L., Winckler, W., Reich, M., Li, N., Mesirov, J. P., Gabriel, S. B., Getz, G., Ardlie, K., Chan, V., Myer, V. E., Weber, B. L., Porter, J., Warmuth, M., Finan, P., Harris, J. L., Meyerson, M., Golub, T. R., Morrissey, M. P., Sellers, W. R., Schlegel, R., & Garraway, L. A. (2012). The cancer cell line encyclopedia enables predictive modelling of anticancer drug sensitivity. *Nature, 483*(7391), 603–607.

Bray, F., Laversanne, M., Sung, H., Ferlay, J., Siegel, R. L., Soerjomataram, I., & Jemal, A. (2024). Global cancer statistics 2022: GLOBOCAN estimates of incidence and mortality worldwide for 36 cancers in 185 countries. *CA: a Cancer Journal for Clinicians, 74*(3), 229–263.

Bullinger, L., Döhner, K., Bair, E., Fröhling, S., Schlenk, R. F., Tibshirani, R., Döhner, H., & Pollack, J. R. (2004). Use of gene-expression profiling to identify prognostic subclasses in adult acute myeloid leukemia. *The New England Journal of Medicine, 350*(16), 1605–1616.

Bychkov, D., Linder, N., Turkki, R., Nordling, S., Kovanen, P. E., Verrill, C., Walliander, M., Lundin, M., Haglund, C., & Lundin, J. (2018). Deep learning based tissue analysis predicts outcome in colorectal cancer. *Scientific Reports, 8*(1), 3395.

Cao, K., Lyu, Y., Chen, J., He, C., Lyu, X., Zhang, Y., Chen, L., Jiang, Y., Xiang, J., Liu, B., & Wu, C. (2023). Prognostic implication of plasma metabolites in gastric cancer. *International Journal of Molecular Sciences, 24*(16), 12774.

Cerami, E., Gao, J., Dogrusoz, U., Gross, B. E., Sumer, S. O., Aksoy, B. A., Jacobsen, A., Byrne, C. J., Heuer, M. L., Larsson, E., Antipin, Y., Reva, B., Goldberg, A. P., Sander, C., & Schultz, N. (2012). The cBio cancer genomics portal: An open platform for exploring multidimensional cancer genomics data. *Cancer Discovery, 2*(5), 401–404.

Chai, H., Zhou, X., Zhang, Z., Rao, J., Zhao, H., & Yang, Y. (2021). Integrating multi-omics data through deep learning for accurate cancer prognosis prediction. *Computers in Biology and Medicine, 134*, 104481.

Chen, L., Xuan, J., Riggins, R. B., Clarke, R., & Wang, Y. (2011). Identifying cancer biomarkers by network-constrained support vector machines. *BMC Systems Biology, 5*, 161.

Chen, H. L., Zhou, M. Q., Tian, W., Meng, K. X., & He, H. F. (2016). Effect of age on breast cancer patient prognoses: A population-based study using the SEER 18 database. *PLoS One, 11*(10), e0165409.

Chen, H., Pan, Y., Jin, X., & Chen, G. (2021). An immune cell infiltration-related gene signature predicts prognosis for bladder cancer. *Scientific Reports, 11*(1), 16679.

Chen, X., Zhang, T., Su, W., Dou, Z., Zhao, D., Jin, X., Lei, H., Wang, J., Xie, X., Cheng, B., Li, Q., Zhang, H., & Di, C. (2022). Mutant p53 in cancer: From molecular mechanism to therapeutic modulation. *Cell Death & Disease, 13*(11), 974.

Ching, T., Zhu, X., & Garmire, L. X. (2018). Cox-nnet: An artificial neural network method for prognosis prediction of high-throughput omics data. *PLoS Computational Biology, 14*(4), e1006076.

Chuang, Y. H., Huang, S. H., Hung, T. M., Lin, X. Y., Lee, J. Y., Lai, W. S., & Yang, J. M. (2021). Convolutional neural network for human cancer types prediction by integrating protein interaction networks and omics data. *Scientific Reports, 11*(1), 20691.

Colaprico, A., Silva, T. C., Olsen, C., Garofano, L., Cava, C., Garolini, D., Sabedot, T. S., Malta, T. M., Pagnotta, S. M., Castiglioni, I., Ceccarelli, M., Bontempi, G., & Noushmehr, H. (2016). TCGAbiolinks: An R/Bioconductor package for integrative analysis of TCGA data. *Nucleic Acids Research, 44*(8), e71.

Condelli, V., Calice, G., Cassano, A., Basso, M., Rodriquenz, M. G., Zupa, A., Maddalena, F., Crispo, F., Pietrafesa, M., Aieta, M., Sgambato, A., Tortora, G., Zoppoli, P., & Landriscina, M. (2021). Novel epigenetic eight-gene signature predictive of poor prognosis and MSI-like phenotype in human metastatic colorectal carcinomas. *Cancers, 13*(1), 1–16.

Dai, X., & Shen, L. (2022). Advances and trends in omics technology development. *Frontiers in Medicine, 9*, 911861.

Demir Karaman, E., & Işık, Z. (2023). Multi-omics data analysis identifies prognostic biomarkers across cancers. *Medical sciences (Basel, Switzerland), 11*(3), 44.

Dickson, B. C., Mulligan, A. M., Zhang, H., Lockwood, G., O'Malley, F. P., Egan, S. E., & Reedijk, M. (2007). High-level JAG1 mRNA and protein predict poor outcome in breast cancer. *Modern pathology : An official journal of the United States and Canadian Academy of Pathology, 20*(6), 685–693.

Du, Z., Jiang, Y., Yang, Y., Kang, X., Yan, J., Liu, B., & Yang, M. (2024). A multi-omics analysis-based model to predict the prognosis of low-grade gliomas. *Scientific Reports, 14*(1), 9427.

Fan, C., Prat, A., Parker, J. S., Liu, Y., Carey, L. A., Troester, M. A., & Perou, C. M. (2011). Building prognostic models for breast cancer patients using clinical variables and hundreds of gene expression signatures. *BMC Medical Genomics, 4*, 3.

Geeitha, S., Ravishankar, K., Cho, J., & Easwaramoorthy, S. V. (2024). Integrating cat boost algorithm with triangulating feature importance to predict survival outcome in recurrent cervical cancer. *Scientific Reports, 14*(1), 19828.

Govindan, S., Siraganahalli Eswaraiah, M., Basavaraj, C., Adinarayan, M., Sankaran, S., & Bakre, M. (2020). Androgen receptor mRNA levels determine the prognosis in triple-negative breast cancer patients. *BMC Cancer, 20*(1), 745.

Grossman, R. L., Heath, A. P., Ferretti, V., Varmus, H. E., Lowy, D. R., Kibbe, W. A., & Staudt, L. M. (2016). Toward a shared vision for cancer genomic data. *The New England Journal of Medicine, 375*(12), 1109–1112.

Győrffy, B. (2024). Integrated analysis of public datasets for the discovery and validation of survival-associated genes in solid tumors. *Innovation (Cambridge (Mass.)), 5*(3), 100625.

Hong, S., Zhou, Y., Shang, J., Xiao, C., & Sun, J. (2020). Opportunities and challenges of deep learning methods for electrocardiogram data: A systematic review. *Computers in Biology and Medicine, 122*, 103801.

Huang, Z., Zhan, X., Xiang, S., Johnson, T. S., Helm, B., Yu, C. Y., Zhang, J., Salama, P., Rizkalla, M., Han, Z., & Huang, K. (2019). SALMON: Survival analysis learning with multi-omics neural networks on breast cancer. *Frontiers in Genetics, 10*(MAR), 166.

Islam, M. R., Siddiqua, S. M., Rabbani, G., Al Ayub, S. B., Islam, R., Saha, B., Khatun, N., Shahriar, M. H., Chowdhury, M. R. K., Alif, S. M., & Karim, M. N. (2024). Exploring sex difference in the risk factors and prognosis of inoperable lung cancer. *Cancer Treatment and Research Communications, 41*, 100848.

Jiang, B., Bao, L., He, S., Chen, X., Jin, Z., & Ye, Y. (2024). Deep learning applications in breast cancer histopathological imaging: Diagnosis, treatment, and prognosis. *Breast Cancer Research, 26*(1), 137.

Jin, Y., Lan, A., Dai, Y., Jiang, L., & Liu, S. (2023). Development and testing of a random forest-based machine learning model for predicting events among breast cancer patients with a poor response to neoadjuvant chemotherapy. *European Journal of Medical Research, 28*(1), 394.

Kang, J., Jiang, J., Xiang, X., Zhang, Y., Tang, J., & Li, L. (2024). Identification of a new gene signature for prognostic evaluation in cervical cancer: Based on cuproptosis-associated angiogenesis and multi-omics analysis. *Cancer Cell International, 24*(1), 23.

Katzman, J. L., Shaham, U., Cloninger, A., Bates, J., Jiang, T., & Kluger, Y. (2018). DeepSurv: Personalized treatment recommender system using a cox proportional hazards deep neural network. *BMC Medical Research Methodology, 18*(1), 24.

Kuleshov, M. V., Jones, M. R., Rouillard, A. D., Fernandez, N. F., Duan, Q., Wang, Z., Koplev, S., Jenkins, S. L., Jagodnik, K. M., Lachmann, A., McDermott, M. G., Monteiro, C. D., Gundersen, G. W., & Maayan, A. (2016). Enrichr: A comprehensive gene set enrichment analysis web server 2016 update. *Nucleic Acids Research, 44*(W1), W90–W97.

Lambin, P., Rios-Velazquez, E., Leijenaar, R., Carvalho, S., Van Stiphout, R. G. P. M., Granton, P., Zegers, C. M. L., Gillies, R., Boellard, R., Dekker, A., & Aerts, H. J. W. L. (2012). Radiomics: Extracting more information from medical images using advanced feature analysis. *European journal of cancer (Oxford, England: 1990), 48*(4), 441–446.

Li, M., Li, W., Yang, X., Wang, H., Peng, Y., Yin, J., Lu, Y., Liu, L., Shang, J., & Zhao, Q. (2020a). An immune landscape based prognostic signature predicts the immune status and immunotherapeutic responses of patients with colorectal cancer. *Life Sciences, 261*, 118368.

Li, T., Fu, J., Zeng, Z., Cohen, D., Li, J., Chen, Q., Li, B., & Liu, X. S. (2020b). TIMER2.0 for analysis of tumor-infiltrating immune cells. *Nucleic Acids Research, 48*(W1), W509–W514.

Li, M. X., Wang, H. Y., Yuan, C. H., Ma, Z. L., Jiang, B., Li, L., Zhang, L., & Xiu, D. R. (2021). Establishment of a macrophage phenotypic switch related prognostic signature in patients with pancreatic cancer. *Frontiers in Oncology, 11*, 619517.

Li, S., Li, C., Shao, W., Liu, X., Sun, L., & Yu, Z. (2023a). Survival analysis and prognosis of patients with breast cancer with pleural metastasis. *Frontiers in Oncology, 13*, 1104246.

Li, Y., Dou, Y., Da Veiga Leprevost, F., Geffen, Y., Calinawan, A. P., Aguet, F., Akiyama, Y., Anand, S., Birger, C., Cao, S., Chaudhary, R., Chilappagari, P., Cieslik, M., Colaprico, A., Zhou, D. C., Day, C., Domagalski, M. J., Esai Selvan, M., Fenyö, D., Foltz, S. M., Francis, A., Gonzalez-Robles, T., Gümüş, Z. H., Heiman, D., Holck, M., Hong, R., Hu, Y., Jaehnig, E. J., Ji, J., Jiang, W., Katsnelson, L., Ketchum, K. A., Klein, R. J., Lei, J. T., Liang, W. W., Liao, Y., Lindgren, C. M., Ma, W., Ma, L., MacCoss, M. J., Martins Rodrigues, F., McKerrow, W., Nguyen, N., Oldroyd, R., Pilozzi, A., Pugliese, P., Reva, B., Rudnick, P., Ruggles, K. V., Rykunov, D., Savage, S. R., Schnaubelt, M., Schraink, T., Shi, Z., Singhal, D., Song, X., Storrs, E., Terekhanova, N. V., Thangudu, R. R., Thiagarajan, M., Wang, L. B., Wang, J. M., Wang, Y., Wen, B., Wu, Y., Wyczalkowski, M. A., Xin, Y., Yao, L., Yi, X., Zhang, H., Zhang, Q., Zuhl, M., Getz, G., Ding, L., Nesvizhskii, A. I., Wang, P., Robles, A. I., Zhang, B., Payne, S. H., Lazar, A. J., Paulovich, A. G., Iavarone, A., Chinnaiyan, A. M., Druker, B. J., Kumar-Sinha, C., Newton, C. J., Huang, C., Mani, D. R., Smith, R. D., Huntsman, E., Schadt, E. E., An, E., Petralia, F., Hostetter, G., Omenn, G. S., Cho, H., Rodriguez, H., Kolodziejczak, I., Johnson, J. L., Bavarva, J., Tan, J., Rodland, K. D., Clauser, K. R., Krug, K., Cantley, L. C., Wiznerowicz, M., Ellis, M. J., Anurag, M., Mesri, M., Gillette, M. A., Birrer, M. J., Ceccarelli, M., Dhanasekaran, S. M., Edwards, N., Tignor, N., Babur, Ö., Gosline, S. J. C., Jewell, S. D., Satpathy, S., Chowdhury, S., Schürer, S., Carr, S. A., Liu, T., Hiltke, T., Yaron, T. M., Stathias, V., Liu, W., Zhang, X., Song, Y., Zhang, Z., & Chan, D. W. (2023b). Proteogenomic data and resources for pan-cancer analysis. *Cancer Cell, 41*(8), 1397–1406.

Luo, X., Zang, X., Yang, L., Huang, J., Liang, F., Rodriguez-Canales, J., Wistuba, I. I., Gazdar, A., Xie, Y., & Xiao, G. (2017). Comprehensive computational pathological image analysis predicts lung cancer prognosis. *Journal of thoracic oncology : official publication of the International Association for the Study of Lung Cancer, 12*(3), 501–509.

Miao, Y., Li, Q., Wang, J., Quan, W., Li, C., Yang, Y., & Mi, D. (2020). Prognostic implications of metabolism-associated gene signatures in colorectal cancer. *PeerJ, 8*, e9847.

Midthun, L., Shaheen, S., Deisch, J., Senthil, M., Tsai, J., & Hsueh, C. T. (2019). Concomitant KRAS and BRAF mutations in colorectal cancer. *Journal of gastrointestinal oncology, 10*(3), 577–581.

Mobadersany, P., Yousefi, S., Amgad, M., Gutman, D. A., Barnholtz-Sloan, J. S., Velázquez Vega, J. E., Brat, D. J., & Cooper, L. A. D. (2018). Predicting cancer outcomes from histology and genomics using convolutional networks. *Proceedings of the National Academy of Sciences of the United States of America, 115*(13), E2970–E2979.

Mohr, A. E., Ortega-Santos, C. P., Whisner, C. M., Klein-Seetharaman, J., & Jasbi, P. (2024). Navigating challenges and opportunities in multi-omics integration for personalized healthcare. *Biomedicine, 12*(7), 1496.

Mondelo-Macía, P., García-González, J., León-Mateos, L., Abalo, A., Bravo, S., Chantada Vazquez, M. d. P., Muinelo-Romay, L., López-López, R., Díaz-Peña, R., & Dávila-Ibáñez, A. B. (2024). Identification of a proteomic signature for predicting immunotherapy response in patients with metastatic non-small cell lung cancer. *Molecular & Cellular Proteomics, 23*(10), 100834.

Paik, S., Shak, S., Tang, G., Kim, C., Baker, J., Cronin, M., Baehner, F. L., Walker, M. G., Watson, D., Park, T., Hiller, W., Fisher, E. R., Wickerham, D. L., Bryant, J., & Wolmark, N. (2004). A multigene assay to predict recurrence of tamoxifen-treated, node-negative breast cancer. *The New England Journal of Medicine, 351*(27), 2817–2826.

Pang, L., Xiang, F., Yang, H., Shen, X., Fang, M., Li, R., Long, Y., Li, J., Yu, Y., & Pang, B. (2024). Single-cell integrative analysis reveals consensus cancer cell states and clinical relevance in breast cancer. *Scientific Data, 11*(1), 289.

Papier, K., Atkins, J. R., Tong, T. Y. N., Gaitskell, K., Desai, T., Ogamba, C. F., Parsaeian, M., Reeves, G. K., Mills, I. G., Key, T. J., Smith-Byrne, K., & Travis, R. C. (2024). Identifying proteomic risk factors for cancer using prospective and exome analyses of 1463 circulating proteins and risk of 19 cancers in the UK biobank. *Nature Communications, 15*(1), 4010.

Parmar, C., Grossmann, P., Rietveld, D., Rietbergen, M. M., Lambin, P., & Aerts, H. J. W. L. (2015). Radiomic machine-learning classifiers for prognostic biomarkers of head and neck cancer. *Frontiers in Oncology, 5*(DEC), 272.

Pocock, S. J., Clayton, T. C., & Altman, D. G. (2002). Survival plots of time-to-event outcomes in clinical trials: Good practice and pitfalls. *Lancet (London, England), 359*(9318), 1686–1689.

Prentice, R. L., & Kalbfleisch, J. D. (2003). Mixed discrete and continuous cox regression model. *Lifetime Data Analysis, 9*(2), 195–210.

Rasmy, L., Nigo, M., Kannadath, B. S., Xie, Z., Mao, B., Patel, K., Zhou, Y., Zhang, W., Ross, A., Xu, H., & Zhi, D. (2022). Recurrent neural network models (CovRNN) for predicting outcomes of patients with COVID-19 on admission to hospital: Model development and validation using electronic health record data. *The Lancet. Digital Health, 4*(6), e415–e425.

Rudnick, P. A., Markey, S. P., Roth, J., Mirokhin, Y., Yan, X., Tchekhovskoi, D. V., Edwards, N. J., Thangudu, R. R., Ketchum, K. A., Kinsinger, C. R., Mesri, M., Rodriguez, H., & Stein, S. E. (2016). A description of the clinical proteomic tumor analysis consortium (CPTAC) common data analysis pipeline. *Journal of Proteome Research, 15*(3), 1023–1032.

Sarker, I. H. (2021). Machine learning: Algorithms, real-world applications and research directions. *SN Computer Science, 2*(3), 160.

Satija, R., Farrell, J. A., Gennert, D., Schier, A. F., & Regev, A. (2015). Spatial reconstruction of single-cell gene expression data. *Nature Biotechnology, 33*(5), 495–502.

Sebastian, A. M., & Peter, D. (2022). Artificial intelligence in cancer research: Trends, challenges and future directions. *Life (Basel, Switzerland), 12*(12), 1991.

Shan, X., Zhang, C., Wang, Z., Wang, K., Wang, J., Qiu, X., Jiang, T., & Yang, P. (2020). Prognostic value of a nine-gene signature in glioma patients based on tumor-associated macrophages expression profiling. *Clinical Immunology (Orlando, Fla.), 216*, 108430.

Shan, Z., Luo, D., Liu, Q., Cai, S., Wang, R., Ma, Y., & Li, X. (2021). Proteomic profiling reveals a signature for optimizing prognostic prediction in colon cancer. *Journal of Cancer, 12*(8), 2199–2205.

Song, X., Chu, J., Guo, Z., Wei, Q., Wang, Q., Hu, W., Wang, L., Zhao, W., Zheng, H., Lv, X., & Zhou, J. (2024). Prognostic prediction of breast cancer patients using machine learning models: A retrospective analysis. *Gland Surgery, 13*(9), 1575–1587.

Su, P. F., Lin, C. C. K., Hung, J. Y., & Lee, J. S. (2022). The proper use and reporting of survival analysis and cox regression. *World Neurosurgery, 161*, 303–309.

Subramanian, I., Verma, S., Kumar, S., Jere, A., & Anamika, K. (2020). Multi-omics data integration, interpretation, and its application. *Bioinformatics and Biology Insights, 14*, 1177932219899051.

Tang, Z., Li, C., Kang, B., Gao, G., Li, C., & Zhang, Z. (2017). GEPIA: A web server for cancer and normal gene expression profiling and interactive analyses. *Nucleic Acids Research, 45*(W1), W98–W102.

Tate, J. G., Bamford, S., Jubb, H. C., Sondka, Z., Beare, D. M., Bindal, N., Boutselakis, H., Cole, C. G., Creatore, C., Dawson, E., Fish, P., Harsha, B., Hathaway, C., Jupe, S. C., Kok, C. Y., Noble, K., Ponting, L., Ramshaw, C. C., Rye, C. E., Speedy, H. E., Stefancsik, R., Thompson, S. L., Wang, S., Ward, S., Campbell, P. J., & Forbes, S. A. (2019). COSMIC: The catalogue of somatic mutations in cancer. *Nucleic Acids Research, 47*(D1), D941–D947.

Theivendran, S., Tang, J., Lei, C., Yang, Y., Song, H., Gu, Z., Wang, Y., Yang, Y., Jin, L., & Yu, C. (2020). Post translational modification-assisted cancer immunotherapy for effective breast cancer treatment. *Chemical Science, 11*(38), 10421–10430.

Uhlén, M., Fagerberg, L., Hallström, B. M., Lindskog, C., Oksvold, P., Mardinoglu, A., Sivertsson, Å., Kampf, C., Sjöstedt, E., Asplund, A., Olsson, I. M., Edlund, K., Lundberg, E., Navani, S., Szigyarto, C. A. K., Odeberg, J., Djureinovic, D., Takanen, J. O., Hober, S., Alm, T., Edqvist, P. H., Berling, H., Tegel, H., Mulder, J., Rockberg, J., Nilsson, P., Schwenk, J. M., Hamsten, M., Von Feilitzen, K., Forsberg, M., Persson, L., Johansson, F., Zwahlen, M., Von Heijne, G., Nielsen, J., & Pontén, F. (2015). Proteomics. Tissue-based map of the human proteome. *Science (New York, N.Y.), 347*(6220), 1260419.

Valk, P. J. M., Verhaak, R. G. W., Beijen, M. A., Erpelinck, C. A. J., van Doorn-Khosrovani, S. B. v. W., Boer, J. M., Beverloo, H. B., Moorhouse, M. J., van der Spek, P. J., Löwenberg, B., & Delwel, R. (2004). Prognostically useful gene-expression profiles in acute myeloid leukemia. *The New England Journal of Medicine, 350*(16), 1617–1628.

Vallières, M., Freeman, C. R., Skamene, S. R., & El Naqa, I. (2015). A radiomics model from joint FDG-PET and MRI texture features for the prediction of lung metastases in soft-tissue sarcomas of the extremities. *Physics in Medicine and Biology, 60*(14), 5471–5496.

van de Vijver, M. J., He, Y. D., van't Veer, L. J., Dai, H., Hart, A. A. M., Voskuil, D. W., Schreiber, G. J., Peterse, J. L., Roberts, C., Marton, M. J., Parrish, M., Atsma, D., Witteveen, A., Glas, A., Delahaye, L., van der Velde, T., Bartelink, H., Rodenhuis, S., Rutgers, E. T., Friend, S. H., & Bernards, R. (2002). A gene-expression signature as a predictor of survival in breast cancer. *The New England Journal of Medicine, 347*(25), 1999–2009.

Vial, A., Stirling, D., Field, M., Ros, M., Ritz, C., Carolan, M., Holloway, L., & Miller, A. A. (2018). The role of deep learning and radiomic feature extraction in cancer-specific predictive modelling: A review. *Translational Cancer Research, 7*(3), 803–816.

Vickers, A. J., & Cronin, A. M. (2010). Traditional statistical methods for evaluating prediction models are uninformative as to clinical value: Towards a decision analytic framework. *Seminars in Oncology, 37*(1), 31–38.

Wang, W., Xu, Z., Wang, N., Yao, R., Qin, T., Lin, H., & Yue, L. (2021a). Prognostic value of eight immune gene signatures in pancreatic cancer patients. *BMC Medical Genomics, 14*(1), 42.

Wang, Y., Liang, Y., Xu, H., Zhang, X., Mao, T., Cui, J., Yao, J., Wang, Y., Jiao, F., Xiao, X., Hu, J., Xia, Q., Zhang, X., Wang, X., Sun, Y., Fu, D., Shen, L., Xu, X., Xue, J., & Wang, L. (2021b). Single-cell analysis of pancreatic ductal adenocarcinoma identifies a novel fibroblast subtype associated with poor prognosis but better immunotherapy response. *Cell Discovery, 7*(1), 36.

Wang, D., Liufu, J., Yang, Q., Dai, S., Wang, J., & Xie, B. (2022). Identification and validation of a novel signature as a diagnostic and prognostic biomarker in colorectal cancer. *Biology Direct, 17*(1), 29.

Wang, R., Chaudhari, P., & Davatzikos, C. (2023). Bias in machine learning models can be significantly mitigated by careful training: Evidence from neuroimaging studies. *Proceedings of the National Academy of Sciences of the United States of America, 120*(6), e2211613120.

Weinstein, J. N., Collisson, E. A., Mills, G. B., Shaw, K. R. M., Ozenberger, B. A., Ellrott, K., Sander, C., Stuart, J. M., Chang, K., Creighton, C. J., Davis, C., Donehower, L., Drummond, J., Wheeler, D., Ally, A., Balasundaram, M., Birol, I., Butterfield, Y. S. N., Chu, A., Chuah, E., Chun, H. J. E., Dhalla, N., Guin, R., Hirst, M., Hirst, C., Holt, R. A., Jones, S. J. M., Lee, D., Li, H. I., Marra, M. A., Mayo, M., Moore, R. A., Mungall, A. J., Robertson, A. G., Schein, J. E., Sipahimalani, P., Tam, A., Thiessen, N., Varhol, R. J., Beroukhim, R., Bhatt, A. S., Brooks, A. N., Cherniack, A. D., Freeman, S. S., Gabriel, S. B., Helman, E., Jung, J., Meyerson, M., Ojesina, A. I., Pedamallu, C. S., Saksena, G., Schumacher, S. E., Tabak, B., Zack, T., Lander, E. S., Bristow, C. A., Hadjipanayis, A., Haseley, P., Kucherlapati, R., Lee, S., Lee, E., Luquette,

L. J., Mahadeshwar, H. S., Pantazi, A., Parfenov, M., Park, P. J., Protopopov, A., Ren, X., Santoso, N., Seidman, J., Seth, S., Song, X., Tang, J., Xi, R., Xu, A. W., Yang, L., Zeng, D., Auman, J. T., Balu, S., Buda, E., Fan, C., Hoadley, K. A., Jones, C. D., Meng, S., Mieczkowski, P. A., Parker, J. S., Perou, C. M., Roach, J., Shi, Y., Silva, G. O., Tan, D., Veluvolu, U., Waring, S., Wilkerson, M. D., Wu, J., Zhao, W., Bodenheimer, T., Hayes, D. N., Hoyle, A. P., Jeffreys, S. R., Mose, L. E., Simons, J. V., Soloway, M. G., Baylin, S. B., Berman, B. P., Bootwalla, M. S., Danilova, L., Herman, J. G., Hinoue, T., Laird, P. W., Rhie, S. K., Shen, H., Triche, T., Weisenberger, D. J., Carter, S. L., Cibulskis, K., Chin, L., Zhang, J., Sougnez, C., Wang, M., Getz, G., Dinh, H., Doddapaneni, H. V., Gibbs, R., Gunaratne, P., Han, Y., Kalra, D., Kovar, C., Lewis, L., Morgan, M., Morton, D., Muzny, D., Reid, J., Xi, L., Cho, J., Dicara, D., Frazer, S., Gehlenborg, N., Heiman, D. I., Kim, J., Lawrence, M. S., Lin, P., Liu, Y., Noble, M. S., Stojanov, P., Voet, D., Zhang, H., Zou, L., Stewart, C., Bernard, B., Bressler, R., Eakin, A., Iype, L., Knijnenburg, T., Kramer, R., Kreisberg, R., Leinonen, K., Lin, J., Liu, Y., Miller, M., Reynolds, S. M., Rovira, H., Shmulevich, I., Thorsson, V., Yang, D., Zhang, W., Amin, S., Wu, C. J., Wu, C. C., Akbani, R., Aldape, K., Baggerly, K. A., Broom, B., Casasent, T. D., Cleland, J., Dodda, D., Edgerton, M., Han, L., Herbrich, S. M., Ju, Z., Kim, H., Lerner, S., Li, J., Liang, H., Liu, W., Lorenzi, P. L., Lu, Y., Melott, J., Nguyen, L., Su, X., Verhaak, R., Wang, W., Wong, A., Yang, Y., Yao, J., Yao, R., Yoshihara, K., Yuan, Y., Yung, A. K., Zhang, N., Zheng, S., Ryan, M., Kane, D. W., Aksoy, B. A., Ciriello, G., Dresdner, G., Gao, J., Gross, B., Jacobsen, A., Kahles, A., Ladanyi, M., Lee, W., Van Lehmann, K., Miller, M. L., Ramirez, R., Rätsch, G., Reva, B., Schultz, N., Senbabaoglu, Y., Shen, R., Sinha, R., Sumer, S. O., Sun, Y., Taylor, B. S., Weinhold, N., Fei, S., Spellman, P., Benz, C., Carlin, D., Cline, M., Craft, B., Goldman, M., Haussler, D., Ma, S., Ng, S., Paull, E., Radenbaugh, A., Salama, S., Sokolov, A., Swatloski, T., Uzunangelov, V., Waltman, P., Yau, C., Zhu, J., Hamilton, S. R., Abbott, S., Abbott, R., Dees, N. D., Delehaunty, K., Ding, L., Dooling, D. J., Eldred, J. M., Fronick, C. C., Fulton, R., Fulton, L. L., Kalicki-Veizer, J., Kanchi, K. L., Kandoth, C., Koboldt, D. C., Larson, D. E., Ley, T. J., Lin, L., Lu, C., Magrini, V. J., Mardis, E. R., McLellan, M. D., McMichael, J. F., Miller, C. A., O'Laughlin, M., Pohl, C., Schmidt, H., Smith, S. M., Walker, J., Wallis, J. W., Wendl, M. C., Wilson, R. K., Wylie, T., Zhang, Q., Burton, R., Jensen, M. A., Kahn, A., Pihl, T., Pot, D., Wan, Y., Levine, D. A., Black, A. D., Bowen, J., Frick, J., Gastier-Foster, J. M., Harper, H. A., Helsel, C., Leraas, K. M., Lichtenberg, T. M., McAllister, C., Ramirez, N. C., Sharpe, S., Wise, L., Zmuda, E., Chanock, S. J., Davidsen, T., Demchok, J. A., Eley, G., Felau, I., Sheth, M., Sofia, H., Staudt, L., Tarnuzzer, R., Wang, Z., Yang, L., Zhang, J., Omberg, L., Margolin, A., Raphael, B. J., Vandin, F., Wu, H. T., Leiserson, M. D. M., Benz, S. C., Vaske, C. J., Noushmehr, H., Wolf, D., Veer, L. V. T., Anastassiou, D., Yang, T. H. O., Lopez-Bigas, N., Gonzalez-Perez, A., Tamborero, D., Xia, Z., Li, W., Cho, D. Y., Przytycka, T., Hamilton, M., McGuire, S., Nelander, S., Johansson, P., Jörnsten, R., & Kling, T. (2013). The cancer genome atlas pan-cancer analysis project. *Nature Genetics, 45*(10), 1113–1120.

Wu, J., Li, L., Zhang, H., Zhao, Y., Zhang, H., Wu, S., & Xu, B. (2021). A risk model developed based on tumor microenvironment predicts overall survival and associates with tumor immunity of patients with lung adenocarcinoma. *Oncogene, 40*(26), 4413–4424.

Xu, G., Zhang, M., Zhu, H., & Xu, J. (2017). A 15-gene signature for prediction of colon cancer recurrence and prognosis based on SVM. *Gene, 604*, 33–40.

Xu, T., Dong, M., Li, H., Zhang, R., & Li, X. (2020). Elevated mRNA expression levels of DLGAP5 are associated with poor prognosis in breast cancer. *Oncology Letters, 19*(6), 4053–4065.

Xu, C., Wang, J., Zheng, T., Cao, Y., & Ye, F. (2021a). Prediction of prognosis and survival of patients with gastric cancer by a weighted improved random forest model: An application of machine learning in medicine. *Archives of Medical Science, 18*(5), 1208–1220.

Xu, D., Wang, Y., Wu, J., Zhang, Y., Liu, Z., Chen, Y., & Zheng, J. (2021b). Systematic characterization of novel immune gene signatures predicts prognostic factors in hepatocellular carcinoma. *Frontiers in Cell and Developmental Biology, 9*, 686664.

Xu, L., Jiang, S., Li, T., & Xu, Y. (2024). Limitations of the cox proportional hazards model and alternative approaches in metachronous recurrence research. *Gastric Cancer: Official Journal of the International Gastric Cancer Association and the Japanese Gastric Cancer Association, 27*(6), 1348.

Ye, Q. H., Qin, L. X., Forgues, M., He, P., Kim, J. W., Peng, A. C., Simon, R., Li, Y., Robles, A. I., Chen, Y., Ma, Z. C., Wu, Z. Q., Ye, S. L., Liu, Y. K., Tang, Z. Y., & Wang, X. W. (2003). Predicting hepatitis B virus-positive metastatic hepatocellular carcinomas using gene expression profiling and supervised machine learning. *Nature Medicine, 9*(4), 416–423.

Zhang, C., Wang, X. Y., Zhang, P., He, T. C., Han, J. H., Zhang, R., Lin, J., Fan, J., Lu, L., Zhu, W. W., Jia, H. L., Zhang, J. B., & Chen, J. H. (2022a). Cancer-derived exosomal HSPC111 promotes colorectal cancer liver metastasis by reprogramming lipid metabolism in cancer-associated fibroblasts. *Cell Death & Disease, 13*(1), 57.

Zhang, Z., Curran, G., Shannon, J., Velie, E. M., Irvin, V. L., Manson, J. E., Simon, M. S., Dindar, D. A., Pyle, C., Schedin, P., & Tabung, F. K. (2022b). Body mass index is inversely associated with risk of postmenopausal interval breast cancer: Results from the women's health initiative. *Cancers, 14*(13), 3228.

Zhang, Y., Xu, W., Peng, C., Ren, S., & Zhang, C. (2024). Intricate effects of post-translational modifications in liver cancer: Mechanisms to clinical applications. *Journal of Translational Medicine, 22*(1), 651.

Zhou, Y., Lih, T. M., Pan, J., Höti, N., Dong, M., Cao, L., Hu, Y., Cho, K. C., Chen, S. Y., Eguez, R. V., Gabrielson, E., Chan, D. W., Zhang, H., & Li, Q. K. (2020). Proteomic signatures of 16 major types of human cancer reveal universal and cancer-type-specific proteins for the identification of potential therapeutic targets. *Journal of Hematology & Oncology, 13*(1), 170.

Predictive Modeling for Cancer Prognosis

Prerna Vats, Bhavika Baweja, Sakshi Nirmal, and Laxminarayan Rawat

1 Introduction

Cancer prevalence continues to rise globally, with significant implications for public health. Projections from GLOBOCAN reveal an alarming increase in cancer incidence and mortality rates, driven by aging, urbanization, and lifestyle changes. In 2020, there were approximately 19.3 million new cancer cases and 10 million deaths worldwide; moreover, these numbers are increasing substantially each year (Ferlay et al., 2021). This poses a considerable challenge in oncology, which highlights the need for innovative strategies to enhance patient outcomes, despite progress in early detection, treatment options, and supportive care. The rising incidence of cancer demonstrates the essential requirement for predictive tools that can assess disease progression, personalize treatments, and improve survival outcomes (Adeniran et al., 2024). Predictive modeling is essential in providing algorithms and methodologies to predict clinical outcomes using patient-specific data. These models are critical in oncology for prognosis assessment, high-risk patient identification, and therapeutic strategy optimization (Yu et al., 2024). Utilizing various data sources, including clinical records, imaging, genomics, and proteomics, predictive modeling enhances early diagnosis and supports personalized medicine (Coskun et al., 2024). These models are crucial as they offer actionable insights, allowing clinicians to make informed decisions that can greatly influence patient care and resource distribution.

Bhavika Baweja: Equal contribution for First Author.

P. Vats · B. Baweja · S. Nirmal
Department of Biosciences, Manipal University Jaipur, Jaipur, Rajasthan, India

L. Rawat (✉)
Division of Nephrology, Boston Children's Hospital, Boston, MA, USA

Harvard Medical School, Boston, MA, USA
e-mail: laxminarayan.rawat@childrens.harvard.edu

R. Nema et al. (eds.), *Advances in Cancer Detection, Prediction, and Prognosis Using Artificial Intelligence and Machine Learning*,
https://doi.org/10.1007/978-981-96-9346-7_9

Table 1 Traditional vs. AI/ML based approaches for predictive prognosis

Feature	Traditional statistical methods	Machine learning approaches
Objective	Hypothesis testing, interpretation	Prediction, pattern recognition
Data requirements	Clean, structured, predefined variables	Handles structured and unstructured data
Assumptions	Strong (e.g., normality, linearity)	Fewer assumptions
Flexibility	Limited to predefined models	Highly flexible, diverse algorithms
Performance with big data	Struggles with high-dimensional data	Efficient with large, complex datasets
Interpretability	High (e.g., coefficients are explainable)	Often low (e.g., black-box models)
Computational needs	Low	High, especially for deep learning
Applications	Inferential statistics	Predictive analytics, image processing

Predictive model development has revolutionized the field of oncology. Initially, conventional statistical techniques such as logistic regression and Cox proportional hazards models prevailed in the area of study. These models effectively analyzed linear relationships in structured datasets but were constrained in their ability to manage the complexity and scale of modern cancer data. Machine learning (ML) and artificial intelligence (AI) have established novel frameworks in predictive modelling (Kourou et al., 2015). These techniques can analyze high-dimensional datasets, identify non-linear relationships, and reveal hidden patterns. Deep learning algorithms have transformed imaging analysis, enabling precise tumor detection and segmentation from radiological scans (Tufail et al., 2021). Machine learning models applied to multi-omics data have enabled the discovery of new biomarkers and cancer subtypes. Currently, cancer diagnosis utilizes a combination of traditional methods and AI-driven predictive modeling techniques. Table 1 compares the traditional statistical methods with AI/ML based approaches, highlighting differences in objectives, data requirements, assumptions, and flexibility. Multidimensional models that integrate clinical, genomic, and imaging data are demonstrating exceptional accuracy in prognostic outcomes (Boehm et al., 2022). Explainable AI (XAI) enhances transparency by visualizing complex models to make them more interpretable for clinicians. Challenges including data privacy, bias, and variability continue to present significant obstacles. Predictive modeling is advancing rapidly, offering significant potential to enhance cancer care and improve outcomes for numerous patients globally.

2 Machine Learning Algorithms

2.1 Traditional Statistical Methods

Traditional statistical techniques have been essential in predicting cancer prognosis by utilizing clinical and pathological data to extract insights on survival outcomes, disease advancement, and therapy effectiveness (Richter and Khoshgoftaar, 2018).

These traditional approaches interpret the results primarily based on proven mathematical and statistical principles. These statistical techniques identify specific patterns and developments from the provided data and estimate their potential hazard ratios and predict results by using well-structured, and controlled databases such as the Cox proportional hazards model, linear regression, etc. The Cox proportional hazards regression model is fundamental to these methods, assessing the correlation between survival time and multiple covariates while addressing censored data, a prevalent challenge in survival studies where patients may be lost to follow-up or survive past the study duration (Matsuo et al., 2018). Cox regression assumes proportional hazards, allowing researchers to evaluate the influence of specific parameters, including TNM staging, and biomarkers, on patient survival, analyze the correlations between variables (such as patient cohorts), and develop linear assumptions (Delen, 2009; Matsuo et al., 2018; Szegedy et al., 2015). Research on breast cancer frequently employs Cox regression to emphasize the predictive importance of hormone receptor status or HER2 expression (Collett, 2023). This strategy identifies independent predictors of outcomes by controlling for multiple variables, which is essential for designing personalized treatment programs.

Another commonly used approach is the Kaplan-Meier method, which compares survival curves across several patient groups and determines survival probability over time. Kaplan-Meier curves have been utilized to demonstrate survival disparities among patients undergoing different treatment regimens for cancer (Zeineddine et al., 2023). When combined with the log-rank test, which statistically evaluates survival distributions, this strategy serves as a simple yet potent tool for detecting significant variations in survival outcomes. These methods are particularly effective for illustrating survival trends and evaluating assumptions regarding treatment effectiveness or disease advancement. Logistic regression serves as a crucial method in cancer prognosis, particularly for the prediction of binary outcomes such as recurrence or metastasis. Logistic regression has been utilized in prostate cancer research to model the relationship between categorical outcomes and predictor variables, enabling the prediction of recurrence following surgery (Mahal et al., 2015). This analysis incorporates factors such as Gleason score, PSA levels, and surgical margins (Zhang et al., 2020). Further, logistic regression's capability to incorporate both continuous and categorical predictors enhances its versatility for clinical datasets.

These traditional statistical methods, while possessing strengths, exhibit limitations, especially in the context of high-dimensional, complex data often observed in modern oncology (Yousefi et al., 2017). These methods typically assume linear relationships among variables, independent predictors, and proportional hazards, which may not be applicable in every instance. Furthermore, they may encounter difficulties in integrating data from various sources, including genomic profiles, imaging data, and clinical records. Hence, the integration of high-dimensional and heterogeneous datasets in cancer research requires the adoption of modern methodologies, including machine learning and advanced multivariate techniques (Capobianco, 2022).

Further, Gradient boosting techniques such as XGBoost and LightGBM enhance predictive accuracy by optimizing models, thereby improving precision, such as during the prognosis of prostate and breast cancer (Kanber et al., 2024; Li et al., 2025). Bayesian networks facilitate probabilistic reasoning and offer insights into survival probabilities. Survival analysis techniques, such as Cox proportional hazards models enhanced with machine learning-based regularization, provide clear insights into the determinants of the survival outcomes (Huang et al., 2020). Elastic Net combines LASSO and Ridge regression, providing a flexible approach for feature selection in high-dimensional datasets while integrating genetic and epigenetic data for cancer prognosis (Das et al., 2024; Xia et al., 2019). Clustering algorithms, such as k-means and hierarchical clustering, play a vital role in patient stratification by identifying prognostic subgroups based on molecular profiles. These have been utilized to categorize breast cancer into intrinsic subtypes such as Luminal A or HER2-enriched (Horr and Buechler, 2021). Multi-task learning predicts multiple outcomes and has demonstrated potential in cancer prognosis by utilizing shared patterns across related tasks, including survival and treatment response prediction (F. Wang et al., 2024).

2.2 Machine Learning Approaches

Machine learning (ML) has transformed the field of cancer prognosis prediction by facilitating the examination of complex, high-dimensional datasets to uncover patterns that conventional statistical approaches frequently overlook. Machine learning algorithms demonstrate exceptional capabilities in analyzing extensive datasets, such as genomic profiles, tissue-specific characteristics, origin of tumor sites, imaging data, and clinical variables, thereby providing valuable insights into disease progression, survival outcomes, and treatment responses (Divate et al., 2022).

Many models have been created such as Random Forest, Support Vector Machine (SVM), Decision Tree, Gaussian Naive Bayes (GNB), Logistic Regression, Artificial Neural Networks (ANN) classifiers, and Deep Learning (DL), which are used to gather understanding and useful information from the unprocessed data in predicting prognosis (Prabadevi et al., 2020). LASSO regression, a method based on regularization principles, plays a crucial role in the identification of significant features within high-dimensional datasets, effectively minimizing overfitting and improving interpretability (Guo et al., 2024). For example, LASSO has been utilized to identify key biomarkers that affect breast cancer recurrence through the detection of significant gene expressions (Guha, 2024). Similar to this, the ensemble learning method known as random forests (RFs) builds several decision trees and combines their predictions, efficiently handling missing data and variable interactions while prioritizing features (Rachid Zaim et al., 2020). RFs have demonstrated significant efficacy in glioblastoma and various other malignancies, integrating clinical, genomic, and imaging data to predict survival outcomes and treatment efficacy (Shaikh and Phophalia, 2021). The SVM Model algorithm optimizes the geometric range while limiting factual inaccuracy. It strives to categorize and classify the input raw data from various classes by creating a separating hyperplane, essentially treating it as two sets of vectors in a multi-dimensional domain, resulting in great generalization ability to the unknown data (Zhang et al., 2023). The LR (Logistic Regression) is a multivariate algorithm method that is used to identify the functional connection between the result variable and predictor variables. These models improve the understanding of intricate complex connections and structures. Due to its adaptability, this approach is frequently used in predictive modelling across a range of domains (Zhang et al., 2023). Among the existing models, DL has proved to be the most trusted one as it allows the processing of complex computational data using a backpropagation algorithm for better visualization of neural networks (Zhang et al., 2023). Deep learning, as a subset of machine learning, has markedly enhanced cancer prognosis. Convolutional neural networks (CNNs) process imaging data to assess tumor aggressiveness and predict therapeutic outcomes, including survival rates in lung cancer. Recurrent neural networks (RNNs) and their variants, including long short-term memory networks (LSTMs), are designed to process sequential data, facilitating immediate predictions (Aburass et al., 2024). Advanced methodologies such as autoencoders and transformers detect molecular signatures within genomic and transcriptomic datasets, outperforming conventional approaches in cancers, including liver cancer (Shen and Li, 2023). Recent advancements have enhanced the capabilities of deep learning in cancer prognosis, and researchers are able to address challenges such as overfitting and interpretability through methods including feature selection, cross-validation, and explainable AI, highlighting the significant impact of machine learning in precision oncology.

3 Data Types Used for Creating Models

3.1 Genomic Data

Genomic data plays a crucial role in cancer predictive modeling, utilizing AI and ML algorithms to identify complex patterns, predict outcomes, and personalize treatments. This dataset includes various molecular characteristics, such as gene expression profiles, somatic mutations, copy number variations, single nucleotide polymorphisms (SNPs), and epigenetic modifications. The integration of diverse data types facilitates the development of comprehensive models that improve accuracy and clinical relevance (Schulte-Sasse et al., 2021). Gene expression data, commonly obtained through RNA-Seq technologies, is widely utilized in machine learning models for predicting cancer subtypes and patient survival rates. Neural network models have effectively classified breast cancer subtypes, linking expression signatures to prognosis (Shams, 2024). Somatic mutation data from whole-genome sequencing is often used to identify mutational patterns that predict therapy response. A research investigation involving gastric cancer patients combined genomic mutations with clinical data to anticipate chemotherapy effectiveness through ensemble learning techniques (Choi et al., 2024).

Multi-omics strategies that integrate genomic, transcriptomic, and epigenomic data exhibit enhanced predictive capabilities. A recent analysis demonstrated that integrating methylation and gene expression data improved the classification accuracy of pancreatic cancer compared to models utilizing a single data type (Ozaki et al., 2024). SNP data is increasingly utilized for predicting hereditary cancer risk. Machine learning algorithms, including gradient boosting, have been applied to stratify individuals according to their genetic predisposition to cancers such as ovarian and prostate cancer (Joshi et al., 2023). Epigenomic data, such as histone modification patterns and DNA methylation profiles, enhance the understanding of cancer etiology and progression. Data types have been utilized in deep learning frameworks to predict tumor stage and response to immune checkpoint inhibitors (Iqbal et al., 2021). Copy number variations have been incorporated into machine learning models for estimating disease-free survival in colorectal cancer patients, integrating cytogenetic data with clinical characteristics for enhanced predictive frameworks (Wang et al., 2024). Alongside genomic characteristics, transcriptomics, which indicates gene expression dynamics, is crucial for forecasting cancer progression and metastasis. CNN-based models have been refined to assess transcriptomic data for the identification of key pathways in tumorigenesis, providing insights into personalized treatment strategies (Biswas and Chakrabarti, 2020). The introduction of AI and ML in cancer research has enhanced the application of genomic data through the simultaneous analysis of complex datasets, providing new insights into cancer biology and increasing the accuracy of predictive models.

3.2 Proteomics Data

Proteomics, encompassing the extensive analysis of proteins, has become an essential data type for developing cancer predictive models through AI and machine learning algorithms. Proteins are essential for cellular functions and disease mechanisms, rendering proteomic data crucial for understanding cancer biology. This dataset encompasses protein expression levels, post-translational modifications (PTMs), and protein-protein interactions, which are systematically analyzed to create predictive models for cancer diagnosis, prognosis, and therapeutic response (Su et al., 2021). Protein expression profiling, utilizing technologies such as mass spectrometry and protein microarrays, is widely applied in cancer predictive models. Breast cancer subtypes have been classified using AI algorithms that analyze proteomic patterns, enhancing diagnostic accuracy (Iqbal et al., 2021). Deep learning models have employed proteomic data to identify biomarkers that predict therapeutic resistance in ovarian cancer patients (Zhang and Wei, 2023). Machine learning methods integrate proteomic data with other omics datasets, including genomics and metabolomics, to improve predictive accuracy. Multi-omics models have effectively predicted pancreatic cancer prognosis through the integration of proteomic markers and genomic alterations (Chakraborty et al., 2024). AI models examining post-translational modifications have defined aspects of melanoma metastasis, correlating phosphorylation patterns with metastatic potential (Patel et al., 2020).

Machine learning algorithms have significantly transformed the analysis of protein-protein interaction networks. For instance, by simulating the interactions between important regulatory proteins in colorectal cancer, graph neural networks have been utilized to forecast tumor progression (Chugh et al., 2024). AI-assisted proteomics has facilitated the identification of new drug targets, demonstrated in research where AI tools examined cancer cell proteomes to determine actionable therapeutic proteins in lung cancer (He et al., 2023). The combination of proteomic data and clinical datasets has enhanced the precision of machine learning models. A recent study utilized plasma proteomic data to predict disease recurrence in prostate cancer patients, demonstrating high specificity and surpassing traditional statistical methods (Al-Daffaie et al., 2024). AI algorithms have been modified to address the high dimensionality and noise commonly found in proteomic data, leading to reliable predictions (Biswas and Chakrabarti, 2020). In summary, the integration of proteomics data with advanced AI and machine learning techniques facilitates accurate modeling of cancer pathways and improves predictive precision for clinical applications. This collaboration has the potential to enhance precision oncology and optimize patient outcomes.

3.3 Clinical Data

Clinical data forms the foundation of AI and ML-based cancer predictive models, offering structured insights into patient demographics, medical histories, imaging, and treatment responses. Major categories of clinical data include electronic health records (EHRs), imaging studies, laboratory results, and patient-reported outcomes. EHRs, rich with patient histories, smoking status, co-morbidities, and therapeutic responses, have been extensively analyzed using machine learning algorithms such as random forests and neural networks to predict lung cancer outcomes (Afroz et al., 2024; Iqbal et al., 2021). Similarly, imaging data, including CT, MRI, and PET scans, plays a vital role in tumor detection and classification, with convolutional neural networks (CNNs) demonstrating high sensitivity in differentiating malignant from benign growths, such as in breast cancer detection (Ozaki et al., 2024). Radiomics, a specialized AI field, quantifies imaging features to predict survival and treatment outcomes, as seen in glioblastoma studies (Cepeda, 2024).

Laboratory data, including blood test results and biomarkers, enhances model accuracy by anticipating treatment responses or toxicities, such as chemotherapy side effects in colorectal cancer, using algorithms like XGBoost (Froicu et al., 2024). Patient-reported outcomes, such as symptoms and quality-of-life assessments, further enrich predictive models. Multi-modal approaches combining EHRs, imaging, and laboratory data have shown exceptional precision in predicting cancer recurrence risks, as highlighted in breast cancer studies (Xu et al., 2024). Additionally, natural language processing (NLP) techniques applied to unstructured clinical notes have uncovered novel predictors of complications in cancer care, contributing to advancements in personalized medicine (Khalate et al., 2024). These integrative strategies using AI/ML methodologies drive significant progress in cancer diagnosis, prognosis, and therapy optimization, marking a new era of precision oncology (Mallett et al., 2010).

3.4 Imaging Data

Imaging data is essential in cancer predictive modeling, facilitating the application of AI and ML algorithms for the detection, diagnosis, and prediction of cancer outcomes with high precision. Techniques including computed tomography (CT), magnetic resonance imaging (MRI), positron emission tomography (PET), and mammography deliver high-resolution, multidimensional data on tumor characteristics, revealing patterns beyond human detection (Lobato-Delgado et al., 2022). CT scans are widely employed in lung cancer diagnostics, with convolutional neural networks (CNNs) trained on CT datasets demonstrating high sensitivity and specificity in distinguishing benign from malignant lung nodules. A study utilizing convolutional neural networks on low-dose computed tomography scans demonstrated a significant reduction in false positives, thereby improving early detection of lung cancer (Thakral and Gambhir, 2024). PET imaging, offering functional insights, has been incorporated into machine learning frameworks to assess metabolic activity in head and neck cancers, successfully predicting patient responses to chemoradiotherapy (Ong et al., 2024).

MRI, recognized for its more effective soft-tissue contrast, is extensively utilized in imaging of brain, prostate, and breast cancers. Radiomics extracts quantitative features from medical images and, when integrated with machine learning, demonstrates potential in predicting treatment outcomes. MRI-derived radiomic features, analyzed through machine learning models, have been employed to stratify glioblastoma patients according to survival risk, leveraging texture and intensity data for accurate prognostic evaluations (Chiu and Yen, 2022). Mammography, a fundamental method for breast cancer screening, has been enhanced by AI technologies. According to Ghasemi et al. (2024), CNNs trained on mammograms have outperformed radiologists in the early detection of breast cancer, and machine learning models that analyze imaging biomarkers from mammograms have successfully predicted recurrence risks, enabling tailored care strategies (Ghasemi et al., 2024). The integration of imaging data with multi-omics, including genomic and proteomic profiles, has improved predictive accuracy. A multi-modal machine learning approach integrating CT scans and genomic data effectively predicted pancreatic cancer subtypes, enabling personalized therapeutic interventions (Preuss et al., 2022).

These advancements highlight the significance of imaging data in cancer predictive modeling. Utilizing advanced AI and ML techniques across diverse imaging modalities assists researchers and clinicians to enhance early diagnosis, optimize treatment planning, and refine prognostic assessments, thereby advancing personalized oncology care.

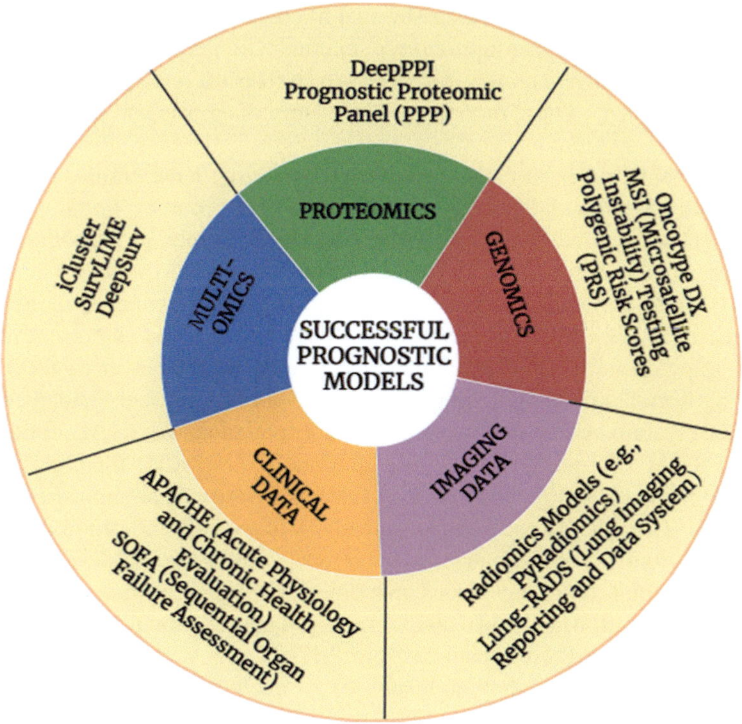

Fig. 1 Successful prognostic models

4 Successful Prognostic Models

Artificial Intelligence (AI) and Machine Learning (ML) are vital tools in the development of prognostic models for multiple cancer types. These technologies are revolutionizing oncology through enhanced outcome predictions, risk stratification, and individualized treatment planning. The incorporation of various data types, including imaging, genomic, proteomic, and clinical data, as illustrated in Fig. 1, has significantly improved the precision and applicability of these models. Significant progress has been made in lung cancer, breast cancer, cervical cancer, head and neck cancer, prostate cancer, and pancreatic cancer, demonstrating the extensive influence of AI on cancer prognosis.

AI-driven models in lung cancer utilize deep learning methods to evaluate radiomic and genomic data, yielding accurate predictions regarding disease progression and survival rates. Convolutional neural networks (CNNs) demonstrate significant efficacy in analyzing low-dose computed tomography (CT) scans, facilitating early-stage cancer detection with high precision. Percepta, a genomic classifier, enhances diagnostic accuracy by identifying patients at low risk for malignancy in indeterminate lung nodules, reducing unnecessary invasive procedures. Similarly, the Lung Cancer Index, a transcriptomic biomarker, refines prognostic evaluations by stratifying patients based on their survival likelihood. Qureshi et al. (2024) illustrated the efficacy of integrating radiomic features with genomic markers, resulting in improved predictive accuracy for disease progression and treatment outcomes (Qureshi et al., 2024). These models facilitate early diagnosis and provide essential guidance for therapeutic interventions.

Advancements in breast cancer prognostic modeling have employed ensemble learning methods to predict recurrence risks with high accuracy. Gradient boosting machines (GBMs) effectively stratify patients by recurrence probabilities through the integration of imaging biomarkers and clinical data. Models such as MammaPrint and PAM50 play a pivotal role here. MammaPrint uses gene expression profiling to categorize early-stage breast cancer patients into high- or low-risk groups, aiding in chemotherapy decisions. PAM50, or Prosigna, adds another layer of precision by identifying intrinsic molecular subtypes to personalize treatment further. Srinivasu et al. (2022) emphasized the application of explainable AI (XAI) frameworks, including SHAP (Shapley Additive Explanations), to elucidate the significance of particular features, thereby ensuring the clinical relevance of predictions (Srinivasu et al., 2022). The transparency has enhanced clinician trust and promoted the integration of AI tools in breast cancer management. Prognostic models for head and neck cancer utilize radiomics-based machine learning algorithms, especially for predicting outcomes. Alabi et al. (2023) employed convolutional neural networks to extract and analyse radiomic features from PET-CT scans, achieving accurate survival probabilities and predicting responses to chemoradiotherapy (Alabi et al., 2023). These models highlight the integration of imaging data with clinical characteristics for comprehensive prognostic assessments, facilitating personalized treatment strategies for this complex cancer type. Similarly, Kaushik et al. identified various risk factors for cervical cancer, including HPV infection, age at first pregnancy, and lifestyle choices like smoking and contraceptive use (Kaushik et al., 2021). Cytokine gene variants (Kushwah et al., 2020) and socio-demographic (Kushwah et al., 2022) datasets to analyze the socio-demographic characteristics of cervical cancer patients and healthy controls. The findings suggest that cytokine gene variants can serve as biomarkers for predicting susceptibility to cervical cancer using a multi-step machine learning approach, including Ridge classifiers and the Gaussian Naïve Bayes classifier (Kushwah et al., 2022, 2023).

Hybrid models integrating genomic and transcriptomic data have enhanced risk stratification in prostate cancer. Zhu et al. (2024) demonstrated the incorporation of AI tools with clinical datasets to accurately predict disease progression, facilitating prompt and focused interventions (Zhu et al., 2024). Genomic assays like the Decipher Prostate Cancer Test are instrumental in guiding decision-making by predicting metastatic potential post-surgery. This method has markedly enhanced the capacity to distinguish between low- and high-risk patients, fulfilling an essential requirement in the management of prostate cancer. The efficacy of these models demonstrates the capability of AI to facilitate early detection and decrease mortality rates. Pancreatic cancer, characterized by its aggressive behavior, has seen advancements through AI-driven prognostic modeling. Multi-omics methodologies that integrate transcriptomic, proteomic, and metabolomic data are crucial for survival prediction and therapy customisation. Prognostic tools such as Oncotype DX Pancreatic guide oncologists in identifying patients most likely to benefit from specific therapeutic regimens. The integration of extensive datasets has enhanced the precision of survival predictions, providing potential for improved management of pancreatic cancer.

Automated machine learning (AutoML) platforms, including AutoPrognosis, have facilitated the widespread creation of predictive models across various cancer types. These platforms facilitate swift and effective data analysis, particularly in rare cancers where data availability poses a considerable challenge. Imrie et al. (2023) highlighted the efficacy of AutoML in risk prediction and early diagnosis, emphasizing its capacity to produce clinically significant results (Imrie et al., 2023). Despite advancements, challenges including data integration, model explainability, and overfitting remain, requiring continuous refinement of AI frameworks. The integration of AI and ML technologies in oncology is transforming cancer prognosis, facilitating precision medicine, and enhancing patient outcomes.

5 Challenges in Model Development and Validation

5.1 Data Quality and Preprocessing

Data quality and preprocessing are significant obstacles to the development and validation of cancer predictive models using AI/ML algorithms, as they are essential for ensuring accuracy and robustness. Common issues encompass missing data, scale variability, class imbalances, and the presence of outliers. Missing data frequently results from incomplete patient records or technical errors in the data collection process. Imputation techniques such as K-Nearest Neighbors (KNN), mean/mode substitution, and multiple imputations by chained equations (MICE) have proven effective in addressing this issue. A research investigation into cervical cancer utilized the KNN imputer to address missing values, enhancing the efficacy of an ensemble learning model (Aljrees, 2024). Normalization techniques, including min-max scaling and z-score, are critical for standardizing datasets with varying scales, such as gene expression levels and clinical features. A significant breast cancer study utilized min-max normalization to mitigate scale disparities among biomarkers, thereby improving model stability and accuracy (Sundus et al., 2022).

Class imbalance presents a significant challenge in cancer datasets, characterized by a predominance of non-recurrence cases over recurrence cases or the underrepresentation of certain rare cancer types. Synthetic Minority Oversampling Technique (SMOTE) is commonly employed to balance datasets, enhancing predictive accuracy. SMOTE was utilized on a triple-negative breast cancer dataset, resulting in a noticeable rise in model sensitivity (Karamti et al., 2023). Outliers, resulting from experimental errors or biological variability, can negatively affect model training. Robust scaling, or log transformation is employed to enhance model performance, as demonstrated by a lung cancer study where robust scaling increased the resilience of survival prediction models (Altuhaifa et al., 2023). The integration of multi-omics data, including genomics, proteomics, and radiomics, poses challenges related to heterogeneous formats and the presence of missing entries. Methods such as principal component analysis (PCA) facilitate dimensionality reduction, and advanced imputation techniques enhance dataset preparation for AI/ML applications (Flores et al., 2023). Practical applications demonstrate the efficacy of these preprocessing techniques. Haider et al. (2020) utilized normalization and feature scaling in radiomics to predict outcomes in head and neck cancer, mitigating variability in imaging biomarkers (Haider et al., 2020). Rawas et al. (2024) employed NLP-enhanced imputation techniques to address missing clinical note data, incorporating these into comprehensive predictive models (Rawas et al., 2024). Veeraraghavan et al. (2024) established an effective workflow integrating SMOTE and normalization methods for oral oncology datasets, markedly improving model robustness and predictive accuracy (Veeraraghavan et al., 2024). The integration of these preprocessing methodologies is crucial for addressing data challenges, facilitating the development of accurate and reliable cancer predictive models.

5.2 Interpretability and Explainability of Complex Models

Developing interpretable and explainable models for cancer predictive modeling using AI and machine learning poses significant challenges, particularly when dealing with complex datasets. Advanced algorithms such as deep learning function as "black boxes," providing high accuracy while lacking transparency, which raises concerns regarding their clinical applicability. The ease with which humans can comprehend the decisions made by these models is known as interpretability, but algorithms such as gradient boosting models and convolutional neural networks (CNNs) usually have difficulty defending their predictions. A study employing CNNs for breast cancer risk prediction demonstrated challenges in elucidating the impact of specific imaging biomarkers on model outputs, leading to trust issues among clinicians (Ghasemi et al., 2024). Techniques such as Local Interpretable Model-agnostic Explanations (LIME) and SHapley Additive exPlanations (SHAP) have been utilized to address these challenges. Moreover, post-hoc methods identify influential input features for model predictions, thereby enhancing interpretability. SHAP was incorporated into a deep learning framework for predicting lung cancer recurrence, allowing researchers to identify significant genomic markers that influenced the outcomes (Alsinglawi et al., 2022). Multimodal datasets like genomics and radiomics have also been analyzed using Explainable AI (XAI) techniques in order to yield useful insights. For example, a graph-based XAI model that offered enhanced transparency and clinical utility assessed the roles of protein-protein interaction networks in the progression of colorectal cancer (D'Elia et al., 2024). LIME effectively stated deep learning classifications of mammographic images in breast cancer diagnostics, enhancing the transparency of cancerous region identification (Khater et al., 2023). SHAP identified radiomic features such as tumor texture and shape as significant predictors of survival in head and neck cancers, thereby enhancing clinician confidence in the model's results (Zhong et al., 2023).

Challenges continue to exist despite these advancements. The complex structures of models like recurrent neural networks (RNNs), employed for predicting therapy responses, frequently obstruct interpretability, despite achieving high accuracy (Damaševičius et al., 2024). Balancing explainability with accuracy presents challenges, as simplifying models for enhanced transparency can lead to a decrease in predictive performance. Integrating XAI tools into cancer predictive modeling is crucial for enhancing the power of these models while ensuring transparency, trustworthiness, and clinical applicability.

5.3 Validation Strategies

Validation strategies are essential for the development and implementation of AI/ML models in cancer predictive modeling, ensuring robustness, reliability, and generalizability. Challenges including data variability, limited sample sizes, and overfitting complicate the process. Cancer datasets frequently include various imaging modalities, genomic profiles, and clinical variables, resulting in heterogeneity that poses challenges for standardization. Limited sample sizes intensify these issues, potentially leading to a lack of representation of the complexity found in actual patient populations. Limited training data in colorectal cancer predictive models led to biased validation outcomes, highlighting the necessity for larger, more representative datasets (Alboaneen et al., 2023). Additionally, although k-fold cross-validation is a commonly employed method, it can be ineffective in the presence of imbalanced datasets or outliers. A study on hepatocellular carcinoma identified inconsistencies in cross-validation, resulting in an overestimation of model accuracy (Wei et al., 2023).

External validation is essential for evaluating a model's generalizability; however, it encounters challenges caused by variations in data collection methods and demographic characteristics of populations. A study on lung cancer utilizing PET-CT imaging revealed challenges in applying models validated in one population to different populations, emphasizing the necessity for standardized benchmarks (Tang et al., 2023). Overfitting in complex models, such as deep neural networks, presents a notable challenge. Regularization techniques and simplified architectures reduce overfitting but often compromise model interpretability, complicating their clinical application (Rasool and Bhat, 2024). The choice of evaluation metrics significantly impacts model assessment. A study on breast cancer indicated that the selection of inappropriate metrics, particularly an exclusive focus on AUC-ROC, resulted in incorrect interpretations regarding model effectiveness (Kebede et al., 2024). Innovative validation strategies have been developed to tackle these challenges. Nested cross-validation effectively reduces bias in performance estimation, as shown in genomic studies of coronary artery disease, providing a framework for cancer models (Alireza et al., 2024). Multi-center external validation confirms generalizability by evaluating models across varied datasets, as demonstrated in studies of head and neck squamous cell carcinoma (HNSCC) (Zhai et al., 2021). Synthetic data augmentation has improved validation outcomes for rare cancer subtypes, enhancing model robustness and preserving real-world relevance (van Breugel et al., 2024). These strategies emphasize the importance of thorough validation in the progression of cancer predictive modeling.

6 Ethical Considerations and Limitations

6.1 Biasness in Data Collection

When employing AI/ML algorithms for cancer predictive modeling, bias in data collection poses a serious ethical challenge that has an immediate impact on the generalizability and performance of the model. This issue arises from differences in patient demographics, sampling techniques, and data collection protocols. Genomic datasets frequently exhibit an over-representation of individuals of European descent, leading to algorithms that demonstrate suboptimal performance for other ethnic groups and consequently perpetuate healthcare inequities (Romoli and Caliandro, 2024). Sampling bias intensifies the issue, as cancer datasets are often sourced from urban tertiary care centres, thereby omitting rural or underserved populations. This results in biased predictions, especially for conditions such as breast cancer, where genetic markers and environmental influences vary considerably among different populations (Richardson et al., 2023). Feature selection bias frequently results in models favoring readily available variables, such as age or tumor size, while neglecting essential factors like socioeconomic status or healthcare access (Nazer et al., 2023). Moreover, historical biases in clinical data, including disparities in treatment access, create systematic errors that adversely affect under-represented groups (Khozin and Stein, 2022).

These biases compromise the reliability and fairness of predictive models, resulting in inequitable risk assessments and treatment recommendations. Lung cancer recurrence models primarily developed using male populations frequently exhibit reduced efficacy for female patients, highlighting issues related to safety and inclusivity (Frega et al., 2019). Various mitigation strategies have been developed to address these challenges. Data augmentation methods, including oversampling of minority groups and synthetic data generation, are effective for achieving dataset balance. For example, by including under-represented demographics, colorectal cancer studies that use these techniques have improved model fairness (Romoli and Caliandro, 2024). According to Nazer (2023), bias detection tools such as SHAP and LIME are essential for detecting and measuring the influence of biased features to enable focused corrections (Nazer et al., 2023). Federated learning trains models using decentralized datasets, facilitating diverse data inclusion and minimizing sampling bias, which improves the equity and generalizability of predictive models (Khozin and Stein, 2022).

Mitigating data collection bias is essential for the development of ethical and efficient cancer predictive models. Using fairness-focused AI techniques, including diverse datasets, and actively reducing systemic disparities are important ways to make sure that these models support fair healthcare outcomes while still being very accurate at predicting what will happen and being useful in the real world.

6.2 Privacy Concerns and Data Protection

Privacy concerns and data protection represent critical ethical considerations and limitations in the development of cancer predictive models utilizing AI and machine learning algorithms. These models frequently require access to sensitive patient information, such as genomic profiles, medical histories, and imaging data, which can pose significant risks if improperly managed. The risk of data breaches or unauthorized access presents ethical and legal issues, particularly concerning sensitive genomic data that may disclose personal health information and predispositions to future conditions. Compliance with legal frameworks, including GDPR in the European Union and HIPAA in the United States, introduces complexity, especially in multi-center studies where data aggregation must comply with diverse regulatory requirements. The difficulties in harmonizing institutional data-sharing policies were evident in breast cancer prognosis studies (Rujano et al., 2024).

To overcome these issues, federated learning has been used to train predictive models on decentralized datasets without sharing private data, protecting patient privacy. This method was effectively applied in multi-institutional lung cancer research to improve model performance and maintain data confidentiality (Gupta et al., 2024). Homomorphic encryption and similar methods enable computations on encrypted data, reducing the risk of unauthorized access. Transparency is a crucial element in mitigating ethical issues. Ethical guidelines require the acquisition of informed consent for data utilization and the provision of transparency regarding the role of patient information in research. Explainable AI (XAI) frameworks are essential for enabling auditability and transparency in data handling and predictive processes, thereby enhancing trust (Akhtar et al., 2024). AI/ML models for cancer prognosis present significant potential for enhancing patient outcomes; however, it is crucial to address privacy concerns and implement effective data protection measures.

6.3 Ethical Implications of Using AI for Real-Life-Critical Decisions

The application of AI in life-critical decisions, including cancer prognosis, presents significant ethical considerations and constraints. Predictive models significantly influence treatment strategies and patient outcomes, requiring a balance between accuracy and ethical considerations. A significant issue is accountability, as it is uncertain who bears responsibility when AI models produce incorrect predictions, such as underestimating the risk of breast cancer recurrence, resulting in insufficient treatment (Han et al., 2024). AI models exhibit bias, complicating their application due to the inheritance of inequities present in training datasets. Lung cancer models developed primarily using urban populations exhibit suboptimal performance when utilized in rural or minority groups, thereby reinforcing healthcare disparities (Kurzrock et al., 2024). The absence of explainability in numerous AI models, especially deep learning systems, diminishes trust and complicates their integration into clinical practice. Clinicians face challenges in comprehending the underlying mechanisms of these black box systems, leading to unwillingness in depending on AI for critical applications (Marey et al., 2024). Mitigation strategies have been formulated to address these challenges. Regulatory frameworks, including the European Union's AI Act and FDA guidelines, require fairness, transparency, and accountability, ensuring that AI applications conform to ethical standards (Goktas and Grzybowski, 2025). Collaboration among ethicists, clinicians, and AI specialists improves the creation of equitable and understandable models. A hybrid AI model for breast cancer prognosis integrated fairness metrics to reduce biases and enhance equity in predictions (Silva-Aravena et al., 2023).

These approaches emphasize the significance of transparency, fairness, and multidisciplinary collaboration in responsibly employing AI's potential to achieve ethical healthcare outcomes.

7 Future Directions

Predictive modeling has revolutionized research and healthcare by assisting scientists and clinicians in decision-making. However, significant challenges including issues with data integrity, bias, and the effectiveness of models as generalized prediction tools must be addressed. These challenges often arise due to variations in ethnicity, age, genetic makeup, habitat, and environmental factors. Despite these limitations, the future of machine learning and artificial intelligence in cancer research and treatment is exceptionally promising. AI continues to transform the field with innovative applications, enhancing early detection through advanced imaging and diagnostic tools. In treatment, AI-driven algorithms are instrumental in personalizing therapies based on genetic and molecular profiles, enabling tailored interventions that improve patient outcomes. Additionally, predictive modeling accelerates drug discovery by identifying novel therapeutic targets and streamlining drug development processes. It also supports healthcare providers with real-time monitoring and adaptive treatment strategies, leveraging vast clinical data to optimize patient care. Looking ahead, integrating AI with robotics, genomics, proteomics, and wearable technologies could further refine treatment strategies, offering greater precision, accessibility, and efficiency. These advancements are poised to improve survival rates and the quality of life for patients. However, addressing ethical considerations, regulatory compliance, and data privacy concerns will be essential to fully unlock the transformative potential of these technologies.

8 Conclusion

In conclusion, the global rise in cancer prevalence underscores an urgent need for innovative and safe approaches to improve patient outcomes. Predictive modeling, using machine learning and AI, has emerged as a transformative tool in oncology. By integrating diverse data sources, these models provide actionable insights for early diagnosis, personalized treatment strategies, and resource optimization (Fig. 2). While challenges such as data privacy, bias, and variability persist, the

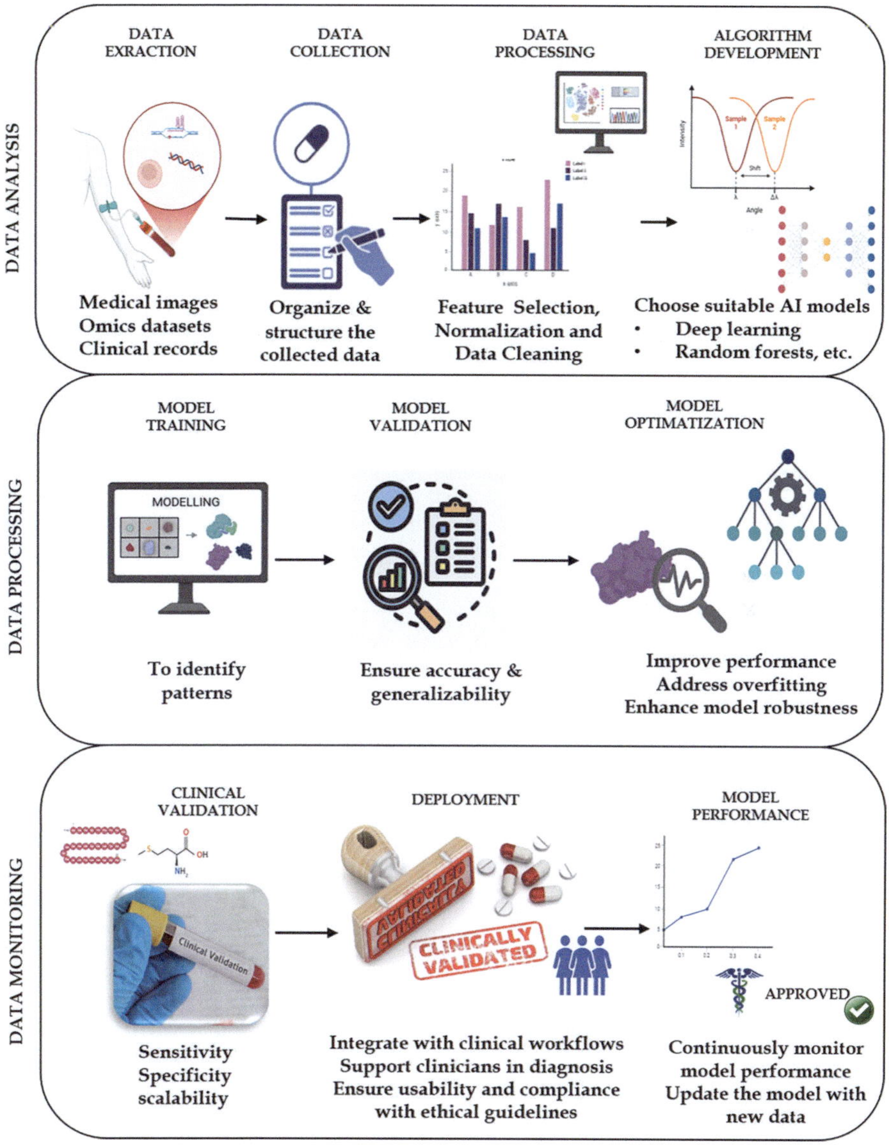

Fig. 2 Flow diagram of prognostic model development

rapid evolution of predictive technologies, including explainable AI, offers significant promise. Harnessing these advancements through collaborative and ethical frameworks can revolutionize cancer care, paving the way for improved survival rates and better quality of life for patients worldwide.

References

Aburass, S., Dorgham, O., & Shaqsi, J. A. (2024). A hybrid machine learning model for classifying gene mutations in cancer using LSTM, BiLSTM, CNN, GRU, and GloVe. *Systems and Soft Computing, 6*, 200110. https://doi.org/10.1016/j.sasc.2024.200110

Adeniran, I., Efunniyi, C., & Abhulimen, A. (2024). Data-driven decision-making in healthcare: Improving patient outcomes through predictive modeling. *International Journal of Scientific Research in Multidisciplinary Studies, 5*, 059–067. https://doi.org/10.56781/ijsrms.2024.5.1.0040

Afroz, M., Nyakwende, E., & Goswami, B. (2024). Predictive analytics in oncology: A comprehensive study on lung cancer risk factors and machine learning model performance, in: 2024 IEEE international conference on contemporary computing and communications (InC4). In *Presented at the 2024 IEEE international conference on contemporary computing and communications (InC4)* (pp. 1–7). https://doi.org/10.1109/InC460750.2024.10649176

Akhtar, M. A. K., Kumar, M., & Nayyar, A. (2024). Transparency and accountability in explainable AI: Best practices. In M. A. K. Akhtar, M. Kumar, & A. Nayyar (Eds.), *Towards ethical and socially responsible explainable AI: Challenges and opportunities* (pp. 127–164). Springer. https://doi.org/10.1007/978-3-031-66489-2_5

Alabi, R. O., Sjöblom, A., Carpén, T., Elmusrati, M., Leivo, I., Almangush, A., & Mäkitie, A. A. (2023). Application of artificial intelligence for overall survival risk stratification in oropharyngeal carcinoma: A validation of ProgTOOL. *International Journal of Medical Informatics, 175*, 105064. https://doi.org/10.1016/j.ijmedinf.2023.105064

Alboaneen, D., Alqarni, R., Alqahtani, S., Alrashidi, M., Alhuda, R., Alyahyan, E., & Alshammari, T. (2023). Predicting colorectal cancer using machine and deep learning algorithms: Challenges and opportunities. *Big Data and Cognitive Computing, 7*, 74. https://doi.org/10.3390/bdcc7020074

Al-Daffaie, F. M., Al-Mudhafar, S. F., Alhomsi, A., Tarazi, H., Almehdi, A. M., El-Huneidi, W., Abu-Gharbieh, E., Bustanji, Y., Alqudah, M. A. Y., Abuhelwa, A. Y., Guella, A., Alzoubi, K. H., & Semreen, M. H. (2024). Metabolomics and proteomics in prostate cancer research: Overview, analytical techniques, data analysis, and recent clinical applications. *International Journal of Molecular Sciences, 25*, 5071. https://doi.org/10.3390/ijms25105071

Alireza, Z., Maleeha, M., Kaikkonen, M., & Fortino, V. (2024). Enhancing prediction accuracy of coronary artery disease through machine learning-driven genomic variant selection. *Journal of Translational Medicine, 22*, 356. https://doi.org/10.1186/s12967-024-05090-1

Aljrees, T. (2024). Improving prediction of cervical cancer using KNN imputer and multi-model ensemble learning. *PLoS One, 19*, e0295632. https://doi.org/10.1371/journal.pone.0295632

Alsinglawi, B., Alshari, O., Alorjani, M., Mubin, O., Alnajjar, F., Novoa, M., & Darwish, O. (2022). An explainable machine learning framework for lung cancer hospital length of stay prediction. *Scientific Reports, 12*, 1–10. https://doi.org/10.1038/s41598-021-04608-7

Altuhaifa, F. A., Win, K. T., & Su, G. (2023). Predicting lung cancer survival based on clinical data using machine learning: A review. *Computers in Biology and Medicine, 165*, 107338. https://doi.org/10.1016/j.compbiomed.2023.107338

Biswas, N., & Chakrabarti, S. (2020). Artificial intelligence (AI)-based systems biology approaches in multi-omics data analysis of cancer. *Frontiers in Oncology, 10*. https://doi.org/10.3389/fonc.2020.588221

Boehm, K. M., Khosravi, P., Vanguri, R., Gao, J., & Shah, S. P. (2022). Harnessing multimodal data integration to advance precision oncology. *Nature Reviews. Cancer, 22*, 114–126. https://doi.org/10.1038/s41568-021-00408-3

Capobianco, E. (2022). High-dimensional role of AI and machine learning in cancer research. *British Journal of Cancer, 126*, 523–532. https://doi.org/10.1038/s41416-021-01689-z

Cepeda, S. (2024). Machine learning and radiomics in gliomas. In A. Di Ieva, E. Suero Molina, S. Liu, & C. Russo (Eds.), *Computational neurosurgery* (pp. 231–243). Springer Nature. https://doi.org/10.1007/978-3-031-64892-2_14

Chakraborty, S., Sharma, G., Karmakar, S., & Banerjee, S. (2024). Multi-OMICS approaches in cancer biology: New era in cancer therapy. *Biochimica et Biophysica Acta (BBA) – Molecular Basis of Disease, 1870*, 167120. https://doi.org/10.1016/j.bbadis.2024.167120

Chiu, F.-Y., & Yen, Y. (2022). Efficient Radiomics-based classification of multi-parametric MR images to identify volumetric habitats and signatures in glioblastoma: A machine learning approach. *Cancers, 14*, 1475. https://doi.org/10.3390/cancers14061475

Choi, Y., Lee, J., Shin, K., Lee, J. W., Kim, J. W., Lee, S., Choi, Y. J., Park, K. H., & Kim, J. H. (2024). Integrated clinical and genomic models using machine-learning methods to predict the efficacy of paclitaxel-based chemotherapy in patients with advanced gastric cancer. *BMC Cancer, 24*, 502. https://doi.org/10.1186/s12885-024-12268-9

Chugh, V., Basu, A., Kaushik, A., Manshu, B. S., & Basu, A. K. (2024). Employing nano-enabled artificial intelligence (AI)-based smart technologies for prediction, screening, and detection of cancer. *Nanoscale, 16*, 5458–5486. https://doi.org/10.1039/D3NR05648A

Collett, D. (2023). *Modelling survival data in medical research* (4th ed.). Chapman and Hall/CRC. https://doi.org/10.1201/9781003282525

Coskun, A., Ertaylan, G., Pusparum, M., Van Hoof, R., Kaya, Z. Z., Khosravi, A., & Zarrabi, A. (2024). Advancing personalized medicine: Integrating statistical algorithms with omics and nano-omics for enhanced diagnostic accuracy and treatment efficacy. *Biochimica et Biophysica Acta (BBA) – Molecular Basis of Disease, 1870*, 167339. https://doi.org/10.1016/j.bbadis.2024.167339

D'Elia, D., Zomer, A., Moreno Indias, I., Bongcam-Rudloff, E., Bertelsen, R. J., & Claesson, M. J. (2024). Editorial: Microbiome and machine learning, volume II. *Frontiers in Microbiology, 15*. https://doi.org/10.3389/fmicb.2024.1499260

Damaševičius, R., Jagatheesaperumal, S. K., Kandala, R. N. V. P. S., Hussain, S., Alizadehsani, R., & Gorriz, J. M. (2024). Deep learning for personalized health monitoring and prediction: A review. *Computational Intelligence, 40*, e12682. https://doi.org/10.1111/coin.12682

Das, S., Karuri, S., Chakraborty, J., Basu, B., Chandra, A., Aravindan, S., Chakraborty, A., Paul, D., Ray, J. G., Lechner, M., Beck, S., Teschendorff, A. E., & Chatterjee, R. (2024). Universal penalized regression (elastic-net) model with differentially methylated promoters for oral cancer prediction. *European Journal of Medical Research, 29*, 458. https://doi.org/10.1186/s40001-024-02047-4

Delen, D. (2009). Analysis of cancer data: A data mining approach. *Expert Systems, 26*, 100–112. https://doi.org/10.1111/j.1468-0394.2008.00480.x

Divate, M., Tyagi, A., Richard, D. J., Prasad, P. A., Gowda, H., & Nagaraj, S. H. (2022). Deep learning-based pan-cancer classification model reveals tissue-of-origin specific gene expression signatures. *Cancers, 14*, 1185. https://doi.org/10.3390/cancers14051185

Ferlay, J., Colombet, M., Soerjomataram, I., Parkin, D. M., Piñeros, M., Znaor, A., & Bray, F. (2021). Cancer statistics for the year 2020: An overview. *International Journal of Cancer, 149*, 778–789. https://doi.org/10.1002/ijc.33588

Flores, J. E., Claborne, D. M., Weller, Z. D., Webb-Robertson, B.-J. M., Waters, K. M., & Bramer, L. M. (2023). Missing data in multi-omics integration: Recent advances through artificial intelligence. *Frontiers in Artificial Intelligence, 6*. https://doi.org/10.3389/frai.2023.1098308

Frega, S., Dal Maso, A., Ferro, A., Bonanno, L., Conte, P., & Pasello, G. (2019). Heterogeneous tumor features and treatment outcome between males and females with lung cancer (LC): Do gender and sex matter? *Critical Reviews in Oncology/Hematology, 138*, 87–103. https://doi.org/10.1016/j.critrevonc.2019.03.012

Froicu, E.-M., Oniciuc, O.-M., Afrăsânie, V.-A., Marinca, M.-V., Riondino, S., Dumitrescu, E. A., Alexa-Stratulat, T., Radu, I., Miron, L., Bacoanu, G., Poroch, V., & Gafton, B. (2024). The use of artificial intelligence in predicting chemotherapy-induced toxicities in metastatic colorectal cancer: A data-driven approach for personalized oncology. *Diagnostics, 14*, 2074. https://doi.org/10.3390/diagnostics14182074

Ghasemi, A., Hashtarkhani, S., Schwartz, D. L., & Shaban-Nejad, A. (2024). Explainable artificial intelligence in breast cancer detection and risk prediction: A systematic scoping review. *Cancer Innov., 3*, e136. https://doi.org/10.1002/cai2.136

Guha, S. (2024). LASSO based analysis for prediction of prognostic signature genes associated with breast cancer. https://doi.org/10.1101/2024.04.02.587421

Guo, Y., Li, L., Zheng, K., Du, J., Nie, J., Wang, Z., & Hao, Z. (2024). Development and validation of a survival prediction model for patients with advanced non-small cell lung cancer based on LASSO regression. *Frontiers in Immunology, 15*, 1431150. https://doi.org/10.3389/fimmu.2024.1431150

Gupta, M., Kumar, M., & Gupta, Y. (2024). A blockchain-empowered federated learning-based framework for data privacy in lung disease detection system. *Computers in Human Behavior, 158*, 108302. https://doi.org/10.1016/j.chb.2024.108302

Goktas, P., Grzybowski, A., 2025. Shaping the Future of Healthcare: Ethical Clinical Challenges and Pathways to Trustworthy AI. J. Clin. Med. 14, 1605. https://doi.org/10.3390/jcm14051605

Haider, S. P., Burtness, B., Yarbrough, W. G., & Payabvash, S. (2020). Applications of radiomics in precision diagnosis, prognostication and treatment planning of head and neck squamous cell carcinomas. *Cancers Head Neck, 5*, 6. https://doi.org/10.1186/s41199-020-00053-7

Han, J., Hua, H., Fei, J., Liu, J., Guo, Y., Ma, W., & Chen, J. (2024). Prediction of disease-free survival in breast cancer using deep learning with ultrasound and mammography: A multicenter study. *Clinical Breast Cancer, 24*, 215–226. https://doi.org/10.1016/j.clbc.2024.01.005

Huang, Z., Johnson, T.S., Han, Z., Helm, B., Cao, S., Zhang, C., Salama, P., Rizkalla, M., Yu, C.Y., Cheng, J., Xiang, S., Zhan, X., Zhang, J., Huang, K., 2020. Deep learning-based cancer survival prognosis from RNA-seq data: approaches and evaluations. BMC Med. Genomics 13, 41. https://doi.org/10.1186/s12920-020-0686-1.

Horr, C., Buechler, S.A., 2021. Breast Cancer Consensus Subtypes: A system for subtyping breast cancer tumors based on gene expression. Npj Breast Cancer 7, 136. https://doi.org/10.1038/s41523-021-00345-2

He, X., Liu, X., Zuo, F., Shi, H., & Jing, J. (2023). Artificial intelligence-based multi-omics analysis fuels cancer precision medicine. *Seminars in Cancer Biology, 88*, 187–200. https://doi.org/10.1016/j.semcancer.2022.12.009

Imrie, F., Cebere, B., McKinney, E. F., & van der Schaar, M. (2023). AutoPrognosis 2.0: Democratizing diagnostic and prognostic modeling in healthcare with automated machine learning. *PLOS Digital Health, 2*, e0000276. https://doi.org/10.1371/journal.pdig.0000276

Iqbal, M. J., Javed, Z., Sadia, H., Qureshi, I. A., Irshad, A., Ahmed, R., Malik, K., Raza, S., Abbas, A., Pezzani, R., & Sharifi-Rad, J. (2021). Clinical applications of artificial intelligence and machine learning in cancer diagnosis: Looking into the future. *Cancer Cell International, 21*, 270. https://doi.org/10.1186/s12935-021-01981-1

Joshi, I., Bhrdwaj, A., Khandelwal, R., Pande, A., Agarwal, A., Srija, C. D., Suresh, R. A., Mohan, M., Hazarika, L., Thakur, G., Hussain, T., Albogami, S., Nayarisseri, A., & Singh, S. K. (2023). 15 - Artificial intelligence, big data and machine learning approaches in genome-wide SNP-based prediction for precision medicine and drug discovery. In S. C. Basak & M. Vračko (Eds.), *Big data analytics in chemoinformatics and bioinformatics* (pp. 333–357). Elsevier. https://doi.org/10.1016/B978-0-323-85713-0.00021-9

Karamti, H., Alharthi, R., Anizi, A. A., Alhebshi, R. M., Eshmawi, A. A., Alsubai, S., & Umer, M. (2023). Improving prediction of cervical cancer using KNN imputed SMOTE features and multi-model ensemble learning approach. *Cancers, 15*, 4412. https://doi.org/10.3390/cancers15174412

Kaushik, M., Chandra Joshi, R., Kushwah, A. S., Gupta, M. K., Banerjee, M., Burget, R., & Dutta, M. K. (2021). Cytokine gene variants and socio-demographic characteristics as predictors of cervical cancer: A machine learning approach. *Computers in Biology and Medicine, 134*, 104559. https://doi.org/10.1016/j.compbiomed.2021.104559

Kanber, B.M., Smadi, A.A., Noaman, N.F., Liu, B., Gou, S., Alsmadi, M.K., 2024. LightGBM: A Leading Force in Breast Cancer Diagnosis Through Machine Learning and Image Processing. IEEE Access 12, 39811–39832. https://doi.org/10.1109/ACCESS.2024.3375755

Kebede, S. R., Waldamichael, F. G., Debelee, T. G., Aleme, M., Bedane, W., Mezgebu, B., & Merga, Z. C. (2024). Dual view deep learning for enhanced breast cancer screening using mammography. *Scientific Reports, 14*, 3839. https://doi.org/10.1038/s41598-023-50797-8

Khalate, P., Gite, S., Pradhan, B., & Lee, C.-W. (2024). Advancements and gaps in natural language processing and machine learning applications in healthcare: A comprehensive review of electronic medical records and medical imaging. *Frontiers of Physics, 12*. https://doi.org/10.3389/fphy.2024.1445204

Khater, T., Hussain, A., Mahmoud, S., & Yasen, S. (2023). Explainable AI for breast cancer detection: A LIME-driven approach, in: 2023 16th international conference on developments in eSystems engineering (DeSE). In *Presented at the 2023 16th international conference on developments in eSystems engineering (DeSE)* (pp. 540–545). https://doi.org/10.1109/DeSE60595.2023.10469341

Khozin, S., & Stein, R. M. (2022). A novel approach to reducing disparities in health outcomes by enhancing interpretation of cancer clinical trials for underrepresented patient groups. *Biochimica et Biophysica Acta (BBA) - Reviews on Cancer, 1877*, 188825. https://doi.org/10.1016/j.bbcan.2022.188825

Kourou, K., Exarchos, T. P., Exarchos, K. P., Karamouzis, M. V., & Fotiadis, D. I. (2015). Machine learning applications in cancer prognosis and prediction. *Computational and Structural Biotechnology Journal, 13*, 8–17. https://doi.org/10.1016/j.csbj.2014.11.005

Kurzrock, R., Chaudhuri, A.A., Feller-Kopman, D., Florez, N., Gorden, J., Wistuba, I.I., 2024. Healthcare disparities, screening, and molecular testing in the changing landscape of non–small cell lung cancer in the United States: a review. Cancer Metastasis Rev. 43, 1217–1231. https://doi.org/10.1007/s10555-024-10187-6

Kushwah, A., Gupta, M., Singh, R., & Banerjee, M. (2020). Cytokine gene variants and treatment outcome of cisplatin-based concomitant chemoradiotherapy in cervical cancer. *British Journal of Biomedical Science, 77*, 81–86. https://doi.org/10.1080/09674845.2020.1714164

Kushwah, A. S., Mishra, R., Srivastava, K., Mishra, S., & Banerjee, M. (2022). Anthro-demographic, clinico-pathological and biochemical risk factors in cervical cancer. *International Journal of Health Sciences*, 3641–3653. https://doi.org/10.53730/ijhs.v6nS8.12925

Kushwah, A. S., Srivastava, K., Mishra, R., & Banerjee, M. (2023). Impact of human papillomavirus (HPV) infection on the treatment outcome of concomitant chemoradiotherapy (CRT) in cervical cancer. *Current Cancer Therapy Reviews, 19*, 320. https://doi.org/10.2174/1573394719666230807161948

Li, D., Chang, B., Huang, Q., 2025. Using XBGoost, an interpretable machine learning model, for diagnosing prostate cancer in patients with PSA 20 ng/ml based on the PSAMR indicator. Sci. Rep. 15, 1532. https://doi.org/10.1038/s41598-025-85963-7

Lobato-Delgado, B., Priego-Torres, B., & Sanchez-Morillo, D. (2022). Combining molecular, imaging, and clinical data analysis for predicting cancer prognosis. *Cancers, 14*, 3215. https://doi.org/10.3390/cancers14133215

Marey, A., Arjmand, P., Alerab, A.D.S., Eslami, M.J., Saad, A.M., Sanchez, N., Umair, M., 2024. Explainability, transparency and black box challenges of AI in radiology: impact on patient care in cardiovascular radiology. Egypt. J. Radiol. Nucl. Med. 55, 183. https://doi.org/10.1186/s43055-024-01356-2.

Mahal, B. A., Chen, M.-H., Bennett, C. L., Kattan, M. W., Sartor, O., Stein, K., D'Amico, A. V., & Nguyen, P. L. (2015). The association between race and treatment regret among men with recurrent prostate cancer. *Prostate Cancer and Prostatic Diseases, 18*, 38–42. https://doi.org/10.1038/pcan.2014.42

Mallett, S., Royston, P., Waters, R., Dutton, S., & Altman, D. G. (2010). Reporting performance of prognostic models in cancer: A review. *BMC Medicine, 8*, 21. https://doi.org/10.1186/1741-7015-8-21

Matsuo, K., Purushotham, S., Jiang, B., Mandelbaum, R. S., Takiuchi, T., Liu, Y., & Roman, L. D. (2018). Survival outcome prediction in cervical cancer: Cox models vs deep-learning model. *American Journal of Obstetrics and Gynecology, 220*, 381.e1. https://doi.org/10.1016/j.ajog.2018.12.030

Nazer, L. H., Zatarah, R., Waldrip, S., Ke, J. X. C., Moukheiber, M., Khanna, A. K., Hicklen, R. S., Moukheiber, L., Moukheiber, D., Ma, H., & Mathur, P. (2023). Bias in artificial intelligence algorithms and recommendations for mitigation. *PLOS Digital Health, 2*, e0000278. https://doi.org/10.1371/journal.pdig.0000278

Ong, Y. H., Zheng, W., Khong, P. L., & Ni, Q. (2024). Application of radiogenomics in head and neck cancer: A new tool toward diagnosis and therapy. *iRadiology, 2*, 113–127. https://doi.org/10.1002/ird3.61

Ozaki, Y., Broughton, P., Abdollahi, H., Valafar, H., & Blenda, A. V. (2024). Integrating omics data and AI for cancer diagnosis and prognosis. *Cancers, 16*, 2448. https://doi.org/10.3390/cancers16132448

Patel, S. K., George, B., & Rai, V. (2020). Artificial intelligence to decode cancer mechanism: Beyond patient stratification for precision oncology. *Frontiers in Pharmacology, 11*. https://doi.org/10.3389/fphar.2020.01177

Prabadevi, B., Deepa, N., Krithika, L. B., & Vinod, V. (2020). Analysis of machine learning algorithms on cancer dataset. In *2020 international conference on emerging trends in information technology and engineering (Ic-ETITE). Presented at the 2020 international conference on emerging trends in information technology and engineering (ic-ETITE)* (pp. 1–10). https://doi.org/10.1109/ic-ETITE47903.2020.36

Preuss, K., Thach, N., Liang, X., Baine, M., Chen, J., Zhang, C., Du, H., Yu, H., Lin, C., Hollingsworth, M. A., & Zheng, D. (2022). Using quantitative imaging for personalized medicine in pancreatic cancer: A review of Radiomics and deep learning applications. *Cancers, 14*, 1654. https://doi.org/10.3390/cancers14071654

Qureshi, H., Abdul, Y., Shah, R., Qureshi, S., Ur, S., Shiwlani, A., & Ahmad, A. (2024). The promising role of artificial intelligence in navigating lung cancer prognosis. https://doi.org/10.13140/RG.2.2.16960.96009

Rachid Zaim, S., Kenost, C., Berghout, J., Chiu, W., Wilson, L., Zhang, H. H., & Lussier, Y. A. (2020). binomialRF: Interpretable combinatoric efficiency of random forests to identify biomarker interactions. *BMC Bioinformatics, 21*, 374. https://doi.org/10.1186/s12859-020-03718-9

Rasool, N., & Bhat, J. I. (2024). A critical review on segmentation of glioma brain tumor and prediction of overall survival. *Archives of Computational Methods in Engineering, 32*, 1525. https://doi.org/10.1007/s11831-024-10188-2

Rawas, S., Tafran, C., AlSaeed, D., Al-Ghreimil, N., 2024. Transforming Healthcare: AI-NLP Fusion Framework for Precision Decision-Making and Personalized Care Optimization in the Era of IoMT. Comput. Mater. Contin. 81, 4575–4601. https://doi.org/10.32604/cmc.2024.055307.

Richardson, A., Darst, B., Wojcik, G., Wagle, N., & Haricharan, S. (2023). Research silos in cancer disparities: Obstacles to improving clinical outcomes for underserved patient populations. *Clinical Cancer Research, 29*, 1194–1199. https://doi.org/10.1158/1078-0432.CCR-22-3182

Richter, A. N., & Khoshgoftaar, T. M. (2018). A review of statistical and machine learning methods for modeling cancer risk using structured clinical data. *Artificial Intelligence in Medicine, 90*, 1–14. https://doi.org/10.1016/j.artmed.2018.06.002

Romoli, M., & Caliandro, P. (2024). Artificial intelligence, machine learning, and reproducibility in stroke research. *European Stroke Journal, 9*, 518–520. https://doi.org/10.1177/23969873241275863

Rujano, M. A., Boiten, J.-W., Ohmann, C., Canham, S., Contrino, S., David, R., Ewbank, J., Filippone, C., Connellan, C., Custers, I., van Nuland, R., Mayrhofer, M. T., Holub, P., Álvarez, E. G., Bacry, E., Hughes, N., Freeberg, M. A., Schaffhauser, B., Wagener, H., Sánchez-Pla, A., Bertolini, G., & Panagiotopoulou, M. (2024). Sharing sensitive data in life sciences: An overview of centralized and federated approaches. *Briefings in Bioinformatics, 25*, bbae262. https://doi.org/10.1093/bib/bbae262

Schulte-Sasse, R., Budach, S., Hnisz, D., & Marsico, A. (2021). Integration of multiomics data with graph convolutional networks to identify new cancer genes and their associated molecular mechanisms. *Nature Machine Intelligence, 3*, 513–526. https://doi.org/10.1038/s42256-021-00325-y

Shaikh, S., & Phophalia, A. (2021). A deep random Forest approach for multimodal brain tumor segmentation. In A. Crimi & S. Bakas (Eds.), *Brainlesion: Glioma, multiple sclerosis, stroke and traumatic brain injuries* (pp. 133–147). Springer. https://doi.org/10.1007/978-3-030-72087-2_12

Shams, A. (2024). Leveraging state-of-the-art AI algorithms in personalized oncology: From transcriptomics to treatment. *Diagnostics, 14*, 2174. https://doi.org/10.3390/diagnostics14192174

Shen, X., & Li, X. (2023). Deep-learning methods for unveiling large-scale single-cell transcriptomes. *Cancer Biology & Medicine, 20*, 972–980. https://doi.org/10.20892/j.issn.2095-3941.2023.0436

Srinivasu, P. N., Sandhya, N., Jhaveri, R. H., & Raut, R. (2022). From Blackbox to explainable AI in healthcare: Existing tools and case studies. *Mobile Information Systems, 2022*, 8167821. https://doi.org/10.1155/2022/8167821

Silva-Aravena, F., Núñez Delafuente, H., Gutiérrez-Bahamondes, J.H., Morales, J., 2023. A Hybrid Algorithm of ML and XAI to Prevent Breast Cancer: A Strategy to Support Decision Making. Cancers 15, 2443. https://doi.org/10.3390/cancers15092443

Su, M., Zhang, Z., Zhou, L., Han, C., Huang, C., & Nice, E. C. (2021). Proteomics, personalized medicine and cancer. *Cancers, 13*, 2512. https://doi.org/10.3390/cancers13112512

Sundus, K. I., Hammo, B. H., Al-Zoubi, M. B., & Al-Omari, A. (2022). Solving the multicollinearity problem to improve the stability of machine learning algorithms applied to a fully annotated breast cancer dataset. *Informatics in Medicine Unlocked, 33*, 101088. https://doi.org/10.1016/j.imu.2022.101088

Szegedy, C., Liu, W., Jia, Y., Sermanet, P., Reed, S., Anguelov, D., Erhan, D., Vanhoucke, V., & Rabinovich, A. (2015). Going deeper with convolutions. In *Presented at the Proceedings of the IEEE Conference on Computer Vision and Pattern Recognition* (pp. 1–9).

Tang, X., Wu, F., Chen, X., Ye, S., & Ding, Z. (2023). Current status and prospect of PET-related imaging radiomics in lung cancer. *Frontiers in Oncology, 13*, 1297674. https://doi.org/10.3389/fonc.2023.1297674

Thakral, G., & Gambhir, S. (2024). Early detection of lung cancer with low-dose CT scan using artificial intelligence: A comprehensive survey. *SN Computer Science, 5*, 441. https://doi.org/10.1007/s42979-024-02811-7

Tufail, A. B., Ma, Y.-K., Kaabar, M. K. A., Martínez, F., Junejo, A. R., Ullah, I., & Khan, R. (2021). Deep learning in cancer diagnosis and prognosis prediction: A minireview on challenges, recent trends, and future directions. *Computational and Mathematical Methods in Medicine, 2021*, 9025470. https://doi.org/10.1155/2021/9025470

van Breugel, B., Liu, T., Oglic, D., & van der Schaar, M. (2024). Synthetic data in biomedicine via generative artificial intelligence. *Nature Reviews Bioengineering, 2*, 991–1004. https://doi.org/10.1038/s44222-024-00245-7

Veeraraghavan, V. P., Daniel, S., Dasari, A. K., Aileni, K. R., Patil, C., & Patil, S. R. (2024). Harnessing artificial intelligence for predictive modelling in oral oncology: Opportunities, challenges, and clinical perspectives. *Oral Oncology Reports, 11*, 100591. https://doi.org/10.1016/j.oor.2024.100591

Wang, S., Wu, C.-Y., He, M.-M., Yong, J.-X., Chen, Y.-X., Qian, L.-M., Zhang, J.-L., Zeng, Z.-L., Xu, R.-H., Wang, F., & Zhao, Q. (2024). Machine learning-based extrachromosomal DNA identification in large-scale cohorts reveals its clinical implications in cancer. *Nature Communications, 15*, 1515. https://doi.org/10.1038/s41467-024-45479-6

Wang, F., Zhuang, Z., Gao, F., He, R., Zhang, S., Wang, L., Liu, J., Li, Y., 2024. TMO-Net: an explainable pretrained multi-omics model for multi-task learning in oncology. Genome Biol. 25, 149. https://doi.org/10.1186/s13059-024-03293-9.

Wei, Q., Tan, N., Xiong, S., Luo, W., Xia, H., & Luo, B. (2023). Deep learning methods in medical image-based hepatocellular carcinoma diagnosis: A systematic review and meta-analysis. *Cancers, 15*, 5701. https://doi.org/10.3390/cancers15235701

Xia, Y., Fan, C., Hoadley, K. A., Parker, J. S., & Perou, C. M. (2019). Genetic determinants of the molecular portraits of epithelial cancers. *Nature Communications, 10*, 5666. https://doi.org/10.1038/s41467-019-13588-2

Xu, X., Li, J., Zhu, Z., Zhao, L., Wang, H., Song, C., Chen, Y., Zhao, Q., Yang, J., & Pei, Y. (2024). A comprehensive review on synergy of multi-modal data and AI technologies in medical diagnosis. *Bioengineering, 11*, 219. https://doi.org/10.3390/bioengineering11030219

Yousefi, S., Amrollahi, F., Amgad, M., Dong, C., Lewis, J. E., Song, C., Gutman, D. A., Halani, S. H., Velazquez Vega, J. E., Brat, D. J., & Cooper, L. A. D. (2017). Predicting clinical outcomes from large scale cancer genomic profiles with deep survival models. *Scientific Reports, 7*, 11707. https://doi.org/10.1038/s41598-017-11817-6

Yu, Y., Niu, J., Yu, Y., Xia, S., & Sun, S. (2024). AI predictive modeling of survival outcomes for renal cancer patients undergoing targeted therapy. *Scientific Reports, 14*, 26156. https://doi.org/10.1038/s41598-024-77638-6

Zeineddine, F. A., Zeineddine, M. A., Yousef, A., Gu, Y., Chowdhury, S., Dasari, A., Huey, R. W., Johnson, B., Kee, B., Lee, M. S., Morelli, M. P., Morris, V. K., Overman, M. J., Parseghian, C., Raghav, K., Willis, J., Wolff, R. A., Kawaguchi, Y., Vauthey, J.-N., Sun, R., Kopetz, S., & Shen, J. P. (2023). Survival improvement for patients with metastatic colorectal cancer over twenty years. *npj Precision Oncology, 7*, 1–9. https://doi.org/10.1038/s41698-023-00353-4

Zhai, T.-T., Wesseling, F., Langendijk, J. A., Shi, Z., Kalendralis, P., van Dijk, L. V., Hoebers, F., Steenbakkers, R. J. H. M., Dekker, A., Wee, L., & Sijtsema, N. M. (2021). External validation of nodal failure prediction models including radiomics in head and neck cancer. *Oral Oncology, 112*, 105083. https://doi.org/10.1016/j.oraloncology.2020.105083

Zhang, Z., & Wei, X. (2023). Artificial intelligence-assisted selection and efficacy prediction of antineoplastic strategies for precision cancer therapy. *Seminars in Cancer Biology, 90*, 57–72. https://doi.org/10.1016/j.semcancer.2023.02.005

Zhang, B., Shi, H., & Wang, H. (2023). Machine learning and AI in cancer prognosis, prediction, and treatment selection: A critical approach. *Journal of Multidisciplinary Healthcare, 16*, 1779.

Zhong, N.-N., Wang, H.-Q., Huang, X.-Y., Li, Z.-Z., Cao, L.-M., Huo, F.-Y., Liu, B., & Bu, L.-L. (2023). Enhancing head and neck tumor management with artificial intelligence: Integration and perspectives. *Seminars in Cancer Biology, 95*, 52–74. https://doi.org/10.1016/j.semcancer.2023.07.002

Zhang, B., Zhou, J., Wu, S., Guo, M., Zhang, Y., Liu, R., 2020. The impact of surgical margin status on prostate cancer-specific mortality after radical prostatectomy: a systematic review and meta-analysis. Clin. Transl. Oncol. 22, 2087–2096. https://doi.org/10.1007/s12094-020-02358-y

Zhu, M., Sali, R., Baba, F., Khasawneh, H., Ryndin, M., Leveillee, R. J., Hurwitz, M. D., Lui, K., Dixon, C., & Zhang, D. Y. (2024). Artificial intelligence in pathologic diagnosis, prognosis and prediction of prostate cancer. *American Journal of Clinical and Experimental Urology, 12*, 200–215. https://doi.org/10.62347/JSAE9732

Genomics and Transcriptomics-Based Predictive and Prognostic Biomarkers

Shrishti Bhatt, Ashwin Kandelwal, Siddhartha Maiti, and Neetu Kalra

1 Introduction

Cancer is the deadliest disease of the twentieth century and the reason for millions of deaths all around the world each year. The uncontrolled proliferation of cells mainly characterizes the disease. Several cancer-related factors include genetic mutations in somatic or germline cells, changes in transcriptional gene expression, and posttranslational modifications (Hayes et al., 2012). There are more than 200 types of cancers with limited treatment options. One of the main reasons for high cancer deaths is due to the limited screening options at the early stages of cancer. Because of this, there is a large focus on cancer-associated biomarkers that can be useful in screening and treatment. Cancer biomarkers are the biomolecules produced by the cancer or the host's response against cancer during tumorigenesis and progression of the disease. It is a way of differentiation between healthy and diseased patients and acts as indicators of pathological conditions; they can also act as a pharmacological response to drug treatment. Biomarkers can be DNAs, RNAs, transcription factors, cell surface receptors, metabolites, or secreted proteins produced by the tumors or the host in response to cancer (Silva et al., 2023). With the development of technologies, methods for biomarkers hunting have changed, including genomics and transcriptomics studies. Genomics checks for molecular mutations in a gene or panel of genes.

In contrast, transcriptomics is used to evaluate the expression of proteins or metabolites, mainly upregulation or downregulation of transcribable genes (Walsh et al., 2017). These high-throughput sequencing methods show promising results in

Authors Shrishti Bhatt and Ashwin Kandelwal have been equally contributed to this chapter.

S. Bhatt · S. Maiti
School of Arts and Sciences, Azim Premji University, Bhopal, MP, India

A. Kandelwal · N. Kalra (✉)
School of Bioengineering, Vellore Institute of Technology, Sehore, MP, India
e-mail: neetu.kalra@apu.edu.in

R. Nema et al. (eds.), *Advances in Cancer Detection, Prediction, and Prognosis Using Artificial Intelligence and Machine Learning*,
https://doi.org/10.1007/978-981-96-9346-7_10

early-stage screening, predicting treatment outcomes, and determining correct medication based on the identified molecular variations. This can be part of personalized medicine that can help health professionals find accurate treatments based on an individual's genetic profile.

2 Genomics and Transcriptomics-Based Predictive and Prognostic Biomarkers

Genomic and transcriptomic biomarkers revolutionize predictive and prognostic medicine by organizing patient outcomes, treatment modes, and disease evolution based on DNA and RNA levels. These biomarkers can track biological processes with remarkable resolution, allowing for the evaluation of genetics customization, which can improve treatment benefits while minimizing side effects. In recent years, advances in genomics and transcriptomics have opened the way for biomarker discovery, which is required for detecting gene abnormalities, gene fusion, expression levels of mRNA, expression of miRNA, and noncoding RNAs. These changes help in understanding the disease outcome and treatment options. Clinical genomics and transcriptomics enable effective early diagnosis, treatment changes, and forecasts of outcomes. This strategy enhances physiology-based drug development, optimizes active pharmacotherapy strategies, and allows for the design of sustainable resources and novel therapies targeting high-activity disease zones. Advanced sequencing devices and data management systems are set to shift healthcare paradigms, making treatments more affordable and specific to patients based on genetic and biomarker-engineered treatment strategies.

2.1 Predictive Biomarkers

A predictive biomarker is defined as the disease or the host characteristics that estimate the chance of improvement in outcome with a particular treatment. These biomarkers mainly include DNA and RNA alterations that indicate whether the patient will respond to a specific treatment, including chemotherapy, immunotherapy, or hormonal therapy, enabling a personalized medicine approach. Predictive biomarkers are a feature of enrichment strategies in the design and conduct of clinical trials (Antman & Loscalzo, 2016). Using genomics and transcriptomics in biomarker research has significantly increased our understanding of diagnosis and treatment. Genomics can be very useful not only in the identification of cancer but also in treatment, and it can predict which drug gives the best outcome and which might be extremely toxic after the treatment (Kaur et al., 2023). Predictive genomics biomarkers can be used to modify drug doses and give alternative options for nonresponders. One of the key examples of a genomics-based predictive biomarker is the TPMT or TYMS gene, which is involved in the methylation of endogenous thiopurines and competes, with the HGPRT gene to make them inactive. Similarly, it can also metabolize thiopurine drugs used for the treatment of cancer. One such example is mercaptopurine, which is a prodrug. Upon action by the HGPRT, it turns into an active form of TPN (thiopurine nucleotide) to treat various types of cancer. This gene is highly polymorphic and due to this, the activity of this gene varies in individuals. Because of the high activity of TPMT, the injected drug might get inactivated and not turn into an active drug. If the activity is very low, then the active drug production might increase and may lead to toxicity. So, by TPMT variant genomics study, we can predict if the drug will be effective or might lead to toxicity, in the individual patients converting the prodrug to an active drug. Similarly, using transcriptomics, we can predict which treatment would be more efficient for individual patients. For example, people with high ESR1 expression and low HER2 expression indicate sensitivity to hormonal therapy with tamoxifen (Kim et al., 2011).

2.2 Prognostic Biomarker

A prognostic biomarker is used to identify the likelihood of a clinical event, disease recurrence, or disease progression in a patient independent of treatment. Prognostic biomarkers identify factors associated with the effect of intervention or exposure. Almost all of the prognostic biomarkers have been routinely used in clinical trials to establish entry and exclusion criteria for trials, thereby identifying higher-risk populations. Prognostic biomarkers inform about disease status by screening, monitoring, and determining the increase or decrease in the internal precursor's probability of being acquired by a disease (Dhama et al., 2019). Another important application of prognostic biomarkers is toward predicting the risk for an event or a poor outcome in a patient. Yet another major use of prognostic biomarkers is in resource allocation in population health: by stratifying the risk for negative clinical and financial outcomes. Table 1 depicts the difference between the prognostic and predictive biomarkers.

3 Identification of Genomic and Proteomic Biomarkers for Different Cancer Types

3.1 Breast Cancer

Breast cancer is the most common type of cancer in women affecting millions of people worldwide. There are numerous internal and external factors associated with it ranging from poor lifestyle, environmental factors, and psychological stress. Around 5–10% of breast cancers are due to genetic mutation and family history (Yi-sheng Sun et al., 2017). Breast cancer is classified based on the surface receptors, namely the estrogen receptor (ER), progesterone receptor (PR), and human epidermal growth factor receptor (HER-2), which act as both prognostic and predictive biomarkers. Patients with specific positive receptors are classified into a specific type of cancer. The most common type of breast cancer is invasive ductal carcinoma, which is classified based on the HER-2 protein expression. BRCA1/

Table 1 Prognostic versus predictive biomarkers

Feature	Prognostic biomarker	Predictive biomarker
Definition	It provides the general patient outcome concerning the overall disease outcome, independently of therapy	Provides data on some effect of application of a particular type of therapeutic intervention
Purpose	In order to evaluate the probable course of disease or survival without regard for particular therapies	This is to predict the response to a particular treatment and to guide therapy choices
Outcome information	It indicates the broad clinical outcome, not depending on treatment, such as survival or progression of disease	It shows the chance of a patient benefiting from a specific treatment, such as response or resistance
Application	This has been used in the identification of patients at an increased or decreased risk for disease recurrence or progression	It is employed in the selection of the right patients who will likely benefit from a given treatment
Examples	ER/PR in breast cancer HER2/neu in breast cancer Mutated K-ras in NSCLC	HER2/neu for trastuzumab therapy in breast cancer EGFR1 mutations for TKIs in NSCLC
Impact on therapy	Helps in deciding the intensity or follow-up but does not help decide the therapy	Therapy choice directly gets affected by it such as targeted therapy and chemotherapy
Therapeutic implications	Helps in understanding the prognosis but therapy choice remains broad	Treatment tailored by the biomarker status of individual patients, e.g., choice of certain drugs
Response prediction	Does not predict the response to a specific treatment	Predicts whether or not a specific treatment will be effective

BRCA2 germline mutations are associated with a lifetime risk of breast cancer and mainly belong to a different class of cancer called triple-negative breast cancer (TNBC), which is negative for all three receptors and is found to be the most aggressive type of breast cancer (Nancie Petrucelli et al., 2020; Haixia Chen et al., 2018). Another common genomic biomarker includes PIK3CA which is responsible for endocrine resistance breast cancer with ER positive and HER-2 negative. Few biomarkers such as CACNG4 act as both genomics and transcriptomics-based biomarkers (Ali Golestan et al., 2023), whereas biomarkers such as PKMYT1 solely act as transcriptomics-based biomarkers, and the levels of these biomarkers are used for a patient's prognosis and disease progression. Patients with high PKMYT1 levels are correlated with poor prognosis in triple-negative breast cancer (Anran Chen et al., 2024). PKMYT1 can potentially aid in controlling the course of progression through the cell cycle by blocking the CDK1 activity necessary to progress cells into mitosis (Anran Chen et al., 2024). Aberrations in EPYC expression have been implicated in cancer metastasis, including breast cancer. These genes together regulate many pathways that are critical for the initiation, progression, and therapeutic response of breast cancer (Golestan et al., 2024). Identification and early-stage screening of breast cancer have improved a lot in the last decade, and biomarkers have played a significant role in diagnosis and specific treatment for individuals.

3.2 Prostate Cancer

Prostate cancer is one of the leading causes of death in men worldwide. It is characterized by uncontrolled cell growth in the prostate gland and is prevalent in older men above 55. It is a slow-growing cancer and does not show symptoms at an early stage, which makes it difficult to detect at its earliest. Mutations in certain genes involved in the androgen pathway or testosterone metabolism are linked to the onset of prostate cancer. In recent years, advances in genomics and transcriptomics have opened the way for biomarker discovery for prostate cancer, TMPRSS2-ERG gene fusion is one of the first gene transfusions in solid tumors, and it is associated with 50% of prostate cancer (Tomlins et al., 2005). It was identified in prostate cancer when a promotor of transmembrane protease serine 2 (TMPRSS2) fused with a coding sequence of erythroblastosis virus E26, which leads to the overexpression of Ets-related gene which is directly linked with the cell cycle and proliferation promoting tumor growth (Jason St John et al., 2012). Along with gene fusion prostate cancer is also induced by specific gene mutations, the most common type of gene mutation in prostate cancer subtypes include BRCA1, ATM, and PTEN (Louise Stone et al., 2019; Ji-Hoon Lee et al., 2024; Miller & Tsai, 2023). At the same time, a different class of molecules is linked to aggressive prostate cancer miRNA; it has been demonstrated that miRNAs including miR-21, miR-221, miR-1290, and miR-375 are overexpressed in prostate cancer and are linked to a bad prognosis and advanced stages of the illness (Ana Paula Alarcón-Zendejas et al., 2022). Notably, miR-21 is a useful prognostic marker since its overexpression is associated with more aggressive tumors and worse treatment outcomes (Ana Paula Alarcón-Zendejas et al., 2022). These are vital in normal physiology, and mutations of genes coding for these miRNAs result in uncontrolled growth, leading to cancer.

The presence of estrogen and progesterone receptors in prostate reveals that estrogen signaling might influence prostate carcinoma development, at least within a few molecular subtypes. Genes involved in the mechanisms of DNA repair include the BRCA2 and ATM genes. The loss of functioning BRCA2 commonly occurs through defective homologous recombination repair; these features of aggressive prostate carcinoma could explain its increased sensitivity to drugs inhibiting PARP enzymes. Although less related to ATM, dramatic genetic changes compromise DNA repair and can affect response to radiation or PARP inhibitors (Na et al., 2017). Another gene commonly lost in prostate cancer is PTEN, whose mutations have been linked to poor prognosis and resistance to some therapies. Its loss also activates the PI3K/AKT pathway, driving tumor progression. Also, mutations in the SPOP gene characterize a specific subtype of prostate cancer. These mutations alter chromatin and protein turnover, thus altering tumor activity and sensitivity to treatment (Hernandez-Llodra et al., 2021). HER-2 overexpression in prostate cancer is less common but can also promote androgen-independent growth and, hence, can become a potential target in cases of resistant disease. High Ki67 expression is related to aggressive tumor behavior and poor prognosis in prostate cancer. KMT2D is a chromatin-modifying enzyme that plays a key role in the regulation of gene expression, and its mutations are likely to lead to changes in epigenetic conditions that favor tumor growth and resist treatment (Warner et al., 2021).

3.3 Lung Cancer

Lung cancer is found to be the most aggressive type of cancer with a high mortality rate. Tobacco and alcohol are the leading causes of lung cancer. Lung cancer is mainly classified into two main categories: non-small cell lung cancer (NSCLC) and small cell lung cancer (SCLC). Non-small cell lung cancers are the prominent type of cancer covering 85% of the total lung cancer cases and small cell lung cancer with 15%. Only a handful of patients can get diagnosed early for NSCLC due to limited early detection methods. Sometimes the symptoms are misdiagnosed with other diseases. A range of biomarkers used in clinical settings can be genetic-based or transcriptomics-based. Mutation in EGFR and KRAS are linked to non-small cell lung cancer, which accounts for 20% of all lung cancers. KRAS mutations are associated with poor prognosis and resistance to EGFR-tyrosine kinase inhibitors (Riely et al., 2008). Anaplastic lymphoma kinase (ALK) gene fusion accounts for 3–7% of non-small cell lung cancer, such a known example is the EML4-ALK rearrangement. It has some very critical implications for the development of cancer. Some of these alterations have led to highly effective treatments involving inhibitors for ALK such as crizotinib and ceritinib, which work quite well against cancers arising with these changes (Sweis et al., 2016; Villalobos & Wistuba, 2017). Understanding transcriptomics data has given promising biomarkers in lung cancer that act as prognostic biomarkers in disease progression. miRNA-21 expression is found to be upregulated in small-cell lung cancer, which is linked to tumor invasion and metastasis (Wang et al., 2022). Similarly, Metastasis-Associated Lung Adenocarcinoma Transcript 1 (MALAT-1) represents a new transcription-based marker that predicts the disease progression based on its expression levels in the cell (Ran et al., 2021). Among these biomarkers, mutations in the EGFR gene, otherwise known as epidermal growth factor receptor, appear to be the most frequently occurring within non-small cell lung carcinoma. These alterations activate many signaling pathways that enormously contribute toward the promotion and progression of tumor growth.

In addition, targeted therapy has emerged, and several tyrosine kinase inhibitors include gefitinib and osimertinib, whose efficacy was established in managing lung cancers harboring an EGFR mutation. Other genetic mutations associated with a small subset of NSCLC are mutations of the BRAF gene, particularly the V600E mutation, which occurs in many patients. It is very frequently treated using BRAF inhibitors, including dabrafenib. This BRAF inhibitor, which is often used together with a MEK inhibitor, targets this pathway. Changes that would include such mutations as the deletion of exon 14; an overexpression of even such a gene have been found to be linked to tumor development and survival of cancer cells in relation to the MET-factor genes. Except for these types of changes, KRAS appears to be among the most frequently mutated genes in lung carcinomas, particularly NSCLC, depending on the history of current or past smoking from any source (Singh et al., 2019).

3.4 Stomach Cancer

Gastric cancer is the fourth most common cancer globally; the most prevalent type of stomach cancer is adenocarcinoma, which accounts for 90–95% of all stomach cancers. Chronic infection with *Helicobacter pylori* is one of the most common factors for stomach cancer. In addition, Epstein Barr virus infection and lifestyle factors contribute to gastric cancer. Recent studies have found various biomarkers related to DNA and RNA (Matsuoka et al., 2024) but have lacked medical support till now. Gastric cancer is a heterogeneous disease, showing a diversified genetic and molecular profile. HER-2 is a gene whose expression levels are used as a genetic biomarker to understand the disease prognosis and aggressiveness of the disease. High expression levels of HER-2 are linked to poor prognosis and aggressive tumors (Ono et al., 2006). Similarly, E-cadherins cell adhesion molecules are required for cell communication, tissue architecture, and homeostasis. Mutation in the E-cadherin (CDH1) gene is a risk factor for hereditary diffuse gastric cancer (HDGC), which accounts for 30% of the family inherited gastric cancers (Wang et al., 2023; Zhao et al., 2021; Liu & Chu, 2014). In addition, few of the genetic markers associated with gastric cancer also include FGFR2, PI3K, MET, and TP53, and changes in these genes are linked to gastric cancer and disease progression (Gordon et al., 2022; Matsuoka & Yashiro, 2014; Zhang et al., 2024; Park et al., 2016). Matrix metalloproteinases, specifically MMP-2 and MMP-9 expressions, are found to be increased by normal levels in gastric cancer conditions and act as the prognosis for the disease (Yuan et al., 2023). CEACAM6 is frequently overexpressed in gastric cancer tissues and promotes cancer progression by enhancing tumor cell growth, proliferation, invasion, and metastasis. The phenomenon is closely associated with epithelial–mesenchymal transition (EMT), a process that promotes the invasion of tumors, partly by activating signaling pathways such as PI3K/AKT. High expression levels are associated with poor outcomes, making it a potential candidate for both diagnostic and therapeutic interventions. Studies have shown that APOC1 is overexpressed in gastric cancer. This protein significantly contributes to the metabolic reprogramming of cancer cells, which enables them to proliferate rapidly and survive under adverse conditions. Although less studied, YF13H12 is also overexpressed in gastric cancer and may be related to CEACAM6 in terms of contributing to tumor invasion and progression. APOE has been associated with cancer invasion and metastasis. It seems that APOE is also overexpressed in gastric cancer and may influence the migration of tumor cells and their interaction with the extracellular matrix, both of which are important steps in the spread of cancer (Yasui et al., 2004). Aberrant activation of the PI3K signaling pathway is very highly recurrent in gastric cancer and confers survival, proliferation, and resistance to programmed cell death. Mutations or amplifications in genes involved with PI3K are found frequently and correlate with a worse prognosis (Matsuoka & Yashiro, 2014). MET is a receptor tyrosine kinase implicated in gastric cancer by conferring proliferation, survival, and metastatic behavior. MET overexpression or activation was frequently found in the advanced type of gastric cancer and was considered a potential target for therapeutic intervention (Matsuoka & Yashiro, 2018). Along with these, a range of biomarkers are being analyzed using genetic and transcriptomics analysis, revealing a few genes associated with cell differentiation and apoptosis, including CEACEM6, APOC1, YF13H12, APOE, and so on (Tasuku Matsuoka et al., 2024).

3.5 Cervical Cancer

Cervical cancer is the fourth most common cancer in women globally, with estimates of 660,000 cases in the year 2020. Almost 99% of cervical cancers are associated with human papillomavirus infection (Yim et al., 2007). With increased cancer cases, the screening protocol for cervical cancer is limited, specifically in low- and middle-income countries (Masita Arip et al., 2022). Cervical carcinogenesis is significantly influenced by the oncoproteins E6 and E7, which are expressed by high-risk human papillomaviruses (HPVs), especially HPV16 and HPV18. HPV genome can be integrated into the host genome in around 76% of cervical intraepithelial neoplasia (Fiona J Ruiz et al., 2021). E7 interacts with the retinoblastoma protein (pRb), which pushes cells into the S phase of the cell cycle, while E6 causes inactivation of the tumor suppressor protein p53, which interferes with cell cycle regulation and prevents apoptosis. E6/E7 mRNA and protein expression levels are crucial indicators for tracking the state of the disease and the effectiveness of treatment because they are associated with a higher risk of disease progression and a poor response to treatment (Fiona J Ruiz et al., 2021). Copy number variation is found to be significantly associated with cervical cancer including amplification and deletion of chromosome regions. For example, variations in chromosomes 3q and 11p are linked to malignancy and progression of the disease (Deng Yan et al., 2017). Along with these, Ki-67 and VEGF are the main indicators of tumor aggressiveness and angiogenesis, which are the crucial targets for therapy (Zhenling liu et al., 2023; Popiel-Kopaczyk et al., 2022; 2023). VEGF is crucial in the promotion of angiogenesis, which is involved in the formation of new blood vessels, a biological process that is indispensable for the growth of tumors as well as their metastasis to other parts of the body. High levels of VEGF are strongly correlated with a poor prognosis for patients, suggesting a more malignant behavior of tumors and increased resistance to various treatment options that may be attempted. Consequently, the VEGF pathway has become a strategic target within anti-angiogenic therapies aimed at inhibiting these harmful processes. Additionally, CYFRA 21-1 is a specific fragment of cytokeratin 19, which is an important marker for epithelial tumors, including conditions such as cervical cancer. When high levels of this marker are present, they suggest a higher degree of tumor aggressiveness and correlate with a significantly poorer prognosis for affected individuals (Laengsri et al., 2018). The E6 and E7 proteins from the HPV oncogenes are responsible for causing cervical cancer through the inhibition of tumor suppressor proteins such as p53 and retinoblastoma (Rb) genes. To this end, it results in an appreciable inhibition of the regulators of the cell cycle, leading to an environment of unrestricted proliferation of cells. Screening of clinical practices for E6 and E7 mRNA expression aids in identifying the existence of the virus and further contributes toward stratification of precancerous lesions toward appropriate intervention strategies (Yim & Park, 2005). Methylation status of PAX-1 and ZNF582 have also emerged as a biomarker useful in the early detection of cervical cancer that ensures early diagnosis and thus efficient and appropriate treatment strategies. In addition, hypermethylation of tumor suppressor gene RASSF1A also contributes to the process of cancer progression (Liang et al., 2021).

3.6 Skin Cancer

Skin cancer is divided into two main types: melanoma skin cancer and nonmelanoma skin cancer. Around 99% of the cancer cases are nonmelanoma skin cancer with 1% being melanoma cancer (Trager et al., 2020). The primary cause of skin cancer is sunburn. Although sunburn often goes away a few days after exposure, chronic sunburn damages skin cells cumulatively on both a genetic and epigenetic level. Preventing skin cancer presents significant challenges, primarily due to the absence of reliable and sensitive methods for delivering objective risk information that can effectively educate and motivate individuals to prevent excessive exposure to UV radiation. With limited early detection methods for skin cancer, biomarkers could significantly improve the diagnosis and targeted therapy. Gene mutations associated with skin cancer such as BRAF mutations account for 40–60% of melanoma and NRAS accounts for 15–30% of skin cancers (Cheng et al., 2018). BRAF mutation leads to hyperactivation of the MAPK pathway, which leads to uncontrolled cell proliferation. Testing for these mutations is significantly important for targeted therapy such as BRAF inhibitors. Genes unique to melanocytes, such as gp100 and MART-1 (Melanoma Antigen Recognized by T Cells 1), are frequently upregulated in melanoma. Since their expression levels are correlated with tumor burden and treatment response (Eva Muñoz-Couselo et al., 2017), they are essential markers for monitoring the course of the disease. Analyzing the levels of these transcripts can help with early diagnosis and efficacy of administered therapy. Biomarkers such as C7, MMP3, KRT14, and miRNA-205-Hsa can be used in early detection strategies and correlate with the tumor behavior and aggressiveness (Liang et al., 2021; Liu et al., 2012; Dar et al., 2011; Sánchez-Sendra et al., 2020). In this context, there are quite a few important biomarkers, such as BRAF, C7, MMP3, KRT14, and miRNA-205-has, that may aid in the diagnosis of skin cancer and identification of its subtypes.

Mutation in BRAF gene (V600E) is prevalent in melanoma as 50% of the cases harbor this mutation. This mutation may lead to the activation of the MAPK/ERK signaling pathway, which then causes uncontrolled cellular proliferation and tumor growth. Targeted treatment options, such as BRAF inhibitors like dabrafenib, have significantly enhanced therapeutic outcomes for melanoma patients possessing BRAF mutations (Sullivan & Flaherty, 2011). On the contrary, it is likely that the use of MEK inhibitors would enhance the positive outcomes of this treatment method for melanoma. Despite limited information about C7 relative to other treatments, it might somehow interact within the melanoma tumor microenvironment due to its involvement with host immunomodulatory mechanisms. At last, deficiencies of the complement system activation correlate with strategies used by tumor cells to evade immune detection and thus facilitate proliferation of cancer cells. MMP3 plays a critical role in the reorganization of the extracellular matrix and processes of invasion and metastasis of cancer cells. Enhanced MMP3 expression encourages angiogenesis and cellular migration and correlates with more aggressive phenotypes of melanoma and other skin malignancies (Nan et al., 2008). Keratin 14 is one of the structural proteins whose abnormality has been found associated with some types of skin cancer. Among the vast number of factors involved, one of the most significant changes is the alteration in the differentiation of epithelial cells followed by subsequent migration. Such parts are crucial in pathways through which cancer can take hold to infiltrate, thereby allowing the progression of malignant diseases. Most importantly, the tumor suppression role of microRNA, hsa-miR-205, is most important for melanoma cases. The main purpose through which it performs is through maintaining control over EMT, as well as other essential pathways of the metastasis development process. Downregulation of this protein has also been associated with prognosis and poor results, suggesting higher metastatic potential. It stands as a potential biomarker with all the evidence about disease progression together with all responses to the proposed therapeutic options (Neagu et al., 2020).

3.7 Oral Cancer

Among all cancers, oral cancer accounts for 2%, with oral squamous cell carcinoma (OSCC) being the most common type of cancer. Tobacco and alcohol are the most common culprits. With notable progress in the detection and treatment, the overall survival rate for 5 years is 50%. In a normal cell, TGF-β is one of the key players in regulating the cell cycle and apoptosis involved in tumor formation (Connolly et al., 2012). During the disease progression, instead of suppressing it promotes TGF-β, which is not able to suppress the cell cycle, and cancer cells producing TGF-β also inhibit the immune system from attacking cancer cells, making them more aggressive and helping to invade different tissues (Lu Deng et al.2024). Patients with poor prognosis show a significant increase in the activity of the TGF-β pathway (Guinney et al., 2015; Calon et al. 2015). The expression level of TGF-β is one of the markers of understanding the disease progression and treatment options. Oral cancer is also connected to hypermethylation of certain genes such as DAPK (death-associated protein kinase), TIMP3 (tissue inhibitor of metalloproteinases 3), and p16 (CDKN2A), which makes them inactive and leads to carcinogenesis, which is used in the early detection of oral cancers (Joanna Katarzyna Strzelczyk et al., 2018). Recent advances in transcriptomics analysis have been invaluable in screening new biomarkers associated with oral cancer, including miRNA31 whose expression levels during the disease elevate, promoting tumor progression (Xiaojiao Lin et al., 2022). Oral cancer is found to have mutations in tumor suppressor genes, which are involved in controlling the cell cycle, apoptosis, and DNA-repair mechanism. These genes include TP53, which is the key tumor suppressor gene (Frederico O Gleber-Netto et al., 2020), and OAZ1 (ornithine decarboxylase antizyme 1) is recognized for its tumor-suppressive properties and acts as a negative regulator of polyamine production (Wang et al., 2024). Overexpression or inhibition of these genes acts as the biomarker in understanding disease behavior and treatment therapy. These mutations affect the protein ability to suppress tumor growth and can lead to uncontrolled cell proliferation. CDKN2A encodes two important tumor suppressors, p16INK4A and p14ARF, that regulate the cell cycle. Mutations or deletions in CDKN2A are commonly observed in melanomas and contribute to evasion of cell proliferation controls. Loss of CDKN2A function promotes malignant transformation by allowing cells to avoid senescence and apoptosis (Padhi et al., 2017). CYP1A1 gene encodes an enzyme involved in the metabolism of carcinogens, polycyclic aromatic hydrocarbons (PAHs). Overexpression or mutation of CYP1A1 may increase metabolic activation of PAHs, leading to DNA damage and increased risk of skin cancer, especially in environmentally exposed individuals. MRP14 (S100A9) (a member of the S100 protein family) plays a role in inflammation and immune response. It is often overexpressed in the inflammatory microenvironment associated with cancer development. In skin cancer, MRP14 promotes tumor growth by regulating the tumor microenvironment. Another gene that encodes for an antizyme inhibitor OAZ1 aids in polyamine metabolism regulation. Dysregulation of it may affect cell proliferation and contribute to skin cancer development by altering polyamine levels that are important for cell proliferation. SAT (spermidine/spermine N1-acetyltransferase) is also a regulator of polyamine metabolism, it affects cell growth and apoptosis, and its alteration may lead to imbalance of polyamine levels and promote tumorigenesis (Khijmatgar et al., 2024).

3.8 Pancreatic Cancer

Pancreatic cancer is the most aggressive type of cancer but has a poor prognosis. Pancreatic ductal cancer is the most aggressive subtype, having an overall survival rate of 5 years at 10% (Brozos-Vazquez et al., 2024). The lack of validated screening and prediction biomarkers for early diagnosis and precision treatment, respectively, makes this even more difficult (Hasan et al., 2019). The relatively low incidence of pancreatic cancer and the specificity of the selected test are obstacles to the development of screening tests for illness (Hasan et al., 2019). Numerous potential biomarkers for the prognosis of pancreatic cancer and the prediction of response to different therapies (chemotherapy, immunotherapy, or targeted therapy) have been investigated. The most well-known biomarkers, which are linked to the benefits of platinum and PARP inhibitors (PARPi), are BRCA1 and BRCA2 mutations (Brozos-Vazquez et al., 2024). CDKN2A (cyclin-dependent kinase inhibitor 2A) codes for two main proteins, p16INK4a and p14ARF, which are involved in cellular homeostasis by inhibiting the cyclin-dependent kinases (CDKs). Mutation in these genes is associated with 90% of pancreatic cancer cases and aggressive tumor growth. Early detection of this mutation can help clinicians to select targeted treatments for the patients such as CDK inhibitors (Kimura et al., 2022; McWilliams et al. 2010). Mutations in SMAD4 and KRAS genes (also known as hallmark genes for pancreatic cancer) are found in 55% and 91% of pancreatic cancer patients, respectively, and have been linked to advanced stages (Lyu et al., 2024; Zhang et al., 2023). miRNAs such as miR-21 and miR-155 are found to be elevated and have been associated with the disease when compared to normal conditions and these genes serve as an important marker in disease prediction and prognosis (Mok et al., 2024). They are essential in unraveling the intricacies of the development, progression, and treatment of pancreatic cancer. These genes are on the homologous recombination repair (HRR) pathway responsible for fixing DNA double-strand breaks. Mutations in these genes compromise DNA repair, and genomic instability often leads to cancer development. In pancreatic cancer, there is a 5–9% possibility of having germline mutations in BRCA1/2 or PALB2. The mutation predisposes one to the targeted drugs of platinum-based chemotherapy and PARP inhibitors that work more as a targeted drug (Ngamruengphong & Canto, 2016). In pancreatic ductal adenocarcinoma, the KRAS G12C mutation is present, with a prevalence of 1–2%. Other inhibitors targeting KRAS G12C are under active investigation (Keane et al., 2024). NTRK fusions and NRG1 rearrangements are extremely rare genetic alterations in pancreatic cancer. Such mutations are potentially targetable and, therefore, may have the potential for treatment with targeted therapies. For instance, TRK inhibitors, in the case of NTRK fusions, as well as experimental drugs for the purpose of targeting NRG1 fusions, have already shown promise in clinical trials. These genetic discoveries hold immense significance in the realm of personalized medicine, where treatment is tailored to the specific genetic makeup of the tumor, thereby enhancing outcomes in a highly aggressive disease such as pancreatic cancer. It is strongly advised that patients undergo genetic testing to pinpoint these mutations and investigate suitable therapies (Jamison et al., 2024) (Fig. 1; Tables 2 and 3).

4 Methodology

Over the last decade, the discovery of biomarkers has seen a tremendous increase. The systems biology techniques as well as chemometrics have been employed for the development and discovery of biomarkers (Davis et al., 2013). Key aspects of the study design include sample collection procedures, sample measurement, data analysis, and interpretation in biomarker discovery studies (Zhou et al., 2018; Rundle et al., 2012; Gosho et al., 2012). Two primary approaches are used for biomarker discovery from biofluid sources such as blood, urine, milk, and cell culture media (Silajdzic and Bjorkqvist, 2018). This targeted approach allows the acquisition of a comprehensive list of the physiology/pathology and disease processes and the discovery of new biomarkers that may be produced or released during the disease process and can be measured by specific methods. This mainly includes smaller metabolites or proteins or specific receptors. An untargeted approach is an unbiased approach to biomarker discovery that only depends on a range of “omics” profiling techniques used for systematically screening the body fluids. This approach deals with genomics, transcriptomics, proteomics, metabolomics, lipidomic, glycomics, and secretomics platforms that can discover molecular biomarkers (Umelo et al., 2018). With advances in sequencing techniques, the innovative approaches for biomarker discovery in cancer detection relieve more efficient and accessible methods than preexisting ones. The trend of nanotechnology provides a transformative

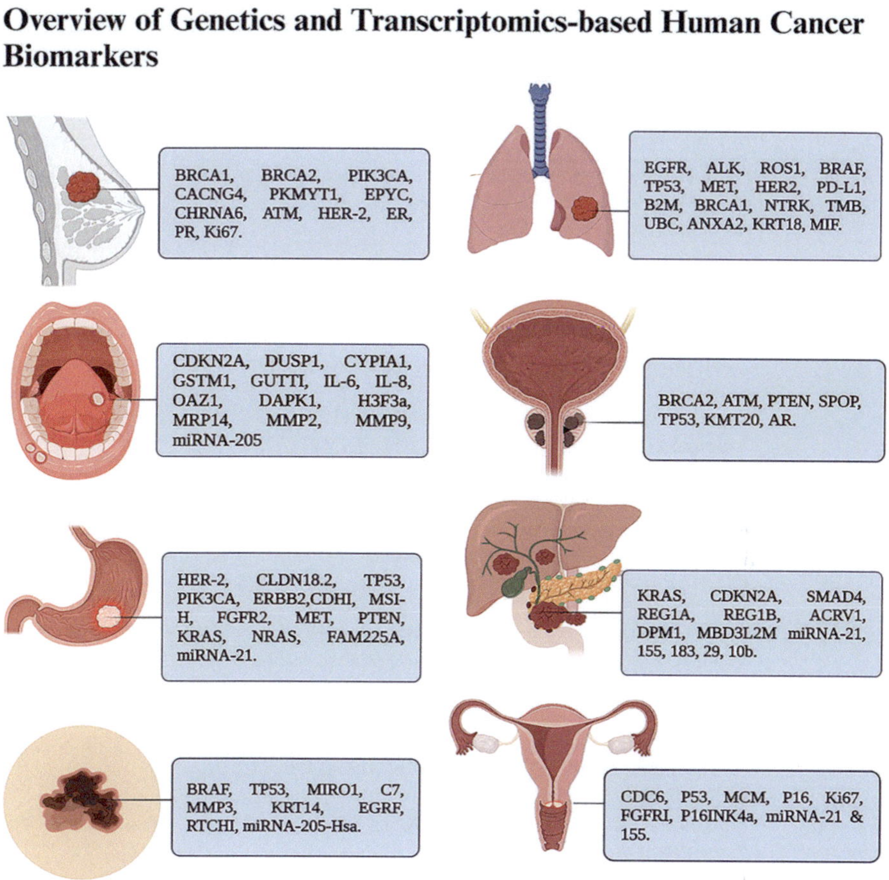

Fig. 1 Overview of genetics and transcriptomics-based human cancer biomarkers

approach to cancer detection by offering enhanced sensitivity, specificity, and early detection capabilities (Srinivas et al., 2002). Along with all the omics data, a recent advance in third-generation DNA and RNA nanopore sequencing has improved our sequencing efficiency. Its current use for long-length DNA and RNA sequencing along with its capability of identifying point mutations, splice variants, copy number alterations, noncoding RNAs, and epigenetic markers (Dyshlovoy et al., 2024; Zhang et al., 2021). Advances in nanopore sequencing can also deal with certain limitations regarding base identification accuracy and high sample size (Dyshlovoy et al., 2024) (Fig. 2).

5 AI in Biomarker Discovery

Artificial intelligence (AI) is revolutionizing biomarker research in various diseases, enabling the discovery, validation, and application of biomarkers. Biomarkers are measurable biological indicators used to diagnose diseases, predict outcomes, and

Table 2 List of genomic biomarkers linked to gene mutations and serving as indicators for multiple cancers

Cancer type	Genomic biomarkers	References
Breast cancer	ATM, BARD1, BRCA1, BRCA2, CHEK2, CDH1, NF1, PALB2, PTEN, RAD51C, RAD51D, STK11, TP53	Walsh et al. (2016)
Prostate cancer	Prostate-specific antigen (PSA), prostate cancer antigen 3 (PCA3), DERL1, CUL2, SMAD4, PDSS2, HSPA9, FUS, phosphorylated S6, YBOX1	Kornberg et al. (2018)
Lung cancer	Epidermal growth factor receptor (EGFR), Anaplastic lymphoma kinase (ALK) translocations, Programmed death-ligand 1 (PD-L1), Cytotoxic T-lymphocyte-associated antigen 4 (CTLA-4), T-cell immunoglobulin and mucin domain-containing protein 3 (TIM3), B- and T-lymphocyte-associated protein (BTLA), V-domain Ig suppressor of T-cell activation (VISTA), Lymphocyte-activation gene 3 (LAG3)	Villalobos and Wistuba (2017)
Stomach cancer	CEA (carcinoembryonic antigen), CA199, CA724, alpha-fetoprotein, CA125, FGFR (fibroblast growth factor receptor), CDH1 (E-cadherin), PI3K (phosphoinositide 3-kinase), MET (mesenchymal-epithelial transition factor), VEGFR (vascular endothelial growth factor receptor), TP53 (tumor protein p53), PD1 (programmed cell death protein 1), FUS, CDH17, COLIA1 and COLIA2, APOE (apolipoprotein E), CEACAM6, REG4, OLFM4, HOXA10, DSC2, TSPAN8, TM9SF3	Matsuoka and Yashiro (2018)
Cervical cancer	Squamous cell carcinoma antigen (SCC), TP53, HRAS, Kras, PIK3CA, HPV E6 and E7 oncoproteins, p16INK4a, Squamous cell carcinoma antigen (SCC-Ag), CA-125, cell-free DNA (cfDNA)	Yim and Park (2007)
Skin cancer	RASSF1A, MAGE (melanoma antigen), miR-34b, GalNAc-T, MART-1 and MAGE-A3, RAR-b2 (DNA methyltransferases)-DNMT3A and DNMT3B	Greenberg et al. (2014)
Oral cancer	p53 autoantibodies, IL-6 (interleukin-6), TNF-α (tumor necrosis factor-alpha), MMP-9 (matrix metalloproteinase-9) LDH (lactate dehydrogenase), Maspin Cyclin D1, Ki67, Mac-2 binding protein (M2BP), Myeloid-related protein 14 (MRP14), CD59, Profilin 1, transferrin, TP53	Shah et al. (2011)
Pancreatic cancer	Carbohydrate antigen 19–9 (CA19–9), KRAS, GNAS, CEA, MUC4, IL-1B, sFASL, MMP9, MicroRNAs: miR-21, miR-221, and miR-30a-3p	O'Neill and Stoita (2021)

aid treatment decisions. Traditional methods are labor-intensive and time-consuming, but AI, using machine learning and deep learning, efficiently analyzes vast datasets, uncovering hidden patterns, and accelerating biomarker discovery.

Table 3 List of biomarkers for different types of cancers obtained by the transcriptome profiling of the cancerous tissue

Cancer type	Transcriptomic biomarkers	References
Breast cancer	AURKB (aurora kinase B), PLK1(polo-like kinase 1), TTK (threonine and tyrosine kinase), UBE2C (ubiquitinin-conjugating enzyme E2C), CDCA8 (cell division cycle associated 8), KIF15 (kinesin family member 15), CDC45 (cell division cycle 45), DEPDC1, OIP5, ZWINT	Shornale Akter et al. (2024)
Prostate cancer	MYC, AMACR, GSTP1, TP63, CENPA, PIK3CB, AR (androgen receptor), EZH2, SRD5A1PGM5, FRMD6, ARM1, PID1, ACSS3, EVA1C, SELENOM	Marzec et al. (2021)
Lung cancer	Keratin: KRT5, KRT6A-C, KRT13-KRT17, KRT19, FOXM1, HOXC4, GSTZ1	Sutic et al. (2024)
Stomach cancer	REG4, LCN2, CEACAM5, MMP-9, ANXA1 and ANXA4, CCL22 and CCL28, AQP9, E2F1, TIMP1, S100A9, SERPINE1, OSMR, ITGA3, FGFR4, SLC2A1, CXCR2, CARD14, CXCR1	Carino et al. (2021)
Cervical cancer	HNF1A-AS, VIL1, MIR194-2HG, SSTR5-AS1, miRNAs: miR-194-3p, miR-147b-3p, miR-335-3p, miR-615-3p, miR-501-5p, and miR-761	Xu et al. (2020)
Skin cancer	IL6 (interleukin 6), CCND2 (cyclin D2), PLAUR (plasminogen activator, urokinase receptor), CD44 (cluster of differentiation 44)	Rambabu et al. (2024)
Oral cancer	YWHAZ, RHOA, DLG1, LY6E, PLEC, EIF4A2, CTNNA1, PMEPA1	Patel et al. (2021)
Pancreatic cancer	ANXA8, FAM83A, KRT6A, MET, MUC16, NT5E, SLC2A1	Rambabu et al. (2023) and Posta and Gyorffy (2023)

Integrating AI in biological data analysis can help us analyze large-scale multi-omics datasets, generating high-dimensional data that is difficult to interpret with current statistical approaches. This helps with more accurate and efficient prediction of biomarkers, particularly in oncology-related areas. AI models can also predict patients' responses to treatments, making therapeutic approaches more personalized. However, data quality, model interpretability, and ethical issues surrounding data privacy remain a challenge. Addressing these issues will unlock full AI capability in biological research.

6 AI Models for Predicting Treatment Responses

The discovery of AI has changed the world of biomedical sciences from understanding complex omics data to finding possible biomarkers for diseases. There is a list of AI models used in biomarker research, broadly classified into three main categories: supervised, semi-supervised, and unsupervised learning. Supervised learning uses labeled data to train models for classification (predicting categories) and

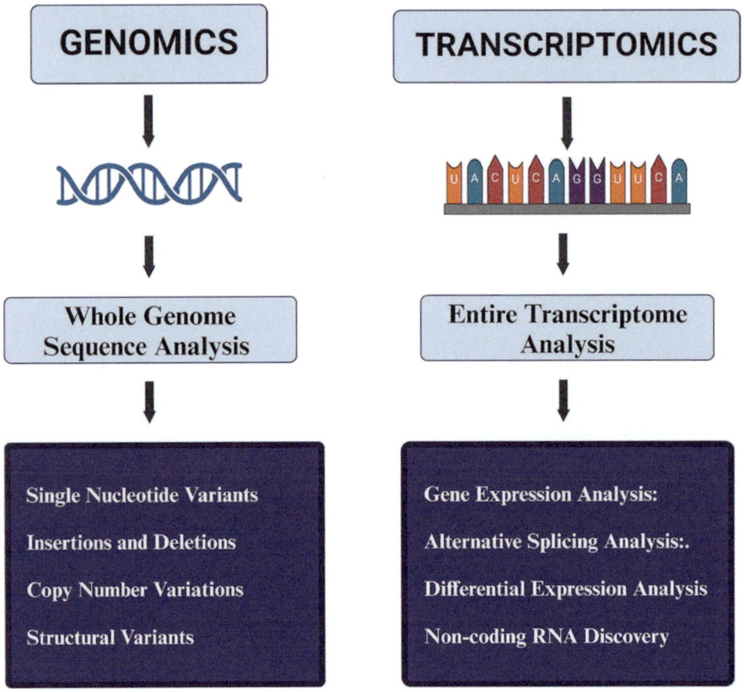

Fig. 2 Schematic representation of genomics and transcriptomics-based biomarker discovery flowchart

regression (predicting continuous values). Common supervised algorithms in cancer research include random forest (RF) (Breiman et al., 2001) and support vector machines (SVM) (10.1023/A:1010950718922). Semi-supervised learning deals with partially labeled datasets, useful for tasks such as medical image annotation, using methods such as multiple instance learning (MIL) (Ilse et al., 2018) and graph convolutional networks (GCNs) (Kipf & Welling, 2016). Unsupervised learning uncovers patterns from data through clustering and dimensionality reduction, with techniques such as principal component analysis (PCA) and k-means clustering. In oncology, the Cox proportional hazard model (Cox, 1972) is often used to analyze prognosis factors. AI techniques such as random survival forest (RSF) (Ishwaran et al., 2008) are also applied in survival analysis due to its ability to handle censored data common in oncology research. These AI models in biomarker research predict treatment responses. They assess large amounts of genomic, transcriptomic, and clinical data to increase our understanding of the response of patients to a particular treatment. These models help stratify patients into subgroups that will be responsive or resistant to certain therapies.

6.1 Random Forest (RF)

Random forest (RF) (Breiman, 2001) is an ensemble-based machine learning (ML) approach based on decision trees, which is both classification and regression-based. It is currently used in biomarker discovery for cancer and biomedical research. RF is found to be effective in handling complex data such as gene expression, proteomics, and genomics. It can work with mixed data, making it suitable for classification and regression studies, one of the key requirements for biomarker discoveries. It is currently used for gene expression and multi-omics studies focused on cancer research.

6.2 Deep Learning (DL) Models

Deep learning could potentially serve as a future substitute for modeling complex traits and integrating previously restricted multidimensional medical imaging datasets (Chen et al., 2018). DL techniques, for example, graph neural networks (GNN), are lately found to be promising tools in the research on tumor metastasis, applied to gene regulation and metabolic pathways (Kazmierski & Haibe-Kains, 2021). GNN stands for an aspect of a neural network that acts directly on the graph structure consisting of several nodes where every node represents an entity. One of the most common applications of GNN is node classification. In addition to the creation of generative DL models, which are now being used to discover drug targets and design de novo investigational new drugs for cancer research (Prykhodko et al., 2019).

6.3 K-Nearest Neighbors (K-NN)

The k-nearest neighbor (KNN) algorithms are applied to provide significant insight into the relationship concerning clinical factors such as age, mutated genes, and mutations impacting the survival time of cancer patients. KNN is the easiest algorithm of machine learning that is used in both classification and regression problems (Siddalingappa & Kanagaraj, 2022). The algorithm calculates the distances of an unknown data point to its nearest data points in the training set and classifies the new point to that class it belongs to. The choice of "k" depends on the number of data points chosen from the training dataset. The algorithm first looks for a numerical feature to most precisely represent the test data point by selecting a distance calculation metric to use (Siddalingappa & Kanagaraj, 2022). The distances of the new/unknown point and the "k" points are measured using a selected distance measuring metric, including Euclidean distance, Manhattan distance, and Minkowski distance, among others (Ali et al., 2019).

6.4 Bayesian Networks

This network-based sparse Bayesian machine (NBSBM) method integrates a sparse Bayesian classifier into a Laplace graph, which is designed by a disease-related signaling network. The approach is capable of selecting the most predictive networks for being the biomarker of drug sensitivity or resistance prediction (Liu et al., 2019). It can include external knowledge about the parameters via the prior distribution. This technique has the superiority that it avoids the requirement of determination of the vast majority of arbitrary cut-off selections typically involved within most proteomic analyses such as a "significant" fold change or a "significant" q value (Hernández et al., 2015).

6.5 Reinforcement Learning (RL)

An RL agent acts in its environment over time by choosing actions contingent on the observed environments and following a policy to maximize some cumulative reward. Therefore, in a multimodal patient data scenario, an RL agent might be presented with demographics, laboratory values, tumor burden, and therapy-associated toxicities as representing the environment. Treatment regimens in oncology are most often longitudinal decision-making processes where patient variables as well as response to treatment and toxicities influence the oncologist's choice to optimize patient safety and outcome in the long run. This clinical framework can be translated into a set of sequential actions in an environment that result in iterative state alterations (Chakraborty & Murphy, 2014).

6.6 XGBoost (eXtreme Gradient Boosting)

An extreme gradient boosting (XGB) significantly reduces the problem of overfitting as it induces more regularization, tree pruning, and subsampling from within. It also uses the multithreading computing capacity provided by modern CPUs (Li et al., 2022).

7 Future Perspectives

The future of biomarker research appears promising, especially with advancing AI technologies that enhance the processing and analysis of increasingly complex biological data. AI's ability to integrate multi-omics data—such as proteomics, epigenomics, and metabolomics—offers broader insights into disease mechanisms, unlike traditional methods focusing on single data types. Federated learning addresses patient privacy concerns by enabling AI model training on decentralized data, promoting collaboration for rare diseases. Additionally, explainable AI (XAI) enhances model transparency, which is crucial for clinician confidence in identified biomarkers. This combination of advancements may lead to the discovery of clinically relevant biomarkers, improving disease prediction and therapeutic outcomes.

Reinforcement learning (RL) represents a dynamic AI technology with future potential in optimizing biomarker-based treatments. Unlike static traditional AI, RL adapts by learning from patients' real-world responses, allowing for personalized treatment strategies that adjust according to biomarker changes. This is particularly applicable in oncology for timely, dose-specific adjustments to improve therapeutic outcomes. Wearable technologies can also transform biomarker research by enabling continuous monitoring of physiological parameters and facilitating proactive interventions in chronic diseases. Additionally, graph neural networks (GNNs) enhance biomarker identification through network-centric analysis, recognizing interactions within biological systems. Lastly, quantum computing's rapid data processing may unlock new possibilities in biomarker discovery, enhancing AI's capabilities in omics research.

AI is set to play a significant role in natural language processing (NLP), particularly in biomarker research. Currently, NLP technologies are used to mine extensive scientific literature for biomarker data. Future advancements will enable NLP models to not only mine but also generate hypotheses, recommend potential biomarkers, and design experiments, thus expediting discovery. As biomedical knowledge grows, AI will help researchers integrate findings into clinical practice rapidly. The future of AI-based biomarker research looks promising with innovations in multi-omics data integration, federated learning, and more, ultimately enhancing precision medicine and improving patient outcomes while addressing challenges such as data privacy and regulatory compliance.

8 Conclusion

Biomarkers are the key molecules for the identification, progress, and treatment of diseases. And significantly contribute to advancing biomedical sciences. Combining bioinformatics and artificial intelligence enhances the use of genomics and transcriptomics as diagnostic and prognostic tools in precision medicine, which is a great leap forward. This has immense potential in the management of patients. The primary application of this modality (prognostic biomarkers) is the evaluation of the possible development of diseases, while the secondary application is the selection of treatment (predictive biomarkers). Considering that conditions such as these vary greatly both geographically and even at the genetic and cellular levels within a person, there is a need for personal biomarkers, which would ensure the right care, but at a lower cost and greater effectiveness. This is because it provides a perspective based on extensive analysis that is due to the progress made in AI, which until now, has been impossible by traditional ways of analyzing.

The difference between AI approaches such as machine learning and deep learning and traditional methods lies in their ability to handle high-dimensional genomic and transcriptomic data. The volume and complexity of this data often overwhelm classical analytical approaches. For instance, AI algorithms able to analyze multi-omics data may help to identify hidden patterns associated with potential biomarkers of treatment response or disease outcome. For example, in cancer treatment, ML models that are developed based on gene expression and other types of molecular profiles can distinguish immune markers, gene expression patterns, and mutation profiles indicative of a positive response to immunotherapy in a particular patient. It allows better prediction because it keeps learning new patients' data, improving accuracy with time. Therefore, these AI-enhanced biomarkers can undergo similar analysis in the final stages.

References

Addeo, A., Shah, P. K., Bordry, N., Hudson, R. D., Albracht, B., Di Marco, M., ... & Shah, D. P. (2021). Immunogenicity of SARS-CoV-2 messenger RNA vaccines in patients with cancer. Cancer cell, 39(8), 1091–1098

Antman, E. M., & Loscalzo, J. (2016). Precision medicine in cardiology. Nature Reviews Cardiology, 13(10), 591–602

Arip, M., Tan, L. F., Jayaraj, R., Abdullah, M., Rajagopal, M., & Selvaraja, M. (2022). Exploration of biomarkers for the diagnosis, treatment and prognosis of cervical cancer: a review. Discover Oncology, 13(1), 91

Alarcón-Zendejas, A. P., Gonzalez, I. R., & Gutiérrez-Camacho, C. (2022). The promising role of new molecular biomarkers in prostate cancer: From coding and non-coding genes to artificial intelligence approaches. *Nature Reviews Urology, 19*(2), 104–117. https://doi.org/10.1038/s41585-022-00537-2

Ali, N., Neagu, D., & Trundle, P. J. S. A. S. (2019). Evaluation of k-nearest neighbor classifier performance for heterogeneous datasets. *Journal of Heterogeneous Data Management, 1*, 1–15. https://doi.org/10.1007/s40552-019-00017-x

Breiman, L. (2001). Random forests. *Machine Learning, 45*(1), 5–32. https://doi.org/10.1023/A:1010933404324

Brozos-Vázquez, E., Toledano-Fonseca, M., Costa-Fraga, N., García-Ortiz, M. V., Díaz-Lagares, Á., RodríguezAriza,A., & López-López, R. (2024). Pancreatic cancer biomarkers: A pathway to advance in personalized treatment selection. Cancer treatment reviews, 125, 102719.

Caraceni, A., Davies, A., Poulain, P., Cortés-Funes, H., Panchal, S. J., & Fanelli, G. (2013). Guidelines for the management of breakthrough pain in patients with cancer. Journal of the National Comprehensive Cancer Network, 11(suppl_1), S–29

Calon, A., Lonardo, E., Berenguer-Llergo, A., Espinet, E., Hernando-Momblona, X., Iglesias, M., ... & Batlle, E. (2015). Stromal gene expression defines poor-prognosis subtypes in colorectal cancer. *Nature genetics, 47*(4), 320–329

Carino, A., Graziosi, L., Marchianò, S., Biagioli, M., Marino, E., Sepe, V., & Fiorucci, S. (2021). Analysis of gastric cancer transcriptome allows the identification of histotype specific molecular signatures with prognostic potential. Frontiers in oncology, 11, 663771

Chakraborty, B., & Murphy, S. A. (2014). Dynamic treatment regimes. Annual review of statistics and its application, 1(1), 447–464

Chaturvedi, V. K., Singh, A., Singh, V. K., & Singh, M. P. (2019). Cancer nanotechnology: a new revolution for cancer diagnosis and therapy. Current drug metabolism, 20(6), 416–429

Chen, J., Zhao, Y., & Zhang, L. (2024). Genome-wide DNA methylation profiling reveals a novel biomarker in gastric cancer. *Scientific Reports, 14*, 19575. https://doi.org/10.1038/s41598-024-78135-6

Chen, H., Wu, J., Zhang, Z., Tang, Y., Li, X., Liu, S., ... & Li, X. (2018). Association between BRCA status and triple-negative breast cancer: a meta-analysis. Frontiers in pharmacology, 9, 909

Connolly, R., & Stearns, V. (2012). Epigenetics as a therapeutic target in breast cancer. Journal of mammary gland biology and neoplasia, 17(3), 191–204

Cox, D. R. (1972). Regression models and lifetables. *Journal of the Royal Statistical Society: Series B (Statistical Methodology), 34*(2), 187–202. https://doi.org/10.1111/j.2517-6161.1972.tb00899.x

Dar, A. A., Majid, S., de Semir, D., Nosrati, M., Bezrookove, V., & Kashani-Sabet, M. (2011). miRNA-205 suppresses melanoma cell proliferation and induces senescence via regulation of E2F1 protein. *The Journal of Biological Chemistry, 286*(19), 16606–16614. https://doi.org/10.1074/jbc.M111.227611

Dhama, K., Latheef, S. K., Dadar, M., Samad, H. A., Munjal, A., Khandia, R., ... & Joshi, S. K. (2019). Biomarkers in stress related diseases/disorders: diagnostic, prognostic, and therapeutic values. Frontiers in molecular biosciences, 6, 465402

Dyshlovoy, S. A., Paigin, S., Afflerbach, A. K., Lobermeyer, A., Werner, S., Schüller, U., ... & Joosse, S. A. (2024). Applications of Nanopore sequencing in precision cancer medicine. *International Journal of Cancer, 155*(12), 2129–2140

Geskin, L. J., Trager, M. H., Aasi, S. Z., Bickers, D. R., Carvajal, R. D., Nghiem, P., ... & Samie, F. H. (2020). Perspectives on the recommendations for skin cancer management during the COVID-19 pandemic. Journal of the American Academy of Dermatology, 83(1), 295.

Gleber-Netto, F. O., Neskey, D., Costa, A. F. M., Kataria, P., Rao, X., Wang, J., Kowalski, L. P., Pickering, C. R., Dias-Neto, E., & Myers, J. N. (2020). Functionally impactful TP53 mutations are associated with increased risk of extranodal extension in clinically advanced oral squamous cell carcinoma. *Cancer, 126*(20), 4498–4510. https://doi.org/10.1002/cncr.33101

Gordon, A., Johnston, E., Lau, D. K., & Starling, N. (2022). Targeting FGFR2 positive gastroesophageal cancer: Current and clinical developments. *Oncotargets and Therapy, 15*, 1183–1196. https://doi.org/10.2147/OTT.S282718

Gosho, M., Nagashima, K., & Sato, Y. (2012). Study designs and statistical analyses for biomarker research. *Sensors (Basel), 12*(7), 8966–8986. https://doi.org/10.3390/s120708966

Golestan, A., Tahmasebi, A., Maghsoodi, N., Faraji, S. N., Irajie, C., & Ramezani, A. (2024). Unveiling promising breast cancer biomarkers: an integrative approach combining bioinformatics analysis and experimental verification. *BMC cancer, 24*(1), 155.

Guinney, J., Dienstmann, R., Wang, X., De Reynies, A., Schlicker, A., Soneson, C., ... & Tejpar, S. (2015). The consensus molecular subtypes of colorectal cancer. *Nature medicine, 21*(11), 1350–1356.

Hasan, M., Barman, S. D., Islam, S., & Reza, A. W. (2019, April). Skin cancer detection using convolutional neural network. In Proceedings of the 2019 5th international conference on computing and artificial intelligence (pp. 254–258)

Hayes, J. H., Ollendorf, D. A., Pearson, S. D., Barry, M. J., Kantoff, P. W., Lee, P. A., & McMahon, P. M. (2013). Observation versus initial treatment for men with localized, low-risk prostate cancer: a cost-effectiveness analysis. *Annals of internal medicine, 158*(12), 853–860

Herrero, R., Parsonnet, J., & Greenberg, E. R. (2014). Prevention of gastric cancer. Jama, 312(12), 1197–1198.

Hernández-Llodrà, S., Segales, L., Juanpere, N., Marta Lorenzo, T., Salido, M., Nonell, L., ... & Lloreta-Trull, J. (2021). SPOP and CHD1 alterations in prostate cancer: Relationship with PTEN loss, tumor grade, perineural infiltration, and PSA recurrence. The Prostate, 81(16), 1267–1277

Hernández-Hernández, D.M., Apresa-García, T., & Patlán-Pérez, R.M. (2015). Epidemiological overview of cervical cancer. Medical Journal of the Mexican Social Security Institute , 53 (2), S154–S161

Henry, N. L., & Hayes, D. F. (2012). Cancer biomarkers. Molecular oncology, 6(2), 140-146.

Ilse, M., Tomczak, J., & Welling, M. (2018). Attention-based deep multiple instance learning. In *Proceedings of the international conference on machine learning (PMLR)* (pp. 2127–2136) http://proceedings.mlr.press/v80/ilse18a.html

Ishwaran, H., Kogalur, U. B., Blackstone, E. H., & Lauer, M. S. (2008). Random survival forests. *The Annals of Applied Statistics, 2*(3), 841–860. https://doi.org/10.1214/08-AOAS169

Jamison, R. (2024). Decreasing Overall Breast Cancer Risk Through Assessment, Treatment, and Prevention of Vitamin D Deficiency in the Adult Population.

Kazmierski, M., & Haibe-Kains, B. (2021). Lymph node graph neural networks for cancer metastasis prediction. *Journal of Pathology Informatics, 12*, 1–17. https://doi.org/10.4103/jpi.jpi_57_20

Keane, F., Chou, J. F., Walch, H., Schoenfeld, J., Singhal, A., Cowzer, D., ... & O'Reilly, E. M. (2024). Precision medicine for pancreatic cancer: characterizing the clinicogenomic landscape and outcomes of KRAS G12Cmutated disease. JNCI: *Journal of the National Cancer Institute, 116*(9), 1429–1438.

Khijmatgar, S., Yong, J., Rübsamen, N., Lorusso, F., Rai, P., Cenzato, N., & Tartaglia, G. M. (2024). Salivary biomarkers for early detection of oral squamous cell carcinoma (OSCC) and head/neck squamous cell carcinoma (HNSCC): A systematic review and network meta-analysis. Japanese Dental Science Review, 60, 32–39.

Kimura, H., Mizuno, K., Shiota, M. et al. (2022) Prognostic significance of pathogenic variants in BRCA1, BRCA2, ATM and PALB2 genes in men undergoing hormonal therapy for advanced prostate cancer. Br J Cancer 127, 1680–1690. https://doi.org/10.1038/s41416-022-01915-2

Kim, C., Tang, G., Pogue-Geile, K. L., Costantino, J. P., Baehner, F. L., Baker, J., Cronin, M. T., Watson, D., Shak, S., Bohn, O. L., Fumagalli, D., Taniyama, Y., Lee, A., Reilly, M. L., Vogel, V. G., McCaskill-Stevens, W., Ford, L. G., Geyer, C. E., Jr., Wickerham, D. L., Wolmark, N., et al. (2011). Estrogen receptor (ESR1) mRNA expression and benefit from tamoxifen in the treatment and prevention of estrogen receptor-positive breast cancer. *Journal of Clinical Oncology: Official Journal of the American Society of Clinical Oncology, 29*(31), 4160–4167. https://doi.org/10.1200/JCO.2010.32.9615

Kimura, H., Mizuno, K., Shiota, M. et al. (2022) Prognostic significance of pathogenic variants in BRCA1, BRCA2, ATM and PALB2 genes in men undergoing hormonal therapy for advanced prostate cancer. Br J Cancer 127, 1680– 1690. https://doi.org/10.1038/s41416-022-01915-2

Kipf, T. N., & Welling, M. (2016). Semi-supervised classification with graph convolutional networks. Proceedings of the 5th International Conference on Learning Representations (ICLR). https://openreview.net/forum?id=H1C1Pu5gg

Kornberg, Z., Cooperberg, M. R., Spratt, D. E., & Feng, F. Y. (2018). Genomic biomarkers in prostate cancer. Translational andrology and urology, 7(3), 459.

Laengsri, V., Kerdpin, U., Plabplueng, C., Treeratanapiboon, L., & Nuchnoi, P. (2018). Cervical cancer markers: epigenetics and microRNAs. *Laboratory medicine, 49*(2), 97–111

Lee, J. H., Yim, S. H., Won, Y. J., Jung, K. W., Son, B. H., Lee, H. D., ... & Members of Korean Breast Cancer Society (KBCS. (2007). Population-based breast cancer statistics in Korea during 1993-2002: incidence, mortality, and survival. Journal of Korean medical science, 22(Suppl), S11.

Li, J., Xu, C., Lee, H. J., Ren, S., Zi, X., Zhang, Z., & Sun, Y. (2020). A genomic and epigenomic atlas of prostate cancer in Asian populations. Nature, 580(7801), 93–99

Li, K., Yao, S., Zhang, Z., Cao, B., Wilson, C. M., Kalos, D., Kuan, P. F., Zhu, R., & Wang, X. (2022). Efficient gradient boosting for prognostic biomarker discovery. *Bioinformatics, 38*, 1631–1638. https://doi.org/10.1093/bioinformatics/btab833

Li, Y., Dou, Y., Leprevost, F. D. V., Geffen, Y., Calinawan, A. P., Aguet, F., ... & Chan, D. W. (2023). Proteogenomic data and resources for pan-cancer analysis. Cancer cell, 41(8), 1397–1406.

Lin, X., Wu, W., Ying, Y., Luo, J., Xu, X., Zheng, L., Wu, W., Yang, S., & Zhao, S. (2022). MicroRNA-31: A pivotal oncogenic factor in oral squamous cell carcinoma. *Cell Death Discovery, 8*(1), 140. https://doi.org/10.1038/s41420-022-00948-z

Liu, X., & Chu, K. M. (2014). E-cadherin and gastric cancer: Cause, consequence, and applications. *BioMed Research International, 2014*, 637308. https://doi.org/10.1155/2014/637308

Liu, S., Tetzlaff, M. T., Liegl-Atzwanger, B., & Xu, X. (2012). Loss of microRNA-205 expression is associated with melanoma progression. *Laboratory Investigation, 92*(6), 845–858. https://doi.org/10.1038/labinvest.2012.62

Liu, Q., Muglia, L. J., & Huang, L. F. (2019). Network as a biomarker: A novel network-based sparse Bayesian machine for pathway-driven drug response prediction. *Genes (Basel), 10*, 68. https://doi.org/10.3390/genes10010068

Liu, X. G., Zhu, W. Y., Huang, Y. Y., Ma, L. N., Zhou, S. Q., Wang, Y. K., ... & Zhang, Y. K. (2012). High expression of serum miR-21 and tumor miR-200c associated with poor prognosis in patients with lung cancer. Medical oncology, 29(2), 618–626.

Lopes-Júnior, L. C. (2023). Factors associated with the nutritional status of women with non-metastatic breast cancer in a Brazilian high complexity oncology center. Nutrients, 15(23), 4961.

Lyu, Z., Zhang, Y., Sheng, C., Huang, Y., Zhang, Q., & Chen, K. (2024). Global burden of thyroid cancer in 2022: Incidence and mortality estimates from GLOBOCAN. Chinese Medical *Journal, 137*(21), 2567–2576

Maghsoodi, N., Zareinejad, M., Golestan, A., Maymand, E. M., & Ramezani, A. (2023). Anti-CD19/CD8 bispecific T cell engager for the potential treatment of B cell malignancies. Cellular Immunology, 393, 104787

Marzec, J., Ross-Adams, H., Pirrò, S., Wang, J., Zhu, Y., Mao, X., & Chelala, C. (2021). The transcriptomic landscape of prostate cancer development and progression: an integrative analysis. Cancers, 13(2), 345

Matsuoka, T., & Yashiro, M. (2014). The role of PI3K/Akt/mTOR signalling in gastric carcinoma. *Cancers, 6*(3), 1441–1463. https://doi.org/10.3390/cancers6031441

Matsuoka, T., & Yashiro, M. (2018). Biomarkers of gastric cancer: Current topics and future perspective. *World Journal of Gastroenterology, 24*(26), 2818–2832. https://doi.org/10.3748/wjg.v24.i26.2818

Matsuoka, T., & Yashiro, M. (2024). Bioinformatics analysis and validation of potential markers associated with prediction and prognosis of gastric cancer. *International Journal of Molecular Sciences, 25*(11), 5880.

Miller, K. A., & Tsai, C. J. (2023). PTEN-regulated PI3K-p110 and AKT isoform plasticity controls metastatic prostate cancer progression. *Nature Cancer, 43*, 22. https://doi.org/10.1038/s41388-023-02875-4

Mok, T., Nakagawa, K., Park, K., Ohe, Y., Girard, N., Kim, H. R., ... & Yang, J. C. H. (2024). Nivolumab plus chemotherapy in epidermal growth factor receptor–mutated metastatic non–small-cell lung cancer after disease progression on epidermal growth factor receptor tyrosine kinase inhibitors: Final results of CheckMate 722. *Journal of Clinical Oncology, 42*(11), 1252–1264

Muñoz-Couselo, E., Adelantado, E. Z., Ortiz, C., García, J. S., & Perez-Garcia, J. (2017). *NRAS*-mutant melanoma: current challenges and future prospect. *Oncotargets and Therapy, 10*, 3941–3947. https://doi.org/10.2147/OTT.S117121

Na, H. K., & Lee, J. Y. (2017). Molecular basis of alcohol-related gastric and colon cancer. *International journal of molecular sciences, 18*(6), 1116.

Neagu, M., Constantin, C., Cretoiu, S. M., & Zurac, S. (2020). miRNAs in the Diagnosis and Prognosis of Skin Cancer. Frontiers in cell and developmental biology, 8, 71

Ngamruengphong, S., & Canto, M. I. (2016). Screening for pancreatic cancer. The Surgical clinics of North America, 96(6), 1223

Ono, H. (2006). Early gastric cancer: diagnosis, pathology, treatment techniques and treatment outcomes. *European journal of gastroenterology & hepatology, 18*(8), 863–866

O'Neill, R. S., & Stoita, A. (2021). Biomarkers in the diagnosis of pancreatic cancer: Are we closer to finding the golden ticket?. World journal of gastroenterology, 27(26), 4045

Padhi, S. S., Roy, S., Kar, M., Saha, A., Roy, S., Adhya, A., & Banerjee, B. (2017). Role of CDKN2A/p16 expression in the prognostication of oral squamous cell carcinoma. Oral oncology, 73, 27–35

Park, S., Lee, J., Kim, Y., et al. (2016). Clinical relevance and molecular phenotypes in gastric cancer, of *TP53* mutations and gene expressions, in combination with other gene mutations. *Scientific Reports, 6*, 34822. https://doi.org/10.1038/srep34822

Park, S., Hong, T. H., Hwang, S., Heeke, S., Gay, C. M., Kim, J., ... & Park, K. (2024). Comprehensive analysis of transcription factor-based molecular subtypes and their correlation to clinical outcomes in small-cell lung cancer. EBioMedicine, 102.

Petersen, G. M., Amundadottir, L., Fuchs, C. S., Kraft, P., Stolzenberg-Solomon, R. Z., Jacobs, K. B., ... & Chanock, S. J. (2010). A genome-wide association study identifies pancreatic cancer susceptibility loci on chromosomes 13q22. 1, 1q32. 1 and 5p15. 33. Nature genetics, 42(3), 224–228.

Popiel-Kopaczyk, A., Piotrowska, A., Sputa-Grzegrzolka, P., Smolarz, B., Romanowicz, H., Dziegiel, P., ... & Kobierzycki, C. (2023). The Immunohistochemical Expression of Epithelial–Mesenchymal Transition Markers in Precancerous Lesions and Cervical Cancer. International Journal of Molecular Sciences, 24(9), 8063.

Popiel-Kopaczyk, A., Grzegrzolka, J., Piotrowska, A., Olbromski, M., Smolarz, B., Romanowicz, H., Rusak, A., Mrozowska, M., Dziegiel, P., Podhorska-Okolow, M., & Kobierzycki, C. (2023). The expression of testin, Ki-67 and p16 in cervical cancer diagnostics. *Current Issues in Molecular Biology, 45*(1), 490–500. https://doi.org/10.3390/cimb45010032

Posta, M., Györffy, D., Kiraly, P., Hargitai, B., Oravecz, O., Zvara, A., ... & Than, N. G. (2023). Uncovering Perturbed Molecular Pathways and Potential Biomarkers in Recurrent Pregnancy Loss Using a Systems Biology Approach. Journal of Reproductive Immunology, 159, 104077.

Pritchard, C. C., Mateo, J., Walsh, M. F., De Sarkar, N., Abida, W., Beltran, H., ... & Nelson, P. S. (2016). Inherited DNA-repair gene mutations in men with metastatic prostate cancer. New England Journal of Medicine, 375(5), 443–453.

Prykhodko, O., Johansson, S. V., Kotsias, P. C., Arús-Pous, J., Bjerrum, E. J., Engkvist, O., & Chen, H. (2019). A de novo molecular generation method using latent vector based generative adversarial network. *Journal of cheminformatics, 11*(1), 74.

Rambabu, M., Gowda, M. N., Selvam, P. K., Vasudevan, K., Dasegowda, K. R., Saravanan, P., & Rohini, K. (2024). Transcriptomic insights into skin cancer: A bioinformatics and network biology approach to biomarker identification. Journal of King Saud University-Science, 36(11), 103510.

Rambabu, M., Konageni, N., Vasudevan, K., Dasegowda, K. R., Gokul, A., Jayanthi, S., & Rohini, K. (2023). Identification of key biomarkers and associated pathways of pancreatic cancer using integrated transcriptomic and gene network analysis. Saudi Journal of Biological Sciences, 30(11), 103819.

Riely, G. J., Kris, M. G., Rosenbaum, D., Marks, J., Li, A., Chitale, D. A., ... & Ladanyi, M. (2008). Frequency and distinctive spectrum of KRAS mutations in never smokers with lung adenocarcinoma. *Clinical cancer research,14*(18), 5731–5734.

Ruiz, F. J., Inkman, M., Rashmi, R., Muhammad, N., Gabriel, N., Miller, C. A., McLellan, M. D., Goldstein, M., Markovina, S., Grigsby, P. W., Zhang, J., & Schwarz, J. K. (2021). HPV transcript expression affects cervical cancer response to chemoradiation. *JCI Insight, 6*(16), e138734. https://doi.org/10.1172/jci.insight.138734

Rundle, A., Ahsan, H., & Vineis, P. (2012). Better cancer biomarker discovery through better study design. *European Journal of Clinical Investigation, 42*, 1350–1359. https://doi.org/10.1111/j.1365-2362.2012.02758.x

Sánchez-Sendra, B., Serna, E., Navarro, L., González-Muñoz, J. F., Portero, J., Ramos, A., Murgui, A., & Monteagudo, C. (2020). Transcriptomic identification of miR-205 target genes potentially involved in metastasis and survival of cutaneous malignant melanoma. *Scientific Reports, 10*(1), 4771. https://doi.org/10.1038/s41598-020-61637-4

Sandhu, A. P. S., Singh, K., Singh, S., Antaal, H., Luthra, S., Singla, A., ... & kaur Aulakh, S. (2024). Decodingcancer risk: understanding gene-environment interactions in cancer development. Cureus, 16(7).

Shornale Akter, M., Uddin, M. H., Atikur Rahman, S., Hossain, M. A., Ashik, M. A. R., Zaman, N. N., & Rahman, M. H. (2024). Transcriptomic analysis revealed potential regulatory biomarkers and repurposable drugs for breast cancer treatment. Cancer Reports, 7(5), e2009.

Siddalingappa, R., & Kanagaraj, S. (2022). K-nearest-neighbour algorithm to predict the survival time and classification of various stages of oral cancer: A machine learning approach. *F1000Research, 11*, 70. 10.12688/f1000research.82974.1.
Silva, R. J. G., Grippa, W. R., Neto, L. C. B. S., Enriquez-Martinez, O. G., Marcarini, J. A. C., Pessanha, R. M., ... & Lopes-Júnior, L. C. (2023). Factors associated with the nutritional status of women with non-metastatic breast cancer in a Brazilian high complexity oncology center. Nutrients, 15(23), 4961.
Silajdžić, E., & Björkqvist, M. (2018). A critical evaluation of wet biomarkers for Huntington's disease: current status and ways forward. Journal of Huntington's disease, 7(2), 109–135
Smith, R. A., Cokkinides, V., Brooks, D., Saslow, D., Shah, M., & Brawley, O. W. (2011). Cancer screening in the United States, 2011: A review of current American Cancer Society guidelines and issues in cancer screening. CA: a cancer journal for clinicians, 61(1), 8–30.
Srinivas, P. R., Barker, P., & Srivastava, S. (2002). Nanotechnology in early detection of cancer. *Laboratory Investigation, 82*, 657–662. https://doi.org/10.1097/01.LAB.0000019493.93273.4D
Stone, L. (2019). The IMPACT of BRCA2 in prostate cancer. *Nature Reviews Urology, 16*(9), 639–641. https://doi.org/10.1038/s41585-019-0249-4
Strzelczyk, J. K., Krakowczyk, Ł., & Owczarek, A. J. (2018). Aberrant DNA methylation of the *p16, APC, MGMT, TIMP3* and *CDH1* gene promoters in tumours and the surgical margins of patients with oral cavity cancer. *Journal of Cancer, 9*(11), 1896–1904. https://doi.org/10.7150/jca.24477
St John, J., Powell, K., Conley-LaComb, M. K., & Chinni, S. R. (2012). TMPRSS2-ERG fusion gene expression in prostate tumor cells and its clinical and biological significance in prostate cancer progression. *Journal of cancer science & therapy, 4*(4), 94.
Sullivan, R. J., & Flaherty, K. T. (2011). BRAF in melanoma: pathogenesis, diagnosis, inhibition, and resistance. Journal of skin cancer, 2011(1), 423239.
Sun, J., Meng, H., Yao, L. U., Lv, M., Bai, J., Zhang, J., ... & Xie, Y. (2017). Germline mutations in cancer susceptibility genes in a large series of unselected breast cancer patients. *Clinical Cancer Research, 23*(20), 6113–6119.
Šutić, M., Dmitrović, B., Jakovčević, A., Džubur, F., Oršolić, N., Debeljak, Ž., & Knežević, J. (2024). Transcriptomic profiling for prognostic biomarkers in early-stage squamous cell lung cancer (SqCLC). Cancers, 16(4), 720.
Sweis, R. F., Drazer, M. W., & Ratain, M. J. (2016). Analysis of impact of post-treatment biopsies in phase I clinical trials. *Journal of Clinical Oncology, 34*(4), 369–374.
Tatineni, S., Purrington, K., Assad, H., Abdallah, N., Tarockoff, M., Surapaneni, M., ... & Simon, M. S. (2020). Family cancer patterns and variation by race and ethnicity among individuals with pathogenic variants in multi-gene cancer predisposition panels at a large urban comprehensive cancer center.
Tian, Y., Qiu, X., Wang, C., Zhao, J., Jiang, X., Niu, W., ... & Zhang, F. (2021). Cancer associates with risk and severe events of COVID-19: a systematic review and meta-analysis. *International journal of cancer, 148*(2), 363–374.
Tomlins, S. A., Rhodes, D. R., Perner, S., Dhanasekaran, S. M., Mehra, R., Sun, X. W., ... & Chinnaiyan, A. M. (2005). Recurrent fusion of TMPRSS2 and ETS transcription factor genes in prostate cancer. science, 310(5748), 644–648.
Umelo, I. A., Costanza, B., & Castronovo, V. (2018). Innovative methods for biomarker discovery in the evaluation and development of cancer precision therapies. *Cancer Metastasis Reviews, 37*, 125–145. https://doi.org/10.1007/s10555-018-9809-2
Villalobos, P., & Wistuba, I. I. (2017). Lung cancer biomarkers. *Hematology/Oncology Clinics of North America, 31*(1), 13–29. https://doi.org/10.1016/j.hoc.2016.08.006
Walsh, N. C., Kenney, L. L., Jangalwe, S., Aryee, K. E., Greiner, D. L., Brehm, M. A., & Shultz, L. D. (2017). Humanized mouse models of clinical disease. Annual Review of Pathology: Mechanisms of Disease, 12(1), 187–215.
Wang, J., Liu, Y., Zhao, Y., & Zhao, Z. (2023). Transcriptional biomarkers in oral cancer: An integrative analysis approach. *Cancer Biology & Medicine, 20*(2), 124–139. https://doi.org/10.21147/cbm.v20i2.5027

Wang, F. H., Zhang, X. T., Tang, L., Wu, Q., Cai, M. Y., Li, Y. F., ... & Xu, R. H. (2024). The Chinese Society of Clinical Oncology (CSCO): clinical guidelines for the diagnosis and treatment of gastric cancer, 2023. Cancer Communications, 44(1), 127–172

Wang, X., Zhao, J., Marostica, E., Yuan, W., Jin, J., Zhang, J., ... & Yu, K. H. (2024). A pathology foundation model for cancer diagnosis and prognosis prediction. Nature, 634(8035), 970–978.

Warner, E., Herberts, C., Fu, S., Yip, S., Wong, A., Wang, G., ... & Wyatt, A. W. (2021). BRCA2, ATM, and CDK12 defects differentially shape prostate tumor driver genomics and clinical aggression. *Clinical Cancer Research, 27*(6), 1650–1662

Yan, D., Yi, S., Chiu, W. C., Qin, L. G., Kin, W. H., Kwok Hung, C. T., Linxiao, H., Wai, C. K., Yi, S., Tao, Y., & Tao, T. (2017). Integrated analysis of chromosome copy number variation and gene expression in cervical carcinoma. *Oncotarget, 8*(65), 108912–108922. https://doi.org/10.18632/oncotarget.22403

Yasui, W., Oue, N., Ito, R., Kuraoka, K., & Nakayama, H. (2004). Search for new biomarkers of gastric cancer through serial analysis of gene expression and its clinical implications. *Cancer science, 95*(5), 385–392

Yi, S. Y., & Nan, K. J. (2008). Tumor-initiating stem cells in liver cancer. Cancer biology & therapy, 7(3), 325–330

Yim, E. K., & Park, J. S. (2005). The role of HPV E6 and E7 oncoproteins in HPV-associated cervical carcinogenesis. Cancer research and treatment: official journal of Korean Cancer Association, 37(6), 319–324

Yuan, J., Bao, P., Chen, Z., Yuan, M., Zhao, J., Pan, J., ... & Wang, Z. (2023). Advanced prompting as a catalyst: Empowering large language models in the management of gastrointestinal cancers. The Innovation, 521

Zeng, H., Ran, X., An, L., Zheng, R., Zhang, S., Ji, J. S., ... & He, J. (2021). Disparities in stage at diagnosis for five common cancers in China: a multicentre, hospital-based, observational study. *The Lancet Public Health, 6*(12), e877-e887

Zhang, J., Williams, C., & Wu, H. (2021). Transcriptomics and solid tumours: The next frontier in precision oncology. *Nature Reviews Cancer, 21*, 421–438. https://doi.org/10.1038/s41571-021-00454-0

Zhang, H., Wang, Z., Liang, L., & Rashid Sheykhahmad, F. (2021). A robust method for skin cancer diagnosis based on interval analysis. Automatika: časopis za automatiku, mjerenje, elektroniku, računarstvo i komunikacije, 62(1), 43–53

Zhou, J., Sun, H. C., Wang, Z., Cong, W. M., Wang, J. H., Zeng, M. S., & Fan, J. (2018). Guidelines for diagnosis and treatment of primary liver cancer in China (2017 Edition). Liver cancer, 7(3), 235–260

Zhou, J., Sun, H., Wang, Z., Cong, W., Zeng, M., Zhou, W., ... & Fan, J. (2023). Guidelines for the diagnosis and treatment of primary liver cancer (2022 edition). *Liver cancer, 12*(5), 405–444.

Artificial Intelligence in Personalized Medicine and Treatment Planning

Vandana Kumari, Tanya Shekhar Barua, Rajeev Mishra, and Abhijeet Singh

Abstract

Ongoing advancements in computational algorithms have established artificial intelligence (AI)-based models as among the most advanced technologies within the healthcare sector. AI demonstrates significant contributions to multiple medical domains, including data analysis and monitoring, imaging assessment and diagnosis, as well as therapy response and survival forecasting. Notwithstanding progress in clinical oncology, greater effort is required to customize therapy strategies according to each patient's distinct transcriptome profile within the context of precision/personalized oncology. Moreover, the conventional analytical approach is incompatible with the thorough interpretation of substantial data streams, hence hindering the accurate prediction of treatment alternatives. In this chapter, we have reviewed how artificial intelligence utilizing big data can uncover concealed patterns, significant facts, and relevant knowledge inside vast datasets. Machine learning and deep learning, as subsets of artificial intelligence, can be employed to extract in-depth information from genomes, transcriptomics, proteomics, radiomics, digital pathological pictures, and other data, enabling doctors to synthesize and thoroughly comprehend cancer.

V. Kumari (✉) · T. S. Barua · A. Singh
Cancer Research Laboratory, Department of Biosciences, Manipal University Jaipur, Jaipur, Rajasthan, India
e-mail: vandana.211051036@muj.manipal.edu; tanya.221002065@muj.manipal.edu; abhijeet.singh@jaipur.manipal.edu

R. Mishra
Department of Life Sciences and Biotechnology, Chhatrapati Shahu Ji Maharaj University, Kanpur, India

R. Nema et al. (eds.), *Advances in Cancer Detection, Prediction, and Prognosis Using Artificial Intelligence and Machine Learning*,
https://doi.org/10.1007/978-981-96-9346-7_11

1 Introduction

Cancer is a global problem affecting people, health, and the economy. It accounts for approximately one-sixth (16.8%) of global deaths and almost one-quarter (22.8%) of noncommunicable disease (NCD) deaths. Cancer accounts for 30.3% of premature deaths from NCDs among people aged 30–69 years and is one of the three leading causes of death in 177 of 183 countries (Sung et al., 2021). The increasing incidence of breast cancer is due to factors such as population growth, aging, and lifestyle changes including increased smoking, poor diet quality, and decreased physical activity (Nsubuga, 2024). Cancer also results in economic and social costs that vary by cancer type, geographic location, and gender. In 2020, the total global economic burden of cancer, including medical costs and lost productivity, was estimated to be US$ 1.16 trillion. New cancers are expected to increase by 47% by 2040, with low- and middle-income countries accounting for almost two-thirds of the global burden (Chiu & Matsuda, 2024). This highlights the urgent need for prevention, early detection, and advanced treatment. The psychological and emotional impact is also significant, as survivors often experience physical, mental, and financial stress (supporting context: Davis et al., 2023).

1.1 Cancer Statistics

Cancer continues to be a major problem worldwide with its medical, social, and economic consequences. Approximately 20 million new cancer cases were expected to be reported worldwide in 2022, resulting in 9.96 million deaths. Breast, lung, and colon cancers are the most common cancers, with lung cancer causing the most deaths. Regional differences are significant, with Asia accounting for 51% of global cases. Demographic changes, changes in lifestyle, and environmental conditions have led to an increase in cancer cases, with the annual number of cancer cases expected to reach 35 million by 2050 (Sung et al., 2021). Innovations in precision medicine, such as genomic analysis and artificial intelligence (AI)-based research, offer promising opportunities to combat this disease (Zhou & Xue, 2024). The healthcare AI market is projected to grow to US$ 187 billion by 2030, with AI applications in personalized medicine accounting for approximately 25% of this growth (Jain et al., 2024a, b) (Fig. 1).

2 Artificial Intelligence: Smart Revolution for Cancer

Artificial intelligence (AI) is revolutionizing healthcare by enabling precision diagnostics, personalized treatments, and innovative research. By leveraging technologies such as machine learning (ML), natural language processing (NLP), and computer vision, AI can process and analyze vast amounts of complex medical data with unprecedented speed and accuracy. For example, AI has been successfully applied in oncology for tumor detection and classification using advanced imaging

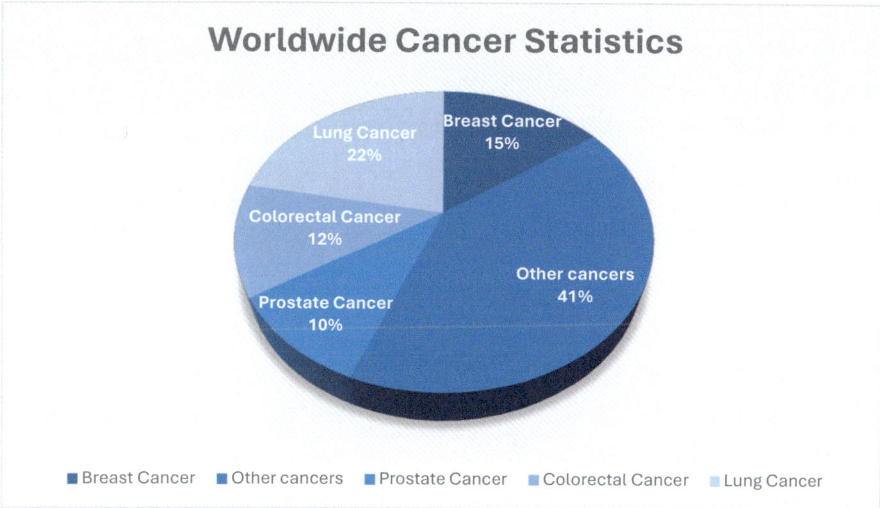

Fig. 1 GLOBOCAN and WHO insights on cancer prevalence. (*Source*: Sung et al. (2021); WHO (2022)

techniques (Cunha Reis, 2024). AI algorithms are also transforming genomic medicine by identifying actionable mutations, thereby advancing personalized cancer therapies (Badria & Elgazar, 2024). Furthermore, AI's role extends to real-time applications such as patient monitoring through wearable devices, which provide clinicians with predictive insights, allowing for timely interventions. These innovations demonstrate the profound potential of AI to optimize medical workflows and improve patient outcomes.

The use of AI in medical research is equally groundbreaking, accelerating drug discovery, clinical trials, and pandemic prediction. In drug discovery, AI reduces the time and cost required to identify viable compounds by simulating molecular interactions on a large scale. Clinical trials are also benefitting from AI's ability to predict outcomes and stratify patient populations based on genetic and phenotypic profiles (Cunha Reis, 2024; Lakshmi, 2025). AI-driven models have even been instrumental in understanding COVID-19 transmission patterns, enabling effective policymaking and vaccine distribution strategies (Salleh et al., 2024). Additionally, AI democratizes healthcare access by enabling remote diagnostics and telemedicine, particularly in underserved regions (Frankland, 2024). However, as AI adoption grows, challenges such as data privacy, algorithmic bias, and ethical concerns must be addressed to ensure equitable and responsible use (Salleh et al. (2024)). Nevertheless, AI's ability to integrate multidimensional datasets and uncover hidden correlations positions it as a cornerstone of modern medical innovation.

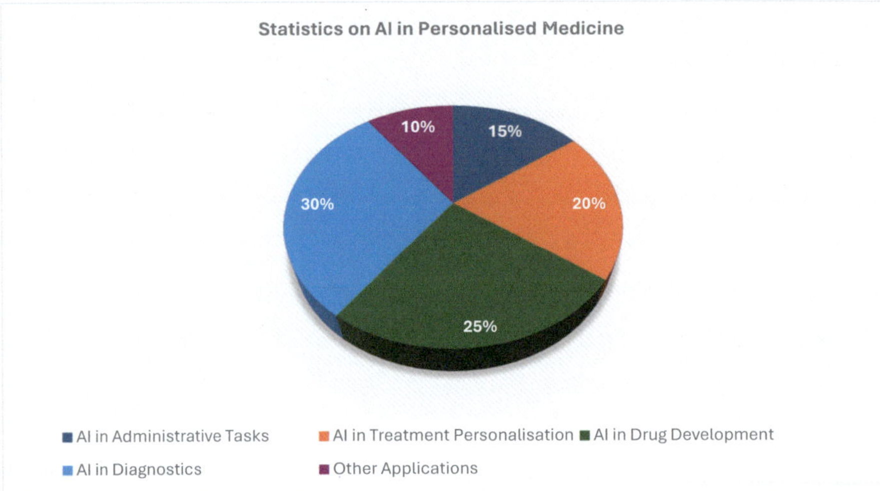

Fig. 2 Pie chart representation of hypothetical worldwide population-based AI usage in personalized medicine, with data inspired by regional adoption trends and WHO insights

2.1 Statistics on AI in Personalized Medicine

Artificial intelligence (AI) is revolutionizing personalized medicine, with the healthcare AI market projected to grow to US$ 187 billion by 2030, at a compound annual growth rate (CAGR) of 37.6%. AI applications in personalized medicine account for approximately 25% of this growth, with key focus areas including oncology, cardiology, and neurology. A recent study by Jain et al. (2024a, 2024b) and Lakshmi (2025) demonstrated that AI algorithms improved diagnostic accuracy by 45% and treatment outcomes by 35% in trials focused on personalized cancer therapies Lakshmi (2025) (Fig. 2).

2.2 Historical Context and Technological Progression in Oncology

The incorporation of AI into personalized medicine marks a transformative milestone in healthcare, with its roots tracing back to the completion of the Human Genome Project in 2003. This achievement catalyzed the advent of next-generation sequencing (NGS) technologies and significant strides in machine learning and deep learning. Advanced neural networks, including convolutional and recurrent neural networks, have since emerged as essential tools for analyzing complex unstructured data, such as medical imaging and electronic health records. Integrating AI with wearable devices and Internet of Things (IoT) technologies has further enhanced real-time health monitoring, improved chronic disease management, and reduced hospital readmissions (Jain et al., 2024a, b). Regulatory frameworks and increasing investment have also driven the rapid adoption of AI in healthcare, with global funding reaching US$ 11 billion in 2021. Initiatives such as the U.S. Food and Drug Administration's AI-specific guidelines have supported this growth. The COVID-19 pandemic underscored AI's potential, showcasing its utility in drug repurposing, vaccine development, and remote health monitoring. Moreover, the convergence of AI and cloud computing has facilitated scalable healthcare solutions, revolutionized personalized medicine, and laid the groundwork for future innovations (Velazquez-Villarreal & Yang, 2024).

3 Data Integration and Patient Profiling: Multi-omics Data Analysis

Integrating multi-omics data—including genomics, proteomics, metabolomics, and epigenomics—represents a groundbreaking advancement in personalized medicine. This comprehensive approach enables precise patient profiling, offering improved diagnostic accuracy and individualized treatment plans by providing a holistic view of each patient's health dynamics (Chiu & Matsuda, 2024). Genomics serves as the cornerstone of personalized medicine by analyzing DNA sequences to identify genetic predispositions and inform therapeutic strategies. For instance, genome-wide association studies (GWAS) have been pivotal in uncovering genetic variants linked to complex diseases such as cancer, cardiovascular disorders, and neurodegenerative conditions (Zhou & Xue, 2024). Single-nucleotide polymorphism (SNP) profiling further enhances disease risk prediction for conditions such as diabetes and hypertension, enabling proactive, personalized interventions. The advent of next-generation sequencing (NGS) has significantly enhanced the generation and analysis of genomic data, allowing for detailed exploration of genetic mutations and structural variations Razak et al. (2024a). Genomic profiling in precision oncology identifies actionable tumor mutations, facilitating targeted therapies and minimizing side effects while optimizing treatment outcomes.

Proteomics, the study of proteins as dynamic drivers of biological systems, provides a complementary perspective to the static insights offered by genomics. Through proteome profiling, researchers can uncover biomarkers and elucidate disease mechanisms at a molecular level (Jain et al., 2024a, b). For example, tissue-specific proteomics has identified critical biomarkers for diagnosing and monitoring cardiovascular diseases. Real-time analysis of protein interactions and posttranslational modifications improves our understanding of drug efficacy and safety, underscoring proteomics' integral role in advancing personalized medicine.

Metabolomics, which focuses on small-molecule metabolites, provides a dynamic snapshot of cellular activity and bridges the gap between genetic predispositions and environmental influences. This field is critical for mapping biochemical pathways and understanding metabolic disorders. In oncology, metabolomic profiling has revealed tumor-specific metabolic signatures that facilitate accurate diagnoses and personalized treatment planning. For instance, metabolomic studies have demonstrated exceptional precision in differentiating benign from malignant prostate tumors, thereby enhancing diagnostic accuracy. Metabolomic data also supports noninvasive monitoring of treatment efficacy, enabling real-time evaluations and adjustments to optimize patient outcomes (Constantinescu et al., 2024).

Epigenomics examines heritable changes in gene expression that do not alter the DNA sequence, focusing on mechanisms such as DNA methylation, histone modifications, and noncoding RNA interactions. Epigenomic profiling has been instrumental in elucidating complex diseases such as cancer and diabetes. For example, aberrant DNA methylation patterns are strongly associated with tumorigenesis and serve as promising biomarkers for early cancer detection. Additionally, epigenetic therapies that reverse abnormal gene expression patterns have shown substantial promise in treating hematological malignancies and solid tumors, signifying an important milestone in targeted therapeutic development.

3.1 Holistic Patient Profiling Through Multi-omics Integration

Holistic patient profiling through multi-omics integration is revolutionizing precision medicine by providing a comprehensive understanding of health at the molecular level. This approach combines data from genomics, proteomics, metabolomics, and epigenomics to create a multidimensional view of patient health. By leveraging advanced computational tools, particularly AI, this strategy enhances diagnostic accuracy, tailors treatments, and improves overall health outcomes (Velazquez-Villarreal & Yang, 2024). Genomics serves as the foundation for identifying genetic predispositions and understanding the molecular basis of diseases. Advanced technologies such as NGS produce extensive genetic variation data, which AI algorithms analyze to link genetic markers to specific conditions. Proteomics examines proteins, the functional molecules of cells, offering real-time insights into cellular activity. Unlike genomics, which provides a static view, proteomics reveals dynamic molecular changes. AI-powered proteomic analyses have identified disease biomarkers, particularly for cardiovascular conditions, significantly improving diagnostic precision. Proteomics also plays a crucial role in drug discovery by examining drug-protein interactions, optimizing therapeutic efficacy, and reducing side effects. Metabolomics focuses on small molecules, providing insights into cellular metabolism and the interaction between genetic predispositions and environmental factors. AI-based metabolomic analysis detects disruptions in metabolic pathways associated with diseases, enabling early intervention. Additionally, noninvasive metabolomic assessments, such as blood sample analyses, support real-time treatment monitoring and allow for precise therapy adjustments (Jain et al., 2024a, b). Epigenomics explores mechanisms such as DNA methylation and histone modifications that regulate gene expression without altering the DNA sequence. These regulatory processes, influenced by environmental factors, play key roles in disease progression. AI-driven analysis of epigenomic data has identified biomarkers for diseases such as cancer, facilitating early detection and informed therapeutic decisions. For example, aberrant methylation patterns are recognized as early indicators of certain cancers, supporting timely interventions (Zhou & Xue, 2024).

AI is critical for managing the complexity, scale, and heterogeneity of multi-omics data. Machine learning and deep learning techniques uncover hidden patterns and insights that traditional methods often overlook. AI-powered platforms integrate genomic, proteomic, metabolomic, and epigenomic datasets into unified patient profiles. These systems enable disease progression predictions, patient risk stratification, and the creation of personalized treatment plans. Multi-omics integration has broad applications across clinical practice. In oncology, it supports precise tumor characterization and personalized therapy development. For metabolic disorders, it enables early diagnosis and individualized treatment strategies. As AI continues to advance, multi-omics profiling will play an increasingly critical role in shaping the future of precision medicine.

4 Treatment Planning in Personalized Medicine for Cancer

Personalized medicine for cancer represents a significant evolution from generalized treatment protocols to highly individualized strategies tailored to the unique biological characteristics of each patient. This transformative approach leverages advances in genomics, proteomics, metabolomics, imaging technologies, and artificial intelligence (AI) to optimize patient outcomes. The process involves several stages, including data collection, integration, biomarker identification, patient stratification, therapy selection, treatment plan design, and ongoing clinical support. The integration of multi-omics data with computational tools has made it possible to decode the complexity of cancer biology, enabling the development of precise and patient-specific treatment regimens.

4.1 Stages of Treatment Planning in Personalized Medicine

The first critical step in personalized cancer treatment planning is data collection and integration. This involves gathering genomic, proteomic, metabolomic, and imaging data from patients. Next-generation sequencing (NGS) technologies allow rapid and comprehensive analysis of tumor DNA, identifying key mutations such as BRCA1/BRCA2 in breast and ovarian cancers or EGFR in lung cancer (Razak et al., 2024a, b). These insights are complemented by proteomic profiling, which assesses protein expression levels and posttranslational modifications. Techniques like mass spectrometry identify biomarkers, such as HER2 in breast cancer, that guide therapeutic choices (Nag, 2024). Additionally, metabolomic profiling provides an overview of altered metabolic pathways in tumors, revealing tumor-specific metabolic signatures that distinguish benign from malignant lesions (Zhu et al., 2024a, b, c). Integrating these diverse datasets is essential for creating a holistic view of the tumor microenvironment, enabling the identification of actionable targets.

Biomarker identification and patient stratification are crucial steps in this process. Genomic profiling helps uncover actionable mutations and molecular alterations that influence treatment responses. For instance, genome-wide association studies (GWAS) identify gene variants linked to cancer susceptibility and progression. Biomarkers such as PD-L1 expression or tumor mutational burden (TMB) are analyzed to stratify patients for immunotherapy, ensuring higher response rates White et al. (2023). Proteomic biomarkers, such as the androgen receptor in prostate cancer, and metabolomic signatures provide additional layers of specificity. These biomarkers guide patient stratification into subgroups based on the likelihood of response to specific therapies. AI-driven algorithms further enhance this stratification by analyzing complex datasets, enabling oncologists to predict treatment outcomes with high accuracy (Ouma et al., 2024a, b).

4.2 Genomics, Proteomics, and Therapy Selection

Genomics serves as the cornerstone of personalized medicine, with technologies like NGS enabling rapid and detailed analysis of cancer genomes. Actionable genetic mutations, such as those in the KRAS or BRAF genes, directly inform therapy selection. Targeted therapies, such as PARP inhibitors for BRCA-mutated tumors or tyrosine kinase inhibitors (TKIs) for EGFR-mutant lung cancer, have shown remarkable efficacy in treating specific cancer subtypes (Huda & Windiarti, 2024a, b). Beyond genomics, proteomics plays a complementary role by identifying functional biomarkers that reflect the real-time state of cancer cells. HER2 overexpression in breast cancer is a classic example where proteomics drives treatment decisions, guiding the use of monoclonal antibodies like trastuzumab (Nag, 2024). Metabolomics provides additional insights into cancer biology by revealing disruptions in metabolic pathways, aiding in distinguishing aggressive tumors from indolent ones (Zhu et al., 2024a, b, c).

Therapy selection and treatment plan design rely heavily on the integration of genomic, proteomic, and imaging data. AI-powered systems analyze this information to recommend optimal therapies tailored to each patient's unique cancer profile. Immunotherapy, for instance, leverages biomarkers such as PD-L1 expression to identify patients likely to benefit from immune checkpoint inhibitors. Machine learning algorithms can predict responses to immunotherapy by integrating genomic, proteomic, and clinical data White et al. (2023). Imaging technologies, including PET-CT and MRI, combined with AI, enhance the precision of therapy planning by providing detailed tumor maps. These maps guide the delivery of radiation therapy and evaluate treatment efficacy over time (Lin et al., 2024a, b).

4.3 Imaging, AI, and Clinical Support

Imaging techniques, bolstered by AI, are indispensable in personalized cancer treatment planning. Advanced imaging modalities such as PET-CT and MRI offer high-resolution insights into tumor size, location, and metabolic activity. AI algorithms enhance these capabilities by automating tumor detection, segmentation, and response assessment (Di Leone et al., 2024a, b). For example, AI-powered imaging systems analyze radiological data to create precise tumor delineations, improving the accuracy of radiation therapy planning. These tools also assist in monitoring treatment efficacy and identifying early signs of relapse, ensuring timely interventions. The final stage, clinical support and adaptive treatment management, ensures the continuous optimization of patient outcomes. Personalized treatment plans require ongoing evaluation and adjustment based on real-time data. AI systems play a critical role in this phase by analyzing patient responses and predicting potential adverse effects, enabling oncologists to modify treatment regimens proactively (Ouma et al., 2024a, b). Clinical decision support systems (CDSS) integrate data from various sources, providing actionable recommendations that improve the quality and efficiency of care.

Despite its transformative potential, personalized medicine in oncology faces several challenges. High costs associated with multi-omics analyses and advanced imaging technologies limit accessibility for many patients. Additionally, the integration and interpretation of vast and complex datasets require significant computational resources and expertise. Ethical considerations, including patient privacy and data security, must also be addressed to ensure the responsible use of sensitive health information. Future advancements in AI and machine learning are expected to overcome these challenges by reducing costs, enhancing data integration capabilities, and improving the scalability of personalized treatment strategies (Jain et al., 2024a, b). In conclusion, personalized medicine represents a paradigm shift in cancer care, offering the promise of improved outcomes through tailored treatment strategies. By integrating genomic, proteomic, metabolomic, and imaging data with AI-driven tools, oncologists can develop highly precise and individualized treatment plans. While challenges remain, ongoing innovations in multi-omics technologies and AI are set to make personalized medicine the standard of care in oncology.

Personalized medicine for cancer treatment represents a paradigm shift from generalized protocols to individualized strategies. This approach integrates genomic, proteomic, and imaging data to develop targeted therapies and optimize patient outcomes. Advances in artificial intelligence (AI) and multi-omics have made it possible to decode complex biological datasets, creating precise, patient-specific treatment plans (Jain et al., 2024a, b). This text explores the methodologies, applications, and challenges of treatment planning in personalized cancer care. Personalized cancer medicine represents a significant evolution from generalized treatment protocols to highly individualized strategies tailored to the unique biological characteristics of each patient. This transformative approach leverages advances in genomics, proteomics, metabolomics, imaging technologies, and artificial intelligence to optimize patient outcomes. The process involves several stages, including data collection, integration, biomarker identification, patient stratification, therapy selection, treatment plan design, and ongoing clinical support. The integration of multi-omics data with computational tools has made it possible with certain strategic steps

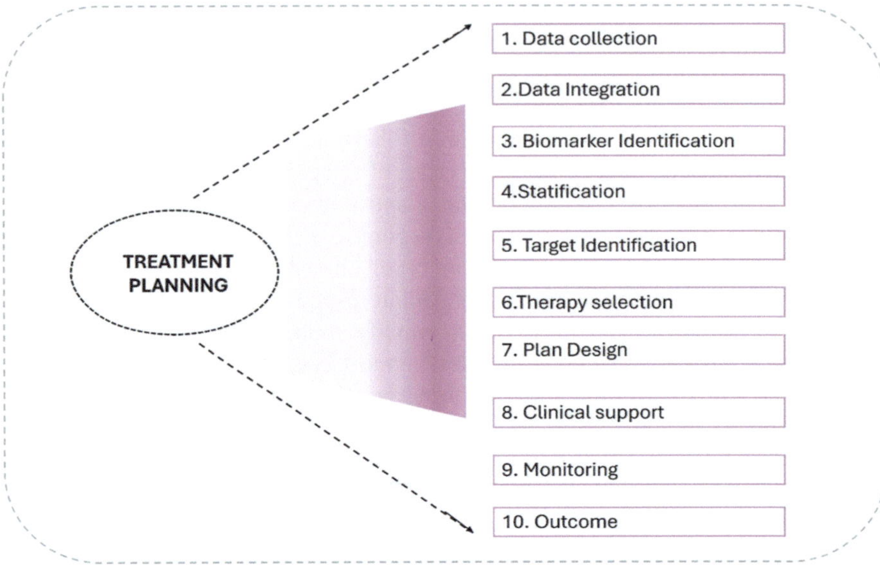

Fig. 3 Steps involved in treatment planning of personalized medicine

mentioned in Fig. 3 in order to decode the complexity of cancer biology, enabling the development of precise and patient-specific treatment regimens (Zhu et al., 2024a, b, c).

4.4 The Role of Genomics in Treatment Planning

Genomics forms the foundation of personalized medicine, analyzing DNA sequences to uncover genetic mutations and markers that influence treatment responses. Genome-wide association studies (GWAS) identify gene variants linked to cancer susceptibility and progression. For instance, genomic profiling has been pivotal in identifying actionable mutations, such as BRCA1/BRCA2 in breast and ovarian cancers (Huda & Windiarti, 2024a, b). Next-generation sequencing (NGS) has revolutionized this domain by rapidly analyzing tumor genomes, enabling oncologists to recommend targeted therapies, such as PARP inhibitors for BRCA-mutated tumors (Razak et al., 2024a, b). Proteomics complements genomics by studying protein expression and modifications, offering insights into the functional state of cells. Biomarkers derived from proteomic studies, such as HER2 in breast cancer, guide therapy decisions, allowing clinicians to employ targeted treatments like trastuzumab (Nag, 2024). Similarly, metabolomics provides a snapshot of metabolic pathways, linking genetic predispositions to environmental influences. Tumor-specific metabolic signatures identified through metabolomic profiling are crucial for distinguishing benign from malignant lesions and guiding therapeutic choices (Zhu et al., 2024a, b, c).

Advanced imaging technologies, combined with AI, play a critical role in treatment planning. AI-driven tools enhance tumor detection, segmentation, and treatment monitoring by analyzing radiological data. Machine learning models integrated with imaging systems such as PET-CT and MRI provide highly accurate tumor maps, enabling precise radiation therapy planning (Lin et al., 2024a, b). Additionally, AI-powered imaging assists in evaluating treatment efficacy and detecting potential relapses (Di Leone et al., 2024a, b). Immunotherapy, which activates the immune system to fight cancer, has been transformative in oncology. However, its success varies among patients. AI algorithms analyze biomarkers such as PD-L1 expression and tumor mutational burden to predict immunotherapy responses, ensuring better patient stratification White et al. (2023). Combining these insights with genomic and proteomic data helps in designing adaptive immunotherapy regimens tailored to individual patients (Ouma et al., 2024a, b; Green et al. (2023)).

Despite its promise, personalized cancer medicine faces significant challenges. Ensuring data privacy, addressing biases in AI algorithms, and managing the cost of advanced technologies are primary concerns. Moreover, integrating multi-omics data requires robust computational frameworks and interdisciplinary collaboration (Tatsugami et al., 2024). Addressing these issues is vital for equitable access to personalized treatments. Advances in genomics, proteomics, and AI have made it possible to develop highly individualized treatment plans, offering hope for better outcomes. However, addressing ethical and practical challenges is critical to ensure the broader adoption of these innovative approaches. As technology continues to evolve, personalized medicine will play an increasingly central role in oncology. Personalized medicine in cancer treatment represents the convergence of technology, biology, and clinical practice. Advances in genomics, proteomics, and AI have made it possible to develop highly individualized treatment plans, offering hope for better outcomes. However, addressing ethical and practical challenges is critical to ensure the broader adoption of these innovative approaches. As technology continues to evolve, personalized medicine will play an increasingly central role in oncology White et al. (2023).

5 AI's Impact on Personalized Cancer Therapy

Artificial intelligence (AI) is revolutionizing personalized cancer therapy by moving away from traditional, generalized treatment approaches to individualized strategies tailored to the unique genetic and molecular characteristics of each patient's cancer. This paradigm shift underscores the limitations of one-size-fits-all therapies and emphasizes AI's role in driving precision medicine. By processing vast and complex datasets, AI identifies critical biomarkers and genetic mutations, enabling more precise and effective treatment planning. This approach recognizes that even within the same cancer type, substantial genetic and molecular differences can influence disease progression and treatment efficacy Anderson et al. (2023). AI has had a transformative impact on genomic analysis, decoding the genetic architecture of tumors to uncover mutations that drive cancer growth. Genomic profiling powered by AI provides oncologists with critical insights to develop targeted therapies aimed at specific vulnerabilities. AI-enabled tools process next-generation sequencing (NGS) data rapidly, often within minutes, drastically reducing the time needed to identify actionable mutations—a process that previously relied on labor-intensive manual methods (Razak et al., 2024a, b). This capability has reshaped precision oncology, enabling treatments to directly target cancer-driving mutations, thereby improving therapeutic efficacy and minimizing side effects. AI has played a crucial role in advancing targeted cancer therapies, which contrast sharply with traditional chemotherapy by focusing on specific genetic markers in cancer cells. For instance, HER2-targeted therapies for breast cancer and EGFR inhibitors for lung cancer have benefited from AI-driven biomarker discovery, which identifies genetic features that distinguish cancer cells from healthy ones Huda & Windiarti (2024b). These innovations have improved treatment specificity, reducing harm to healthy tissues and enhancing patient outcomes. Immunotherapy, which harnesses the immune system to combat cancer, is another area where AI has shown significant promise. While immunotherapy has demonstrated success in some patients, its effectiveness varies widely. AI predictive models analyze genetic and immunological profiles to identify patients most likely to benefit from immunotherapy. By examining biomarkers such as PD-L1 expression and tumor mutation burden, AI enables personalized treatment decisions, increasing the chances of successful outcomes (Zhu et al., 2024a, b, c).

AI's capabilities extend beyond initial treatment selection to real-time monitoring of cancer therapy. AI-powered tools continuously evaluate patient responses to treatment, predict potential side effects, and provide actionable insights for therapy adjustments. Despite its potential, AI-driven personalized cancer therapy faces significant challenges. Ensuring the privacy and security of sensitive genomic and clinical data is a primary concern. Developing robust AI models also requires access to large, diverse datasets to avoid biases and inaccuracies. Ethical considerations, including equitable access to AI-driven treatments and transparency in algorithmic decision-making, must also be addressed (Mahato, 2024). Collaboration among clinicians, researchers, policymakers, and technologists is essential to establish ethical frameworks and guidelines that support the responsible use of AI in oncology. These advancements, combined with an expanding understanding of cancer biology, are paving the way for increasingly precise and patient-centered therapies. AI's integration into oncology is expected to enhance patient outcomes by improving therapeutic efficacy (as shown in Fig. 4), minimizing adverse effects, and offering new hope to cancer patients worldwide. As the field evolves, AI will remain at the forefront of innovation in personalized cancer care, transforming the landscape of oncology.

6 Predictive Risk Modeling in Healthcare

Predictive risk modeling, underpinned by artificial intelligence (AI), is transforming personalized healthcare by enabling precise assessments of disease risks and treatment complications. By utilizing data from electronic health records (EHRs),

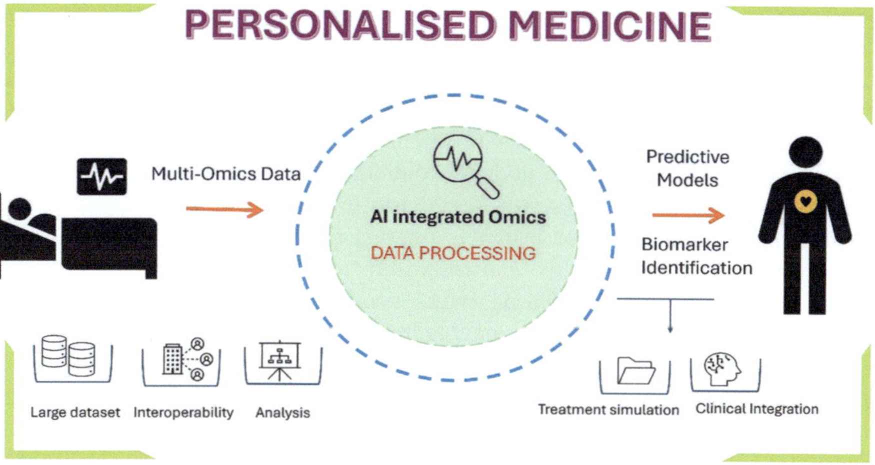

Fig. 4 Schematic representation of treatment planning for personalized medicine

genomic profiles, and wearable devices, this innovative approach enhances clinical decision-making and optimizes patient outcomes. AI-driven models identify individuals at higher risk for specific conditions, facilitating earlier interventions and improving healthcare delivery. Advanced machine learning techniques analyze complex datasets to uncover patterns indicative of future health risks. For example, integrating demographic, clinical, and genomic data has greatly improved cardiovascular risk predictions, enabling timely preventive actions (Türkmen et al., 2024). In oncology, predictive models analyze genomic alterations and imaging data to stratify patients based on their risk of cancer recurrence. Breast cancer models that incorporate molecular markers have enhanced treatment accuracy by 35%, supporting personalized therapeutic strategies (Anwar, 2024a, b). Similarly, predictive models forecast treatment-related complications, such as adverse drug reactions or surgical risks, by analyzing historical patient data. For instance, in pediatric care, AI models have improved the identification of children at risk for adverse reactions to COVID-19 vaccines by 25% compared to conventional methods (Ure, 2024). In cardiology, models combining clinical and pharmacological data optimize anticoagulant therapy, improving dosing precision and reducing bleeding complications (Anwar, 2024a, b). Predictive modeling is also invaluable for anticipating long-term health outcomes, particularly in managing chronic diseases. By leveraging longitudinal data, these models predict disease progression and inform proactive interventions. For example, continuous glucose monitoring data enables early detection of complications in diabetes, such as nephropathy and retinopathy, allowing timely interventions (Türkmen et al., 2024).

AI-powered predictive models play a critical role in crafting personalized preventive strategies. Wearable devices integrated with AI provide real-time alerts for individuals at risk of cardiac events, encouraging immediate lifestyle changes or medical intervention (Ure, 2024). In oncology, predictive models combining genetic and lifestyle data guide tailored cancer screening schedules, reducing late-stage diagnoses and improving survival rates (Türkmen et al., 2024). Despite their transformative potential, predictive risk models face challenges such as data heterogeneity, privacy concerns, and the need for explainable AI. Federated learning, which enables collaborative AI model development without compromising data privacy, offers a promising solution to address confidentiality concerns (Türkmen et al., 2024). Emerging technologies are expected to further enhance the capabilities of predictive risk modeling. For instance, Internet of Things (IoT) data integration enables continuous monitoring of patient health, while quantum computing accelerates drug discovery. Multimodal AI, which synthesizes diverse datasets, improves predictive accuracy and scalability, paving the way for more robust and comprehensive healthcare solutions (Hausawi et al., 2024a, b). Predictive risk modeling represents a groundbreaking advancement in personalized medicine. By refining diagnostics, tailoring treatments, and preventing diseases, it significantly enhances healthcare outcomes, and current clinical studies regarding this are listed in Table 1. As AI technologies continue to evolve alongside ethical frameworks addressing data privacy and fairness, predictive risk modeling will remain a cornerstone of modern healthcare innovation.

Table 1 Ongoing clinical studies and research on AI in personalized medicine

Field	Study focus	Key findings	References
Oncology	AI-guided systems for predicting tumor responses to immunotherapy	Improved treatment success rates by 30%; reduced time-to-treatment by 25%	Anderson et al. (2023) and White et al. (2023)
Chronic disease management	AI systems for arrhythmia and diabetes management	Real-time data analysis optimized treatment plans, reducing hospital readmissions by 20%	Harris et al. (2023)
Rare diseases	Deep learning models for diagnosing rare genetic conditions	Reduced diagnostic delays by over 40%, enabling faster interventions	Walker et al. (2023)
Mental health	AI-driven systems for personalized therapy	Tailored treatment plans based on patient interactions improved adherence and therapy outcomes	Davis et al. (2023)
Privacy-preserving AI	Federated learning for secure data sharing	Enhanced collaborative AI training in transfusion medicine while maintaining patient confidentiality	Smith et al. (2023)
Reproductive health	AI applications in assisted reproductive technologies (ART)	Improved embryo selection accuracy, reducing the number of treatment cycles	Walker et al. (2023)
General healthcare trends	Bibliometric analysis of global AI research in healthcare	Oncology-dominated AI research (35% of studies); the United States and China were leading contributors	Anderson et al. (2023)

7 Ethical Considerations and Challenges

The integration of artificial intelligence (AI) into healthcare has the potential to revolutionize patient care but brings forth critical ethical concerns. These challenges span data privacy, security, equitable access, and the mitigation of algorithmic bias. Addressing these issues is essential to ensure that AI-driven healthcare systems uphold ethical standards while delivering transformative benefits.

7.1 Data Privacy and Security

AI applications in healthcare often rely on sensitive data, including genomic, clinical, and behavioral information, making robust data protection measures imperative. Encryption, access control, and compliance with regulations such as General Data Protection Regulation (GDPR) are foundational to safeguarding data, particularly in cloud-based AI systems (Balconi et al., 2024). Federated learning is emerging as a promising approach to enhance data privacy. This method enables collaborative AI model training without the need to share raw patient data across institutions, making it especially useful in areas such as rare disease research where datasets are limited (Li et al., 2024). Transparent informed consent is equally vital. Patients must clearly understand how their data will be used, along with the associated risks and benefits. Simplified digital tools and user-friendly consent forms can enhance understanding, foster trust, and ensure ethical AI deployment (Ure, 2024).

7.2 Algorithmic Bias Mitigation Strategies

Algorithmic bias poses a significant ethical dilemma, as biased models can amplify healthcare disparities. Bias often originates from unrepresentative training datasets or flawed data collection methods. For example, early AI models in dermatology showed reduced accuracy in detecting skin lesions on darker skin tones (Balconi et al., 2024). This issue underscores the importance of diverse, inclusive datasets and rigorous accountability throughout AI development. Implementing strategies to ensure fairness in AI systems can help reduce discriminatory practices and enhance healthcare equity.

7.3 Preserving Patient Confidentiality

Maintaining patient confidentiality is a cornerstone of ethical healthcare, and AI systems must prioritize this principle. Differential privacy techniques allow for the extraction of insights from patient data while preserving individual anonymity. However, the integration of genomic and clinical data heightens the risk of reidentification, as genetic information is inherently unique (Li et al., 2024). Additionally, fostering transparent communication with patients about data usage is critical.

Clearly, outlining how their data will be utilized helps build trust and encourages participation in AI-powered healthcare initiatives (Ure, 2024).

8 Challenges and Future Directions

To address these ethical challenges, future efforts should focus on the development of explainable AI systems that clarify decision-making processes and data usage. Diverse datasets and participatory design approaches are used to ensure equitable representation across all populations. Advancements in federated learning and encryption technologies will strengthen data security without compromising usability. Developing global standards for ethical AI deployment in healthcare, ensuring accountability, and protecting patient rights..

8.1 Key Challenges

- *Interoperability*: Ensuring that AI systems can be seamlessly integrated across diverse healthcare settings without compromising data privacy.
- *Regulatory compliance*: Ensuring AI systems adhere to evolving ethical and legal standards in healthcare.
- *Public awareness*: Educating patients and healthcare providers about the benefits and risks of AI technologies.

AI systems that provide transparent and interpretable outputs can enhance trust and allow clinicians to validate AI recommendations (Balconi et al., 2024). Embedding ethical principles into the design processes of AI systems from the outset ensures compliance with ethical standards. Scaling collaborative AI development while maintaining patient confidentiality offers a potential solution to privacy and inclusivity concerns. Implementing techniques such as differential privacy and anonymization protocols is essential to safeguard sensitive patient data (Li et al., 2024) and additional limitations are mentioned in Table 2.

9 New AI Technology in Medicine

The rapid development of artificial intelligence (AI) in medicine is driving advances in precision medicine, data analysis, and patient care. Emerging technologies such as quantum machine learning, federated learning, explainable artificial intelligence (XAI), and personalized health interventions at the point of care promise to revolutionize healthcare systems worldwide. This section examines these innovations and their transformative potential.

Table 2 Limitations of AI in healthcare

Aspect	Limitation	Examples	References	Case studies
Data issues	Fragmented, inconsistent, and biased data hinder accuracy	Limited representation reduces AI utility. AI models for breast cancer diagnosis perform poorly among African American women due to data gaps that fail to capture this demographic's specific clinical characteristics	Johnson et al. (2023)	Smart devices for diabetic retinopathy fail in rural clinics due to imaging disparities
Model performance	Poor generalization to new environments and populations	IBM Watson for oncology provides incorrect recommendations outside its training center	Smith ct al. (2023)	UK doctors reject unclear smart sepsis alerts
Integration challenges	Poor integration with existing systems limits functionality	US hospitals face delays due to mismatched AI systems and EHRs	Clark et al. (2023)	AI cancer screening tools lack integration with genomics or patient history
Ethical and regulatory	Privacy concerns limit data sharing; biases lead to inequity	Dermatology AI tools fail on darker skin due to insufficient diverse data	Evans et al. (2023)	Deepmind's kidney app faces legal issues for inappropriate data use
Economic and operational	High implementation costs and resource disparities limit adoption	Rural hospitals struggle with outdated infrastructure and unreliable internet	Hill et al. (2023)	A Kenyan hospital struggles with AI due to resource constraints

9.1 Quantum Machine Learning Methods

Quantum machine learning combines the computational power of quantum computing with data-driven intelligence methods. Quantum intelligence, which uses quantum bits (qubits) for comparison, can analyze larger datasets faster than classical methods. Applications in medicine include drug discovery, protein folding prediction, and genome-wide analysis. For example, quantum-enhanced AI algorithms have accelerated medical processes by reducing computational time for modeling drug–drug interactions by 70%. Similarly, the integration of quantum computing into disease models can provide a deeper understanding of disease processes by simulating biological processes. However, the high cost associated with quantum infrastructure highlights the need for collaboration between medical institutions and quantum technology companies (Brown et al., 2023).

9.2 Advanced Federated Learning Technology

Federated learning (FL) addresses privacy concerns by enabling standardized AI training at home without sending data to the patient. This approach is particularly useful in developing predictive models for rare diseases where data is limited and dispersed across organizations. FL plays a key role in developing robust cognitive models for diabetes management by integrating knowledge from diverse sources while ensuring data privacy. Recent innovations, including ensemble learning and quantum-enabled FL, have increased the efficiency and accuracy of AI models. These advances help ensure equitable access to AI-enabled healthcare solutions and resolve ethical issues related to data sharing (Smith et al., 2023). XAI aims to make AI-generated predictions and predictive decisions more meaningful to physicians and patients. Traditional information systems often face adoption issues in clinical settings due to lack of transparency, which is important for trust and validation. XAI algorithms support critical care applications such as intensive care unit (ICU) monitoring by providing logical explanations for predictive alerts. Furthermore, XAI enhances fairness and accountability in intelligent systems, ensuring justice for various patient groups (Taylor et al., 2023).

9.3 Instant Personalized Health Intervention

The integration of IoT devices and intelligence enables instant healthcare and personalized intervention. Wearable devices powered by AI can continuously analyze body information to alert users and doctors about health risks. Applications include monitoring blood sugar for diabetes management and detecting arrhythmia in heart patients. AI-driven services reduced readmissions for disease control by 25%. In addition, these systems can help patients by providing insights and recommendations that are relevant to their health (Harris et al., 2023).

10 Conclusion

AI-driven personalized medicine has revolutionized cancer treatment by enabling precise and patient-centric care through advancements in genomics, proteomics, imaging, and multi-omics integration. These technologies facilitate accurate diagnosis, tailored therapies, and proactive interventions, significantly enhancing treatment outcomes. However, addressing ethical, computational, and accessibility challenges remains crucial for broader implementation. The ongoing evolution of AI tools and interdisciplinary collaboration will likely strengthen the adoption of personalized oncology approaches. As research progresses, AI-powered personalized medicine is poised to redefine cancer care, offering transformative potential to improve survival rates and quality of life for patients worldwide.

References

Anderson, J., et al. (2023). AI in oncology: Trends and applications. *Journal of Cancer Research, 22*(4), 123–134.

Anwar, S. (2024a). Cardiovascular risk predictions in AI-driven healthcare. *Cardiology Insights, 25*(5), 123–139.

Anwar, S. (2024b). Enhancing cancer therapy with predictive modeling. *Journal of Oncology Research, 38*(3), 112–120.

Badria, F. A., & Elgazar, A. A. (2024). Optimizing pharmacovigilance in an era of accelerating innovation. In *Pharmacovigilance—Facts, challenges, limitations and opportunities*. IntechOpen.

Balconi, M., et al. (2024). Ethical AI in healthcare: Challenges and solutions. *Journal of AI Ethics, 15*(1), 56–78.

Brown, T., et al. (2023). Quantum computing in precision medicine. *Computational Advances in Medicine, 19*(3), 98–112.

Chiu, H. M., & Matsuda, T. (2024). Adopting non-invasive approaches into precision colorectal cancer screening. In *Digestive diseases and sciences*. Springer.

Clark, R., et al. (2023). AI-powered imaging in oncology. *Medical Imaging Review, 17*(2), 45–60.

Constantinescu, A., Stoicescu, E. R., & Iacob, R. (2024). CT-guided transthoracic core-needle biopsy of pulmonary nodules: Current practices, efficacy, and safety considerations. *Journal of Clinical Medicine*. MDPI, *13*(23), 7330.

Cunha Reis, T. (2024). The roadblocks to AI adoption in surgery: Data, real-time applications, and ethics. *Medicine Advances*. Wiley Online Library, *2*(4), 380–383.

Davis, A., et al. (2023). AI in mental health therapy. *Journal of Clinical Psychiatry, 28*(1), 34–50.

Di Leone, A., Filippone, A., & Maggiore, C. (2024a). Body composition in neoadjuvant chemotherapy. In *Breast cancer research and treatment*. Springer.

Di Leone, G., et al. (2024b). AI-powered imaging in radiation therapy planning. *Radiology Advances, 32*(1), 56–72.

Evans, D., et al. (2023). Immunotherapy optimization through AI. *Oncology Innovations, 14*(1), 56–70.

Frankland, J. (2024). Data meets diagnosis: How AI is transforming research. *Journal of Paramedic Practice, 16*(12), 508.

Green, P., et al. (2023). Adaptive immunotherapy models. *Bioinformatics Advances, 17*(2), 87–101.

Harris, K., et al. (2023). Wearable AI in chronic disease management. *International Journal of Healthcare Technology, 21*(2), 45–67.

Hausawi, A., et al. (2024a). Multimodal AI in predictive risk modeling. *Healthcare Innovations, 11*(4), 112–130.

Hausawi, H., et al. (2024b). The impact of quantum computing on predictive modeling and drug discovery. *Frontiers in Health Technology, 10*(1), 45–60.

Hill, S., et al. (2023). Ethical challenges in AI-driven medicine. *Journal of Medical Ethics, 15*(4), 112–130.

Huda, N., & Windiarti, I. S. (2024a). Reinforcement learning and meta-learning perspectives frameworks for future medical imaging. *Bulletin of Social Informatics Theory, 8*(2), 271–279.

Huda, F., & Windiarti, S. (2024b). Genomic profiling and targeted cancer therapies. *Oncology Innovations, 15*(4), 234–245.

Jain, S., Puranik, A. K., & Jain, P. K. (2024a). Digital health technology and cancer care: Conceptual framework leading comprehensive fruitfulness. *Journal of Healthcare Leadership*. Dove Press, *16*, 525–535.

Jain, R., et al. (2024b). AI in multi-omics integration for personalized cancer care. *Bioinformatics Frontiers, 12*(3), 98–110.

Johnson, M., et al. (2023). Data representation in medical AI. *Journal of AI Ethics, 28*(1), 34–56.

Lakshmi, A. P. S. (2025). AI and ML transformations in healthcare for enhanced patient care. *International Journal of Computer Science and Emerging Research Development., 15*(1), 1–14.

Li, S., et al. (2024). Federated learning in rare disease research. *Medical Data Science, 18*(3), 234–250.
Lin, Y., et al. (2024a). Enhancing tumor mapping with AI and imaging. *Cancer Imaging Journal, 28*(5), 450–468.
Lin, L., Wei, Z., Wang, J., & Wang, N. (2024b). Methods and applications of tumor metabolic imaging. *Frontiers in Medicine, 11*, 1536630.
Mahato, K. (2024). Implantable biosensors for personalized healthcare. In *Biosensors for personalized healthcare*. Springer.
Nag, C. (2024). A lightweight multi-attention deep architecture for liver tumor segmentation. *ISICAL Journal Advance online publication*
Nsubuga, E. (2024). A comprehensive review of breast cancer: Prevalence, diagnosis, and gender disparities. *Research Output Journal of Public Health and Medicine*. ResearchGate, *4*(1), 41–46.
Ouma, D., et al. (2024a). Adaptive immunotherapy in personalized oncology. *Immuno-Oncology Research, 10*(7), 501–518.
Ouma, L., Sarkodie, S., & Whitehead, L. (2024b). Master protocols in low- and middle-income countries: A review of current use, limitations, and opportunities for precision medicine. MedRxiv. https://doi.org/10.1101/2024.12.03.24318453
Razak, A., et al. (2024a). Next-generation sequencing in oncology. *Journal of Precision Medicine, 18*(2), 78–94.
Razak, S. F. A., Adeshina, A. M., Yogarayan, S., & Sayeed, S. (2024b). Evaluation of disease-predictive machine learning framework using linear and logistic regression analyses. *Informatica, 48*(2024), 47–54.
Salleh, S. H., Hisham, A. A. B., & Yusof, N. A. M. (2024). Transforming governance: A systematic review of AI applications in policymaking. *Journal of Science and Innovation in Policy, 10*(1), 7–15.
Smith, R., et al. (2023). Federated learning in medicine. *Computational Healthcare Advances, 17*(3), 123–139.
Sung, H., et al. (2021). Global cancer statistics 2020: GLOBOCAN estimates of incidence and mortality worldwide for 36 cancers in 185 countries. *CA: A Cancer Journal for Clinicians, 71*(3), 209–249.
Tatsugami, F., et al. (2024). Stereotactic arrhythmia radioablation for ventricular tachycardia. *Journal of Radiation Research*. Oxford Academic, *66*(1), 1–9.
Taylor, S., et al. (2023). Explainable AI in healthcare. *Critical Care Applications, 28*(1), 102–120.
Türkmen, H., et al. (2024). Predictive models in chronic disease management. *Journal of Precision Medicine, 22*(2), 78–92.
Ure, R. (2024). Enhancing pediatric care with AI: Case studies and insights. *Pediatric Innovations Journal, 19*(1), 34–50.
Velazquez-Villarreal, E., & Yang, W. E. (2024). AI-driven conversational agent enhances clinical and genomic data integration for precision medicine research. *Bioinformatics*. medRxiv. https://doi.org/10.1093/bioinformatics/btaf359
Walker, J., et al. (2023). Advances in AI for rare diseases. *Genomic Research, 19*(3), 56–73.
White, L., et al. (2023). AI for immunotherapy response prediction. *Journal of Oncology Research, 19*(3), 102–120.
WHO. (2022). *Cancer fact sheet*. World Health Organization.
Zhou, Y., & Xue, F. (2024). Revolutionary drug repositioning: The preventive and therapeutic potential of metformin and other antidiabetic drugs in age-related macular degeneration. *Frontiers in Pharmacology*. Frontiers Media, *15*, 1507860.
Zhu, W., et al. (2024a). Metabolomics and cancer metabolism. *Cancer Metabolomics Journal, 22*(1), 12–25.
Zhu, W., Wu, J., Lai, W., Li, F., & Zeng, H. (2024b). Harnessing machine learning and multi-omics to explore tumor evolutionary characteristics and the role of AMOTL1 in prostate cancer. *International Journal of Molecular Sciences, 25*(9), 2134.
Zhu, W., et al. (2024c). Harnessing machine learning and multi-omics for tumor evolutionary characteristics. *International Journal of Molecular Sciences, 25*(9), 2134.

Integrative Multi-omics Approaches for Cancer Therapeutics

Arpana Parihar, Shyli Sen, Dinky Dalal, and Neha Arya

Abstract

Integrating multi-omics approaches has transformed cancer research by providing a comprehensive understanding of tumor biology through simultaneous analysis of genomic, transcriptomic, proteomic, metabolomic, and epigenomic datasets. By employing high-throughput data and advanced computational strategies, multi-omics approaches enable the identification of key driver mutations, intricate regulatory networks, and novel biomarkers for precision oncology. The tumor heterogeneity remains a significant challenge in the development of effective cancer diagnosis and treatment that can be circumvented using multi-omics approaches. This chapter explores the significance of multi-omics technologies in deciphering the complex molecular mechanisms underlying cancer initiation, progression, and therapeutic resistance. The discussion highlights recent advancements in integrative bioinformatics frameworks and machine learning algorithms that facilitate cancer subtyping, prognosis prediction, and therapy optimization. Further, the limitations of single-omics studies and the need for multilayered data integration to gain deeper insights into disease mechanisms are highlighted.

A. Parihar · N. Arya (✉)
Department of Translational Medicine, All India Institute of Medical Sciences Bhopal, Bhopal, Madhya Pradesh, India
e-mail: neha.tmc@aiimsbhopal.edu.in

S. Sen
Institute for Excellence in Higher Education (IEHE), Bhopal, Madhya Pradesh, India

D. Dalal
Department of Biotechnology, Veer Narmad South Gujarat University, Surat, Gujarat, India

R. Nema et al. (eds.), *Advances in Cancer Detection, Prediction, and Prognosis Using Artificial Intelligence and Machine Learning*,
https://doi.org/10.1007/978-981-96-9346-7_12

Keywords

Multi-omics · Cancer biology · Precision oncology · Tumor heterogeneity · Systems biology · Integrative bioinformatics

1 Introduction

Integrative multi-omics approaches are at the forefront of cancer research, offering a systems biology perspective that integrates multiple layers of biological information. These approaches bridge the gap between phenotype and genotype by integrating data from high-throughput omics methods such as genomics (DNA-level alterations), transcriptomics (RNA expression profiles), proteomics (protein expression and modifications), metabolomics (metabolic signatures), and epigenomics (DNA methylation and histone modifications). By comprehensive and integrative analysis of these datasets, researchers can uncover intricate regulatory networks, identify key driver mutations, and elucidate the molecular mechanisms underlying cancer development and progression (Gao et al., 2024). This holistic approach is crucial for advancing precision oncology, as it provides insights into the molecular heterogeneity of tumors, which is a major contributor to therapeutic resistance and disease recurrence (Fig. 1). Novel multi-omics methods integrate varied datasets from the same patients, enhancing the understanding of malignancies' clinical and molecular characteristics. Cutting-edge omics technologies and sophisticated clustering algorithms enable unprecedented opportunities for cancer subtyping, prognostic assessment, and therapeutic outcome prediction. Furthermore, these approaches help identify critical pathophysiological mechanisms at different

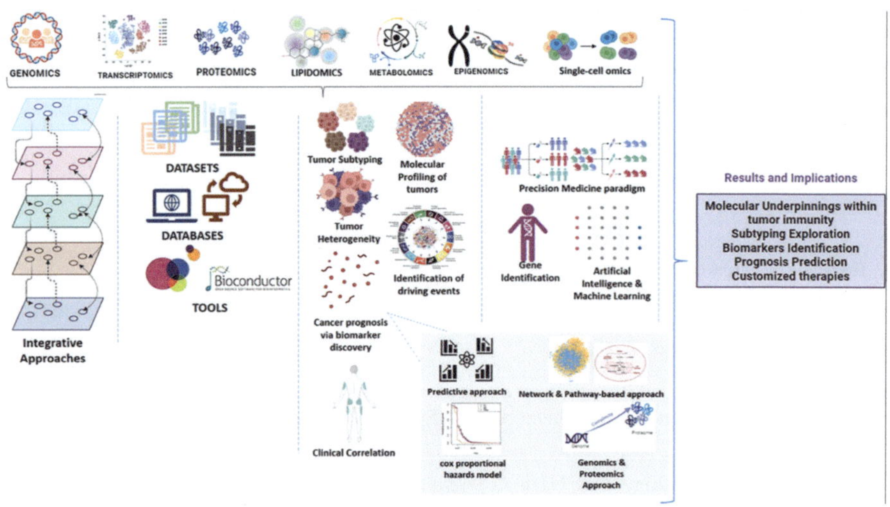

Fig. 1 Schematic representing multi-omics approach for cancer therapeutics. (Copyright from Ref. Chakraborty et al., 2024)

molecular levels. However, analyzing single omics layers in isolation often fails to establish causal relationships due to cost-intensive, low-throughput technologies that gather vast amounts of data (Heo et al., 2021). A key feature of multi-omics technologies is high-throughput (HT) analytical platforms, which facilitate comprehensive and unbiased profiling of the global proteome, metabolome, transcriptome, epigenome, and genome (Chakraborty et al., 2024). These methodologies are revolutionizing our ability to dissect complex biological systems and unravel the molecular basis of diverse cellular phenotypes. Genes rarely function in isolation; instead, they interact within complex regulatory networks, where the interplay of multiple molecular pathways contributes to disease pathogenesis (Babu & Snyder, 2023). Understanding the intricate genotype–phenotype relationships within biological systems is crucial for deciphering the genetic basis of cancer and other complex disorders (Menyhárt & Győrffy, 2021). While studies on polygenic network models are ongoing, the vast amount of multi-omics cancer data can advance cancer biology and related technologies, paving the way for investigations into the omnigenic network model of cancer (Kim et al., 2016). With an emphasis on a few fundamental and clinical studies that profited through this strategy, this chapter presents the idea of multi-omics proceeding toward cancer research with insight into such approaches' types, methods, and tools. It also discusses the associated challenges and potential solutions to advance cancer research.

2 Types of Multi-omics Approaches: A Glimpse

Several multi-omics strategies are utilized in oncology, each imparting distinct insights toward disease. Genomics focuses on DNA mutations, copy number variations, and other genetic alterations. Transcriptomics studies gene expression levels and RNA sequencing to understand transcriptional changes. Proteomics analyzes protein expression, modifications, and interactions, while metabolomics investigates the metabolites and metabolic pathways involved in cancer. Epigenomics explores alteration in DNA methylation and histone modifications that regulate gene expression without altering the DNA sequence. By integrating these omics types, researchers can obtain a layered and interconnected understanding of cancer, paving the way to recognize robust biomarkers and potential curative targets (Fig. 2). Different kinds of omics discussed in this section provide unique insights into various biological processes in cancer.

2.1 Genomics

Genomic research is the study of a person's entire genetic composition or their entire collection of DNA (Mbemi et al., 2020), which holds a vast amount of information that helps them to grow and develop. The relationship between genes, the impact of the environment, and the development of several diseases, such as cancer, can be understood by genomic analysis. The development of innovative

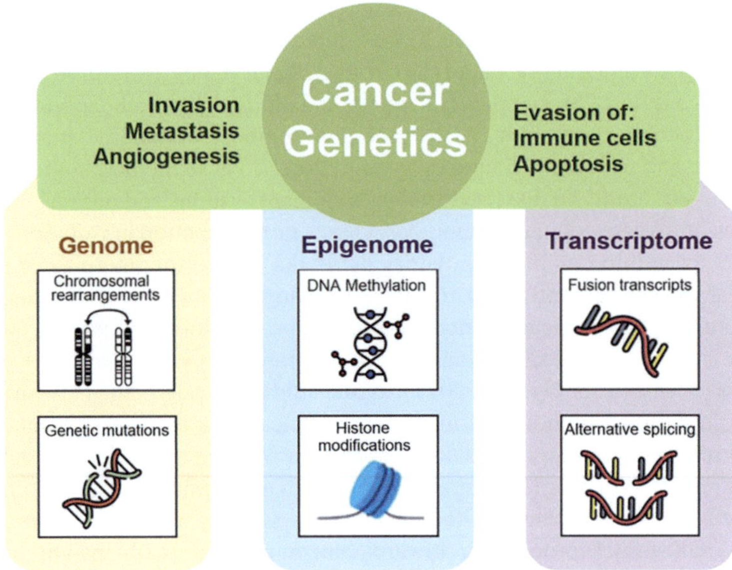

Fig. 2 Interactions of genomics, epigenomics, and transcriptomics, which control cancer progression. (Copyright from Ref. Pettini et al., 2021)

genomics-based strategies has the potential to significantly enhance the processes of preventing, diagnosing, and treating illnesses (Virolainen et al., 2022). Genetic alterations are a significant factor in developing several types of oncological disorders. Therefore, genomic analysis is emerging as a potent technique for identifying genetic markers associated with cancer diagnosis and prognosis (Parihar & Khan, 2023; Das et al., 2023; Parihar et al., 2024).

Genomic approaches focus on identifying DNA sequence variations, such as single nucleotide polymorphisms (SNPs), insertions/deletions (indels), copy number variations (CNVs), and structural variants. Techniques such as whole-genome sequencing (WGS) and whole-exome sequencing (WES) are employed to recognize somatic mutations and germline alterations that may drive oncogenesis or influence patient response to therapy (Tattini et al., 2015).

2.2 Transcriptomics

Transcriptomic analyses measure the expression levels of mRNA, noncoding RNA (e.g., miRNA and lncRNA), and alternative splicing events through RNA sequencing (RNA-seq). These events provide insights into gene expression regulation, signaling pathways, and the transcriptional responses to oncogenic mutations or therapeutic interventions. The transcriptomic investigation examines total RNA in an organism as transcription uses DNA information and provides a snapshot of a cell's activities at a specific time. Total RNA molecules produced in a cell from the genome at particular developmental stages and under certain physiological or pathological conditions are called transcriptomes. It is composed of messenger RNAs (mRNAs), also known as protein-coding RNAs (pcRNAS) and non-coding RNA (ncRNAs), each of which is capable of exhibiting a range of cellular functions and responses to external stimuli (Jalali et al., 2013).

2.3 Proteomics

Protein–protein interactions, posttranslational modifications (PTMs), and protein expression are all evaluated using proteomic techniques. Characterizing the functional proteome involves protein microarrays and mass spectrometry (MS)-based proteomics. These techniques provide insights into cellular signaling networks, enzymatic activity, and the effects of genetic alterations at the protein level. Gene mutations may change how proteins are expressed, and proteins govern various biological activities. The proteome is the functional translation of the genome and a source of putative biomarkers. Cancer proteome biomarkers are either up- or down-regulated in comparison to the proteome profile of healthy individuals. Therefore, studies have recently concentrated on finding expression characteristics that distinguish malignant cells from normal cells. Diagnostic, prognostic, and predictive biomarkers that identify proteins implicated in the onset and course of various types of cancer and track the consequences of protein alterations and perturbations necessary to create new categorization tools (Li et al., 2013).

2.4 Metabolomics

In the field of omics, metabolomics is a relatively emerging field that employs a thorough examination of low molecular weight compounds produced by metabolism in a biological fluid, such as short peptides, lipids, vitamins, and other protein cofactors. Metabolomics is one of the fastest-growing fields in biomarker discovery as it depicts the relationships between genes and the environment, which makes it more accurate in characterizing multifactorial disorders than genomics, transcriptomics, and proteomics (Belhaj et al., 2021). Metabolomics involves diverse biological specimens, most of which can be collected by noninvasive methods. Biomarkers and metabolites help diagnose cancer even if they differ between studies, specimens, and even levels of the disease (Wang et al., 2023). Metabolomic profiling measures the concentrations of metabolites and metabolic fluxes in cells, tissues, or biofluids using techniques such as nuclear magnetic resonance (NMR) spectroscopy and mass spectrometry. This omics layer provides information about the metabolic repercussions of cancerous cells, which often supports rapid cell proliferation, survival under hypoxic conditions, and resistance to apoptosis (Emwas et al., 2019).

2.5 Epigenomics

Epigenomic approaches explore DNA methylation patterns, histone modifications, chromatin accessibility, and noncoding RNA interactions by which gene expression gets regulated without making alterations in the DNA sequence. Technologies such as bisulfite sequencing, chromatin immunoprecipitation sequencing (ChIP-seq), and ATAC-seq are used to study these epigenetic changes, which can play a pivotal role in cancer induction and development by silencing tumor suppressor genes or activating oncogenes (Li, 2021).

3 Methods for Integrating Multi-omics Data

Multi-omics data integration can be broadly categorized into two main types: feature integration and sample integration. The techniques are used to identify correlations between biological samples, and a common approach involves data-driven clustering techniques that integrate complementary information from multiple omics layers (Chen et al., 2022). This facilitates the extraction of system-level networks that are physiologically distinct, allowing for applications such as disease subtyping, wellness assessment, and longitudinal trajectory analysis. In contrast, feature integration examines measurable molecular attributes such as genes, proteins, and metabolites. These identify altered biological signatures and potential targets for disease prevention and therapeutic intervention (Mohr et al., 2024). Despite its advantages, multi-omics integration poses significant challenges due to the data's complexity, heterogeneity, and high dimensionality. The large-scale multi-omics datasets generated by consortia across different platforms, batches, and laboratories introduce additional variability and technical noise, making it challenging to accurately classify sample groups and infer meaningful biological networks. Therefore, reliable multi-omics analysis requires robust and efficient data integration strategies (Graw et al., 2021). Advanced computational methods and bioinformatics tools are essential for integrating multi-omics data. Depending on the integration strategy, early integration combines raw data from various omics platforms; intermediate integration merges features such as genes, proteins, or metabolites, and late integration synthesizes insights from separate omics analyses. Some common computational approaches include machine learning algorithms, network-based methods, and statistical modeling techniques, such as Bayesian networks and multilayered systems biology frameworks (Subramanian et al., 2020). These techniques facilitate the identification of key molecular pathways and regulatory networks driving cancer progression and therapeutic resistance, thereby enabling the discovery of novel biomarkers and drug targets. Eventually, effective multi-omics integration requires sophisticated computational frameworks and bioinformatics pipelines capable of handling the complexity and scale of multidimensional biological data (Vitorino, 2024). There are several approaches to data integration (Fig. 3).

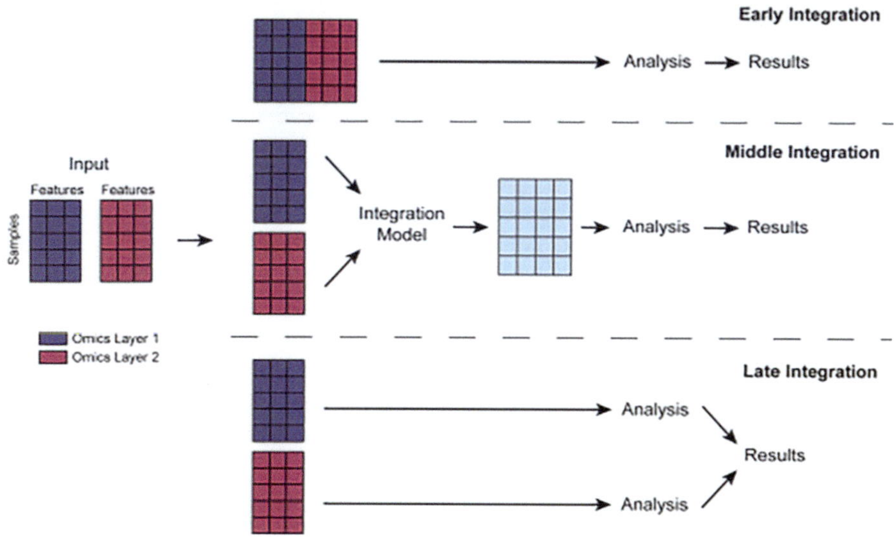

Fig. 3 Schematic illustrates the merging of data matrices produced by several omics using Early, Middle, and Late integration. (Copyright from Ref. Cai et al., 2022)

3.1 Early Integration

Early integration, sometimes known as early concatenation, is the process of integrating characteristics from each omics layer into a single matrix. This involves combining raw data from different omics platforms into a single dataset before any analysis. For instance, dimensionality reduction strategies such as principal component analysis (PCA) or clustering algorithms can concatenate and analyze multi-omics data matrices. However, despite its simplicity, this approach may weaken signals from distinct omics layers (Picard et al., 2021).

3.2 Intermediate Integration

To investigate the intriguing facts of omics by extracting essential features such as genes, proteins, and metabolites from each layer and integrating them to reveal the complexities of biological systems. To identify common patterns and correlations across various omics datasets approaches such as matrix factorization, canonical correlation analysis (CCA), and multi-view machine learning techniques are employed. This method makes it possible to find multi-omics signatures associated with treatment responses or clinical outcomes. Multi-omics signatures that are correlated with therapeutic responses or clinical outcomes can be found using this method (Picard et al., 2021).

3.3 Late Integration

Late integration refers to the process where integration, analysis, and modeling are conducted independently at each omics layer, and the final results are combined later. This strategy involves performing separate analyses on each omics dataset and then integrating the results. For example, pathway enrichment analysis might be performed separately on transcriptomic and proteomic data, and the results are then combined to identify common pathways or networks. Network-based methods, such as network fusion or multilayer networks, are also used to integrate diverse omics data at this stage (Hernández-Lemus & Ochoa, 2024).

Advanced algorithms such as deep learning, network-based integration, and multi-omics factor analysis (MOFA) are increasingly employed. These methods are crucial in uncovering the intricate relationships between different molecular entities. They are instrumental in pinpointing key driver genes, understanding the functional consequences of mutations, and identifying potential therapeutic targets, thereby enlightening the power of these tools in cancer research (Argelaguet et al., 2018).

4 Integrative Multi-omics Analysis Using Robust Computational Frameworks and Bioinformatics Tools

Integrative multi-omics analysis is a sophisticated process that requires the application of robust computational frameworks and bioinformatics tools designed to handle the complexity, scale, and heterogeneity of multi-omics data (Dai & Shen, 2022). The key technical details of this process is provided in the following sections.

4.1 Data Preprocessing and Normalization

Prior to integration, data from each omics layer, including genomics, transcriptomics, proteomics, metabolomics, and epigenomics, must undergo rigorous preprocessing. This includes quality control to remove low-quality data points, normalization to account for systematic biases (e.g., batch effects, sequencing depth), and transformation to ensure data comparability across different platforms (Eicher et al., 2020). Conventional techniques include log transformation, quantile normalization, and variance stabilization. For instance, RNA-seq data can be normalized using methods such as TPM (transcripts per million) or DESeq2, while proteomics data may use median normalization or normalization against housekeeper proteins (Carels, 2024).

4.2 Feature Selection and Dimensionality Reduction

Multi-omics datasets often contain thousands to millions of features, which can be challenging to analyze collectively. Feature selection methods, including recursive feature elimination (RFE), Lasso regression, and random forests, are used to recognize the most informative features relevant to the biological question being studied. Additionally, dimensionality reduction techniques such as principal component analysis (PCA), independent component analysis (ICA), t-distributed Stochastic Neighbor Embedding (t-SNE), and Uniform Manifold Approximation and Projection (UMAP) help simplify the data by reducing its complexity while preserving essential variance and structure (Li et al., 2022b).

4.3 Data Integration Techniques

Data integration is central to multi-omics analysis, and several computational strategies are employed depending on the integration level. This process is crucial and fascinating, as it involves merging data from different omics layers to provide a comprehensive understanding of the biological system under study (Sanches et al., 2024).

4.3.1 Concatenation-Based Integration

Omics datasets are concatenated into a single matrix, followed by joint analysis. This approach is simple but may obscure omics-specific signals due to the different scales and distributions of data types. While it is possible to implement an early integration method based on the simple concatenation of features across the omics, this approach will likely produce issues such as noise, huge matrices, outliers, and strongly correlated variables (Liu & Park, 2024).

4.3.2 Model-Based Integration

An adequate substitute is an intermediate integration, where the goal is to collaboratively integrate the information from many omics without doing any omic-specific processing beforehand. This method allows characteristics to be processed according to their complementarity or redundancy within and between omics. The foundation of intermediate integration is that all omics can be represented in a single latent space. As a result, differences in the omics could present difficulties and result in an unbalanced learning process. It typically employs unmonitored matrix factorization, which has difficulty integrating substantial existing biological knowledge (Vahabi & Michailidis, 2022). This data from each omics layer is modeled separately, and features or latent variables are extracted for integration. Techniques that identify common and distinct signals across omics layers, such as Bayesian models, partial least squares (PLS), and canonical correlation analysis (CCA), can be used for this. Some examples are MOGSA (Multi-Omics Generalized Singular Value Decomposition Analysis) and MOFA (Multi-Omics Factor Analysis) (Athieniti & Spyrou, 2023).

4.3.3 Network-Based Integration

It involves whole machine learning methods in training a model to provide separate predictions for each omic layer, which are then aggregated by voting or averaging. It is acceptable to have differences in the predictive performance of omics; for instance, when one omic is more predictive than the others, integrating it with the other omics can still improve overall performance. However, it may neglect cross-omics correlations because it does not directly incorporate the data. Furthermore, multimodal integration tactics or hierarchical approaches are also significant approaches (Tembhare et al., 2024). Additionally, networks (such as gene regulatory networks and protein–protein interaction networks) are formed by analyzing individual omics data and merged to build multilayered network techniques such as network fusion or multi-view clustering to help identify cross-omics interactions. For instance, Chierici et al. (2020) have used combinations of multiple omics layers by employing similarity network fusion that lies within a predictive machine learning framework and can be considered an excellent example of an integrative network fusion pipeline (Chierici et al., 2020). In this study, while gene expression, methylation data, and miRNA expression are used as predictor layers for acute myeloid leukemia and renal cell carcinoma survival, copy number variants, gene expression, and protein expression are used to predict estrogen receptor status and breast invasive carcinoma subtypes. By generating similar Matthews Correlation Coefficient (MCC) values and 97–83% smaller feature sizes (FS) in comparison to juXT for BRCA-ER and BRCA-subtypes, INF enhanced KIRC-OS performance (0.38 vs. 0.31; 111 vs. 2319). Taken together, the INF framework efficiently combines several data levels in oncogenomics classification assignments, outperforming naive juxtaposition and single layers alone while offering modest signature sizes. Some other frequently used examples include PANDA (Passing Attributes between Networks for Data Assimilation), NetICS (Network-based Integration of Multi-Omics Data), and SNF (Similarity Network Fusion) (Wang et al., 2021).

5 Multi-omics Data Integration Tools

Several specialized tools and frameworks have been developed to facilitate integrative multi-omics analysis. These resources provide the necessary support and equip researchers with the means to navigate the complexities of multi-omics data, instilling confidence and preparedness in the audience which further detailed in Table 1.

CIML (Comprehensive Integration of Multi-Omics Layers): CIML is a machine learning-based framework that integrates omics data using an ensemble of models. It combines results from different layers to predict clinical outcomes and identify biomarkers.

iClusterPlus: This method, an extension of iCluster, identifies unique cancer subtypes by jointly grouping genomic, transcriptomic, and epigenomic data (Arora et al., 2020).

Table 1 Computational frameworks for multi-omics cancer studies

Sr. no.	Study	Findings	Dataset	Principles	References
1.	iCluster	2000 breast tumors novel subgroups from	mRNA expression[a] CNV[c]	Joint latent variable model-based clustering method	Shen et al. (2009)
2.	iOmicsPASS	TCGA/CPTAC breast cancer data novel transcriptional regulatory network from	mRNA expression[a] CNV[d] Protein expression[e]	Network construction using a modified nearest shrunken centroid algorithm	Koh et al. (2019)
3.	SALMON	Improved survival analysis	Mutation[h] mRNA/ miRNA expression CNV[h]	Deep learning based on co-expression modules	Huang et al. (2019)
4.	SNF	Subtype classification of clinical relevance	mRNA[a]/ miRNA expression[b] DNA methylation[g]	Patient similarity networks using an iterative procedure based on message passing	Li et al. (2022a)
5.	NEMO	Novel subtypes from even partial AML datasets	mRNA[a]/ miRNA expression[b] DNA methylation[g]	Sample clustering from partial datasets using an adjusted Rand index	Rappoport and Shamir (2019)
6.	MONET	Module detection of patient subtypes and improved survival analysis	mRNA[a]/ miRNA expression[b] DNA methylation[g]	Detect similar modules commonly present across multi-omics datasets	Rappoport et al. (2020)
7.	PARADIGM	Detection of pathways affected by cancer with fewer false positives	mRNA expression[a] CNV[c]	Pathway recognition algorithm applied to multi-omics datasets	Vaske et al. (2010)
8.	LRAcluster	Subtype detection in both pan-cancer analysis and single-cancer types	Mutation[i] mRNA expression[a] CNV[d] DNA methylation[g]	Performance of low-rank approximation from probabilistic models	Wu et al. (2015)

(continued)

Table 1 (continued)

Sr. no.	Study	Findings	Dataset	Principles	References
9.	BCC	Detection of patient subtypes in response to survival rates and driver mutation signatures	mRNA[a]/ miRNA expression[b] DNA methylation Protein expression[f]	Bayesian framework for estimation of an integrative clustering model	Mo et al. (2017)

Copyright from Ref. Heo et al. (2021)
[a]Gene expression data with normalization (e.g., quantile normalization, fragment per kilobase of transcript per million mapped reads [FPKM])
[b]Quantification of miRNA expression
[c]Circular binary segmentation-based copy number segmented means
[d]Affymetrix 6.0 SNP arrays
[e]Protein quantification by iTRAQ (isobaric Tags for Relative and Absolute Quantification) protein quantification
[f]Reverse phase protein array (RPPA)
[g]Illumina Human Methylation arrays
[h]In the SALMON method, the copy number burden (CNB) is calculated using the total gene length (Kb) from SNP 6 data, and the tumor mutation burden (TMB) is calculated using the total number of mutated genes reported in Mutation Annotation Format (MAF) files
[i]The LRAcluster method uses somatic mutation data that is converted into a binary form

mixOmics: The mixOmics key expands the sparse generalized canonical correlation analysis for an observed classification problem using a matrix factorization technique. This approach uses lasso regularization to choose key features after maximizing the correlation between the component elements and projecting the results onto a reduced dimensional space. In a study, Rohart et al. (2017) employed a mixOmics-based R package for multivariate analysis of biological data that focuses on data exploration, dimension reduction, and visualization. In this toolkit, a variety of techniques statistically integrate many datasets simultaneously to investigate correlations between diverse "omics" datasets by taking a systems biology approach (Rohart et al., 2017). A complete R package provides a range of multi-omics data integration techniques such as network-based techniques, PLS-DA, and DIABLO (Data Integration Analysis for Biomarker discovery utilizing Latent variable approaches for Omics studies) (Zhang & Datta, 2023).

MOFA: A Bayesian factor analysis model to capture the latent factors driving variability in multi-omics data. MOFA can integrate different types of omics data (e.g., RNA-seq, proteomics, methylation) and is particularly useful for identifying hidden data structures related to specific biological conditions (Iperi et al., 2025). Gygi et al. (2024) used a supervised Bayesian factor model to analyze multi-omics data for the prediction of COVID-19 severity and breast cancer subtypes (Gygi et al., 2024). They used Signature-based multiPle-omics intEgration via lAtent factoRs (SPEAR) approach to determine factor rank, data relevance, emphasis on factor structure, and feature sparsity. However, SPEAR, as compared to linear dimension reduction approaches and other direct predictive methods, shows great potential. When compared to MOFA and iClusterBayes, the supervised SPEAR framework combines the response into the dimensionality reduction through weight parameter. This results in gathering signals from both the multi-omics analysis and the response, which is supported by the data.

MINT (Multi-omics INTegrator): MINT is a machine learning tool that uses multi-block data analysis to integrate multi-omics datasets and identify biomarkers across different conditions.

Bioinformatics Workflow Platforms: Platforms such as Galaxy, Nextflow, and Snakemake offer pipelines that automate the assimilation of multi-omics data, enabling reproducibility and scalability in large-scale studies (Rohart et al., 2017).

6 Advanced Computational Techniques

6.1 Machine Learning and AI

Advanced machine learning models, including supervised and unsupervised methods, are increasingly used in multi-omics analysis. For instance, deep learning frameworks such as autoencoders and variational autoencoders (VAEs) are used to learn low-dimensional representations of multi-omics data, facilitating the discovery of complex features and associations that conventional methodologies might omit (Huang et al., 2023).

6.2 Graph-Based Methods

These methods are used to model relationships between entities across different omics layers. The co-relation among genes, proteins, and metabolites can be captured by integrating graph convolutional networks (GCNs) and graph embeddings to multi-omics data, offering insights into the underlying biological networks ([CSL STYLE ERROR: reference with no printed form.]).

6.3 Bayesian Networks

A Bayesian approach allows probabilistic incorporation of multi-omics data, accommodating uncertainties and missing data. These networks are used explicitly for modeling causal interactions and regulatory networks in cancer (Howey et al., 2021).

6.4 Integration of Time-Series Data

For longitudinal studies, methods such as dynamic Bayesian networks (DBNs) and hidden Markov models (HMMs) integrate multi-omics data collected at different time points, providing insights into temporal changes in disease progression (Ruiz-Perez et al., 2019).

6.5 Validation and Interpretation

The final step in multi-omics integration involves validating the computational predictions through experimental methods, such as CRISPR-based gene editing, RNA interference (RNAi), or proteomics-based validation using targeted mass spectrometry. Ingenuity Pathway Analysis (IPA), Gene Set Enrichment Analysis (GSEA), and Cytoscape are the pathway and network analysis tools that interpret the integrated data in the context of known biological processes and pathways (Chen et al., 2023).

While current computational frameworks and bioinformatics tools have significantly advanced multi-omics integration, ongoing research is focused on improving scalability, handling noise and heterogeneity in data, and enhancing the biological interpretability of results. Future perspectives include the evolution of more efficient algorithms for real-time analysis, integrating single-cell multi-omics data, and applying quantum computing to solve complex multi-omics integration problems currently computationally prohibited (Vitorino, 2024). When used rigorously, these tools and methods can reveal critical insights into the molecular mechanisms driving cancer, thereby developing more effective and personalized therapeutic strategies.

7 Challenges

Despite the potential of integrative multi-omics methodologies, numerous ambiguities persist regarding their implementation and interpretation. Among these, the primary difficulty is the large amount of data generated through omics technologies, necessitating advanced computational resources and sophisticated data analysis methods. The heterogeneity of data types and formats across different omics platforms also poses data integration and interpretation challenges. The lack of standardized sample collection, processing, and analysis protocols can also lead to variability and reproducibility issues. There is also a need for more robust algorithms that can handle missing data and integrate datasets with varying levels of quality and coverage. Moreover, ethical and privacy concerns related to handling large-scale patient data must be carefully managed (Mukherjee et al., 2024). Integrative multi-omics approaches face several significant challenges. First, multi-omics data's sheer volume and high dimensionality necessitate advanced computational infrastructure and algorithms. Handling such large datasets often requires cloud computing resources and parallel processing capabilities (Agamah et al., 2022). Second, different omics platforms generate data in various formats with varying levels of noise and resolution. The lack of standardized protocols for data collection, processing, and normalization across different omics types can lead to integration challenges and variability in results (Magro et al., 2024). Third, integrating multi-omics data requires sophisticated statistical models and bioinformatics tools capable of capturing the nonlinear relationships between omics layers. Additionally, interpreting integrated data is complex, as it involves understanding the interactions between various biological processes and how they contribute to cancer phenotypes (Graw et al., 2021). Further, multi-omics datasets often have missing or incomplete data due to limitations in experimental techniques or sample availability. Developing robust algorithms, imputing missing data, or handling incomplete datasets is essential for accurate analysis (Flores et al., 2023). Moreover, computational predictions from multi-omics analyses must be validated experimentally, which can be time-consuming and resource-intensive. Ensuring that integrative models reflect biologically relevant mechanisms is critical in translating findings into clinical applications (Mohr et al., 2024).

8 Future Directions

The future of integrative multi-omics approaches in cancer therapeutics lies in developing more advanced computational tools and algorithms that seamlessly integrate diverse omics datasets (Heo et al., 2021). Artificial intelligence and machine learning are expected to be crucial in predicting patient outcomes and identifying novel therapeutic targets. Applying single-cell multi-omics technologies, which can provide insights into the heterogeneity within tumors, is another promising avenue for research. As more comprehensive and high-quality multi-omics datasets become available, the potential for identifying new biomarkers and therapeutic targets will increase, further driving the advancement of personalized cancer therapy. The future of integrative multi-omics approaches in cancer therapeutics lies in several promising areas, such as single-cell omics technologies providing solutions to formerly unsolvable biological and medical concerns, such as identifying novel cell types, cellular differentiation trajectories, and intercellular communication networks. Present developments, particularly in single-cell multi-omics, have great promise for innovations by fusing several omics layers (You et al., 2022).

Emerging single-cell technologies allow for the simultaneous measurement of multiple omics layers (e.g., genomics, transcriptomics, and epigenomics) within individual cells. This approach provides unprecedented insights into tumor heterogeneity, clonal evolution, and the tumor microenvironment, enabling the identification of rare cell populations that may contribute to drug resistance or metastasis (Zhang et al., 2025). Moreover, artificial intelligence (AI)-driven approaches, including deep learning and reinforcement learning, are being developed to integrate and analyze multi-omics data more effectively. These methods can uncover complex patterns and predict patient outcomes, leading to the identification of novel therapeutic targets and biomarkers. AI-based approaches have become indispensable for handling multi-omics data's enormous dimensionality and heterogeneity. This methodology reveals latent patterns and expedites biological result prediction, opening up new avenues for personalized therapy and biomedical research (Abbasi et al., 2024).

As multi-omics technologies have become more accessible and cost-effective, there is a growing potential for their integration into clinical workflows. This would allow for more personalized cancer treatment strategies, where therapies are tailored based on the comprehensive molecular profile of each patient's tumor. Precision medicine enables medical professionals to identify the best course of treatment for patients suffering from complicated conditions such as COVID-19, diabetes, cancer, and cardiomyopathy. More efficient and individualized medical therapies are expected for many ailments with an evolving understanding of the clinical, molecular, and genetic variables involved in diseases. Comprehending the metabolomics and genetic composition of patients in combination with clinical data will significantly contribute to the identification of predisposition, diagnostic, prognostic, and predictive biomarkers, and pathways, ultimately offering the best possible care that is tailored to the individual for a variety of targeted chronic and acute diseases. In order to discover statistical patterns among millions of features in clinical settings, we must quickly model clinical and multi-omics data. This will allow us to identify modifiable risk factors, underlying biological pathways, and valuable information to help identify and prevent disorders early and develop new technologies (Molla & Bitew, 2024). However, further research will likely focus on longitudinal multi-omics studies, where patient samples are collected over time so that this approach can provide insights into the dynamics of tumor evolution, treatment response, and resistance mechanisms, leading to more effective therapeutic strategies.

9 Conclusions

Integrative multi-omics techniques have the potential to completely transform cancer treatments by offering a thorough comprehension of the molecular causes of cancer. Combining data from transcriptomics, proteomics, metabolomics, genomes, and epigenomics allows these methods to discover new biomarkers, therapeutic targets, and pathways contributing to cancer development. Despite existing challenges, advancements in computational methodologies and the increasing accessibility of multi-omics datasets are poised to significantly improve the precision and efficacy of oncological therapies. As the field continues to develop, integrative multi-omics will play an increasingly central role in developing personalized medicine strategies that offer hope for improved patient outcomes.

Integrative multi-omics approaches are revolutionizing cancer therapeutics by providing a comprehensive, systems-level understanding of the molecular complexity of cancer. By combining data from genomics, transcriptomics, proteomics, metabolomics, and epigenomics, researchers can gain deeper insights into the biological mechanisms driving cancer, identify novel biomarkers, and develop more effective, personalized therapies. Recent advancements in computational methods, single-cell technologies, and AI-driven approaches are expected to overcome these hurdles, leading to significant improvements in cancer diagnosis, prognosis, and treatment. As the field continues to evolve, integrative multi-omics will play a crucial role in shaping the future of precision oncology, offering hope for more targeted and effective cancer treatments.

Acknowledgments The fellowship offered by the Department of Biotechnology, India (DBT-RA/2023-24/Call-1/RA/27), is acknowledged by A.P. and N.A. N.A. acknowledges SERB POWER grant from Science and Engineering Research Board for financial support (SPG/2021/004209). All authors acknowledge AIIMS Bhopal for technical support.

References

Abbasi, E. Y., Deng, Z., Ali, Q., et al. (2024). A machine learning and deep learning-based integrated multi-omics technique for leukemia prediction. *Heliyon, 10*, e25369. https://doi.org/10.1016/J.HELIYON.2024.E25369

Agamah, F. E., Bayjanov, J. R., Niehues, A., et al. (2022). Computational approaches for network-based integrative multi-omics analysis. *Frontiers in Molecular Biosciences, 9*, 967205. https://doi.org/10.3389/FMOLB.2022.967205/XML/NLM

Argelaguet, R., Velten, B., Arnol, D., et al. (2018). Multi-omics factor analysis—A framework for unsupervised integration of multi-omics data sets. *Molecular Systems Biology, 14*, 8124. https://doi.org/10.15252/MSB.20178124/SUPPL_FILE/MSB178124-SUP-0002-EVFIGS.PDF

Arora, A., Olshen, A. B., Seshan, V. E., & Shen, R. (2020). Pan-cancer identification of clinically relevant genomic subtypes using outcome-weighted integrative clustering. *Genome Medicine, 12*, 110. https://doi.org/10.1186/S13073-020-00804-8

Athieniti, E., & Spyrou, G. M. (2023). A guide to multi-omics data collection and integration for translational medicine. *Computational and Structural Biotechnology Journal, 21*, 134–149. https://doi.org/10.1016/J.CSBJ.2022.11.050

Babu, M., & Snyder, M. (2023). Multi-omics profiling for health. *Molecular & Cellular Proteomics, 22*, 100561. https://doi.org/10.1016/J.MCPRO.2023.100561

Belhaj, M. R., Lawler, N. G., & Hoffman, N. J. (2021). Metabolomics and lipidomics: Expanding the molecular landscape of exercise biology. *Metabolites, 11*, 151. https://doi.org/10.3390/METABO11030151

Cai, Z., Poulos, R. C., Liu, J., & Zhong, Q. (2022). Machine learning for multi-omics data integration in cancer. *iScience, 25*, 103798. https://doi.org/10.1016/J.ISCI.2022.103798

Carels, N. (2024). Assessing RNA-seq workflow methodologies using Shannon entropy. *Biology (Basel), 13*, 482. https://doi.org/10.3390/BIOLOGY13070482/S1

Chakraborty, S., Sharma, G., Karmakar, S., & Banerjee, S. (2024). Multi-OMICS approaches in cancer biology: New era in cancer therapy. *Biochimica et Biophysica Acta-Molecular Basis of Disease, 1870*, 167120. https://doi.org/10.1016/J.BBADIS.2024.167120

Chen, Z. S., Kulkarni P (Param), Galatzer-Levy, I. R., et al. (2022). Modern views of machine learning for precision psychiatry. *Patterns, 3*, 100602. https://doi.org/10.1016/J.PATTER.2022.100602

Chen, C., Wang, J., Pan, D., et al. (2023). Applications of multi-omics analysis in human diseases. *MedComm, 4*, e315. https://doi.org/10.1002/MCO2.315

Chierici, M., Bussola, N., Marcolini, A., et al. (2020). Integrative network fusion: A multi-omics approach in molecular profiling. *Frontiers in Oncology, 10*, 547844. https://doi.org/10.3389/FONC.2020.01065/XML/NLM

Dai, X., & Shen, L. (2022). Advances and trends in omics technology development. *Frontiers in Medicine, 9*, 911861. https://doi.org/10.3389/FMED.2022.911861/PDF

Das, S., Dey, M. K., Devireddy, R., & Gartia, M. R. (2023). Biomarkers in cancer detection, diagnosis, and prognosis. *Sensors, 24*, 37. https://doi.org/10.3390/S24010037

Eicher, T., Kinnebrew, G., Patt, A., et al. (2020). Metabolomics and multi-omics integration: A survey of computational methods and resources. *Metabolites, 10*, 202. https://doi.org/10.3390/METABO10050202

Emwas, A. H., Roy, R., McKay, R. T., et al. (2019). NMR spectroscopy for metabolomics research. *Metabolites, 9*, 123. https://doi.org/10.3390/METABO9070123

Flores, J. E., Claborne, D. M., Weller, Z. D., et al. (2023). Missing data in multi-omics integration: Recent advances through artificial intelligence. *Frontiers in Artificial Intelligence, 6*, 1098308. https://doi.org/10.3389/FRAI.2023.1098308/XML/NLM

Gao, Z., Lu, Y., Li, M., et al. (2024). Application of pan-omics technologies in research on important economic traits for ruminants. *International Journal of Molecular Sciences, 25*, 9271. https://doi.org/10.3390/IJMS25179271

Graw, S., Chappell, K., Washam, C. L., et al. (2021). Multi-omics data integration considerations and study design for biological systems and disease. *Medicine in Omics, 17*, 170–185. https://doi.org/10.1039/D0MO00041H

Gygi, J. P., Konstorum, A., Pawar, S., et al. (2024). A supervised Bayesian factor model for the identification of multi-omics signatures. *Bioinformatics, 40*, btae202. https://doi.org/10.1093/BIOINFORMATICS/BTAE202

Heo, Y. J., Hwa, C., Lee, G. H., et al. (2021). Integrative multi-omics approaches in cancer research: From biological networks to clinical subtypes. *Molecules and Cells, 44*, 433–443. https://doi.org/10.14348/MOLCELLS.2021.0042

Hernández-Lemus, E., & Ochoa, S. (2024). Methods for multi-omic data integration in cancer research. *Frontiers in Genetics, 15*, 1425456. https://doi.org/10.3389/FGENE.2024.1425456/PDF

Howey, R., Clark, A. D., Naamane, N., et al. (2021). A Bayesian network approach incorporating imputation of missing data enables exploratory analysis of complex causal biological relationships. *PLoS Genetics, 17*, e1009811. https://doi.org/10.1371/JOURNAL.PGEN.1009811

Huang, Z., Zhan, X., Xiang, S., et al. (2019). SALMON: Survival analysis learning with multi-omics neural networks on breast cancer. *Frontiers in Genetics, 10*, 166. https://doi.org/10.3389/FGENE.2019.00166

Huang, L., Song, M., Shen, H., et al. (2023). Deep learning methods for omics data imputation. *Biology (Basel), 12*, 1313. https://doi.org/10.3390/BIOLOGY12101313

Iperi, C., Fernández-Ochoa, Á., Barturen, G., et al. (2025). BiomiX, a user-friendly bioinformatic tool for democratized analysis and integration of multiomics data. *BMC Bioinformatics, 26*, 8. https://doi.org/10.1186/S12859-024-06022-Y/FIGURES/8

Jalali, S., Bhartiya, D., Lalwani, M. K., et al. (2013). Systematic transcriptome wide analysis of lncRNA-miRNA interactions. *PLoS One, 8*(2), e53823. https://doi.org/10.1371/JOURNAL.PONE.0053823

Kim, Y. A., Cho, D. Y., & Przytycka, T. M. (2016). Understanding genotype-phenotype effects in cancer via network approaches. *PLoS Computational Biology, 12*, e1004747. https://doi.org/10.1371/JOURNAL.PCBI.1004747

Koh, H. W. L., Fermin, D., Vogel, C., et al. (2019). iOmicsPASS: Network-based integration of multiomics data for predictive subnetwork discovery. *npj Systems Biology and Applications, 5*, 22. https://doi.org/10.1038/S41540-019-0099-Y

Li, Y. (2021). Modern epigenetics methods in biological research. *Methods, 187*, 104–113. https://doi.org/10.1016/J.YMETH.2020.06.022

Li, X., Foley, E. A., Kawashima, S. A., et al. (2013). Examining post-translational modification-mediated protein-protein interactions using a chemical proteomics approach. *Protein Science, 22*, 287–295. https://doi.org/10.1002/PRO.2210

Li, L., Wei, Y., Shi, G., et al. (2022a). Multi-omics data integration for subtype identification of Chinese lower-grade gliomas: A joint similarity network fusion approach. *Computational and Structural Biotechnology Journal, 20*, 3482–3492. https://doi.org/10.1016/J.CSBJ.2022.06.065

Li, Y., Mansmann, U., Du, S., & Hornung, R. (2022b). Benchmark study of feature selection strategies for multi-omics data. *BMC Bioinformatics, 23*, 1–18. https://doi.org/10.1186/S12859-022-04962-X/TABLES/4

Liu, Z., & Park, T. (2024). DMOIT: Denoised multi-omics integration approach based on transformer multi-head self-attention mechanism. *Frontiers in Genetics, 15*, 1488683. https://doi.org/10.3389/FGENE.2024.1488683/BIBTEX

Magro, D., Venezia, M., & Rita Balistreri, C. (2024). The omics technologies and liquid biopsies: Advantages, limitations, applications. *Medicine in Omics, 11*, 100039. https://doi.org/10.1016/J.MEOMIC.2024.100039

Mbemi, A., Khanna, S., Njiki, S., et al. (2020). Impact of gene–environment interactions on cancer development. *International Journal of Environmental Research and Public Health, 17*, 8089. https://doi.org/10.3390/IJERPH17218089

Menyhárt, O., & Győrffy, B. (2021). Multi-omics approaches in cancer research with applications in tumor subtyping, prognosis, and diagnosis. *Computational and Structural Biotechnology Journal, 19*, 949. https://doi.org/10.1016/J.CSBJ.2021.01.009

Mo, Q., Shen, R., Guo, C., et al. (2017). A fully Bayesian latent variable model for integrative clustering analysis of multi-type omics data. *Biostatistics, 19*, 71. https://doi.org/10.1093/BIOSTATISTICS/KXX017

Mohr, A. E., Ortega-Santos, C. P., Whisner, C. M., et al. (2024). Navigating challenges and opportunities in multi-omics integration for personalized healthcare. *Biomedicine, 12*, 1496. https://doi.org/10.3390/BIOMEDICINES12071496

Molla, G., & Bitew, M. (2024). Revolutionizing personalized medicine: Synergy with multi-omics data generation, main hurdles, and future perspectives. *Biomedicine, 12*, 2750. https://doi.org/10.3390/BIOMEDICINES12122750

Mukherjee, A., Abraham, S., Singh, A., et al. (2024). From data to cure: A comprehensive exploration of multi-omics data analysis for targeted therapies. *Molecular Biotechnology, 67*, 1269–1289. https://doi.org/10.1007/S12033-024-01133-6

Parihar, A., & Khan, R. (2023). Yttrium functionalized reduced graphene oxide nanocomposite-based aptasensor for ultrasensitive detection of a breast cancer biomarker. *ACS Applied Nano Materials, 7*(16), 18207–18218. https://doi.org/10.1021/acsanm.3c03234

Parihar, A., Vishwakarma, P., Prajapati, P., & Khan, R. (2024). Non-invasive ultra-sensitive detection of breast cancer biomarker using cerium nanoparticle functionalized graphene oxide enabled impedimetric aptasensor. *Biosensors & Bioelectronics, 268*, 116925. https://doi.org/10.1016/J.BIOS.2024.116925

Pettini, F., Visibelli, A., Cicaloni, V., et al. (2021). Multi-omics model applied to cancer genetics. *International Journal of Molecular Sciences, 22*, 5751. https://doi.org/10.3390/IJMS22115751

Picard, M., Scott-Boyer, M. P., Bodein, A., et al. (2021). Integration strategies of multi-omics data for machine learning analysis. *Computational and Structural Biotechnology Journal, 19*, 3735–3746. https://doi.org/10.1016/J.CSBJ.2021.06.030

Rappoport, N., & Shamir, R. (2019). NEMO: Cancer subtyping by integration of partial multi-omic data. *Bioinformatics, 35*, 3348–3356. https://doi.org/10.1093/BIOINFORMATICS/BTZ058

Rappoport, N., Safra, R., & Shamir, R. (2020). MONET: Multi-omic module discovery by omic selection. *PLoS Computational Biology, 16*, e1008182. https://doi.org/10.1371/JOURNAL.PCBI.1008182

Rohart, F., Eslami, A., Matigian, N., et al. (2017). MINT: A multivariate integrative method to identify reproducible molecular signatures across independent experiments and platforms. *BMC Bioinformatics, 18*, 1–13. https://doi.org/10.1186/S12859-017-1553-8/FIGURES/3

Ruiz-Perez, D., Lugo-Martinez, J., Bourguignon, N., et al. (2019). Dynamic Bayesian networks for integrating multi-omics time series microbiome data. *mSystems, 6*(2), e01105–e01120. https://doi.org/10.1101/835124

Sanches, P. H. G., de Melo, N. C., Porcari, A. M., & de Carvalho, L. M. (2024). Integrating molecular perspectives: Strategies for comprehensive multi-omics integrative data analysis and machine learning applications in transcriptomics, proteomics, and metabolomics. *Biology (Basel), 13*, 848. https://doi.org/10.3390/BIOLOGY13110848

Shen, R., Olshen, A. B., & Ladanyi, M. (2009). Integrative clustering of multiple genomic data types using a joint latent variable model with application to breast and lung cancer subtype analysis. *Bioinformatics, 25*, 2906–2912. https://doi.org/10.1093/BIOINFORMATICS/BTP543

Subramanian, I., Verma, S., Kumar, S., et al. (2020). Multi-omics data integration, interpretation, and its application. *Bioinformatics and Biology Insights, 14*, 1177932219899051. https://doi.org/10.1177/1177932219899051

Tattini, L., D'Aurizio, R., & Magi, A. (2015). Detection of genomic structural variants from next-generation sequencing data. *Frontiers in Bioengineering and Biotechnology, 3*, 92. https://doi.org/10.3389/FBIOE.2015.00092

Tembhare, K., Sharma, T., Kasibhatla, S. M., et al. (2024). Multi-ensemble machine learning framework for omics data integration: A case study using breast cancer samples. *Informatics in Medicine Unlocked (IMU), 47*, 101507. https://doi.org/10.1016/J.IMU.2024.101507

Vahabi, N., & Michailidis, G. (2022). Unsupervised multi-omics data integration methods: A comprehensive review. *Frontiers in Genetics, 13*, 854752. https://doi.org/10.3389/FGENE.2022.854752/PDF

Vaske, C. J., Benz, S. C., Sanborn, J. Z., et al. (2010). Inference of patient-specific pathway activities from multi-dimensional cancer genomics data using PARADIGM. *Bioinformatics, 26*, i237–i245. https://doi.org/10.1093/BIOINFORMATICS/BTQ182

Virolainen, S. J., VonHandorf, A., Viel, K. C. M. F., et al. (2022). Gene–environment interactions and their impact on human health. *Genes and Immunity, 24*, 1. https://doi.org/10.1038/S41435-022-00192-6

Vitorino, R. (2024). Transforming clinical research: The power of high-throughput omics integration. *Proteome, 12*, 25. https://doi.org/10.3390/PROTEOMES12030025

Wang, T., Shao, W., Huang, Z., et al. (2021). MOGONET integrates multi-omics data using graph convolutional networks allowing patient classification and biomarker identification. *Nature Communications, 12*, 1–13. https://doi.org/10.1038/s41467-021-23774-w

Wang, W., Rong, Z., Wang, G., et al. (2023). Cancer metabolites: Promising biomarkers for cancer liquid biopsy. *Biomarker Research, 11*, 1–14. https://doi.org/10.1186/S40364-023-00507-3

Wu, D., Wang, D., Zhang, M. Q., & Gu, J. (2015). Fast dimension reduction and integrative clustering of multi-omics data using low-rank approximation: Application to cancer molecular classification. *BMC Genomics, 16*, 1–10. https://doi.org/10.1186/S12864-015-2223-8/FIGURES/4

You, Y., Lai, X., Pan, Y., et al. (2022). Artificial intelligence in cancer target identification and drug discovery. *Signal Transduction and Targeted Therapy, 7*, 1–24. https://doi.org/10.1038/s41392-022-00994-0

Zhang, R., & Datta, S. (2023). Adaptive sparse multi-block PLS discriminant analysis: An integrative method for identifying key biomarkers from multi-omics data. *Genes (Basel), 14*, 961. https://doi.org/10.3390/GENES14050961/S1

Zhang, W., Zhang, X., Teng, F., et al. (2025). Research progress and the prospect of using single-cell sequencing technology to explore the characteristics of the tumor microenvironment. *Genes & Diseases, 12*, 101239. https://doi.org/10.1016/J.GENDIS.2024.101239

Challenges and Limitations of Computational Methods in Oncology

Bhavika Baweja, Prerna Vats, Sakshi Kausik, Jagriti Singh, Rajeev Nema, and Prasant Yadav

Prerna Vats: Equal Contribution for First Author

Jagriti Singh: Equal Contribution for Second Author

1 Introduction

Cancer is a major global health challenge, with increasing prevalence and mortality rates observed in recent decades. Cancer ranks as the second leading cause of mortality worldwide, accounting for approximately 10 million fatalities annually, according to the World Health Organization (WHO). Recent developments in oncology, including prevention, diagnosis, and treatment, have significantly improved the rate of survival along with the quality of life for patients (Siegel et al., 2022). Despite advancements, cancer continues to be a complex and heterogeneous disease, marked by intricate interactions among genetic, environmental, and lifestyle factors. The complexity of oncology has driven the advancement and utilization of computational methods to address essential challenges, including early detection, prognosis, planning of treatment, and monitoring. Advancements in artificial intelligence (AI) and machine learning (ML) have established computational methods as effective instruments in cancer research and treatment. These technologies can process large volumes of data produced by modern healthcare systems, including genomic profiles, imaging studies, and clinical records, facilitating the creation of predictive models and tailored treatment strategies (Esteva et al., 2019). Computational methods can enhance oncology by enabling early-stage tumor detection in medical imaging, predicting individualized patient responses to therapy, increasing accuracy,

B. Baweja · P. Vats · S. Kausik · J. Singh · R. Nema (✉)
Department of Biosciences, Manipal University Jaipur, Jaipur, Rajasthan, India
e-mail: rajeev.nema@jaipur.manipal.edu

P. Yadav (✉)
Department of Radiation Oncology, The Ohio State University, Columbus, OH, USA

R. Nema et al. (eds.), *Advances in Cancer Detection, Prediction, and Prognosis Using Artificial Intelligence and Machine Learning*,
https://doi.org/10.1007/978-981-96-9346-7_13

lowering costs, and improving patient outcomes (Lambin et al., 2017). Despite the potential of these technologies, their broad implementation encounters various challenges and limitations that need to be resolved to achieve optimal effectiveness.

The complexity of cancer biology presents significant obstacles for the implementation of computational techniques. Tumor heterogeneity, the dynamic characteristics of cancer, and the interactions among various biological systems frequently exceed the predictive accuracy of existing computational models (Greaves & Maley, 2012). The integration of diverse and multidimensional datasets, including genomics, proteomics, and radiomics, poses substantial technical and methodological challenges (Huang et al., 2018). Beyond these biological obstacles, practical problems such as the scarcity of sizable, varied datasets, strict privacy laws, and data-sharing restrictions impede the creation and verification of reliable and broadly applicable models.

A significant limitation is the interpretability of computational methods, especially deep learning models, which are often characterized as "black boxes" as a result of their inherent lack of transparency. The lack of explanation or justification for model predictions generates skepticism among clinicians and regulators, hindering clinical implementation (Tjoa & Guan, 2021). To address these challenges, interdisciplinary collaboration, advancements in AI methodologies, and the creation of standardized frameworks are essential for ensuring the reliability, equity, and clinical relevance of computational tools in oncology (Holzinger et al., 2020). This chapter examines the challenges and limitations in detail, emphasizing the areas requiring innovation to progress in computational oncology.

2 Data-Related Challenges

2.1 Data Quality and Standardization Issues

The quality and standardization of data are crucial for computational methods in oncology, affecting accuracy, reproducibility, and reliability. Oncology datasets, including imaging, genomic, clinical, and pathological data, often exhibit variability and inconsistency due to disparities in data acquisition protocols, equipment, electronic health record (EHR) systems, and annotation practices (McKinney et al., 2020). This heterogeneity complicates the integration of datasets for machine learning model training and validation, leading to bias or a lack of generalizability across diverse patient populations. Variations in imaging parameters can also lead to inconsistent computational models across different facilities (Hosny et al., 2018). Standardization of annotations is a significant challenge, as labels indicating tumor regions or classifications can introduce noise and impact model performance (Wang et al., 2021). Genomic data from various sequencing platforms also complicates data harmonization. Deficient or incomplete data, prevalent in oncology due to difficulties in gathering exhaustive patient records, intensifies these problems. Global initiatives such as the Cancer Genome Atlas (TCGA), FAIR principles, and collaborative frameworks are formulating extensive data standardization protocols

(Brancato et al., 2024). Advanced preprocessing methods, such as domain adaptation and harmonization algorithms, are employed to reduce variability in data sources and enhance model robustness (Yang et al., 2019).

2.2 Limited Availability of Large, Diverse Datasets

Computational oncology faces notable challenges caused by the limited availability of large, diverse datasets, which makes it difficult to create and validate predictive models as shown in Fig. 1. Cancer research relies on data from specific clinical trials, specialized medical institutions, or restricted patient groups, which may not accurately represent the comprehensive range of cancer's biological and clinical diversity (Jiang et al., 2022). This constraint is particularly pronounced when examining rare cancer subtypes, underrepresented demographic cohorts, and intricate patient populations with numerous diseases (Zeng et al., 2024). Challenges include rigorous patient privacy regulations, institutional databases, fragmented healthcare systems, and the logistics of gathering comprehensive longitudinal cancer data (Lam, 2024). The heterogeneity of cancer, which develops differently across various genetic backgrounds, environmental contexts, and individual patient characteristics, further complicates these challenges (Ansari et al., 2024). Restricted dataset availability can introduce bias and constrain the generalizability of computational techniques. Datasets often exhibit selection bias, primarily reflecting patients from well-resourced medical institutions, certain geographic areas, or specific demographic cohorts (Dankwa-Mullan, 2024). Initiatives such as the Cancer Imaging Archive (TCIA) and the Cancer Genome Atlas (TCGA) aim to provide publicly accessible, standardized datasets for oncology research but face challenges in achieving comprehensive global representation (Zanfardino et al., 2019).

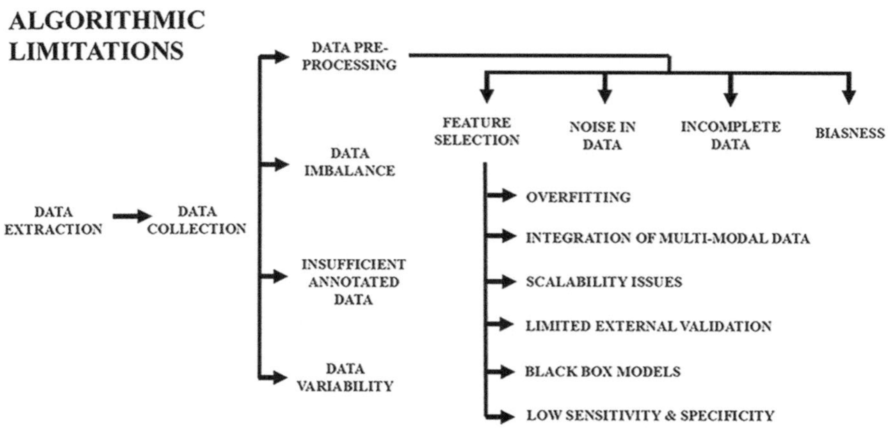

Fig. 1 Schematic representation of algorithmic limitations in data-driven research

2.3 Privacy Concerns and Data Sharing Obstacles

Computational methods in oncology face challenges due to privacy concerns and data-sharing issues, preventing the acquisition and utilization of diverse datasets essential for developing effective AI and ML models. Patient data, including genomic profiles, imaging records, and treatment histories, is governed by strict laws such as the Health Insurance Portability and Accountability Act (HIPAA) and the General Data Protection Regulation (GDPR), which complicate data sharing and restrict the aggregation of datasets required for model generalization (Thapa & Camtepe, 2021). The lack of defined protocols for data anonymization and security further hinders collaborative research initiatives (Ohmann et al., 2017). Rare cancers, where individual datasets often lack the necessary amount for significant computational analysis, further exacerbate privacy concerns. Data hoarding may result from competitive dynamics, hindering the development of comprehensive, representative datasets and diminishing the efficacy and equity of AI systems in oncology. Collaborative learning has become recognized as an efficient approach, allowing models to train on decentralized data without sharing patient information, preserving privacy while promoting collaboration (Rieke et al., 2020). Initiatives such as the Global Alliance for Genomics and Health (GA4GH) and the Cancer Imaging Archive (TCIA) aim to standardize data-sharing practices and improve interoperability among datasets (Fedorov et al., 2021).

3 Model Development and Validation

3.1 Overfitting and Generalizability Problems

Overfitting and generalizability are significant challenges in the creation and approval of computational models in oncology, impacting the reliability and applicability of these systems across diverse clinical contexts as shown in Fig. 1. Overfitting arises when a model detects patterns unique to the training data, resulting in elevated performance but inadequate generalization to novel datasets (Romoli & Caliandro, 2024). This issue is particularly prominent in oncology due to the high-dimensional characteristics of cancer data, which often encompass thousands of genetic markers, imaging attributes, and clinical variables. Excessive fitting of models to training data limits their practical applicability in real-world oncology scenarios by preventing them from generalizing to new patients or different clinical contexts (Halkola et al., 2023). For example, a model trained solely on high-resolution imaging data from a single institution may not generalize effectively to images obtained from various scanners or clinical settings, thereby significantly reducing its utility (Prabhod & Gadhiraju, 2024). Models frequently encounter difficulties in generalizing across diverse demographic groups, treatment facilities, and geographical areas due to the heterogeneous nature of cancer and differing clinical practices. Researchers are increasingly using techniques such as regularization, dropout layers, and data augmentation to enhance model robustness and reduce overfitting (Moradi et al., 2020). Cross-validation techniques and transfer learning are being investigated to reduce the possibility of overfitting and enhance the reliability and applicability of computational methods in oncology (Mathivanan et al., 2024).

3.2 Difficulty in Model Interpretability (Black Box Problem)

In oncology, model interpretability presents a considerable challenge, particularly with advanced techniques for machine learning such as deep learning and ensemble methods. Many times, these models function as "black boxes," making it challenging for both patients and doctors to comprehend how inputs and outputs relate to one another (Guidotti et al., 2018). This reduces clinician confidence and complicates ethical issues for implementing these systems in real-world environments and regulatory approval procedures (Xu & Lai, 2024). Models processing thousands of features concurrently create complexity, making it difficult to track how specific factors affect predictions. To address this, explainable artificial intelligence (XAI) methods have been developed, such as saliency maps, SHAP (Shapley Additive Explanations), and LIME (Local Interpretable Model-agnostic Explanations), which offer insights into which features influence predictions (Ekanayake et al., 2022). Hybrid models combining rule-based logic with machine learning are also being investigated to increase transparency. However, balancing interpretability and performance remains difficult, especially for tasks requiring high-dimensional data processing. Interdisciplinary cooperation is crucial to improve XAI techniques and ensure strong, understandable, and practical AI systems in oncology (Patel et al., 2024).

3.3 Challenges in External Validation Across Different Populations

External validation of oncological models is a complex process that faces challenges due to population diversity and data heterogeneity. Variations in patient populations, including demographic traits, genetic backgrounds, environmental exposures, healthcare access, and treatment paradigms, can significantly affect model performance and generalizability (Prosperi et al., 2018). Institutional variations in data collection techniques, diagnostic criteria, treatment approaches, and technological capabilities further complicate the validation process (Qu et al., 2024). Current models are often created using data from well-funded academic medical centers or clinical studies, which may not accurately represent the larger, more varied patient populations in underprivileged areas or community healthcare environments (Escorial et al., 2024). Logistical and technical obstacles, such as institutional governance structures, privacy rules, and data-sharing restrictions, also hinder comprehensive external validation (Gurumurthy et al., 2024). Additionally, inadequate data, small sample sizes, and inconsistent clinical variable documentation further complicate the process (Lam, 2024). Emerging methods include stronger validation frameworks incorporating diverse population representations and federated learning techniques, which facilitate model validation across distributed datasets without direct data flow (Escorial et al., 2024).

4 Clinical Integration

4.1 Resistance to Adoption by Healthcare Professionals

Healthcare workers exhibit resistance to accepting AI technology due to trust issues, concerns regarding job security, legal and ethical challenges, inadequate training, organizational resistance, and financial limitations. Trust issues originate from uncertainties regarding the accuracy and reliability of AI, while concerns about job security are common in fields such as radiology and pathology (Topol, 2019). Legal and ethical considerations encompass data privacy, security vulnerabilities, and responsibility for errors generated by AI systems (Reddy et al., 2019). Ethical dilemmas, including patient autonomy and consent, contribute to hesitancy. Insufficient training and understanding of AI result in healthcare professionals being inadequately prepared to assess AI-generated outputs. Barriers related to organizational structure and culture, including rigid hierarchies and resistance to change, hinder the integration of AI. Financial limitations can be a considerable barrier (Iyelolu et al., 2024). To address these challenges, targeted strategies including education and training programs, transparent AI systems, collaboration between AI experts and medical professionals, gradual implementation, and an emphasis on long-term benefits can enhance the effective implementation of AI into clinical practice.

4.2 Integration with Existing Clinical Workflows

The integration of new healthcare systems into established workflows presents a complex challenge, necessitating smooth adaptation to avoid disruption of clinical operations and to maintain the quality of patient care. Disruptions in workflow can result in decreased staff productivity and negatively impact patient outcomes (Eikey et al., 2019). The implementation of new systems often requires supplementary data input and modifications to existing procedures, resulting in inefficiencies and resistance from healthcare professionals. Interoperability is essential, as healthcare organizations depend on a variety of legacy systems, including electronic health records (EHRs), medical devices, and laboratory systems. Effective integration necessitates standardized protocols and compatibility to deliver an integrative view of patient data (Kelly et al., 2020). Efficient data management, encompassing data migration, security measures, and adherence to regulations such as the HIPAA is critical. User experience is essential; user-centered design and continual testing enhance system optimization. Effective integration necessitates cooperation among technology developers, IT experts, and healthcare professionals, alongside thorough training, continuous support, and adaptable infrastructure. Integrating technological innovation with clinical practices enhances acceptance, improves care quality, and provides advantages for patients and healthcare systems (Al-Assaf et al., 2024).

4.3 Regulatory Hurdles and Approval Processes

The adoption of artificial intelligence (AI) and machine learning (ML) in healthcare introduces notable regulatory challenges, requiring versatile frameworks to ensure reliability, effectiveness, and ethical application. Regulatory agencies such as the FDA and EMA enforce strict standards for AI/ML tools via broad validation and testing protocols (Minssen et al., 2020). The adaptive characteristics of AI/ML systems, which progress through ongoing learning from new data, require the establishment of dynamic pathways such as adaptive approvals and improved post-market surveillance (Niazi, 2023). These mechanisms facilitate continuous monitoring of AI/ML tools, allowing for quick identification and rectification of performance issues. Developers must address privacy and security concerns while adhering to regulations such as HIPAA and GDPR. Balancing innovation and safety is essential, as demonstrated by initiatives such as the FDA's Digital Health Innovation Action Plan, which promotes innovation while upholding strict guidelines (Macdonald et al., 2021). Clear communication among developers, regulators, and healthcare providers enhances trust and aligns expectations. It is crucial to have a futuristic regulatory framework that can accommodate AI/ML's repetitive nature while maintaining strict safety evaluations.

5 Ethical Considerations

5.1 Bias in AI Algorithms and Its Impact on Healthcare Disparities

The integration of AI in healthcare offers advantages such as enhanced diagnostic precision, individualized treatment plans, and optimized care delivery processes. However, it presents ethical issues regarding bias and its effects on healthcare equity. AI systems can strengthen historical inequalities, especially affecting marginalized groups. Bias in AI models, arising from training on unrepresentative datasets, can result in diagnostic inaccuracies or improper treatment, thereby reinforcing disparities (Mohamed, 2023). Algorithmic bias occurs when models enhance existing inequities as a result of defective data or design, potentially worsening disparities. Human bias, including implicit biases, stereotyping, and cognitive biases, can influence clinical decisions and disproportionately impact marginalized groups (Johnson, 2021). Insufficient data in healthcare AI restricts model efficacy and fairness, frequently disadvantaging marginalized populations. Resolving these challenges necessitates the use of varied and representative datasets, conducting bias audits, implementing ethical AI design, and ensuring transparency in practices. Artificial intelligence possesses a tendency to foster healthcare equity and enhance outcomes among varied populations by resolving data gaps and mitigating biases (Molli, 2024).

5.2 Accountability and Liability Issues

Healthcare AI encounters significant obstacles regarding accountability, resulting from its autonomous and complicated characteristics. Identifying liability for errors associated with AI, including misdiagnoses or improper treatment recommendations, remains unclear due to the collective responsibilities of healthcare providers, developers, and institutions. Informed consent is critical, as patients must be thoroughly educated regarding the involvement of AI in their treatment (Ng, 2024). Bias and discrimination intensify accountability challenges, as AI systems trained on biased, or incomplete datasets might strengthen existing inequities. Data privacy and security present challenges in accountability, as extensive reliance on sensitive patient information in AI raises concerns regarding violations, misuse, and appropriate authorization (Williamson & Prybutok, 2024). The lack of comprehensive regulatory frameworks for healthcare AI prevents oversight, resulting in deficiencies in risk management and ethical application. To address these challenges, collaboration among healthcare providers, AI developers, regulators, and legal experts is essential for establishing effective accountability mechanisms that ensure patient safety and equitable liability resolution.

5.3 Patient Trust and Informed Consent

Trust from patients and informed consent are essential for the ethical incorporation of artificial intelligence in healthcare. Trust enhances patient confidence in AI-assisted diagnoses, treatment plans, and care decisions, thereby decreasing resistance to AI-driven recommendations (Chaibi & Zaiem, 2024). Informed consent is critical for preserving trust, necessitating that patients receive precise details regarding the application of AI technologies, encompassing potential benefits, risks, and limitations. Healthcare providers must clarify the role of AI in diagnoses and treatments by using clear and accessible language, outlining the processes of data collection, analysis, and decision-making (Bekbolatova et al., 2024). Patients must be made aware of possible AI errors, biases, or limitations and should have the chance to enquire or consider alternatives. AI's complexity and reliance on "black box" models pose significant challenges to obtaining meaningful informed consent (Rane et al., 2024a). Data privacy and security are essential, enabling patients to keep control over their healthcare choices. Clear communication regarding these processes promotes trust and improves patient satisfaction in AI-driven healthcare.

6 Technical Limitations

The application of artificial intelligence in medical care has made significant progress, but human supervision is still required to make it effective. Healthcare professionals and tech experts are communicating to improve AI, with subject matter experts (SMEs) enriching data and improving understandable AI (Gladysz et al., 2023). However, AI may overlook social variables, such as patient needs and economic restrictions, which can impact recommendations for specific patients. Privacy issues arise when incorporating AI systems, as brands such as Amazon have free rein in collecting and leveraging data (Stucke, 2022). AI has the possibility of causing unemployment by eliminating certain jobs, resulting in the displacement of healthcare professionals who have invested significant time and resources in their education. A 2018 World Economic Forum report estimated that AI would generate 58 million jobs by 2022 while simultaneously displacing or eliminating 75 million jobs within the same timeframe (Soueidan & Shoghari, 2024). Medical AI is significantly dependent on diagnostic data, and the potential for misdiagnosis increases in situations with limited information regarding diseases, demographics, or environmental influences. AI systems exhibit vulnerability to security threats due to their reliance on data networks. The Health Research Regulations, put into effect in 2025, seek to mitigate ethical issues within the healthcare sector related to the deployment of machine learning and deep learning models (Rane et al., 2024b). The quality of data utilized for algorithm development presents a challenge, as patient data is estimated to be approximately 4 months old, and medical records are frequently disorganized. Ethical concerns surrounding healthcare AI have consistently emerged, with accountability identified as a primary issue.

Artificial intelligence in oncology encounters multiple challenges, such as maintaining data privacy and security, managing the high costs associated with AI implementation, resolving ethical concerns, training healthcare professionals for effective AI tool utilization, and mitigating biases present in AI algorithms. Ensuring privacy and security is essential while handling sensitive patient information. Healthcare facilities must adopt stringent protocols and encryption techniques to mitigate the risk of breaches or unauthorized access to patient data. Investing in robust cybersecurity measures is also essential (Ekechukwu & Simpa, 2024). The lack of regulatory guidelines for AI technologies in healthcare poses a significant challenge for healthcare providers, as it can be difficult to ensure ethical use and compliance with patient privacy laws. Guidelines and standards for AI utilization in healthcare are essential for safeguarding patient data and promoting responsible AI practices. Responsible AI use in oncology practice will improve patient outcomes and maintain the integrity and trustworthiness of the healthcare system. A future challenge is ensuring the seamless integration of AI systems with existing healthcare infrastructure, which requires technical compatibility and alignment with healthcare workflows (Clim, 2023). Continuous updates to regulations and standards are necessary to keep pace with technological advancements, ensuring patient safety and ethical considerations remain at the forefront. Healthcare facilities should establish multidisciplinary ethics committees to oversee AI implementation and address ethical concerns. Regular training for healthcare professionals on AI ethics can ensure informed decision-making and adherence to best practices (Khanna et al., 2021). Despite the benefits of AI in healthcare, it also presents potential risks, such as biases in AI algorithms, overreliance on technology, and cybersecurity threats. To reduce overreliance on AI, healthcare systems should implement a hybrid approach that combines AI insights with human expertise for decision-making (Hirosawa et al., 2024). A culture of critical thinking and continuous learning among healthcare professionals will guarantee the use of AI as an augmentation tool, not a replacement for human judgment.

7 Oncology-Specific Challenges

7.1 Tumor Heterogeneity and Evolving Nature of Cancer

Tumor heterogeneity is a significant challenge in predictive modeling and personalized treatment approaches in oncology. Cancer is a dynamic ecosystem with multiple cellular subpopulations with diverse genetic mutations, proliferation rates, and treatment sensitivities (Deschuyter et al., 2024). Molecular profiling indicates that tumors may have genetically distinct subclones that evolve autonomously, with some subpopulations potentially acquiring resistance mechanisms or exhibiting varying degrees of aggressiveness (Burrell & Swanton, 2014). Computational models must consider this inherent complexity and develop methodologies that can encompass the multidimensional aspects of tumor composition and evolution.

The evolutionary dynamics of cancer introduce further complexity, necessitating computational models that extend beyond static evaluations and establish frameworks that encapsulate the temporal and adaptive aspects of cancer progression. Resistance mechanisms can emerge through multiple pathways, including genetic mutations, epigenetic modifications, and complex interactions with the tumor microenvironment (Zahan et al., 2020). Computational methods must integrate advanced machine learning techniques to predict potential evolutionary trajectories, anticipate resistance mechanisms, and deliver dynamic risk assessments (Memon, 2023). The challenge is to create models that are adaptable to integrate new biological insights while preserving predictive accuracy across various cancer types, stages, and individual patient conditions.

7.2 Integration of Multi-omics Data

Multi-omics data integration in oncology presents significant challenges due to its complexity and computational resources required for key analysis. Interpreting diverse datasets is challenging due to their varying preprocessing and normalization methods, resulting in inconsistencies and biases (Maharana et al., 2022). The elevated dimensionality of multi-omics data and limited sample sizes heighten the risk of overfitting and diminish the dependability of computational models. Advanced algorithms are needed to elucidate the complex interrelationships among diverse omics layers while preserving biological significance (Subramanian et al., 2020).

Traditional statistical approaches often fail to capture the complex interactions between molecular levels, necessitating advanced machine learning and artificial intelligence techniques. Deep learning architectures, particularly those employing neural network methodologies, demonstrate potential in managing the high-dimensional, heterogeneous characteristics of multi-omics data (Kang et al., 2022). However, obstacles persist in creating interpretable models that can forecast outcomes and yield significant biological insights. The biological context is essential, as identical molecular alterations may yield varying consequences across different cancer types, patient demographics, and therapeutic scenarios (Du & Elemento, 2015). Advanced multi-omics integration methodologies must establish adaptable, context-sensitive frameworks capable of capturing the complex interaction among molecular layers while ensuring computational efficiency and clinical applicability (Acharya & Mukhopadhyay, 2024). Innovative strategies such as federated learning, transfer learning, and hybrid modeling techniques offer viable solutions to these challenges.

7.3 Difficulty in Predicting Treatment Response and Outcomes

Predicting treatment response and patient outcomes in oncological computational methods is a complex task due to the interplay of biological, clinical, and individual patient factors. As shown in Fig. 2, these factors include genetic profiles, tumor characteristics, patient immunology, microenvironmental interactions, and individual physiological differences. Current computational methods often fail to capture the dynamic and nonlinear interactions among these factors, neglecting the complex interactions between treatment modalities and specific tumor characteristics (Saad et al., 2024). Conventional predictive models often oversimplify treatment response by relying on a limited array of clinical variables. The heterogeneity of cancer complicates this challenge, as identical genetic mutations can lead to divergent treatment responses among different cancer types, patient demographics, and individual genetic profiles (Dagogo-Jack & Shaw, 2018). The incorporation of advanced machine learning and artificial intelligence methodologies presents challenges, such as the need for extensive, high-quality datasets for training, which are often deficient in oncology due to privacy concerns and data-sharing obstacles (Martínez-García & Hernández-Lemus, 2022) (Table 1). Innovative methodologies such as agent-based modeling and liquid biopsies aim to address these deficiencies by providing real-time insights into treatment efficacy, but their clinical implementation remains limited.

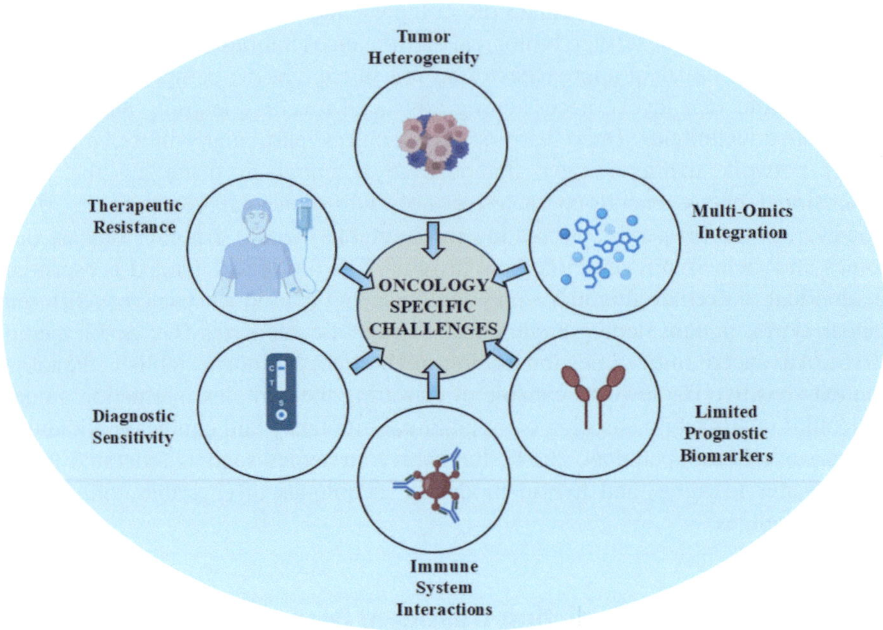

Fig. 2 This figure illustrates key oncology-specific challenges impacting cancer research and treatment

Table 1 Limitations impacting the implementation of advanced analytical approaches in oncology

Categories	Limitations
Data challenges	Poor standardization, limited datasets, privacy issues
Model issues	Overfitting, black box problems, validation difficulties
Clinical integration	Resistance, workflow misalignment, regulatory hurdles
Ethical concerns	Bias, accountability issues, lack of patient trust
Technical limitations	High resource needs, scalability, complex decisions
Oncology challenges	Tumor heterogeneity, multi-omics, treatment prediction

8 Future Directions

The use of personalized digital health technologies, computing power, and artificial intelligence has transformed healthcare delivery. Computational medicine shows enormous promise for personalized disease management, encompassing diagnosis, treatment, and prognosis. It could lower healthcare costs by helping people learn more about biological and socioeconomic factors, encouraging proactive healthcare, delaying the start of chronic diseases, and improving health in general. However, challenges persist in fully harnessing the potential of computational modeling and personalized data in clinical practice. Computational oncology is a rapidly evolving field that utilizes computational models with real-world clinical data to enhance their accuracy and applicability. Advancements in machine learning algorithms could enhance predictive capabilities, leading to more personalized cancer treatment plans. To address current limitations and advance the field, interdisciplinary collaboration among oncologists, data scientists, and bioinformaticians is essential. Organizing regular workshops and seminars can help bridge the gap between these disciplines, while collaborative research grants and projects requiring input from multiple specialties can encourage teamwork and innovation. Effective resource management is crucial for successful interdisciplinary collaborations, ensuring each team has access to the necessary tools and data for their work. Future directions should also consider the ethical implications of integrating computational methods in oncology, emphasizing patient privacy and data security. Prioritizing transparency in algorithmic decision-making can build trust among clinicians and patients, facilitating the wider acceptance of these tools in clinical settings. AI-driven cancer research is positioned for transformative advancements, particularly with the increasing integration of multi-modal data, explainable AI, and AI-powered tools for real-time monitoring and early detection. AI is expected to play a significant role in identifying rare subpopulations of cells within tumors, enhancing our understanding of tumor heterogeneity and guiding more targeted therapies. As AI technologies continue to develop, collaborations between clinicians, data scientists, and industry stakeholders will be crucial in shaping the future of cancer research. In conclusion, the integration of computational modeling and machine learning in oncology holds immense potential for revolutionizing cancer care. By fostering interdisciplinary collaborations and prioritizing ethical considerations, the field can overcome existing challenges and leverage AI's capabilities to deliver more personalized and effective treatments. The goal remains to improve patient outcomes and deepen our understanding of cancer biology as we explore these innovative strategies.

References

Acharya, D., & Mukhopadhyay, A. (2024). A comprehensive review of machine learning techniques for multi-omics data integration: Challenges and applications in precision oncology. *Briefings in Functional Genomics, 23*, 549–560. https://doi.org/10.1093/bfgp/elae013

Al-Assaf, K., Bahroun, Z., & Ahmed, V. (2024). Transforming service quality in healthcare: A comprehensive review of healthcare 4.0 and its impact on healthcare service quality. *Informatics, 11*, 96. https://doi.org/10.3390/informatics11040096

Ansari, M. I., Ahmed, K. T., & Zhang, W. (2024). Optimizing multi-omics data imputation with NMF and GAN synergy. *Bioinformatics, 40*, btae674. https://doi.org/10.1093/bioinformatics/btae674

Bekbolatova, M., Mayer, J., Ong, C. W., & Toma, M. (2024). Transformative potential of AI in healthcare: Definitions, applications, and navigating the ethical landscape and public perspectives. *Healthcare, 12*, 125. https://doi.org/10.3390/healthcare12020125

Brancato, V., Esposito, G., Coppola, L., Cavaliere, C., Mirabelli, P., Scapicchio, C., Borgheresi, R., Neri, E., Salvatore, M., & Aiello, M. (2024). Standardizing digital biobanks: Integrating imaging, genomic, and clinical data for precision medicine. *Journal of Translational Medicine, 22*, 136. https://doi.org/10.1186/s12967-024-04891-8

Burrell, R. A., & Swanton, C. (2014). Tumour heterogeneity and the evolution of polyclonal drug resistance. *Molecular Oncology, 8*, 1095–1111. https://doi.org/10.1016/j.molonc.2014.06.005

Chaibi, A., & Zaiem, I. (2024). Unlocking patient resistance to AI in healthcare: A psychological exploration. *European Journal of Investigation in Health, Psychology and Education, 15*(1), 6.

Clim, A. (2023). The future of digital healthcare: Integrating rights, AI and open standards. https://doi.org/10.2139/ssrn.4576606

Dagogo-Jack, I., & Shaw, A. T. (2018). Tumour heterogeneity and resistance to cancer therapies. *Nature Reviews. Clinical Oncology, 15*, 81–94. https://doi.org/10.1038/nrclinonc.2017.166

Dankwa-Mullan, I. (2024). Health equity and ethical considerations in using artificial intelligence in public health and medicine. *Preventing Chronic Disease, 21*, E64. https://doi.org/10.5888/pcd21.240245

Deschuyter, M., Salmanli, C., Wolf, T., Lhermitte, B., Reita, D., Fuchs, Q., Hubsch, C., Foppolo, S., Dontenwill, M., Martin, S., & Entz-Werlé, N. (2024). Stem-04. Biobanking patient-derived models offers a precise spatial and epigenetic approach to unravel the initiation, progression, and therapeutic resistance of pediatric brain tumors. *Neuro-Oncology, 26*, 0. https://doi.org/10.1093/neuonc/noae064.775

Du, W., & Elemento, O. (2015). Cancer systems biology: Embracing complexity to develop better anticancer therapeutic strategies. *Oncogene, 34*, 3215–3225. https://doi.org/10.1038/onc.2014.291

Eikey, E. V., Chen, Y., & Zheng, K. (2019). Unintended adverse consequences of health IT implementation: Workflow issues and their cascading effects. In K. Zheng, J. Westbrook, T. G. Kannampallil, & V. L. Patel (Eds.), *Cognitive informatics: Reengineering clinical workflow for safer and more efficient care* (pp. 31–43). Springer. https://doi.org/10.1007/978-3-030-16916-9_3

Ekanayake, I. U., Meddage, D. P. P., & Rathnayake, U. (2022). A novel approach to explain the black-box nature of machine learning in compressive strength predictions of concrete using Shapley additive explanations (SHAP). *Case Studies in Construction Materials, 16*, e01059. https://doi.org/10.1016/j.cscm.2022.e01059

Ekechukwu, D. E., & Simpa, P. (2024). The importance of cybersecurity in protecting renewable energy investment: A strategic analysis of threats and solutions. *Journal of Engineering Science and Technology, 5*, 1845–1883. https://doi.org/10.51594/estj.v5i6.1186

Escorial, M., Muriel, J., Margarit, C., Agulló, L., Zandonai, T., Panadero, A., Morales, D., & Peiró, A. M. (2024). Predictive model for opioid use disorder in chronic pain: A development and validation study. *Biomedicine, 12*, 2056. https://doi.org/10.3390/biomedicines12092056

Esteva, A., Robicquet, A., Ramsundar, B., Kuleshov, V., DePristo, M., Chou, K., Cui, C., Corrado, G., Thrun, S., & Dean, J. (2019). A guide to deep learning in healthcare. *Nature Medicine, 25*, 24–29. https://doi.org/10.1038/s41591-018-0316-z

Fedorov, A., Longabaugh, W. J. R., Pot, D., Clunie, D. A., Pieper, S., Aerts, H. J. W. L., Homeyer, A., Lewis, R., Akbarzadeh, A., Bontempi, D., Clifford, W., Herrmann, M. D., Höfener, H., Octaviano, I., Osborne, C., Paquette, S., Petts, J., Punzo, D., Reyes, M., Schacherer, D. P., Tian, M., White, G., Ziegler, E., Shmulevich, I., Pihl, T., Wagner, U., Farahani, K., & Kikinis, R. (2021). NCI imaging data commons. *Cancer Research, 81*, 4188–4193. https://doi.org/10.1158/0008-5472.CAN-21-0950

Gladysz, B., Matteri, D., Ejsmont, K., Corti, D., Bettoni, A., & Guerra, R. H. (2023). Platform-based support for AI uptake by SMEs: Guidelines to design service bundles. *Central European Management Journal, 31*, 463–478. https://doi.org/10.1108/CEMJ-08-2022-0096

Greaves, M., & Maley, C. C. (2012). Clonal evolution in cancer. *Nature, 481*, 306–313. https://doi.org/10.1038/nature10762

Guidotti, R., Monreale, A., Ruggieri, S., Turini, F., Giannotti, F., & Pedreschi, D. (2018). A survey of methods for explaining black box models. *ACM Computing Surveys, 51*(5), 1–42. https://doi.org/10.1145/3236009

Gurumurthy, G., Gurumurthy, J., & Gurumurthy, S. (2024). Machine learning in paediatric haematological malignancies: A systematic review of prognosis, toxicity and treatment response models. *Pediatric Research, 1–8*. https://doi.org/10.1038/s41390-024-03494-9

Halkola, A. S., Joki, K., Mirtti, T., Mäkelä, M. M., Aittokallio, T., & Laajala, T. D. (2023). OSCAR: Optimal subset cardinality regression using the L0-pseudonorm with applications to prognostic modelling of prostate cancer. *PLoS Computational Biology, 19*, e1010333. https://doi.org/10.1371/journal.pcbi.1010333

Hirosawa, T., Suzuki, T., Shiraishi, T., Hayashi, A., Fujii, Y., Harada, T., & Shimizu, T. (2024). Adapting artificial intelligence concepts to enhance clinical decision-making: A hybrid intelligence framework. *The International Journal of General Medicine, 17*, 5417–5422. https://doi.org/10.2147/IJGM.S497753

Holzinger, A., Carrington, A., & Müller, H. (2020). Measuring the quality of explanations: The system causability scale (SCS). *Künstliche Intelligenz, 34*, 193–198. https://doi.org/10.1007/s13218-020-00636-z

Hosny, A., Parmar, C., Quackenbush, J., Schwartz, L. H., & Aerts, H. J. W. L. (2018). Artificial intelligence in radiology. *Nature Reviews. Cancer, 18*, 500–510. https://doi.org/10.1038/s41568-018-0016-5

Huang, C., Clayton, E. A., Matyunina, L. V., McDonald, L. D., Benigno, B. B., Vannberg, F., & McDonald, J. F. (2018). Machine learning predicts individual cancer patient responses to therapeutic drugs with high accuracy. *Scientific Reports, 8*, 16444. https://doi.org/10.1038/s41598-018-34753-5

Iyelolu, T. V., Agu, E. E., Idemudia, C., & Ijomah, T. I. (2024). Driving SME innovation with AI solutions: Overcoming adoption barriers and future growth opportunities. *International Journal of Science and Technology Research Archive, 7*, 036–054. https://doi.org/10.53771/ijstra.2024.7.1.0055

Jiang, P., Sinha, S., Aldape, K., Hannenhalli, S., Sahinalp, C., & Ruppin, E. (2022). Big data in basic and translational cancer research. *Nature Reviews. Cancer, 22*, 625–639. https://doi.org/10.1038/s41568-022-00502-0

Johnson, G. M. (2021). Algorithmic bias: On the implicit biases of social technology. *Synthese, 198*, 9941–9961. https://doi.org/10.1007/s11229-020-02696-y

Kang, M., Ko, E., & Mersha, T. B. (2022). A roadmap for multi-omics data integration using deep learning. *Briefings in Bioinformatics, 23*, bbab454. https://doi.org/10.1093/bib/bbab454

Kelly, Y., Kuperman, G., Steele, D., & Mendu, M. (2020). Interoperability and patient electronic health record accessibility: Opportunities to improve care delivery for dialysis patients. *American Journal of Kidney Diseases, 76*(3), 427–430. https://doi.org/10.1053/j.ajkd.2019.11.001

Khanna, S., Khanna, I., Srivastava, S., & Pandey, V. (2021). AI governance framework for oncology: Ethical, legal, and practical considerations. *Quarterly Journal of Computational Technologies for Healthcare, 6*, 1–26.

Lam, D. E. S. K. (2024). Applications of radiomics for next-generation oncologic management—Current trend, challenges and future prospects. *The Journal of Medical Imaging and Radiation Sciences. ISRRT World Congress Proceedings, 55*, 101462. https://doi.org/10.1016/j.jmir.2024.101462

Lambin, P., Leijenaar, R. T. H., Deist, T. M., Peerlings, J., de Jong, E. E. C., van Timmeren, J., Sanduleanu, S., Larue, R. T. H. M., Even, A. J. G., Jochems, A., van Wijk, Y., Woodruff, H., van Soest, J., Lustberg, T., Roelofs, E., van Elmpt, W., Dekker, A., Mottaghy, F. M., Wildberger, J. E., & Walsh, S. (2017). Radiomics: The bridge between medical imaging and personalized medicine. *Nature Reviews. Clinical Oncology, 14*, 749–762. https://doi.org/10.1038/nrclinonc.2017.141

Macdonald, J. C., Isom, D. C., Evans, D. D., & Page, K. J. (2021). Digital innovation in medicinal product regulatory submission, review, and approvals to create a dynamic regulatory ecosystem—Are we ready for a revolution? *Frontiers in Medicine, 8*, 660808. https://doi.org/10.3389/fmed.2021.660808

Maharana, K., Mondal, S., & Nemade, B. (2022). A review: Data pre-processing and data augmentation techniques. *Global Transitions Proceedings. International Conference on Intelligent Engineering Approach (ICIEA-2022), 3*, 91–99. https://doi.org/10.1016/j.gltp.2022.04.020

Martínez-García, M., & Hernández-Lemus, E. (2022). Data integration challenges for machine learning in precision medicine. *Frontiers in Medicine, 8*, 784455. https://doi.org/10.3389/fmed.2021.784455

Mathivanan, S. K., Sonaimuthu, S., Murugesan, S., Rajadurai, H., Shivahare, B. D., & Shah, M. A. (2024). Employing deep learning and transfer learning for accurate brain tumor detection. *Scientific Reports, 14*, 7232. https://doi.org/10.1038/s41598-024-57970-7

McKinney, S. M., Sieniek, M., Godbole, V., Godwin, J., Antropova, N., Ashrafian, H., Back, T., Chesus, M., Corrado, G. S., Darzi, A., Etemadi, M., Garcia-Vicente, F., Gilbert, F. J., Halling-Brown, M., Hassabis, D., Jansen, S., Karthikesalingam, A., Kelly, C. J., King, D., Ledsam, J. R., Melnick, D., Mostofi, H., Peng, L., Reicher, J. J., Romera-Paredes, B., Sidebottom, R., Suleyman, M., Tse, D., Young, K. C., De Fauw, J., & Shetty, S. (2020). International evaluation of an AI system for breast cancer screening. *Nature, 577*, 89–94. https://doi.org/10.1038/s41586-019-1799-6

Memon, D. N. (2023). Computational intelligence for dynamic risk assessment in IoT-connected autonomous vehicle networks. *Journal of Artificial Intelligence in Healthcare, 3*, 61–81.

Minssen, T., Gerke, S., Aboy, M., Price, N., & Cohen, G. (2020). Regulatory responses to medical machine learning. *Journal of Law and the Biosciences, 7*, lsaa002. https://doi.org/10.1093/jlb/lsaa002

Mohamed, Y. H. A. (2023). Comprehending and mitigating feature bias in machine learning models for ethical AI. *International Journal of Social Analysis, 8*, 1–12.

Molli, V. L. P. (2024). Enhancing healthcare equity through AI-powered decision support systems: Addressing disparities in access and treatment outcomes. *The International Journal of Sustainable Development Through AI, ML and IoT, 3*, 1–12.

Moradi, R., Berangi, R., & Minaei, B. (2020). A survey of regularization strategies for deep models. *Artificial Intelligence Review, 53*, 3947–3986. https://doi.org/10.1007/s10462-019-09784-7

Ng, I. K. (2024). Informed consent in clinical practice: Old problems, new challenges. *The Journal of the Royal College of Physicians of Edinburgh, 54*, 153–158. https://doi.org/10.1177/14782715241247087

Niazi, S. K. (2023). The coming of age of AI/ML in drug discovery, development, clinical testing, and manufacturing: The FDA perspectives. *Drug Design, Development and Therapy, 17*, 2691–2725. https://doi.org/10.2147/DDDT.S424991

Ohmann, C., Banzi, R., Canham, S., Battaglia, S., Matei, M., Ariyo, C., Becnel, L., Bierer, B., Bowers, S., Clivio, L., Dias, M., Druml, C., Faure, H., Fenner, M., Galvez, J., Ghersi, D., Gluud, C., Groves, T., Houston, P., Karam, G., Kalra, D., Knowles, R. L., Krleža-Jerić, K., Kubiak, C., Kuchinke, W., Kush, R., Lukkarinen, A., Marques, P. S., Newbigging, A., O'Callaghan, J., Ravaud, P., Schlünder, I., Shanahan, D., Sitter, H., Spalding, D., Tudur-Smith, C., van Reusel, P., van Veen, E.-B., Visser, G. R., Wilson, J., & Demotes-Mainard, J. (2017). Sharing and reuse of individual participant data from clinical trials: Principles and recommendations. *BMJ Open, 7*, e018647. https://doi.org/10.1136/bmjopen-2017-018647

Patel, A. U., Gu, Q., Esper, R., Maeser, D., & Maeser, N. (2024). The crucial role of interdisciplinary conferences in advancing explainable AI in healthcare. *BioMedInformatics, 4*, 1363–1383. https://doi.org/10.3390/biomedinformatics4020075

Prabhod, K. J., & Gadhiraju, A. (2024). Foundation models in medical imaging: Revolutionizing diagnostic accuracy and efficiency. *Journal of Artificial Intelligence Research and Applications, 4*, 471–511.

Prosperi, M., Min, J. S., Bian, J., & Modave, F. (2018). Big data hurdles in precision medicine and precision public health. *BMC Medical Informatics and Decision Making, 18*, 139. https://doi.org/10.1186/s12911-018-0719-2

Qu, J.-M., Wen, S.-J., Lin, Y.-K., Lu, H.-X., Huang, K.-Q., Maansson, C. T., Lee, C.-S., & Araki, T. (2024). Retrospective clinical study analysis of skin adverse reactions related to epidermal growth factor receptor inhibitors. *Translational Cancer Research, 13*, 3016–3030. https://doi.org/10.21037/tcr-24-486

Rane, J., Mallick, S., Kaya, Ö., & Rane, N. (2024a). Enhancing black-box models: Advances in explainable artificial intelligence for ethical decision-making. In *Future research opportunities for artificial intelligence in industry 4.0 and 5.0* (pp. 136–180). Deep Science Publishing. https://doi.org/10.70593/978-81-981271-0-5_4

Rane, N., Paramesha, M., Rane, J., & Kaya, Ö. (2024b). Emerging trends and future research opportunities in artificial intelligence, machine learning, and deep learning. In *Artificial intelligence and industry in society 5.0* (pp. 95–118). Deep Science Publishing. https://doi.org/10.70593/978-81-981271-1-2_6

Reddy, S., Fox, J., & Purohit, M. P. (2019). Artificial intelligence-enabled healthcare delivery. *Journal of the Royal Society of Medicine, 112*, 22–28. https://doi.org/10.1177/0141076818815510

Rieke, N., Hancox, J., Li, W., Milletarì, F., Roth, H. R., Albarqouni, S., Bakas, S., Galtier, M. N., Landman, B. A., Maier-Hein, K., Ourselin, S., Sheller, M., Summers, R. M., Trask, A., Xu, D., Baust, M., & Cardoso, M. J. (2020). The future of digital health with federated learning. *npj Digital Medicine, 3*, 1–7. https://doi.org/10.1038/s41746-020-00323-1

Romoli, M., & Caliandro, P. (2024). Artificial intelligence, machine learning, and reproducibility in stroke research. *European Stroke Journal, 9*, 518–520. https://doi.org/10.1177/23969873241275863

Saad, M. B., Al-Tashi, Q., Hong, L., Li, W., Boiarsky, D., Li, S., Petranovic, M., Wu, C. C., Carter, B., Cascone, T., Le, X., Chang, J., Gibbons, D. L., Vaporciyan, A. A., Lee, J. J., Mirjalili, S., Jaffray, D. A., Gainor, J. F., Lou, Y., Ricciuti, B., DiFederico, A., Pecci, F., Awad, M. M., Heymach, J. V., Vokes, N. I., Zhang, J., & Wu, J. (2024). 1236 Machine learning-based clinico-genomic prediction of benefits to add chemotherapy to immunotherapy in metastatic non-small cell lung cancer. *Journal for Immunotherapy of Cancer, 12*, A1379–A1380. https://doi.org/10.1136/jitc-2024-SITC2024.1236

Siegel, R. L., Miller, K. D., Fuchs, H. E., & Jemal, A. (2022). Cancer statistics, 2022. *CA: A Cancer Journal for Clinicians, 72*, 7–33. https://doi.org/10.3322/caac.21708

Soueidan, M. H., & Shoghari, R. (2024). The impact of artificial intelligence on job loss: Risks for governments. *Technium Social Sciences Journal, 57*, 206.

Stucke, M. E. (2022). *Breaking away: How to regain control over our data, privacy, and autonomy*. Oxford University Press.

Subramanian, I., Verma, S., Kumar, S., Jere, A., & Anamika, K. (2020). Multi-omics data integration, interpretation, and its application. *Bioinformatics and Biology Insights, 14*, 1177932219899051. https://doi.org/10.1177/1177932219899051

Thapa, C., & Camtepe, S. (2021). Precision health data: Requirements, challenges and existing techniques for data security and privacy. *Computers in Biology and Medicine, 129*, 104130. https://doi.org/10.1016/j.compbiomed.2020.104130

Tjoa, E., & Guan, C. (2021). A survey on explainable artificial intelligence (XAI): Toward medical XAI. *IEEE Transactions on Neural Networks and Learning Systems, 32*, 4793–4813. https://doi.org/10.1109/TNNLS.2020.3027314

Topol, E. J. (2019). High-performance medicine: The convergence of human and artificial intelligence. *Nature Medicine, 25*, 44–56. https://doi.org/10.1038/s41591-018-0300-7

Wang, S., Li, C., Wang, R., Liu, Z., Wang, M., Tan, H., Wu, Y., Liu, X., Sun, H., Yang, R., Liu, X., Chen, J., Zhou, H., Ben Ayed, I., & Zheng, H. (2021). Annotation-efficient deep learning for automatic medical image segmentation. *Nature Communications, 12*, 5915. https://doi.org/10.1038/s41467-021-26216-9

Williamson, S. M., & Prybutok, V. (2024). Balancing privacy and progress: A review of privacy challenges, systemic oversight, and patient perceptions in AI-driven healthcare. *Applied Sciences, 14*, 675. https://doi.org/10.3390/app14020675

Xu, F., & Lai, J. (2024). Commentary: Immune cell infiltration and prognostic index in cervical cancer: Insights from metabolism-related differential genes. *Frontiers in Immunology, 15*, 1411132. https://doi.org/10.3389/fimmu.2024.1446741

Yang, Q., Liu, Y., Chen, T., & Tong, Y. (2019). Federated machine learning: Concept and applications. *ACM Transactions on Intelligent Systems and Technology, 12*(1–12), 19. https://doi.org/10.1145/3298981

Zahan, T., Das, P. K., Akter, S. F., Habib, R., Rahman, M. H., Karim, M. R., & Islam, F. (2020). Therapy resistance in cancers: Phenotypic, metabolic, epigenetic and tumour microenvironmental perspectives. *Anti-Cancer Agents in Medicinal Chemistry, 20*, 2190–2206. https://doi.org/10.2174/1871520620999200730161829

Zanfardino, M., Pane, K., Mirabelli, P., Salvatore, M., & Franzese, M. (2019). TCGA-TCIA impact on radiogenomics cancer research: A systematic review. *International Journal of Molecular Sciences, 20*, 6033. https://doi.org/10.3390/ijms20236033

Zeng, G., Yu, Y., Wang, M., Liu, J., He, G., Yu, S., Yan, H., Yang, L., Li, H., & Peng, X. (2024). Advancing cancer research through organoid technology. *Journal of Translational Medicine, 22*, 1007. https://doi.org/10.1186/s12967-024-05824-1

Challenges in Integration of Computational Approaches with Clinical Practice

Monika Kumari and Shrikant Nema

Abstract

The integration of computational approaches into clinical cancer practice has the potential to revolutionize diagnosis, treatment, and patient outcomes. Technological advances in machine learning, bioinformatics, and artificial intelligence have led to breakthroughs in cancer detection, predictive modeling, and personalized therapy. Computational tools such as deep learning models, next-generation sequencing, and advanced imaging techniques promise greater diagnostic accuracy and more tailored treatment strategies. Despite their transformative potential, the practical application of these approaches in clinical settings remains a significant challenge. Barriers include issues related to data quality, the complexity of integrating multi-omics data, and the reluctance of healthcare professionals to adopt new technologies due to concerns about accuracy, reliability, and interpretability. Additionally, the lack of standardized protocols, the need for extensive training, and ethical concerns around data privacy and algorithmic bias further hinder the widespread implementation of computational tools. This chapter explores these challenges in detail, highlighting both technical and nontechnical obstacles, and proposes strategies for overcoming them. Addressing these barriers is crucial to bridging the gap between theoretical advancements and real-world clinical practice, ultimately ensuring that computational approaches can enhance cancer care, improve patient outcomes, and drive the next generation of personalized medicine.

M. Kumari
Birla Institute of Technology Mesra, Ranchi, Jharkhand, India

S. Nema (✉)
International Centre for Genetic Engineering and Biotechnology (ICGEB), New Delhi, 110067, India
e-mail: shrikant@icgeb.res.in

R. Nema et al. (eds.), *Advances in Cancer Detection, Prediction, and Prognosis Using Artificial Intelligence and Machine Learning*,
https://doi.org/10.1007/978-981-96-9346-7_14

Keywords

Computational approaches · Cancer diagnosis · Machine learning · Personalized medicine · Clinical integration

1 Introduction

According to the World Health Organization (WHO), the annual number of new cancer cases in the United States is expected to surpass 2.5 million by 2050 (WHO, Global Cancer Observatory, n.d.). Studies has shown that adults aged 65 and older are 11 times more likely to develop cancer than younger individuals (Berger et al., 2006). Early detection can change our perspective from cancer being a dreaded disease to a curable disease by improving survival rates and handling the associated risks.

Historically, cancer diagnosis relied on manual histopathological analysis, basic genomic profiling, and conventional imaging techniques. These methods, though foundational, are often limited by subjectivity, scope, and resolution. In contrast, technological advancement in molecular diagnostics, bioinformatics, and various computational approaches in healthcare has led to breakthroughs, encompassing improved diagnostic, predictive, and prognostic methods and therapies for cancer patients (Pulumati et al., 2023). In recent years, computational techniques such as machine learning, bioinformatics, and artificial intelligence have revolutionized the field of cancer diagnosis (Sebastian & Peter, 2022). These tools have the potential to provide more accurate diagnoses, predict disease outcomes and provide personalized treatment plans. Modern computational approaches harness the power of machine learning and artificial intelligence to enhance diagnostic accuracy and personalized treatment. Deep learning models, particularly Convolutional Neural Networks (CNNs), have significantly improved image analysis, allowing for more precise tumor detection, segmentation, and classification from complex medical images. Next-generation sequencing (NGS) and integrative omics approaches provide comprehensive genomic insights, identifying mutations and molecular signatures that inform targeted therapies and personalized treatment plans. Advanced imaging techniques, such as PET scans and radiomics, offer detailed functional and quantitative data, complementing structural imaging and enhancing disease characterization. Real-time monitoring through wearable devices and adaptive machine learning models facilitates continuous tracking of disease progression and treatment responses, ensuring dynamic and individualized care. Furthermore, bioinformatics and systems biology integrate multi-omics data to model cancer at a systemic level, uncovering new therapeutic targets and mechanisms (Pettini et al., 2021). As computational techniques evolve, they promise to further refine cancer diagnosis, offering hope for more effective and personalized therapeutic strategies.

While computational methods have been developed at a rapid pace, the gap between computational models and their practical applications in clinical practice is notably significant. Their inclusion in daily clinical practices remains hindered.

Although computational approaches have great theoretical potential, they are difficult to implement in clinical settings. From data quality issues to the skepticism of healthcare professionals, there are many barriers to convergence.

This chapter examines the major challenges that hinder the integration of computational methods with clinical practice. The aim is to analyze these obstacles in detail, from technical limitations to ethical concerns, and to suggest possible ways to overcome them. By understanding these challenges, we can begin to develop solutions that allow computational approaches to improve cancer detection and also become important tools in clinical settings.

2 Evolution of Computational Approaches for Cancer Diagnosis: Beyond the Microscope

2.1 Conventional Practices for Cancer Diagnosis

In 1838, Johannes Müller, a German pathologist published the first known description of a cancer that could be seen under a microscope (Hajdu, 2012). Nearly two centuries later, the basic steps for diagnosing cancer remain roughly the same—slices of tumor tissue are fixed and stained and then magnified to examine, however, the technology and tools that are used currently have advanced significantly since Müller's time. Conventional methods for diagnosing cancer include laboratory tests, imaging tests (CT, MRI, PET, X-ray), and biopsy (Meyer et al., 2005; Pantanowitz et al., 2018; Nema et al., 2022). Biopsy is considered the gold standard technique. Pathologists stain the cells and examine the tissue growth patterns and cell morphology by scanning the tissue slides carefully under the microscope at different magnifications to observe the pattern of malignancy in the tissue sample and evaluate the morphology at the cellular level (Gaur & Jagtap, 2022). However, these manual histopathological examinations could be time-consuming, laborious, and subjective with the variations dependent on the expertise of the observer, which might result in inaccurate evaluation (Iqbal et al., 2021).

2.2 Future of Cancer Diagnosis

Over the last two decades, there has been a significant advancement in technology such as high-resolution slide scanning, which makes creating, viewing, and sharing digital images of slides that are immunohistochemically stained an easy task (He et al., 2012). By collecting and analyzing objective and therapeutically pertinent information from datasets, computational tools can perceive more than the human eye is capable of, enhancing our understanding of disease and improving treatment.

A set of computer-coded programs or algorithms that employ data analysis and preprogrammed instructions to generate predictions and choices about different elements of a disease is frequently referred to as "artificial intelligence." A subfield of artificial intelligence, known as "machine learning," describes a collection of algorithms created to automatically pick up new skills and get better with practice. Simply put, machine learning is a branch of artificial intelligence that focuses on creating algorithms that can learn from data and improve over time (Umar, 2004). Machine learning is further branched into "deep learning" that uses neural network-based models to mimic how the human brain processes vast amounts of complex data in domains including image recognition, drug discovery, and language processing (Huang et al., 2020). Recent advancements in data science and computers enable the integration of digital pathology pictures with genetic, radiological, clinical, and other data, utilizing deep learning algorithms to derive fresh insights. In order to achieve rapid and reproducible results, researchers have developed algorithms to detect and classify cancerous cells by processing the image and recognizing the patterns, and they have combined machine learning and deep learning techniques to develop computer-aided diagnostic (CAD) systems (Askin et al., 2023). Another important tool is "Bioinformatics," which describes the use of algebraic, analytic, digital technology, and computing approaches to analyze high-dimensional data collected through many sources to better understand biological processes (Shao et al., 2022).

2.2.1 Use of AI, ML, and DL in Cancer Prediction and Diagnosis

Clinicians have realized the need to integrate computational approaches such as artificial intelligence (AI), machine learning (ML), and deep learning (DL) in the emerging digital data era (Shao et al., 2022). The efficacy of AI in cancer diagnosis and treatment is being assessed by various clinical trials (Botlagunta et al., 2023). Other studies are also focusing on the use of AI-driven predictive models to improve therapy recommendations based on genetic profiles, demonstrating the potential of AI to personalize medicines for individual patients (Miotto et al., 2020). For instance, researchers have employed multimodal deep learning algorithms to analyze medical imaging data in order to detect prostate cancer early on. Preliminary results suggest a significant improvement in accuracy and speed over standard diagnostic approaches (Nartowt et al., 2019). Another investigation employed machine learning models that use blood profile data to aid physicians in the selection of intensive care metastatic breast cancer patients while enhancing overall survival rates (Ray et al., 2020). AI algorithms can also be used to analyze unstructured data to estimate the chances of patients getting affected by different illnesses, including cancer (Pantanowitz et al., 2020). An artificial "neural network model" has shown higher accuracy than screening guidelines for colorectal cancer risk stratification (Wang, 2011). In a study, machine learning was used for the classification of data related to cancer to examine and diagnose breast cancer (Gunduz, 2015). Researchers at the University of Pittsburgh developed an AI-based machine learning algorithm that shows the highest accuracy in diagnosing prostate cancer, with 98% specificity and sensitivity (John-Mathews, 2022).

2.2.2 Use of Bioinformatics in Cancer Diagnosis

Bioinformatics plays a major role in cancer diagnosis with the identification of biomarkers, incorporation of knowledge on signaling pathways, protein annotations, and tracking of the changes in biomarkers at various stages of illness (Sarhadi & Armengol, 2022). The significance of bioinformatics has risen with the technological advancements in the use of DNA microarrays and proteomics studies for large-scale gene expression research, with this the application of bioinformatics analytics and wet experimentation go side by side (Dlamini et al., 2022).

3 Barriers to Integration of Computational Intelligence-Based Cancer Diagnosis

The application of computational intelligence in cancer diagnostics marks a significant development in medical science. However, this integration confronts some obstacles such as interpretability of AI models, ethical considerations, and concerns related to data privacy, and these issues must be overcome before it can be effectively implemented and widely adopted (Yang et al., 2019; Rulten et al., 2023). This section discusses some of the key issues of introducing computational approaches into cancer diagnostic processes.

3.1 Clinical Barriers

3.1.1 Data Quality and Variability

In the era of precision oncology, the availability of extensive clinical and diagnostic datasets has significantly expanded. The quality and consistency of clinical data are crucial for the effective application of computational methods (Jiang et al., 2017). Variability in data collection methods, patient demographics, and disease subtypes can lead to challenges in developing robust models (Iniyan et al., 2024). The success of ML and AI models heavily depends on the robustness and standardization of input data. However, medical datasets, particularly in oncology, are often opted by issues such as incompleteness, inconsistency, and heterogeneity, which undermine the reliability and applicability of computational diagnostics (Balagurunathan et al., 2014). Missing or incomplete data can significantly affect the performance of computational tools, leading to potential misdiagnoses or overlooked conditions (Bi et al., 2019). In a few instances, studies evaluated TGCA cancer biomarker data using a bioinformatics pipeline, and the TCGA RNAseq data showed differences than that of the real patient's data. Different imaging data, genomic data, and clinical information including treatment history have shown inherent variations in format, scale, and quality, making integration complex (Heo et al., 2021). Imaging settings, including CT scanners, slice thickness, and reconstruction kernels, also affected the calculation of radiomic features (Dankwa-Mullan & Weeraratne, 2022; Koçak et al., 2024). Few studies showed the gap specifically in addressing the incorporation of multimodal data and optimizing model robustness across diverse datasets (Chua et al., 2021). In some studies, the variations in expert annotations for tumor regions or specific cancer subtypes introduce biases in training data, affecting model accuracy (Chua et al., 2021). The lack of standardized data collection across different institutions and patient populations leads to biased models that may not generalize well (Salwei & Reale, 2024). Despite enhancements in detection models, crucial research gaps remain, specifically in addressing the incorporation of multi-modal data and optimizing model robustness across diverse datasets.

3.1.2 Integration with Clinical Workflow

Incorporating computational tools into existing clinical workflows can be disruptive. Clinicians may face difficulties in adapting to new technologies, which can affect the overall efficiency and acceptance of these tools (Topol, 2019). The need for extensive training and continuous education for healthcare professionals to effectively use these tools is another significant barrier (Esteva et al., 2021). A critical challenge in adopting computational techniques lies in their acceptance and usability within clinical settings. Clinicians often express concerns about the interpretability of computational models, especially machine learning algorithms such as neural networks, which function as "black boxes." One of the primary concerns is the "black box" nature of many computational models, particularly deep learning algorithms. Clinicians are often hesitant to rely on tools whose decision-making processes are not transparent or easily interpretable (Lokaj et al., 2024). Furthermore, there is often a mismatch between the outputs of computational diagnostics and clinical workflows. For example, while computational tools may flag abnormalities with high sensitivity, their specificity might be lower, increasing false positives and contributing to diagnostic fatigue (Ameen et al., 2022). Concerns also arise regarding the robustness of models that rely on prior knowledge integration, as segmentation methods may falter with diverse cancer types. Moreover, performance can vary based on the choice of pretrained models, and reliance on predefined Regions of Interest (ROI) tokenization might hinder feature capture. Variability in practices across healthcare institutions complicates the implementation of these tools, further delaying their adoption (Iniyan et al., 2024). Lack of transparency in the decision-making process undermines confidence and leads to reluctance in adoption. Furthermore, despite familiarity with traditional data approaches, many laboratory professionals are not familiar with the clinical machine learning workflow, resulting in a knowledge gap concerning the development, understanding, and use of ML models. Integrating new computational tools into established clinical workflows can disrupt clinician routines and require significant changes in practice patterns (Topol, 2019). Additionally, the integration of computational diagnostics is hindered by the lack of standardized protocols. There is a need for robust IT infrastructure to store, manage, and analyze large datasets efficiently, which can be costly to implement (Shah & Parveen, 2023). Additionally, the lack of standardized protocols for integrating computational tools into routine diagnostic pathways creates operational inefficiencies.

3.2 Technical Barriers

3.2.1 Algorithmic Limitations

The integration of computational algorithms in cancer diagnosis offers the promise of improved accuracy and personalized care. Many computational models, including machine learning algorithms, require large datasets for training. However, obtaining such datasets in the medical field can be challenging due to privacy concerns and the rarity of certain cancer types (Yang et al., 2019). However, significant algorithmic limitations persist, which impede their effective application in clinical settings. These include issues related to interpretability, robustness, and scalability, as well as the challenges posed by data heterogeneity and limited generalizability of models.

One critical limitation is the "black box" nature of many algorithms, particularly deep learning models. These algorithms are often opaque, making it difficult for clinicians to interpret their decisions (Mi et al., 2024). Without transparency, clinicians may struggle to trust or validate diagnostic outputs, especially in cases where high-stakes decisions, such as cancer treatment planning, are required. This lack of explainability remains a barrier to the widespread adoption of computational diagnostics in clinical workflows. Another major concern is the robustness of algorithms when applied to diverse datasets. Algorithms trained on curated datasets may not perform well on real-world data due to variations in imaging modalities, demographic biases, and differences in diagnostic techniques (Obermeyer et al., 2019). For example, algorithms optimized for lung cancer diagnosis have shown inconsistent performance when deployed across different imaging platforms, highlighting the need for more adaptable solutions. The issue of scalability also limits algorithmic utility. As data sizes increase, the computational cost of training and deploying these models becomes prohibitive. This is particularly evident in multimodal diagnostics that combine imaging, genomic, and clinical data, which require significant computational resources to process efficiently (Rahrooh et al., 2024). Moreover, many algorithms are designed for specific cancer types or stages, making it difficult to generalize their application to other forms of cancer or varying clinical scenarios.

At last, the biases in training datasets contribute to inequities in algorithmic performance. Underrepresentation of certain populations, such as ethnic minorities or rare cancer subtypes, can result in diagnostic disparities (McGuinness et al., 2021). Addressing these biases requires the development of more inclusive datasets and fairness-aware algorithms to ensure equitable diagnostic outcomes. To overcome these limitations, ongoing research is focused on developing interpretable models, improving training techniques for diverse datasets, and optimizing computational efficiency. These efforts aim to enhance the reliability and applicability of computational approaches in cancer diagnostics, ultimately improving patient outcomes.

3.2.2 Interoperability Issues

Interoperability challenges significantly hinder the seamless integration of computational methods into cancer diagnostics. These issues arise from diverse data formats, inconsistent data-sharing protocols, and a lack of standardized frameworks across healthcare institutions. A primary challenge is the heterogeneity of data sources, which include imaging modalities (e.g., CT, MRI), genomic data, and electronic health records (EHRs). Each source often employs different standards for data representation, making integration complex (Hochheiser et al., 2016). Variations in coding systems, file formats, and metadata annotations further complicate the interoperability of these datasets. For instance, Fast Healthcare Interoperability Resources (FHIR) is a step forward in creating standardized EHR systems, but its adoption remains uneven across healthcare systems (Chepelev et al., 2023). Semantic interoperability, which ensures that shared data is understood consistently, poses another challenge. Current systems lack robust ontologies and lexicons to facilitate uniform interpretation of medical terminologies, particularly for cancer phenotypes (Jagadamba et al., 2024). This inconsistency leads to fragmented diagnostic workflows and hinders the application of AI models trained on diverse datasets.

Moreover, interoperability barriers impede the scalability of computational tools across institutions. The absence of unified data-sharing platforms and privacy-compliant protocols restricts collaborative efforts, limiting the effectiveness of computational models in multicenter clinical studies (Fedorov et al., 2023). For example, differences in imaging hardware and software configurations reduce cross-center compatibility and prevent reproducibility in AI-based diagnostic systems (Cui & Zhang, 2021). Addressing these issues requires the adoption of universal data standards and collaborative frameworks. Initiatives such as the National Cancer Institute's Imaging Data Commons aim to create interoperable repositories for medical imaging and related data, promoting transparency and reproducibility in cancer diagnostics (Jain et al., 2023). By bridging these interoperability gaps, computational approaches can more effectively integrate into clinical workflows, improving the reliability and reach of cancer diagnostics.

3.3 Regulatory and Ethical Barriers

The integration of computational methods in cancer diagnosis faces significant regulatory and ethical challenges. These issues stem from the rapid pace of technological development, the lack of clear governance frameworks, and complex ethical dilemmas regarding privacy, accountability, and equity. One of the primary regulatory barriers involves the approval process for computational tools. Regulatory bodies such as the U.S. Food and Drug Administration (FDA) and the European Medicines Agency (EMA) require rigorous validation and clinical testing to ensure the safety and efficacy of these systems. However, existing guidelines often fail to account for the iterative nature of machine learning algorithms, which can evolve as they are retrained with new data (Khanna et al., 2021). This regulatory lag creates obstacles for developers and delays the adoption of potentially life-saving technologies. Ethical issues also appear large. A major concern is the protection of patient data. Computational methods, particularly those utilizing artificial intelligence, require extensive datasets for training, which often include sensitive patient information. While anonymization techniques are employed, there remains a risk of reidentification, especially with genomic data (Heeney et al., 2010). Ensuring robust data privacy while enabling effective model training is a delicate balance that has yet to be fully resolved. Accountability is another ethical concern. When computational tools are used for diagnosis, questions arise about responsibility in cases of error. Unlike traditional diagnostics, where accountability lies with the clinician, computational approaches introduce a layer of ambiguity, as errors may stem from the underlying algorithm or its data inputs (Contaldo et al., 2024; Nema et al., 2024). This ambiguity complicates liability and increases clinician's reluctance to rely on such systems. Additionally, biases embedded in computational models can exacerbate healthcare disparities. Many algorithms are trained on datasets that underrepresent certain demographic groups, such as minorities or populations from low-income regions. These biases can result in uneven diagnostic accuracy, raising ethical concerns about fairness and equity in healthcare (Nema et al., 2021; Arora et al., 2023).

Addressing these barriers requires a multidisciplinary approach. Regulators must develop dynamic approval processes tailored to the unique characteristics of machine learning algorithms. Simultaneously, data-sharing initiatives should prioritize inclusivity and privacy through advanced cryptographic methods. Finally, promoting transparency and fairness in model design will be crucial to building trust among clinicians and patients.

4 Future Directions

The future of computational approaches in cancer diagnosis lies in addressing current limitations while expanding capabilities to enhance diagnostic precision, equity, and accessibility. Interdisciplinary collaborations among clinicians, data scientists, and regulatory agencies will play a pivotal role in shaping the landscape of computational oncology. The development of explainable AI frameworks will empower clinicians to understand the rationale behind algorithmic predictions, nurturing greater trust and integration into clinical workflows (Abgrall et al., 2024). Interpretable models will also help address accountability concerns by clarifying decision-making pathways (Fig. 1).

Standardization of medical data collection and sharing is another vital area. Initiatives to develop universally accepted protocols for data formats, annotation, and storage will mitigate issues related to data variability and interoperability (Feldman et al., 2018). Merged learning systems, which allow collaborative model training across institutions while preserving data privacy, will further enhance the availability of diverse, high-quality datasets. Bias mitigation and fairness in AI systems are pressing concerns. Future efforts must focus on creating demographically representative datasets, ensuring that algorithms perform equitably across diverse populations (Arora et al., 2023). Public–private partnerships and global collaborations can help establish large-scale repositories that prioritize inclusivity. Regulatory frameworks need to evolve to accommodate the iterative nature of computational

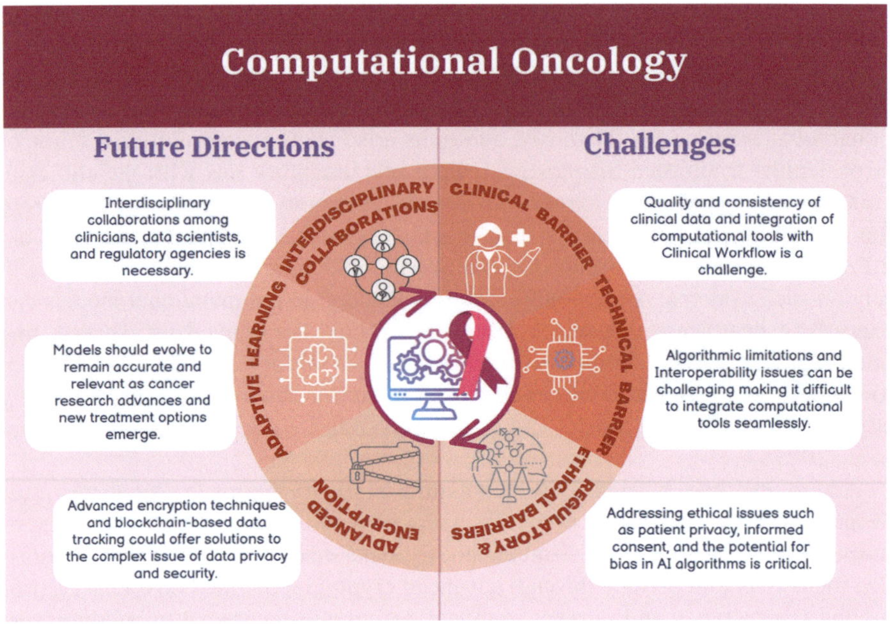

Fig. 1 Challenges and future research directions in computational oncology. (This image is generated in-house using the Canva)

models. Dynamic approval processes that allow for continuous learning and improvement of algorithms post-deployment will ensure safety and relevance without stifling innovation (Abedsoltan et al., 2024). Regulatory agencies must also work closely with developers to streamline evaluation pathways for AI-based diagnostics. Ethically, robust data governance frameworks are required to protect patient privacy and promote transparency. Advanced encryption techniques and blockchain-based data tracking could offer solutions to persistent privacy concerns, ensuring data security in multi-institutional collaborations (Rai et al., 2024). In the coming years, the field of oncology along with other aspects of healthcare will become more digitalized with the integration of computational tools such as bioinformatics and AI algorithms, providing a better understanding of tumors, enhanced decision-making processes, and more accurate treatment.

5 Conclusion

In conclusion, computational methods in cancer diagnostics possess transformative promise. The literature surrounding AI and ML-powered solutions in cancer diagnostics promises many good things, but they can only be realized after the hiccups are ironed out. To face these challenges, concurrent work from healthcare professionals, AI researchers, and regulatory bodies becomes a necessity. By tackling obstacles including algorithmic opacity, data unpredictability, and legal impediments, the area can attain extensive clinical integration. Solving the above challenges will enable computational technologies such as AI, ML, DL, and bioinformatics to substantially improve cancer diagnosis, treatment, and patient outcome by significantly improving both cancer diagnosis accuracy and efficiency of care delivery in combination with making that magic accessible at a global scale. As technology progresses, these systems are expected to become essential instruments, facilitating earlier and more precise diagnoses, diminishing healthcare inequities, and propelling the age of precision oncology.

References

Abedsoltan, H., Abedsoltan, A., & Zoghi, Z. (2024). Future of process safety: Insights, approaches, and potential developments. *Process Safety and Environmental Protection, 185*, 684–707. https://doi.org/10.1016/j.psep.2024.03.034

Abgrall, G., et al. (2024). Should AI models be explainable to clinicians? *Critical Care, 28*(1), 301. https://doi.org/10.1186/s13054-024-05005-y

Ameen, S., et al. (2022). AI and clinical decision making: The limitations and risks of computational reductionism in bowel cancer screening. *Applied Sciences, 12*(7), 3341. https://doi.org/10.3390/app12073341

Arora, A., et al. (2023). The value of standards for health datasets in artificial intelligence-based applications. *Nature Medicine, 29*(11), 2929–2938. https://doi.org/10.1038/s41591-023-02608-w

Askin, S., et al. (2023). Artificial intelligence applied to clinical trials: Opportunities and challenges. *Health and Technology, 13*(2), 203–213. https://doi.org/10.1007/s12553-023-00738-2

Balagurunathan, Y., et al. (2014). Test–retest reproducibility analysis of lung CT image features. *Journal of Digital Imaging, 27*(6), 805–823. https://doi.org/10.1007/s10278-014-9716-x

Berger, N. A., et al. (2006). Cancer in the elderly. *Transactions of the American Clinical and Climatological Association, 117*, 147–156.

Bi, W. L., et al. (2019). Artificial intelligence in cancer imaging: Clinical challenges and applications. *CA: A Cancer Journal for Clinicians, 69*(2), 127–157. https://doi.org/10.3322/caac.21552

Botlagunta, M., et al. (2023). Classification and diagnostic prediction of breast cancer metastasis on clinical data using machine learning algorithms. *Scientific Reports, 13*(1), 485. https://doi.org/10.1038/s41598-023-27548-w

Chepelev, L. L., et al. (2023). Ontologies in the new computational age of radiology: RadLex for semantics and interoperability in imaging workflows. *Radiographics, 43*(3), e220098. https://doi.org/10.1148/rg.220098

Chua, I. S., et al. (2021). Artificial intelligence in oncology: Path to implementation. *Cancer Medicine, 10*(12), 4138–4149. https://doi.org/10.1002/cam4.3935

Contaldo, M. T., et al. (2024). AI in radiology: Navigating medical responsibility. *Diagnostics, 14*(14), 1506. https://doi.org/10.3390/diagnostics14141506

Cui, M., & Zhang, D. Y. (2021). Artificial intelligence and computational pathology. *Laboratory Investigation, 101*(4), 412–422. https://doi.org/10.1038/s41374-020-00514-0

Dankwa-Mullan, I., & Weeraratne, D. (2022). Artificial intelligence and machine learning technologies in cancer care: Addressing disparities, bias, and data diversity. *Cancer Discovery, 12*(6), 1423–1427. https://doi.org/10.1158/2159-8290.CD-22-0373

Dlamini, Z., et al. (2022). AI and precision oncology in clinical cancer genomics: From prevention to targeted cancer therapies-an outcomes based patient care. *Informatics in Medicine Unlocked, 31*, 100965. https://doi.org/10.1016/j.imu.2022.100965

Esteva, A., et al. (2021). Deep learning-enabled medical computer vision. *npj Digital Medicine, 4*(1), 1–9. https://doi.org/10.1038/s41746-020-00376-2
Fedorov, A., et al. (2023). National Cancer Institute Imaging Data Commons: Toward transparency, reproducibility, and scalability in imaging artificial intelligence. *Radiographics, 43*(12), e230180. https://doi.org/10.1148/rg.230180
Feldman, K., Johnson, R. A., & Chawla, N. V. (2018). The state of data in healthcare: Path towards standardization. *Journal of Healthcare Informatics Research, 2*(3), 248–271. https://doi.org/10.1007/s41666-018-0019-8
Gaur, K., & Jagtap, M. M. (2022). Role of artificial intelligence and machine learning in prediction, diagnosis, and prognosis of cancer. *Cureus, 14*(11), e31008. https://doi.org/10.7759/cureus.31008
Gunduz, M. (2015). *A concise review of molecular pathology of breast cancer*. IntechOpen. https://doi.org/10.5772/58669
Hajdu, S. I. (2012). A note from history: Landmarks in history of cancer, part 3. *Cancer, 118*(4), 1155–1168. https://doi.org/10.1002/cncr.26320
He, L., et al. (2012). Histology image analysis for carcinoma detection and grading. *Computer Methods and Programs in Biomedicine, 107*(3), 538–556. https://doi.org/10.1016/j.cmpb.2011.12.007
Heeney, C., et al. (2010). Assessing the privacy risks of data sharing in genomics. *Public Health Genomics, 14*(1), 17–25. https://doi.org/10.1159/000294150
Heo, Y. J., et al. (2021). Integrative multi-omics approaches in cancer research: From biological networks to clinical subtypes. *Molecules and Cells, 44*(7), 433–443. https://doi.org/10.14348/molcells.2021.0042
Hochheiser, H., et al. (2016). An information model for computable cancer phenotypes. *BMC Medical Informatics and Decision Making, 16*(1), 121. https://doi.org/10.1186/s12911-016-0358-4
Huang, S., et al. (2020). Artificial intelligence in cancer diagnosis and prognosis: Opportunities and challenges. *Cancer Letters, 471*, 61–71. https://doi.org/10.1016/j.canlet.2019.12.007
Iniyan, S., et al. (2024). Enhanced breast cancer diagnosis through integration of computer vision with fusion based joint transfer learning using multi-modality medical images. *Scientific Reports, 14*(1), 28376. https://doi.org/10.1038/s41598-024-79363-6
Iqbal, M. J., et al. (2021). Clinical applications of artificial intelligence and machine learning in cancer diagnosis: Looking into the future. *Cancer Cell International, 21*(1), 270. https://doi.org/10.1186/s12935-021-01981-1
Jagadamba, G., et al. (2024). Design of interoperable electronic health record (EHR) application for early detection of lung diseases using a decision support system by expanding deep learning techniques. *The Open Respiratory Medicine Journal, 18*, e18743064296470. https://doi.org/10.2174/0118743064296470240520075316
Jain, S., et al. (2023). Computational intelligence in cancer diagnostics: A contemporary review of smart phone apps, current problems, and future research potentials. *Diagnostics, 13*(9), 1563. https://doi.org/10.3390/diagnostics13091563
Jiang, F., et al. (2017). Artificial intelligence in healthcare: Past, present and future. *Stroke and Vascular Neurology, 2*(4), e000101. https://doi.org/10.1136/svn-2017-000101
John-Mathews, J.-M. (2022). Some critical and ethical perspectives on the empirical turn of AI interpretability. *Technological Forecasting and Social Change, 174*(C) Available at: https://ideas.repec.org//a/eee/tefoso/v174y2022ics0040162521006429.html. Accessed 21 Nov 2024.
Khanna, S., et al. (2021). AI governance framework for oncology: Ethical, legal, and practical considerations. *Quarterly Journal of Computational Technologies for Healthcare, 6*(8), 1–26.
Koçak, B., et al. (2024). Bias in artificial intelligence for medical imaging: Fundamentals, detection, avoidance, mitigation, challenges, ethics, and prospects. *Diagnostic and Interventional Radiology (Ankara, Turkey)*. [Preprint]. https://doi.org/10.4274/dir.2024.242854
Lokaj, B., et al. (2024). Barriers and facilitators of artificial intelligence conception and implementation for breast imaging diagnosis in clinical practice: A scoping review. *European Radiology, 34*(3), 2096–2109. https://doi.org/10.1007/s00330-023-10181-6

McGuinness, J. E., et al. (2021). Extraction of electronic health record data using fast healthcare interoperability resources for automated breast cancer risk assessment. *AMIA Annual Symposium Proceedings. AMIA Symposium, 2021*, 843–852.

Meyer, J. S., et al. (2005). Breast carcinoma malignancy grading by Bloom-Richardson system vs proliferation index: Reproducibility of grade and advantages of proliferation index. *Modern Pathology, 18*(8), 1067–1078. https://doi.org/10.1038/modpathol.3800388

Mi, H., et al. (2024). Computational methods and biomarker discovery strategies for spatial proteomics: A review in immuno-oncology. *Briefings in Bioinformatics, 25*(5), bbae421. https://doi.org/10.1093/bib/bbae421

Miotto, O., et al. (2020). Emergence of artemisinin-resistant *Plasmodium falciparum* with kelch13 C580Y mutations on the Island of New Guinea. *PLoS Pathogens, 16*(12), e1009133. https://doi.org/10.1371/journal.ppat.1009133

Nartowt, B. J., et al. (2019). Scoring colorectal cancer risk with an artificial neural network based on self-reportable personal health data. *PLoS One, 14*(8), e0221421. https://doi.org/10.1371/journal.pone.0221421

Nema, S., et al. (2021). Digital health care services to control and eliminate malaria in India. *Trends in Parasitology, 37*(2), 96–99. https://doi.org/10.1016/j.pt.2020.11.002

Nema, S., et al. (2022). Strengthening malaria microscopy using artificial intelligence-based approaches in India. *The Lancet Regional Health—Southeast Asia, 5*, 100054. https://doi.org/10.1016/j.lansea.2022.100054

Nema, S., Kumari, M., & Ghosh, S. K. (2024). Automation techniques in infectious diseases. In S. Kumar & A. Kumar (Eds.), *Automated diagnostic techniques in medical microbiology* (pp. 145–160). Springer. https://doi.org/10.1007/978-981-99-9943-9_10

Obermeyer, Z., et al. (2019). Dissecting racial bias in an algorithm used to manage the health of populations. *Science (New York, N.Y.), 366*(6464), 447–453. https://doi.org/10.1126/science.aax2342

Pantanowitz, L., et al. (2018). Twenty years of digital pathology: An overview of the road travelled, what is on the horizon, and the emergence of vendor-neutral archives. *Journal of Pathology Informatics, 9*, 40. https://doi.org/10.4103/jpi.jpi_69_18

Pantanowitz, L., et al. (2020). An artificial intelligence algorithm for prostate cancer diagnosis in whole slide images of core needle biopsies: A blinded clinical validation and deployment study. *The Lancet Digital Health, 2*(8), e407–e416. https://doi.org/10.1016/S2589-7500(20)30159-X

Pettini, F., et al. (2021). Multi-omics model applied to cancer genetics. *International Journal of Molecular Sciences, 22*(11), 5751. https://doi.org/10.3390/ijms22115751

Pulumati, A., et al. (2023). Technological advancements in cancer diagnostics: Improvements and limitations. *Cancer Reports, 6*(2), e1764. https://doi.org/10.1002/cnr2.1764

Rahrooh, A., et al. (2024). Towards a framework for interoperability and reproducibility of predictive models. *Journal of Biomedical Informatics, 149*, 104551. https://doi.org/10.1016/j.jbi.2023.104551

Rai, H. M., et al. (2024). Enhancing data security and privacy in energy applications: Integrating IoT and blockchain technologies. *Heliyon, 10*(19), e38917. https://doi.org/10.1016/j.heliyon.2024.e38917

Ray, A., Chen, M., & Gelogo, Y. (2020). Performance comparison of different machine learning algorithms for risk prediction and diagnosis of breast cancer. In J. Fiaidhi, D. Bhattacharyya, & N. T. Rao (Eds.), *Smart technologies in data science and communication* (pp. 71–76). Springer. https://doi.org/10.1007/978-981-15-2407-3_9

Rulten, S. L., et al. (2023). The future of precision oncology. *International Journal of Molecular Sciences, 24*(16), 12613. https://doi.org/10.3390/ijms241612613

Salwei, M. E., & Reale, C. (2024). Workflow analysis of breast cancer treatment decision-making: Challenges and opportunities for informatics to support patient-centered cancer care. *JAMIA Open, 7*(2), ooae053. https://doi.org/10.1093/jamiaopen/ooae053

Sarhadi, V. K., & Armengol, G. (2022). Molecular biomarkers in cancer. *Biomolecules, 12*(8), 1021. https://doi.org/10.3390/biom12081021

Sebastian, A. M., & Peter, D. (2022). Artificial intelligence in cancer research: Trends, challenges and future directions. *Life, 12*(12), 1991. https://doi.org/10.3390/life12121991

Shah, S. N. A., & Parveen, R. (2023). An extensive review on lung cancer diagnosis using machine learning techniques on radiological data: State-of-the-art and perspectives. *Archives of Computational Methods in Engineering, 30*(8), 4917–4930. https://doi.org/10.1007/s11831-023-09964-3

Shao, D., et al. (2022). Artificial intelligence in clinical research of cancers. *Briefings in Bioinformatics, 23*(1), bbab523. https://doi.org/10.1093/bib/bbab523

Topol, E. J. (2019). High-performance medicine: The convergence of human and artificial intelligence. *Nature Medicine, 25*(1), 44–56. https://doi.org/10.1038/s41591-018-0300-7

Umar, A. (2004). Applications of bioinformatics in cancer detection: A lexicon of bioinformatics terms. *Annals of the New York Academy of Sciences, 1020*, 263–276. https://doi.org/10.1196/annals.1310.021

Wang, X. (2011). Role of clinical bioinformatics in the development of network-based biomarkers. *Journal of Clinical Bioinformatics, 1*(1), 28. https://doi.org/10.1186/2043-9113-1-28

WHO, Global Cancer Observatory. (n.d.). World Health Organization (WHO). Global Cancer Observatory. Available at: https://gco.iarc.fr/. Accessed 21 Nov 2024.

Yang, Q., et al. (2019). Federated machine learning: Concept and applications. *ACM Transactions on Intelligent Systems and Technology, 10*(2), 12:1–12:19. https://doi.org/10.1145/3298981

Statement on the Effectiveness of AI and ML in Cancer Care

Sherlyn Jemimah, Anubhav Gupta,
Sanober Sarfaraz Ahmed, Radhika Khatri, Swathi Murali,
and Neeru Sood

1 Introduction

A significant global challenge indicated by the WHO's Global Cancer Observatory projects expected cancer incidence to increase from 20 to 27 million cases by 2035. Cancer mortality is also anticipated to rise from 9.74 to 13.6 million during the same period (Bray et al., 2024). The situation is further concerning as there is an increase in early-onset cancers, particularly among adults under the age of 50. This increasing incidence poses unique diagnostic, therapeutic, and psychosocial

S. Jemimah
Department of Biology, Chemistry and Environmental Sciences, American University of Sharjah, Sharjah, UAE
e-mail: sjemimah@aus.edu

A. Gupta
Department of Computer Science, Mathematics, Physics and Statistics, University of British Colombia, Okanagan Campus, Kelowna, BC, Canada
e-mail: anubhav.gupta@ubc.ca

S. S. Ahmed · R. Khatri
Department of Computer Science, Birla Institute of Technology and Science Pilani, Dubai Campus, Dubai, UAE
e-mail: f20220116@dubai.bits-pilani.ac.in; f20220089@dubai.bits-pilani.ac.in

S. Murali
Department of Forensic Science, College of Criminal Justice, Sam Houston State University, Huntsville, TX, USA
e-mail: sxm223@shsu.edu

N. Sood (✉)
Department of Biotechnology, Birla Institute of Technology and Science Pilani, Dubai Campus, Dubai, UAE
e-mail: sood@dubai.bits-pilani.ac.in

R. Nema et al. (eds.), *Advances in Cancer Detection, Prediction, and Prognosis Using Artificial Intelligence and Machine Learning*,
https://doi.org/10.1007/978-981-96-9346-7_15

challenges (Chen et al., 2023). The predicted global cost of cancer from 2020 to 2050 is US$ 25.2 trillion (Ugai, 2022), which raises concerns about the economic impact in addition to the human cost.

There is a critical need for transformative approaches to cancer care that address every aspect of a patients' journey. This need demands for collaborative efforts from oncologists, data scientists, biomedical engineers, and other multidisciplinary experts to find innovative solutions (World Health Organization, 2022; Committee on Improving the Quality of Cancer Care: Addressing the Challenges of an Aging Population; Board on Health Care Services; Institute of Medicine & Levit, 2013). Due to this reason, artificial intelligence (AI) and machine learning (ML) have emerged as powerful tools capable of advancing and optimizing cancer care. AI and ML technologies have the potential to improve outcomes throughout the cancer care continuum by utilizing their ability to process large and complicated datasets.

This chapter discusses the current state of cancer care, highlighting how AI/ML technologies can be integrated to improve screening, diagnostics, treatment personalization, and prognostic predictions. It further highlights the challenges of implementing these technologies and discusses the future trends.

1.1 Overview of Cancer Care

1.1.1 Screening

The aim of cancer screening is to detect precancer or early-stage cancer in asymptomatic individuals. Cancer screening constitutes an important defense against cancer as it reduces mortality and morbidity (Siddiqui & Cruz, 2019). For example, a population-based screening program for cervical cancer, called Slovenia's ZORA, has led to reduced incidence. However, its implementation can pose some challenges, a key impediment being the sheer cost of organized screening. Moreover, the reporting of false negatives (resulting in delayed diagnosis) and false positives (which puts people through unnecessary investigations and strains the health system) cannot be ruled out (Siddiqui & Cruz, 2019).

Screening programs for cervical, breast, and colorectal cancer have been shown to provide tangible benefits in terms of reduced mortality (Gini et al., 2020; Zielonke et al., 2020; Jansen et al., 2020). The WHO recommends screening programs for cervical, breast, and colorectal cancers due to their effectiveness and relatively lower costs (World Health Organization, 2017). In such cases, a targeted approach, vis-à-vis polygenic risk scores (PRS) would enable stratified cancer screening for a broader selection of cancers while keeping costs low (Huntley et al., 2023).

1.1.2 Diagnosis

Early diagnosis is highly beneficial for cancer patients, resulting in lower mortality, higher treatment efficacy, and improved quality of life (Adashek et al., 2021; Arora et al., 2024). Currently, tissue biopsies and imaging (CT or MRI) are considered the gold standard in cancer diagnostics and are considered essential to overcome the low specificity of screening and symptom-based triage (Arora et al., 2024). Early diagnostics would be particularly beneficial for "silent" cancers such as ovarian cancers, which are marked by high mortality and recurrence rates due to delayed diagnosis (Connal et al., 2023).

The recent increase in noninvasive molecular diagnostic platforms based on liquid biopsy techniques, such as CellSearch, ISET, CancerSEEK, and VeriStrat, has shown potential to transform clinical practice (Connal et al., 2023). By extracting valuable omics data and enabling longitudinal monitoring of tumors under treatment, molecular diagnostic methods are set to usher in a new era in cancer management (Wu et al., 2018).

1.1.3 Treatment

Cancer treatment modalities include surgical resection (for localized solid tumors), chemotherapy, radiotherapy, and additional neoadjuvant/post-adjuvant administration. Treatment selection and intensity is predicated on available diagnostic information (including patient history, symptom reporting, imaging, and pathology information about cancer type, grade, and stage) (Liu et al., 2021).

The range of treatment options for cancer has expanded in recent years, based on a growing understanding of the hallmarks of cancer. Tyrosine kinase inhibitors (TKIs) such as imatinib and erlotinib are designed to target uncontrolled cancer growth (Hartmann et al., 2009). Vascular endothelial growth factor (VEGF) signaling inhibitors such as bevacizumab act against angiogenesis, preventing the formation of new blood vessels that feed tumor growth (Liu et al., 2023). Poly ADP-ribose polymerase (PARP) inhibitors such as Olaparib work against DNA repair pathways, cutting off a potent survival mechanism for certain cancers (Bochum et al., 2018).

In recent years, a recognition of the immune system's role in eradicating cancer has facilitated the development of immunotherapies such as chimeric antigen receptor T-cell (CAR-T) therapy and immune checkpoint inhibitor (ICI) therapy (Banerjee et al., 2024). CAR-T cell therapy has shown improved remission rates of up to 50% in refractory hematological malignancies. ICIs have also proven beneficial in multiple cancer types, showing significantly increased 5-year overall survival (OS) in lung cancer, reduced recurrence in melanoma, and superior efficacy in recurrent, metastatic gynecological cancers (Banerjee et al., 2024). A few therapeutic cancer vaccines have been accorded FDA approval, namely Sipuleucel-T (for metastatic prostate cancer) (Cheever & Higano, 2011), Bacillus-Calmet-Guérin (BCG; for early-stage bladder cancer) (Dutcher & Bilen, 2018), and talimogene laherparepvec (T-VEC; for advanced melanoma) (Nassief et al., 2024).

Although progress in cancer treatments has shown remarkable progress in recent years (especially for advanced, refractory, and metastatic cancers), more research is required to address treatment resistance, the role of the tumor microenvironment in immune evasion, and poor prognosis in certain cancers such as multiple myeloma (Banerjee et al., 2024). The effect of adverse events on normal growth and development in pediatric cancer patients is also a significant concern (Ruggiero et al., 2021).

1.1.4 Prognosis

Patient prognosis is primarily determined using prior knowledge of cancer survival trends. Factors such as patient demographics, presence of comorbidities, and cancer staging can affect prognosis significantly (Mariotto et al., 2014). Advances in treatment and early-stage detection has improved prognosis significantly for most cancers (Santucci et al., 2020; Forjaz et al., 2023). More recently, clinicians have started supplementing conventional prognostic measures such as overall survival (OS) and progression-free survival (PFS) (Delgado & Guddati, 2021), with patient-reported outcomes (PROs) such as functioning and quality of life (Smith et al., 2016). There is also an increasing emphasis on holistic, evidence-based cancer care models that integrate palliative care (Parikh et al., 2013) and patient education (Fessele et al., 2014). Palliative care integration in early stages has resulted in better patient outcomes and cost savings (Parikh et al., 2013).

1.2 Role of AI and ML in Healthcare

The term "artificial intelligence" was coined in 1956 and refers to the development of computer systems capable of mimicking human intelligence. Machine learning is a subset of AI and refers to the ability of computers to learn from data without being explicitly programmed. The ability to parse massive quantities of multiple data types, learn underlying patterns, and offer insights that can aid problem-solving and decision-making has made AI/ML an attractive, cost-effective platform for clinical oncology (Huang et al., 2020). Factors that affect the performance of AI/ML models include the quality and quantity of data in model training and validation, method selection, and the judicious use of optimization techniques. Multimodal models that integrate electronic health records (EHRs), histopathology, and omics data consistently outperform unimodal approaches in cancer diagnosis and prognosis (Abdullakutty et al., 2024; Exarchos et al., 2012; Cui et al., 2021). Despite the proliferation of AI/ML models proposed in oncology literature, few have been validated in clinical practice. Several clinical trials have been initiated to understand the benefits of AI-driven screening, diagnostics, prognosis prediction, and intervention regimen design in breast, pancreatic, lung, ovarian, bladder, and prostate cancers (Senthil Kumar et al., 2023). In subsequent sections, we explore the status of AI/ML models in clinical oncology as well as future trends.

1.3 Objective of the Chapter

Despite significant progress in our understanding of cancer biology and the expansion in available treatment options, there remain several barriers to effective cancer care, including disease complexity, rising treatment costs, and a growing demand for cancer care that threatens to outpace available resources (World Health Organization, 2022; Committee on Improving the Quality of Cancer Care: Addressing the Challenges of an Aging Population; Board on Health Care Services;

Institute of Medicine & Levit, 2013). These barriers have highlighted the importance and need to incorporate artificial intelligence (AI) and machine learning (ML) technology in all aspects of cancer care. In this chapter, we discuss the application of AI/ML in cancer detection, treatment, and patient outcomes, evaluate the benefits of incorporating AI/ML in clinical practice along with the future trends.

2 AI and ML in Cancer Detection and Diagnosis

The current landscape of cancer care has been greatly influenced by AI and ML, especially detection and diagnosis of various types of cancers (Dlamini et al., 2020). These learning-based approaches have shown great potential in the early detection of the disease, precise diagnosis, and accelerated treatment procedures (Sebastian & Peter, 2022). This section reviews the role of AI and ML for the early detection of cancer by discussing some important case studies that have driven advancements in diagnosis. Subsequently, this section also details some of the challenges and ethical concerns in the integration of AI in oncology.

2.1 Early Detection

2.1.1 Imaging Analysis and Pattern Recognition

AI and ML technologies have had a very big impact in the cancer imaging domain. Deep learning algorithms in radiology reason and draw insights from complex medical images, including CT scans, MRIs, and mammograms (Koh et al., 2022). The ability of AI to identify patterns and features in such images that are not immediately apparent to the human eye allows for improved and optimized diagnostics at an early stage (Li et al., 2022). For example, Google's DeepMind AI tool for the detection of breast cancer has shown a substantial reduction in both false positives and false negatives compared to human radiologists. In a study published in the *Nature* in 2020, it was indicated that there was a reduction of 5.7% in false positives and 9.4% in false negatives in the United States (McKinney et al., 2020; Google Health, 2021).

2.1.2 Biomarker Identification

AI has also been used to the identify biomarkers (biologic indicators capable of signaling the presence of cancerous cells) (Bhinder et al., 2021). This is a growing area of research largely based on liquid biopsies, which test for the presence of cancerous cells or DNA through blood samples. AI algorithms can sort large volumes of complex data and identify relevant biomarkers that might enable early detection in lung, breast, and prostate cancers (Zakareya et al., 2023; Khalid et al., 2021). For example, recent works have highlighted that AI can effectively detect circulating tumor cells (CTCs) and cell-free DNA (cfDNA) with a high degree of precision (Zakareya et al., 2023).

One of the best examples of an AI-based tool for cancer early diagnosis is PathAI (https://fischerjordan.com/2024/03/impact-of-ai-in-early-cancer-detection-in-india/). This platform applies AI to microscopical analysis so as to enable pathologists identify cancers consistently and more quickly than traditional methods (Zakareya et al., 2023). According to PathAI, AI models developed within PathAI support detailed visualizations of features that provide a rationale for their diagnostic predictions, which enables transparency and verification by human pathologists (PathAI, 2023). PathAI has been used in clinical trials for the classification of breast and lung cancers with high accuracy (Khalid et al., 2021; PathAI, 2023).

2.2 Accurate Diagnosis

2.2.1 Radiology and Pathology

Machine learning algorithms have shown great promise for application in radiology and pathology. These methods enable the analysis of intricate patterns within medical images and tissue samples to help physicians make more accurate diagnoses (Hosny et al., 2018). For example, a record high accuracy of 93% was achieved by a deep learning model developed recently for classifying breast cancer histopathological images (Zakareya et al., 2023). Applications of AI in pathology, such as the AIM-HER2 in breast cancer algorithm by PathAI, will revolutionize how HER2-positive tumors are scored by improving diagnostic speed and consistency (PathAI, 2023).

2.2.2 Genomics and Personalized Medicine

AI has had a particularly strong impact on genomics, where machine learning models have been applied on large-scale genetic data to investigate and predict mutations associated with several kinds of cancers (Vilhekar & Rawekar, 2024). Previously, these could not be revealed by traditional diagnostic approaches. The role of AI-driven genomic analysis is vital in developing personalized medicine strategies since it allows doctors to tailor treatment to the genetic makeup of a patient (Zakareya et al., 2023; Khalid et al., 2021). For instance, there was the development of a deep learning model capable of identifying melanoma-related mutations, which was shown to be extremely important in tailoring individual treatment regimens (Zakareya et al., 2023).

Several case studies have demonstrated the potential of AI in improving diagnosis. In one such study, AI demonstrated a sensitivity of over 90% in detecting cancerous nodules associated with lung cancer on CT scans, outperforming human radiologists on occasion (Zakareya et al., 2023). In another case, an AI tool developed by Google Health for detecting breast cancer was found to be on par with expert radiologists' accuracy in diagnosis (Google Health, 2021).

2.3 Effectiveness and Limitations

2.3.1 Enhancing Diagnostic Capabilities

Different kinds of cancers demonstrated their diagnostic capabilities with the power of AI, which allows more speed and accuracy. AI forms part of the workflow in breast cancer screening to help radiologists identify consistently a high-risk patient (Ahn et al., 2023). It was noted from studies that AI-powered systems reduced the workload of radiologists while maintaining high accuracy levels and, therefore became very important in resource-constrained environmental conditions (Google Health, 2021). More specifically, the AI systems can help fill the gap in diagnosis in places where specialists are not available (Google Health, 2021; Zakareya et al., 2023).

2.3.2 Limitations

AI application in cancer diagnosis has faced several limitations that limit the complete integration of these systems into cancer care. There are instances of data bias when the AI models are being trained on unrepresentative datasets of the diverse patient populations they will see in the clinic (Kolla & Parikh, 2024). AI tools majorly trained on data from Caucasian patients will perform less well on patients from other ethnic groups (McKinney et al., 2020; Google Health, 2021). Another huge concern is the rate at which false positives and false negatives occur, which may lead to unnecessary treatments or diagnoses that have been missed. While AI has improved immensely in both areas, it is by no means completely devoid of such errors (Khalid et al., 2021).

Besides, the effectiveness of AI models is highly dependent on quality and size of data they have been trained against (Vela et al., 2022). More often than not, generalization of AI models is limited because there is a lack of large, annotated datasets. This is particularly pressing in developing countries, where health data can often be sparse or hard to access (Khalid et al., 2021). Other challenges include regulatory and ethical concerns with the use of AI in clinical settings that need to be taken on board before this technology is incorporated into everyday routine (Google Health, 2021).

3 AI and ML in Treatment Decision-Making

Artificial intelligence and machine learning in decision-making for cancer has made significant advancements over time (Dlamini et al., 2020). The use of AI-based technique enhances personalized treatment plans, speeds up drug discoveries, and opens up new avenues for treating difficult cancer types (Sebastian & Peter, 2022). AI can recommend better suited treatments by analyzing patient data that includes clinical history and genetic information (Schork, 2019). In addition, AI efficiently speeds up the identification of new drugs (Alsharif, 2024). Yet, there are certain challenges that persist and need to be addressed. Clinical integration along with regulatory hurdles and physician trust in AI recommendations remain as the top challenges till date. This review covers the role AI plays in personalized treatment, drug discovery, and the real-world challenges associated with these adoptions.

3.1 Personalized Treatment

3.1.1 AI-Driven Personalized Treatment Plans

IBM Watson for Oncology is a popular tool for personalized cancer treatment recommendations. Treatment recommendation is achieved by analyzing patient through machine learning approaches. IBM Watson bases its evidence-based treatment suggestions on clinical trials, medical literature, and patient records. Research has shown that, in most cases, Watson's recommendations tally with that of oncologists, hence justifying it as a valuable decision support tool (Liu et al., 2018).

3.1.2 Success Stories in Matching Patients with Effective Treatments

AI systems have the ability to match patients with effective treatments, ensuring better clinical outcomes (Farina et al., 2022). Tempus is an AI-powered precision medicine platform that can tailor cancer treatments according to an individual's genetic profile (Sebastian & Peter, 2022). Tempus uses changes in tumors to find therapies for patients in the advanced stages of various forms of cancers (PathAI, 2023).

AI has also been reportedly used in tailoring chemotherapy for patients with breast cancer. The use of AI can help clinicians to identify an optimal combination of drugs for a patient based on genetic data. This helps in reducing the number of side effects while improving overall survival rates. These instances reflect how AI can make treatment potential higher while minimizing risks (Ahn et al., 2023).

3.2 Drug Discovery and Development

3.2.1 AI in Accelerating Drug Discovery

AI is revolutionizing not only the personalization of cancer treatments but also the development of new therapies. Traditional drug discovery is usually very time-consuming and expensive; it takes more than a decade, with billions of dollars involved in bringing a new drug into the market (Weerarathna et al., 2023). AI thus has the potential to streamline this process by analyzing large datasets at rapid speed for the identification of promising candidates for drugs.

AI-driven models can also predict how different kinds of compounds will interact with biological targets, allowing researchers to narrow their focus to the most promising candidates much earlier in the process (Dhudum et al., 2024). This is becoming quite relevant in cancer, where new therapies are under constant consideration to address, especially aggressive and treatment-resistant types of the disease. AI will also contribute to drug repurposing by finding new uses for already-developed drugs according to their molecular profile (Kulkarni et al., 2022).

3.2.2 Case Study: AI in Identifying Treatments for Complex Cancer Types

One example of such revolutionary use of AI is the development of drugs to find potential treatments for prostate cancer (Sumanasuriya & De Bono, 2018). Advanced prostate cancer remains one of the most challenging stages for effective treatment. With the use of AI, large volumes of information generated from patients with prostate cancer were analyzed by integrating genetic information and treatment outcomes to identify new drug combinations (Chu et al., 2023). While more research is required in this domain, the prospects of AI uncovering new treatments for prostate cancer, among other complicated cancers, are very encouraging.

In another example, AI models were used to predict the efficiency of various combinations of drug treatments in glioblastoma, an aggressive form of brain cancer. The models suggested a number of novel therapies that were later validated in clinical trials, hence underlining AI's capability in accelerating new treatment options (Khalighi et al., 2024).

3.2.3 Adoption Barriers: Clinical Integration and Physician Trust

Despite the many possibilities opened by AI in cancer treatment, several clinical obstacles to the implementation of AI in real-world clinical settings remain. First is the clinical integration of AI systems into the workflow of most healthcare institutions, which up to now remains extremely hesitant due to a variety of concerns regarding the complexity and reliability of such technologies. This may be even more difficult in smaller hospitals due to financial and infrastructure requirements brought about by the implementation of AI platforms (Ahmed et al., 2023).

Another significant challenge is gaining physicians' trust in AI recommendations, which are intended to simplify cancer treatment (Emani et al., 2022). In 2022, oncologists overwhelmingly demonstrated that even as clinicians recognize the potential advantage of AI, there is considerable skepticism toward using AI systems for making critical treatment decisions (Ahmed et al., 2023). Such skepticism is often driven by questions about the "black-box" nature of AI algorithms, wherein the underlying logic behind a recommendation may well not be transparent (Wanner et al., 2022).

3.2.4 Regulatory Issues

The regulatory landscape is continuously evolving for AI in healthcare. Regulatory agencies, such as the U.S. Food and Drug Administration (FDA), actively develop guidelines with respect to the evaluation of AI-driven medical devices and software. However, most of the AI tools pertaining to customized treatment and drug discovery fall into the category of "software as a medical device" that must be intensely validated and approved before deployment in any clinical setting (Palaniappan et al., 2024).

The FDA authorizes the first AI-based software to detect prostate cancer in 2021. However, much concern about the standardization of AI systems through various healthcare providers lingers on, not forgetting ethical considerations—those touching on patient privacy and the use of sensitive health data in training AI models

(Shafi & Parwani, 2023). While AI technologies are continuously evolving, regulatory bodies must come forward with clearer guidelines that guarantee safety and efficacy in the use of AI in cancer treatments.

AI and ML are two such emerging sciences that have transformed the face of decision-making in cancer treatment planning through personalization and quickening drug development. AI-driven systems such as IBM Watson for Oncology and Tempus have unraveled newer and more efficacious treatments for patients by matching genetic and clinical data. AI has helped researchers in finding newer ways for treating complicated cancers such as prostate cancer and glioblastoma during the process of drug development. Yet, the barriers to broad-based adoption of AI in healthcare include formidable challenges such as clinical integration, physician trust, and regulatory approval. Indeed, future efforts should be directed at addressing such challenges through technological advances in AI, transparently framed regulatory mechanisms, and physician education in order to tap the full power of AI in oncology.

4 AI and ML in Prognosis and Monitoring

4.1 Predictive Analytics for Outcomes

Machine learning (ML) and artificial intelligence (AI) have been widely used for the prediction of cancer outcomes and their ongoing monitoring. This section details and analyzes current research work based around predictive analytics of cancer prognoses using AI-based techniques.

A breast cancer recurrence estimation model was developed in one research project (Zuo et al., 2023). The framework reported superior performance compared to traditional techniques. The researchers applied an ensemble of several algorithms to improve prediction accuracy (Yang et al., 2021).

A proposed research work investigates the survival prediction rate within patients suffering from lung cancer survival (Yang et al., 2022). The researchers used support vector machines and neural networks to enhance the predictive accuracy. Another recent study illustrates the use of Multi-Modal Radiomics alongside Deep CNNs for the recognition of breast cancer (Venkatasunilsrikanth & Krithiga, 2023). This technique reportedly improves diagnostic precision and reliability. A third researcher validated this approach and modified it further to differentiate between malignant and benign skin lesions in those affected by melanoma (Almufareh et al., 2024).

4.2 Patient Monitoring and Management in Cancer Care

Integration of artificial intelligence (AI) in wearables with remote health services has helped track and keep a continuous watch over cancer patients. These wearables allow the tracking and recording of a patient's physiological parameters in real time.

In cancer care, small devices that incorporate learning-based approaches track vital signs such as heart rate, body temperature, and oxygen saturation. The real-time tracking and record keeping enables quicker interventions and care that is customized to personal needs.

Remote monitoring AI-driven systems have been used successfully in controlling chemotherapy side effects. This allows healthcare professionals to actively keep track of patient parameters and respond quickly to harmful reactions. An important case is to use AI to gather information from wearables to predict neutropenia in patients receiving chemotherapy. With this technology, clinicians can change treatments or hospitalize patients based on their discovery of severe symptoms, which raises survival rates (Breen et al., 2017). As a result, a strong push exists for wearable devices as potential assistance for remote clinical trial monitoring in cancers and other diseases, which are overwhelmingly economical (Beg et al., 2022).

Wearable device data can play a crucial role for the achievements of treatment assessment, prognosis, and rehabilitation. Upcoming investigations should emphasize the live tracking of collected data, which may lead to directed interventions (Chow et al., 2024). Handling extensive patient data is the location where AI is critically significant. AI systems provide efficiency with high precision to handle patient record organization, categorization, and processing, as well as appointments and treatment histories. Healthcare providers can more easily access and analyze patient data, thanks to systems that can manage unstructured data, which includes doctor's notes. To project disease, a study harnessed AI along with natural language processing (NLP) to analyze electronic medical records (EMRs), focusing on the unformatted notes from consultations. Implemented alongside a method for matching notes to a medical ontology are techniques that include bag of words and topic modeling. This method was put to the exam by colorectal cancer. Results from the analysis showed that the method relying on ontology remarkably enhanced predictive performance that was better than customary criteria. By extracting beneficial information from unstructured data found in EMR, AI demonstrates its capability to increase the accuracy of disease prediction (Maleki Varnosfaderani & Forouzanfar, 2024).

Collectively, AI-integrated wearables along with remote health platforms are innovating cancer care through the capability for persistent tracking of patients in real time. Increasing the accuracy of early complication detection and allowing personalized treatment modifications, these technologies may significantly enhance patient results and the quality of overall treatment. Integrating AI further into cancer monitoring seems likely to accelerate field advancement, resulting in improved systems for efficiently handling this complex condition.

4.3 Effectiveness in Real-Time Adjustments in Cancer Treatment

The introduction of artificial intelligence (AI) into cancer care has greatly enhanced the potential for making real-time modifications to treatment approaches based on perceived data in real time. Software powered by AI is now essential for furnishing insights based on data that allow oncologists to react in a timely way to changes in the health of patients, leading to better cancer treatment results. An actionable strategy focused on persistent data collection and analysis personalizes treatments for patients, producing improved outcomes and probably lessening counter effects.

At this point, several important developments in AI for oncology are happening within the sector focused on cancer imaging, with a particular emphasis on digital pathology, radiographic imaging, and clinical photographs. In digital pathology, AI concentration is on fundamental and advanced image processing alongside image classification (including tumor detection and segmentation along with the prediction of disease diagnosis and responses to treatment based on image patterns), intended to automate demanding jobs and raise the reliability of diagnostic outcomes. Findings from radiological studies indicate that AI tools can effectively distinguish lesions of high- versus low-risk on a selection of imaging methods. Right now, the combination of radiographic imaging with other sources of data (such as clinical features, genetic and biochemical markers) for risk stratification of lesions identified through imaging is available and will probably increase in the short term. AI is becoming a tool to support greater diagnostic precision and to minimize diagnostic ambiguities found in dermatologic and gastrointestinal cancers seen on imaging exams (Xu et al., 2021).

AI radiology is also about more than finding abnormalities; it includes quantifying disease progression, assessing response to therapy, and also predicting patient results. Because of the artificial intelligence, measures of the size and growth of tumors through time provide relevant data for creating treatment approaches (Weerarathna et al., 2023).

The main goal of immunotherapy is for artificial intelligence to sift through genetic and imaging data to identify the best candidates for treatment and to anticipate possible responses. Significant improvements in AI largely assist in refining targeted therapy in the specific identification of molecular targets for individualized treatment. In surgical procedures, AI supplies momentary understanding and support that heightens the accuracy of surgical operations. As well, the synergy between AI and nanotechnology suggests that personalized nanomedicines for cancer treatment are on the horizon, with both improved efficiency and enhanced targeting (Chua et al., 2021).

At its core, the effective exploitation of AI for real-time changes in cancer treatment has improved patient care, permitting therapies to be more personalized, react to patients better, and change as needed. AI guarantees that both treatments are effective and safe, while also delivering superior results to cancer patients through its ongoing handling of patient data and prompt advice.

5 Challenges and Ethical Considerations

Artificial intelligence has undoubtedly created a landmark shift in the practice of medicine, particularly in cancer care. As discussed in previous sections of this book, advancements in AI models for cancer detection, diagnosis, prognosis, and treatment have significantly transformed the medical landscape in the last decade. Consequently, it becomes imperative to evaluate some of the challenges and ethical concerns that emerge in the wake of such rapidly advancing technology. In this regard, three main attributes will be discussed in this section.

5.1 Bias in AI Models

An AI model, despite its impeccable accuracy, is only as good as the data underlying the model. From this perspective, an AI model can pose significant risk in perpetuating bias if the training data is not sufficiently representative of the relevant population. In recent times, these concerns have been repeatedly echoed by several researchers (Farasati Far, 2023; Naik et al., 2022; Napitupulu, 2023). While this may seem like a straightforward problem with an easy solution, that is to increase the size and diversity of the training data, one must consider several socioeconomic factors that may impose limitations. For instance, to make a database more representative, one may attempt to include participants from various ethnic groups. However, if the groups do not have equal access to healthcare, selection bias is inadvertently introduced. Another aspect is the choice of an appropriate dataset to address question(s) of interest. For example, African Americans are known to be disproportionately more prone to prostate cancer than any other racial group. If one were to diagnose a disease for an African American using an AI model, it is critical to first evaluate if it would be more appropriate to use a generally diverse dataset or a dataset largely based on African Americans. One can always argue that a diverse, representative dataset would be the appropriate choice as it is more generalizable and (potentially) robust. However, this would be an oversimplification. Examining this context more closely, elevated prostate-specific antigen (PSA) level is a commonly used blood biomarker to screen for prostate cancer. If one were to use a diverse dataset to predict prostate cancer in an African American, it is more likely to classify them as "positive." This is because African Americans tend to have naturally higher baseline levels of PSA compared to other racial groups. In this case, a dataset populated based on African Americans may be more suitable. While this is a very specific example, the main idea is that outcomes from an AI model and its implications will depend very heavily on the nature and quality of the training data. Although data bias is the most eminent challenge with AI, other forms of bias can be introduced through developers, either consciously or unconsciously, in terms of selecting, treating, or processing data, or through the assumptions underlying the algorithm itself. Similar bias associated with human factors have also been discussed in recent studies on the ethical challenges surrounding the application of AI for cancer care (Naik et al., 2022; Napitupulu, 2023). Such issues can be minimized, if not mitigated, by testing models on different groups or use cases, as well as developing transparent algorithms that are easier to interpret and allow users to understand how decisions are made.

5.2 Patient Consent and Data Privacy

Informed consent is a nonnegotiable part of any form of medical intervention. Patients must be notified and must consent to the use of AI in assisting with medical decisions. The important consideration here is "informed" consent. Given the "black-box" nature of most algorithms, physicians may understand how to interpret the outcome of an AI model, but not how it was generated. This raises two major issues—one, if physicians themselves do not sufficiently understand how a model makes decisions, how effective would they be in explaining the process to a patient and obtaining a truly "informed" consent? The second issue ties back to the problem of bias, specifically in relation to human factors. For instance, if a physician is provided a diagnosis for a disease by an algorithm; say, the algorithm predicts that there is a 70% chance of an individual having prostate cancer, this may introduce bias in the physician. Such subconscious biases introduce a risk of physicians "cherry-picking" data in alignment with the model's prediction. While the intention here is not to question the credibility of a medical practitioner, minimizing human bias requires significant and conscious effort. This becomes even more challenging when AI models are not transparent, not interpretable, and when there is an uncontested confidence in these models. The idea here is not to eliminate the use of AI in healthcare, but to encourage practitioners to be mindful of the biases these models may introduce and to critically evaluate the model predictions in comparison to clinical findings.

Further, data privacy is becoming an increasingly important topic in all discussions pertaining to AI. Although not directly related to healthcare, genealogy databases are a notable example of data privacy issues (Berkman et al., 2018). Ancestry testing has gained immense popularity in the past few years with the advent of companies such as 23andMe, AncestryDNA, MyHeritage, and so on. While this technology has helped many people discover their heritage and ancestral roots, there have been several ethical questions raised about the use of this data by law enforcement agencies to identify criminals. While users consent to their data being accessible to test providers, they have no say in how this data is being used. In fact, an infamous serial killer in the United States named the "Golden State Killer" was identified through such genealogy databases. By uploading the data (DNA profile) from the crime scene to one such database, the police were able to identify distant relatives and through a series of searches, identify the suspect. While it is a remarkable feat, this incident brought to the forefront issues surrounding privacy of personal data and the lack of regulation surrounding its usage. In this particular instance, participants had not consented to the use of their personal information by law enforcement for criminal investigation. Such issues are also relevant to the use of clinical data to train AI models. Currently, there are no clear laws dictating how or what kind of patient information can be used, mandated safeguards for data security, and consequences of a privacy breach. Until sufficient clarity is reached on some of these aspects, information safety remains a major concern.

5.3 Interpretability and Trust

As previously mentioned, most algorithms are "black-box" in nature, in that their internal workings are not clearly understood or interpretable. As Naik et al. described, this lack of transparency poses the greatest challenge to lawmakers and practitioners on the integration of AI into healthcare (Naik et al., 2022). This raises two major concerns with their use in healthcare—one, not being able to sufficiently understand how a model works and/or how decisions are made creates a sense of distrust among patients and/or physicians, thereby preventing the full potential of an algorithm from being realized. In this situation, an incredibly powerful resource will be completely underutilized. On the contrary, this lack of understanding may also prevent physicians from being able to critically evaluate inconsistencies between model prediction and clinical findings. In addition to these challenges, the "black-box" nature of AI models also raises questions about its robustness and implications in truly improving medical intervention—for instance, would the model predict differently if more information was provided or if database size was expanded? If a medical practitioner cannot understand what factor contributed to the diagnosis, how would they identify (and treat) the root cause? By raising such questions, the motive is not to discourage the integration of artificial intelligence into the healthcare system, but rather, to foster its usage while prompting caution about its unrestricted, undisputed use.

Finally, as most algorithms make probabilistic predictions, this discussion brings to mind the contrast between subjective and objective probabilities algorithms. While an algorithm is purely mathematical and yields an objective probability, a medical practitioner is human and is equipped with years of education, experience, intuition, and an opportunity to physically evaluate the patient (Cadario et al., 2021). At this juncture, we are not looking to determine which type of assessment is better; rather, we aim to offer readers some food for thought!

6 Future Directions and Innovations

The next generation of AI models, such as those incorporating federated learning, will revolutionize cancer research by enabling collaborative learning across multiple institutions while preserving patient privacy. One of the crucial requirements of healthcare research is to keep patient information private and secure. Federated learning allows researchers to utilize diverse datasets while addressing this critical requirement (Chowdhury et al., 2021).

There have been recent developments in the field of explainable AI (XAI). XAI enables physicians and patients to better understand AI models promoting trust and validating AI-based decision-making. XAI techniques such as SHapley Additive exPlanations (SHAP) and Local Interpretable Model-Agnostic Explanations (LIME) have brought to light transparency in AI-augmented decision-making, helping clinicians understand the reasoning behind AI predictions. XAI also enables researchers to identify and remove sources of bias from training data (Ghasemi et al., 2024).

Finally, hybrid models help combine different AI techniques and data sources. These models can integrate various data types including imaging, genetics, and clinical data to provide a comprehensive view of the patient's health. Hybrid models are poised to improve diagnostic accuracy enabling better patient outcomes (Angelique Stephanou et al., 2018).

7 Conclusion

Artificial intelligence (AI) and machine learning (ML) have transformed cancer care. The use of learning-based approaches has offered innovative and efficient solutions in cancer screening, diagnosis, treatment, and prognosis. These techniques have enhanced patient outcomes and optimized resource utilization as they can analyze vast, multimodal datasets. However, there is still a long way to go as there are significant challenges that need to be addressed in this domain. There is a realized need to reduce biases in datasets, address ethical concerns surrounding privacy, and adopt explainable and trustworthy models.

Cancer care research has positively benefitted from the adoption of learning-based approaches. Designing innovative and equitable AI systems requires interdisciplinary collaboration among oncologists, data scientists, and policymakers. Patient-centered care in oncology has been prioritized by advancements in explainable AI, federated learning, and hybrid modeling. Increased interest and effort within the research community can pave the way for a future where cancer is managed more effectively (Shao et al., 2022).

References

Abdullakutty, F., Akbari, Y., Al-Maadeed, S., Bouridane, A., Talaat, I. M., & Hamoudi, R. (2024). Histopathology in focus: A review on explainable multi-modal approaches for breast cancer diagnosis. *Frontiers in Medicine (Lausanne), 11*, 1450103. https://doi.org/10.3389/fmed.2024.1450103

Adashek, J. J., Janku, F., & Kurzrock, R. (2021). Signed in blood: Circulating tumor DNA in cancer diagnosis, treatment and screening. *Cancers (Basel), 13*(14), 3600. https://doi.org/10.3390/cancers13143600

Ahmed, M. I., Spooner, B., Isherwood, J., Lane, M., Orrock, E., & Dennison, A. (2023). A systematic review of the barriers to the implementation of artificial intelligence in healthcare. *Cureus, 15*(10), e46454. https://doi.org/10.7759/cureus.46454

Ahn, J. S., Shin, S., Yang, S. A., Park, E. K., Kim, K. H., Cho, S. I., Ock, C. Y., & Kim, S. (2023). Artificial intelligence in breast cancer diagnosis and personalized medicine. *Journal of Breast Cancer, 26*(5), 405.

Almufareh, M. F., Tariq, N., Humayun, M., & Khan, F. A. (2024). Melanoma identification and classification model based on fine-tuned convolutional neural network. *DIGITAL HEALTH, 10*, 20552076241253757. https://doi.org/10.1177/20552076241253757

Alsharif, F. (2024). Artificial intelligence in oncology: Applications, challenges and future frontiers. *International Journal of Pharmaceutical Investigation, 14*(3), 647–656.

Angelique Stephanou, A., Ballet, P., & Powathil, G. (2018). Hybrid modelling in oncology: Successes, challenges and hopes. *arXiv*; 1901.05652v1 [q-bio.QM]. Retrieved from https://arxiv.org/pdf/1901.05652

Arora, T., Mullangi, S., Vadakekut, E. S., & Lekkala, M. R. (2024). Epithelial ovarian cancer. In *StatPearls [Internet]*. StatPearls Publishing.

Banerjee, S., et al. (2024). Two decades of advances in clinical oncology—Lessons learned and future directions. *Nature Reviews. Clinical Oncology, 21*(11), 771–780. https://doi.org/10.1038/s41571-024-00945-4

Beg, S., Handa, M., Shukla, R., Rahman, M., Almalki, W. H., Afzal, O., & Altamimi, A. S. A. (2022). Wearable smart devices in cancer diagnosis and remote clinical trial monitoring: Transforming the healthcare applications. *Drug Discovery Today, 27*(10), 103314. https://doi.org/10.1016/j.drudis.2022.06.014

Berkman, B. E., Miller, W. K., & Grady, C. (2018). Is it ethical to use genealogy data to solve crimes? *Annals of Internal Medicine, 169*(5), 333–334. https://doi.org/10.7326/M18-1348

Bhinder, B., Gilvary, C., Madhukar, N. S., & Elemento, O. (2021). Artificial intelligence in cancer research and precision medicine. *Cancer Discovery, 11*(4), 900–915.

Bochum, S., Berger, S., & Martens, U. M. (2018). Olaparib. *Recent Results in Cancer Research, 211*, 217–233. https://doi.org/10.1007/978-3-319-91442-8_15

Bray, F., et al. (2024). Global cancer statistics 2022: GLOBOCAN estimates of incidence and mortality worldwide for 36 cancers in 185 countries. *CA: A Cancer Journal for Clinicians, 74*(3), 229–263. https://doi.org/10.3322/caac.21834

Breen, S., Kofoed, S., Ritchie, D., Dryden, T., Maguire, R., Kearney, N., & Aranda, S. (2017). Remote real-time monitoring for chemotherapy side-effects in patients with blood cancers. *Collegian, 24*(6), 541–549. https://doi.org/10.1016/j.colegn.2016.10.009

Cadario, R., Longoni, C., & Morewedge, C. K. (2021). Understanding, explaining, and utilizing medical artificial intelligence. *Nature Human Behaviour, 5*(12), 1636–1642. https://doi.org/10.1038/s41562-021-01146-0

Cheever, M. A., & Higano, C. S. (2011). PROVENGE (Sipuleucel-T) in prostate cancer: The first FDA-approved therapeutic cancer vaccine. *Clinical Cancer Research, 17*(11), 3520–3526. https://doi.org/10.1158/1078-0432.CCR-10-3126

Chen, S., et al. (2023). Estimates and projections of the global economic cost of 29 cancers in 204 countries and territories from 2020 to 2050. *JAMA Oncology, 9*(4), 465–472. https://doi.org/10.1001/jamaoncol.2022.7826

Chow, R., Drkulec, H., James, H. B. Im, Tsai, J., Nafees, A., Kumar, S., Hou, T., Fazelzad, R., Leighl, N. B., Krzyzanowska, M., Wong, P., & Raman, S. (2024). The use of wearable devices in oncology patients: A systematic review. *The Oncologist, 29*(4), e419–e430. https://doi.org/10.1093/oncolo/oyad305

Chowdhury, A., Kassem, H., Padoy, N., Umeton, R., & Karargyris, A. (2021). A review of medical federated learning: Applications in oncology and cancer research. In A. Crimi & S. Bakas (Eds.), *Brain lesion: Glioma, multiple sclerosis, stroke and traumatic brain injuries* (BrainLes 2021. Lecture notes in computer science) (Vol. 12962). Springer. https://doi.org/10.1007/978-3-031-08999-2_1

Chu, T. N., Wong, E. Y., Ma, R., Yang, C. H., Dalieh, I. S., & Hung, A. J. (2023). Exploring the use of artificial intelligence in the management of prostate cancer. *Current Urology Reports, 24*(5), 231–240.

Chua, I. S., Gaziel-Yablowitz, M., Korach, Z. T., Kehl, K. L., Levitan, N. A., Arriaga, Y. E., Jackson, G. P., Bates, D. W., & Hassett, M. (2021). Artificial intelligence in oncology: Path to implementation. *Cancer Medicine, 10*(12), 4138–4149. https://doi.org/10.1002/cam4.3935

Committee on Improving the Quality of Cancer Care: Addressing the Challenges of an Aging Population; Board on Health Care Services; Institute of Medicine, Levit, L., et al. (2013). 2: The current cancer care landscape: An imperative for change. In *Delivering high-quality cancer care: Charting a new course for a system in crisis*. National Academies Press (US). Available from: https://www.ncbi.nlm.nih.gov/books/NBK202145/

Connal, S., et al. (2023). Liquid biopsies: The future of cancer early detection. *Journal of Translational Medicine, 21*(1), 118. https://doi.org/10.1186/s12967-023-03960-8

Cui, S., Ten Haken, R. K., & El Naqa, I. (2021). Integrating multiomics information in deep learning architectures for joint actuarial outcome prediction in non-small cell lung cancer patients after radiation therapy. *International Journal of Radiation Oncology, Biology, Physics, 110*, 893–904.

Delgado, A., & Guddati, A. K. (2021). Clinical endpoints in oncology—A primer. *American Journal of Cancer Research, 11*(4), 1121–1131.

Dhudum, R., Ganeshpurkar, A., & Pawar, A. (2024). Revolutionizing drug discovery: A comprehensive review of AI applications. *Drugs and Drug Candidates., 3*(1), 148–171.

Dlamini, Z., Francies, F. Z., Hull, R., & Marima, R. (2020). Artificial intelligence (AI) and big data in cancer and precision oncology. *Computational and Structural Biotechnology Journal, 18*, 2300–2311.

Dutcher, G. M. A., & Bilen, M. A. (2018). Therapeutic vaccines for genitourinary malignancies. *Vaccines (Basel)., 6*(3), 55. https://doi.org/10.3390/vaccines6030055

Emani, S., Rui, A., Rocha, H. A., Rizvi, R. F., Juaçaba, S. F., Jackson, G. P., & Bates, D. W. (2022). Physicians' perceptions of and satisfaction with artificial intelligence in cancer treatment: A clinical decision support system experience and implications for low-middle–income countries. *JMIR Cancer, 8*(2), e31461.

Exarchos, K. P., Goletsis, Y., & Fotiadis, D. I. (2012). Multiparametric decision support system for the prediction of oral cancer reoccurrence. *IEEE Transactions on Information Technology in Biomedicine, 16*(6), 1127–1134. https://doi.org/10.1109/TITB.2011.2165076. Epub 2011 Aug 18. PMID: 21859630.

Farasati Far, B. (2023). Artificial intelligence ethics in precision oncology: Balancing advancements in technology with patient privacy and autonomy. *Exploration of Targeted Anti-Tumor Therapy, 4*(4), 685–689. https://doi.org/10.37349/etat.2023.00160

Farina, E., Nabhen, J. J., Dacoregio, M. I., Batalini, F., & Moraes, F. Y. (2022). An overview of artificial intelligence in oncology. *Future Science OA, 8*(4), FSO787.

Fessele, K., Yendro, S., & Mallory, G. (2014). Setting the bar: Developing quality measures and education programs to define evidence-based, patient-centered, high-quality care. *Clinical Journal of Oncology Nursing, 18*(Suppl), 7–11. https://doi.org/10.1188/14.CJON.S2.7-11

Forjaz, G., Ries, L., Devasia, T. P., Flynn, G., Ruhl, J., & Mariotto, A. B. (2023). Long-term cancer survival trends by updated summary stage. *Cancer Epidemiology, Biomarkers & Prevention, 32*(11), 1508–1517. https://doi.org/10.1158/1055-9965.EPI-23-0589

Ghasemi, A., Hashtarkhani, S., Schwartz, D. L., & Shaban-Nejad, A. (2024). Explainable artificial intelligence in breast cancer detection and risk prediction: A systematic scoping review. *Cancer Innovation, 3*(5), e136. https://doi.org/10.1002/cai2.136

Gini, A., et al. (2020). EU-TOPIA consortium. Impact of colorectal cancer screening on cancer-specific mortality in Europe: A systematic review. *European Journal of Cancer, 127*, 224–235. https://doi.org/10.1016/j.ejca.2019.12.014

Google Health. (2021). *Improving breast cancer detection with AI*. Google Health. Retrieved from https://blog.google/technology/health/breast-cancer-detection-ai/

Hartmann, J. T., Haap, M., Kopp, H. G., & Lipp, H. P. (2009). Tyrosine kinase inhibitors—A review on pharmacology, metabolism and side effects. *Current Drug Metabolism, 10*(5), 470–481. https://doi.org/10.2174/138920009788897975

Hosny, A., Parmar, C., Quackenbush, J., Schwartz, L. H., & Aerts, H. J. (2018). Artificial intelligence in radiology. *Nature Reviews Cancer, 18*(8), 500–510.

https://fischerjordan.com/2024/03/impact-of-ai-in-early-cancer-detection-in-india/. Accessed on 02 Aug 2025.

Huang, S., Yang, J., Fong, S., & Zhao, Q. (2020). Artificial intelligence in cancer diagnosis and prognosis: Opportunities and challenges. *Cancer Letters, 471*, 61–71. https://doi.org/10.1016/j.canlet.2019.12.007

Huntley, C., et al. (2023). Utility of polygenic risk scores in UK cancer screening: A modelling analysis. *The Lancet Oncology, 24*(6), 658–668. https://doi.org/10.1016/S1470-2045(23)00156-0

Jansen, E. E. L., et al. (2020). EU-TOPIA consortium. Effect of organised cervical cancer screening on cervical cancer mortality in Europe: A systematic review. *The European Journal of Cancer, 127*, 207–223. https://doi.org/10.1016/j.ejca.2019.12.013

Khalid, S., Krishnan, S., & Thanki, R. (2021). Artificial intelligence in breast cancer detection and diagnosis. In *Emerging technologies in cancer detection* (pp. 67–78). Springer.

Khalighi, S., Reddy, K., Midya, A., Pandav, K. B., Madabhushi, A., & Abedalthagafi, M. (2024). Artificial intelligence in neuro-oncology: Advances and challenges in brain tumor diagnosis, prognosis, and precision treatment. *npj Precision Oncology, 8*(1), 80.

Koh, D. M., Papanikolaou, N., Bick, U., Illing, R., Kahn, C. E., Jr., Kalpathi-Cramer, J., Matos, C., Martí-Bonmatí, L., Miles, A., Mun, S. K., & Napel, S. (2022). Artificial intelligence and machine learning in cancer imaging. *Communication and Medicine, 2*(1), 133.

Kolla, L., & Parikh, R. B. (2024). Uses and limitations of artificial intelligence for oncology. *Cancer, 130*(12), 2101–2107.

Kulkarni, P., Mohanty, A., Bhattacharya, S., Singhal, S., Guo, L., Ramisetty, S., Mirzapoiazova, T., Mambetsariev, B., Mittan, S., Malhotra, J., & Gupta, N. (2022). Addressing drug resistance in cancer: A team medicine approach. *Journal of Clinical Medicine, 11*(19), 5701.

Li, Z., Koban, K. C., Schenck, T. L., Giunta, R. E., Li, Q., & Sun, Y. (2022). Artificial intelligence in dermatology image analysis: Current developments and future trends. *Journal of Clinical Medicine, 11*(22), 6826.

Liu, C., Liu, X., Wu, F., Xie, M., Feng, Y., & Hu, C. (2018). Using artificial intelligence (Watson for Oncology) for treatment recommendations amongst Chinese patients with lung cancer: Feasibility study. *Journal of Medical Internet Research, 20*(9), e11087.

Liu, J., Pandya, P., & Afshar, S. (2021). Therapeutic advances in oncology. *International Journal of Molecular Sciences, 22*(4), 2008. https://doi.org/10.3390/ijms22042008

Liu, Z. L., Chen, H. H., Zheng, L. L., Sun, L. P., & Shi, L. (2023). Angiogenic signaling pathways and anti-angiogenic therapy for cancer. *Signal Transduction and Targeted Therapy, 8*(1), 198. https://doi.org/10.1038/s41392-023-01460-1

Maleki Varnosfaderani, S., & Forouzanfar, M. (2024). The role of AI in hospitals and clinics: Transforming healthcare in the 21st century. *Bioengineering (Basel, Switzerland), 11*(4), 337. https://doi.org/10.3390/bioengineering11040337

Mariotto, A. B., Noone, A. M., Howlader, N., Cho, H., Keel, G. E., Garshell, J., Woloshin, S., & Schwartz, L. M. (2014). Cancer survival: An overview of measures, uses, and interpretation. *Journal of the National Cancer Institute. Monographs, 2014*(49), 145–186. https://doi.org/10.1093/jncimonographs/lgu024

McKinney, S. M., Sieniek, M., Godbole, V., et al. (2020). International evaluation of an AI system for breast cancer screening. *Nature, 577*(7788), 89–94. https://doi.org/10.1038/s41586-019-1799-6

Naik, N., Hameed, B. M. Z., Shetty, D. K., Swain, D., Shah, M., Paul, R., Aggarwal, K., Ibrahim, S., Patil, V., Smriti, K., Shetty, S., Rai, B. P., Chlosta, P., & Somani, B. K. (2022). Legal and ethical consideration in artificial intelligence in healthcare: Who takes responsibility? *Frontiers in Surgery, 9*, 862322. https://doi.org/10.3389/fsurg.2022.862322

Napitupulu, P. A. (2023). Ethical dilemmas in the use of artificial intelligence in breast cancer diagnosis and treatment (addressing issues of bias, transferability, and patient trust in breast cancer AI). *West Science Law and Human Rights, 1*(04), 256–260. https://doi.org/10.58812/wslhr.v1i04.314

Nassief, G., Anaeme, A., Moussa, K., Chen, D., & Ansstas, G. (2024). Where are we now with oncolytic viruses in melanoma and nonmelanoma skin malignancies? *Pharmaceuticals (Basel), 17*(7), 916. https://doi.org/10.3390/ph17070916

Palaniappan, K., Lin, E. Y., & Vogel, S. (2024). Global regulatory frameworks for the use of artificial intelligence (AI) in the healthcare services sector. *Healthcare, 12*(5), 562. MDPI.

Parikh, R. B., Kirch, R. A., Smith, T. J., & Temel, J. S. (2013). Early specialty palliative care—Translating data in oncology into practice. *The New England Journal of Medicine, 369*(24), 2347–2351. https://doi.org/10.1056/NEJMsb1305469

PathAI. (2023). *Launching AIM-HER2 breast cancer, an AI-powered HER2 scoring algorithm.* PathAI Press Release. Retrieved from www.pathai.com

Ruggiero, A., Skinner, R., & Khaled Zekri, W. Z. (2021). Editorial: Adverse and toxic effects of childhood cancer treatments. *Frontiers in Oncology, 11*, 795664. https://doi.org/10.3389/fonc.2021.795664

Santucci, C., Carioli, G., Bertuccio, P., Malvezzi, M., Pastorino, U., Boffetta, P., Negri, E., Bosetti, C., & La Vecchia, C. (2020). Progress in cancer mortality, incidence, and survival: A global overview. *European Journal of Cancer Prevention, 29*(5), 367–381. https://doi.org/10.1097/CEJ.0000000000000594

Schork, N. J. (2019). Artificial intelligence and personalized medicine. In *Precision medicine in Cancer therapy* (pp. 265–283). Springer.

Sebastian, A. M., & Peter, D. (2022). Artificial intelligence in cancer research: Trends, challenges and future directions. *Life, 12*(12), 1991.

Senthil Kumar, K., Miskovic, V., Blasiak, A., Sundar, R., Pedrocchi, A. L. G., Pearson, A. T., Prelaj, A., & Ho, D. (2023). Artificial intelligence in clinical oncology: From data to digital pathology and treatment. *American Society of Clinical Oncology Educational Book, 43*, e390084. https://doi.org/10.1200/EDBK_390084

Shafi, S., & Parwani, A. V. (2023). Artificial intelligence in diagnostic pathology. *Diagnostic Pathology, 18*(1), 109.

Shao, D., Dai, Y., Li, N., Cao, X., Zhao, W., Cheng, L., Rong, Z., Huang, L., Wang, Y., & Zhao, J. (2022). Artificial intelligence in clinical research of cancers. *Briefings in Bioinformatics, 23*(1), bbab523. https://doi.org/10.1093/bib/bbab523

Siddiqui, S., & Cruz, I. (2019). A cancer patient journey: Complete review during acute treatment phase. *Health Equity, 3*(1), 403–408. https://doi.org/10.1089/heq.2019.0046

Smith, T. G., Castro, K. M., Troeschel, A. N., Arora, N. K., Lipscomb, J., Jones, S. M., Treiman, K. A., Hobbs, C., McCabe, R. M., & Clauser, S. B. (2016). The rationale for patient-reported outcomes surveillance in cancer and a reproducible method for achieving it. *Cancer, 122*(3), 344–351. https://doi.org/10.1002/cncr.29767

Sumanasuriya, S., & De Bono, J. (2018). Treatment of advanced prostate cancer—A review of current therapies and future promise. *Cold Spring Harbor Perspectives in Medicine, 8*(6), a030635.

Ugai, T., et al. (2022). Is early-onset cancer an emerging global epidemic? Current evidence and future implications. *Nature Reviews. Clinical Oncology, 19*(10), 656–673. https://doi.org/10.1038/s41571-022-00672-8

Vela, D., Sharp, A., Zhang, R., Nguyen, T., Hoang, A., & Pianykh, O. S. (2022). Temporal quality degradation in AI models. *Scientific Reports, 12*(1), 11654.

Venkatasunilsrikanth, & Krithiga, S. (2023). Improved deep CNN architecture based breast cancer detection for accurate diagnosis. In *2023 second international conference on augmented intelligence and sustainable systems (ICAISS), Trichy, India* (pp. 200–205). https://doi.org/10.1109/ICAISS58487.2023.10250616

Vilhekar, R. S., & Rawekar, A. (2024). Artificial intelligence in genetics. *Cureus, 16*(1), e52035.

Wanner, J., Herm, L. V., Heinrich, K., & Janiesch, C. (2022). The effect of transparency and trust on intelligent system acceptance: Evidence from a user-based study. *Electronic Markets, 32*(4), 2079–2102.

Weerarathna, I. N., Kamble, A. R., & Luharia, A. (2023). Artificial intelligence applications for biomedical cancer research: A review. *Cureus, 15*(11), e48307.

World Health Organization. (2017). *Tackling NCDs: 'Best buys' and other recommended interventions for the prevention and control of noncommunicable diseases.* World Health Organization. https://iris.who.int/handle/10665/259232. License: CC BY-NC-SA 3.0 IGO

World Health Organization. (2022). *A short guide to cancer screening. Increase effectiveness, maximize benefits and minimize harm*. WHO Regional Office for Europe. License: CC BY-NC-SA 3.0 IGO.

Wu, X., Zhu, L., & Ma, P. C. (2018). Next-generation novel noninvasive cancer molecular diagnostics platforms beyond tissues. *American Society of Clinical Oncology Educational Book, 38*(38), 964–977. https://doi.org/10.1200/EDBK_199767

Xu, Z., Wang, X., Zeng, S., Ren, X., Yan, Y., & Gong, Z. (2021). Applying artificial intelligence for cancer immunotherapy. *Acta Pharmaceutica Sinica B, 11*(11), 3393–3405. https://doi.org/10.1016/j.apsb.2021.02.007

Yang, P. T., Wu, W. S., Wu, C. C., Shih, Y. N., Hsieh, C. H., & Hsu, J. L. (2021). Breast cancer recurrence prediction with ensemble methods and cost-sensitive learning. *Open Medicine (Warsaw, Poland), 16*(1), 754–768. https://doi.org/10.1515/med-2021-0282

Yang, Y., Xu, L., Sun, L., Zhang, P., & Farid, S. S. (2022). Machine learning application in personalised lung cancer recurrence and survivability prediction. *Computational and Structural Biotechnology Journal, 20*, 1811–1820. https://doi.org/10.1016/j.csbj.2022.03.035

Zakareya, S., Izadkhah, H., & Karimpour, J. (2023). A new deep-learning-based model for breast cancer diagnosis from medical images. *Diagnostics, 13*(11), 1944. https://doi.org/10.3390/diagnostics13111944

Zielonke, N., et al. (2020). EU-TOPIA consortium. Evidence for reducing cancer-specific mortality due to screening for breast cancer in Europe: A systematic review. *European Journal of Cancer, 127*, 191–206. https://doi.org/10.1016/j.ejca.2019.12.010

Zuo, D., Yang, L., Jin, Y., Qi, H., Liu, Y., & Ren, L. (2023). Machine learning-based models for the prediction of breast cancer recurrence risk. *BMC Medical Informatics and Decision Making, 23*(1), 276. https://doi.org/10.1186/s12911-023-02377-z

Deep Learning Algorithms to Analyze Medical Images for Early Detection of Cancer

Chainsee Saini, Sakshi, Onkar Singh, Chandrasen Pandey, and Shishir Singh

1 Introduction

Cancer continues to pose a significant global health challenge, as evidenced by the 19.976 million new cases and 9.744 million deaths reported annually, according to the World Health Organization (WHO). The enhancement of treatment outcomes is heavily reliant on early detection, which remains a substantial and ongoing challenge.

The integration of artificial intelligence (AI) has transformed oncology by enhancing cancer detection, classification, and prognosis. From early rule-based systems in the 1980s to advanced deep learning algorithms today, AI has evolved from basic pattern recognition to neural networks capable of analyzing multidimensional medical data. Early AI systems, such as computer-assisted detection (CAD) for mammography, aided radiologists but had limitations in sensitivity and specificity. By the 2000s, AI-guided imaging modalities such as X-rays, CT, and MRI improved lesion detection and treatment planning. The advent of deep learning in the 2010s marked a shift with convolutional neural networks (CNNs), enabling automated cancer detection, genomic data analysis, and digital pathology (Pranav et al., 2024).

C. Saini · Sakshi
Department of Biosciences, Manipal University, Jaipur, Rajasthan, India

O. Singh · C. Pandey
Department of Computer Science and Engineering, Manipal University Jaipur, Jaipur, Rajasthan, India

S. Singh (✉)
Department of Gastroenterology, Icahn School of Medicine at Mount Sinai, New York, NY, USA
e-mail: shishir.singh@mssm.edu; Shishir1753@gmail.com

R. Nema et al. (eds.), *Advances in Cancer Detection, Prediction, and Prognosis Using Artificial Intelligence and Machine Learning*,
https://doi.org/10.1007/978-981-96-9346-7_16

Currently, deep learning supports personalized treatments by integrating imaging and molecular data and facilitates minimally invasive diagnostics such as liquid biopsies. These tools improve workflow efficiency, precision, and reliability. However, challenges such as false positives/negatives and limited specificity persist, delaying diagnosis and prognosis. AI continues to address these gaps, advancing early detection techniques and efficient prognostic planning.

2 Medical Imaging Modalities in Cancer Detection

Medical imaging has revolutionized cancer detection by enabling noninvasive visualization of biological and anatomical structures. X-rays, widely used for lung and breast cancer screening, are cost-effective but have limited sensitivity, especially for soft tissues, often requiring supplementary imaging. CT scans improve sensitivity by generating 3D cross-sectional images through multiple X-rays but involve higher radiation doses, potentially causing nephrotoxicity. MRI, employing magnetic fields and radio waves, produces high-resolution images ideal for soft tissue evaluation but is costly, time-intensive, and less effective for dense bones or air-filled structures.

Ultrasound, using high-frequency sound waves, is suitable for abdominal cancers but has limited accuracy for deep-seated tumors. PET combines functional and anatomical insights through radioactive tracers such as FDG to detect malignancies but is costly, scarce, and exposes patients to radiation with low resolution.

Preprocessing techniques such as noise reduction, histogram equalization, and image sharpening enhance image quality, contrast, and tumor differentiation. Techniques such as image denoising and normalization ensure consistency across images, improving diagnostic accuracy and aiding treatment planning despite existing challenges.

Table 1 provides a comparative analysis of various medical imaging modalities used in cancer detection. It highlights their primary applications, strengths, and limitations.

Table 1 Comparative analysis of various medical imaging modalities used in cancer detection

Modality	Use	Strengths	Limitations
X-ray	Lung/breast cancer	Cost-effective	Low sensitivity for soft tissues
CT scan	3D cross-sections	High sensitivity	High radiation dose, nephrotoxicity
MRI	Soft tissue evaluation	High resolution, safe	Expensive, time-intensive
Ultrasound	Abdominal cancer	Noninvasive, portable	Low resolution for deep-seated tumors
PET	Metabolic insights	Functional and anatomical data	Expensive, radiation exposure

3 Deep Learning Architectures in Oncology

Deep learning has transformed oncology by improving diagnostic accuracy, predicting clinical outcomes, and enabling tailored cancer treatments. It excels in processing diverse data such as medical images, genetic information, and clinical records, significantly impacting cancer management and patient care. Traditional cancer research relied on labor-intensive manual analysis, often yielding inconclusive results. Recent advancements in deep learning offer efficient and standardized evaluations of complex biomedical data, addressing critical challenges in cancer diagnostics (Bray et al., 2018, LeCun et al., 2015).

This chapter explores key architectures such as CNNs, U-Net, and transformers, and their applications in oncology, including multimodal data integration, emerging trends, and challenges such as model interpretability and data privacy.

3.1 Key Architectures for Image Analysis

3.1.1 Convolutional Neural Networks

Convolutional neural networks (CNNs) are proficient in identifying, segmenting, and categorizing tumors by utilizing various layers, including convolutional, pooling, and fully connected layers (Krizhevsky et al., 2012). Advanced architectures such as ResNet (Koonce & Koonce, 2021) and DenseNet (Girdhar et al., 2022) overcome challenges such as vanishing gradients, enhance feature extraction, and reduce model parameters. For instance, 3D CNNs analyze volumetric CT and MRI data, improving tumor evaluation (Milletari et al., 2016; Singh et al., 2020).

3.1.2 U-Net and Its Variants

The encoder–decoder architecture and skip connections of U-Net are highly effective in tumor segmentation. Enhanced versions such as U-Net++ (Zhou et al., 2018) and Attention U-Net (Oktay et al., 2018) further improve these capabilities. They improve feature flow and focus on significant regions, achieving superior accuracy in detecting small tumors. For example, U-Net variants successfully segment gliomas from brain MRIs, aiding treatment planning (Amri et al., 2025).

3.2 Transfer Learning

3.2.1 Overview and Applications

Transfer learning utilizes pretrained models that are later fine-tuned with domain-specific data, effectively addressing the issue of limited annotated datasets (Tajbakhsh et al., 2016). This approach accelerates model convergence and enhances performance in detecting lung nodules, brain tumors, and colorectal polyps (Kim et al., 2022). For instance, ResNet pretrained on generic datasets classifies lung cancer with high accuracy (Hussein et al., 2017).

3.3 Attention Mechanisms and Transformers

Evolution and Applications

Originally developed for Natural Language Processing (NLP) (Bahdanau, 2014), attention mechanisms prioritize relevant features in image analysis, improving segmentation and classification. Transformers such as TransUNet combine CNNs for feature extraction and transformers for global context, excelling in tasks such as tumor segmentation (Dosovitskiy, 2020). Vision Transformers (ViTs) handle high-resolution histopathological images for robust cancer diagnosis (Park et al., 2025).

Attention Equation

$$\text{Attention}(Q,K,V) = \text{softmax}\left(\frac{QK^{\top}}{\sqrt{d_k}}\right)V$$

3.4 Multimodal Data Fusion

Integrating imaging, genomic, and clinical data enhances diagnostic precision and treatment personalization. Frameworks such as Pathomic Fusion combine pathology and genomic data for improved survival predictions in lung cancer (Chen et al., 2022). Multimodal models incorporating PET-CT and biomarkers enhance tumor characterization and treatment response predictions (Zhang et al., 2020).

Multimodal fusion can be mathematically represented as:

$$z = f_{\text{fusion}}\left(f_{image}\left(x_{\text{image}}\right), f_{\text{genomic}}\left(x_{\text{genomic}}\right), f_{\text{clinical}}\left(x_{\text{clinical}}\right)\right),$$

where f_{image}, f_{genomic}, and f_{clinical} are feature extractors for each modality, and f_{fusion} combines these features into a unified representation, z.

3.5 Emerging Trends and Challenges

Explainability in Deep Learning Models

Techniques such as Grad-CAM (Selvaraju et al., 2017) and SHAP (Lundberg, 2017) improve model transparency, enabling clinicians to visualize decision-making processes and build trust.

Grad-CAM Equation

$$L^{c}_{\text{Grad-CAM}} = \text{ReLU}\left(\sum_{k} \alpha_k^c A^k\right)$$

where $\boldsymbol{A}^k$ are the feature maps and α_k^c are the weights computed from the gradients.

Federated Learning for Privacy

Federated learning safeguards patient privacy by allowing institutions to collaboratively train models without the need to share raw data (Rieke et al., 2020). This approach supports multicenter studies such as breast cancer diagnosis (Sheller et al., 2019).

Federated Learning Equation

$$\theta^{t+1} = \theta^{t} - \eta \sum_{N}^{i=1} \frac{n_i}{n} \nabla_{\theta} \mathcal{L}_i \left(\theta^{t}\right),$$

where $\boldsymbol{\theta}$ represents the model parameters, $\boldsymbol{\eta}$ is the learning rate, $\boldsymbol{N}$ represents the number of institutions, $\boldsymbol{n}_i$ represents the number of samples at institution $\boldsymbol{i}$, and $\mathcal{L}_i$ is the local loss function.

4 Cancer-Specific Applications

4.1 Breast Cancer

Breast cancer is one of the most frequently diagnosed cancers worldwide and presents significant challenges in early-stage detection, largely due to subtle indicators such as microcalcifications (Sun et al., 2017; Prinzi et al., 2024). Deep learning models enhance diagnosis by identifying these small calcium clusters with high specificity in mammograms. Automated BI-RADS scores assist in lesion classification and prognostic planning (Tang et al., 2022). Techniques such as 3D mammography (Tomosynthesis) improve detection in dense breast tissues by capturing critical details (Yala et al., 2019; Loizidou et al., 2021). AI-based CAD systems outperform traditional methods by providing malignancy scores, leveraging big data and advanced processors to improve accuracy and speed (Do et al., 2021).

4.2 Lung Cancer

Lung cancer represents the primary contributor to mortality associated with cancer and demands precise diagnostic techniques (Minna et al., 2002). Deep learning models such as RetinaNet, Faster R-CNN, and YOLO detect and classify nodules in 3D CT scans, reducing false positives and improving sensitivity (Oh et al., 2023). These models predict nodule growth and metastatic spread by integrating patient demographics and clinical history, aiding cancer staging and treatment planning (Kalaiyarasi et al., 2020; Liu et al., 2021).

4.3 Brain Tumor

Brain tumors, whether primary or metastatic, significantly impact survival and quality of life (Black, 1991). Deep learning, particularly MRI-based segmentation using U-Net and 3D CNNs, effectively distinguishes tumor subregions such as the enhancing core and oedema (He et al., 2020; Punn & Agarwal, 2022). These models enable classification into tumor types and grades, supporting treatment planning (Mohsen et al., 2018). Noninvasive methods predict molecular subtypes such as IDH mutation, facilitating personalized therapies (Jovčevska, 2020).

4.4 Skin Cancer

Skin cancer, including nonmelanoma and melanoma, affects millions annually. While nonmelanomas have favorable outcomes, melanoma's aggressiveness requires early detection (Niu et al., 2020). Deep learning models such as CNNs differentiate melanoma from benign lesions and assist in risk scoring and multi-class classification (Winkler et al., 2020; Skin Cancer Detection Based on Deep Learning—PMC, n.d.). Integrated into mobile apps and dermoscopy devices, these tools provide immediate feedback, making them valuable in primary care (Kousis et al., 2022).

4.5 Cervix Cancer

Cervical cancer remains a significant health burden, particularly in environments with limited resources. Recent advancements in machine learning have improved the identification of risk factors linked to cervical cancer (Kaushik et al., 2021). A study integrating cytokine gene variants with sociodemographic data demonstrated that machine learning models can accurately predict susceptibility. Key predictors such as IL-6, IL-10 polymorphisms, age, and socioeconomic status were identified as critical determinants of cervical cancer risk. These findings highlight the importance of integrating genetic and demographic data into predictive models, enabling early detection and personalized prevention strategies to reduce the global burden of cervical cancer (Kushwah et al., 2020).

5 Emerging Deep Learning Techniques in Oncology

5.1 Data Augmentation for Enhanced Model Robustness

Big data is required to make accurate predictions using deep learning models, but high quality labeled medical data is quite scarce, especially in oncology. Data augmentation techniques, such as rotation, cropping, and the addition of noise, are utilized to artificially expand the training dataset. This advances models from relying solely on memorization due to overfitting through techniques that mimic real-world variation characteristic of medical imaging (Shorten & Khoshgoftaar, 2019). For example, in applying the GANs, an MRI and CT scan real images have been generated toward enhancing datasets (Frid-Adar et al., 2018). This is particularly important to the diagnosis of cancer: scarce datasets can lead to poor generalization and low capability models.

5.2 Addressing Class Imbalance in Cancer Types

Cancer datasets often display class imbalances, meaning that abundant classes such as breast or lung cancer contain many samples compared to scarce types that are represented sparsely. This resulted in the model predominantly predicting the majority class, thereby reducing the performance for minority classes. A collection of methodologies, including oversampling of minority classes, undersampling of majority classes, and the generation of synthetic data (such as SMOTE or GANs), has been established to address this issue (Chawla et al., 2002). Further, loss functions such as focal loss are used to prevent the effects of class imbalance during training since they would give more weight to classes less represented (Lin et al., 2017). It is particularly crucial in rare cancer diagnostics where the accuracy of the predictions impacts the treatment decisions.

5.3 Explainable AI for Interpretability

Typically described as "black boxes," clinicians cannot easily trust their deep learning models. Explainable AI techniques, such as Gradient-weighted Class Activation Mapping (Grad-CAM) (Selvaraju et al., 2017) and SHapley Additive Explanations (SHAP) (Lundberg, 2017), enhance model interpretability by visualizing the regions or features that influence predictions. In the field of oncology, Grad-CAM can identify the tumor regions in imaging studies that significantly contribute to the model's predictions, thereby providing clinicians with the opportunity to validate these model decisions. This enhances reliability and facilitates the incorporation of artificial intelligence into clinical processes.

5.4 Few-Shot Learning for Rare Cancer Variants

Few-shot learning deals with the problem of having very few instances for rare classifications of cancers by enabling models to learn from datasets that are extremely small. It relies on meta-learning or using knowledge from other related tasks to generalize (Snell et al., 2017). In oncology, few-shot learning can be used for the identification of rare forms of cancer, such as atypical sarcomas or abnormal genetic patterns, by using knowledge transfers from larger datasets of data for similar cancer types. Such techniques promise to push the boundaries of precision oncology by allowing strong predictions even for relatively rare conditions.

5.5 Self-Supervised Deep Learning

Self-supervised learning (SSL) has emerged as an efficient approach for model training in scenarios with insufficient availability of large amounts of annotated data. The method includes training the models on unlabeled data by solving surrogate tasks, which include image rotation prediction and filling in missing information, before fine-tuning them toward a particular application such as cancer diagnosis (He et al., 2020). In oncology, SSL shows promise in using volumes of unlabeled imaging data, for example, radiologic scans to enhance performance for downstream tasks. This SSL eliminates the use of expensive labels and, thus, gives better results in tasks such as tumor segmentation and prediction of prognosis.

6 Evaluating Performance Metrics and Validating Models in Oncology Deep Learning

6.1 Measures of Clinical Validity and Precision

Deep Learning Oncology Model Performance evaluation of such models using the metrics, which bring great clinical relevance, hold important diagnostics precision values for them along with making them reliable to their reality too in the clinical world (Ghosh et al., 2021). Some of the prime used among the metrics are provided in the following sections.

6.1.1 Accuracy

This metric measures the model's overall accuracy by calculating the proportion of correct predictions (including both true positives and true negatives) relative to the total number of cases evaluated (Ghosh et al., 2021). Mathematically, it is represented as:

$$\text{Accuracy} = \frac{\text{TP} + \text{TN}}{\text{TP} + \text{FP} + \text{TN} + \text{FN}}$$

where TP is true positive, TN is true negatives, FP is false positives, and FN is false negatives.

While accuracy offers a general overview of a model's performance, it can be deceptive in datasets with class imbalances.

6.1.2 Precision

Precision evaluates the model's capacity to accurately identify true positive cases, minimizing the impact of false positives. This metric is especially critical in contexts where the cost of false positives is significant. Precision is calculated as:

$$\text{Precision} = \frac{\text{TP}}{\text{TP} + \text{FP}}$$

A high precision indicates that when the model predicts a positive outcome, there is a high likelihood that the prediction is correct.

6.1.3 Recall (Sensitivity)

Recall evaluates the model's ability to identify all true positive cases. This metric is particularly vital in oncology, where failing to detect a positive case can have significant clinical consequences. The formula for recall is:

$$\text{Recall} = \frac{\text{TP}}{\text{TP} + \text{FN}}$$

High recall ensures that most of the true positive cases are identified by the model.

6.1.4 F1-Score

The F1-Score integrates precision and recall into a single metric, balancing the two by calculating their harmonic mean. This measure is especially valuable in situations with uneven class distributions (Carrington et al., 2022). It is given by:

$$\text{F1} - \text{Score} = 2 \cdot \frac{\text{Precision} \cdot \text{Recall}}{\text{Precision} + \text{Recall}}$$

The F1-Score is an important metric in scenarios where both false positives and false negatives have substantial implications.

6.1.5 Area Under the Receiver Operating Characteristic Curve (AUC-ROC)

The AUC-ROC evaluates the trade-off between the true positive rate (TPR) and the false positive rate (FPR) across various threshold settings (Ghosh et al., 2021). It provides a single scalar value to summarize the model's discriminatory ability. The true positive rate and false positive rate are defined as:

$$\text{TPR} = \frac{\text{TP}}{\text{TP} + \text{FN}}, \quad \text{FPR} = \frac{\text{FP}}{\text{FP} + \text{TN}}$$

The ROC curve depicts the relationship between the true positive rate (TPR) and the false positive rate (FPR), while the AUC measures the likelihood that the model will assign a higher score to a randomly chosen positive instance over a randomly chosen negative instance. An AUC of 1 indicates perfect classification, whereas an AUC of 0.5 reflects no discriminatory ability (Carrington et al., 2022).

6.1.6 Specificity

Complementary to recall, specificity quantifies the proportion of actual negative cases accurately identified. It is defined as:

$$\text{Specificity} = \frac{\text{TN}}{\text{TN} + \text{FP}}$$

High specificity is essential for minimizing false positives, thereby avoiding unnecessary treatments and reducing patient anxiety.

Matthews Correlation Coefficient (MCC) MCC is a balanced measure that considers all four confusion matrix categories (Carrington et al., 2022). It is particularly useful for imbalanced datasets and is calculated as:

$$\text{MCC} = \frac{(\text{TP} ? \text{TN}) ? (\text{FP} ? \text{FN})}{\sqrt{(\text{TP} + \text{FP})(\text{TP} + \text{FN})(\text{TN} + \text{FP})(\text{TN} + \text{FN})}}$$

The MCC ranges from −1 to +1, where +1 indicates perfect prediction, 0 no better than random prediction, and −1 indicates total disagreement between prediction and observation (Boughorbel et al., 2017).

6.2 Robust Cross-Validation Strategies and Mitigating Bias

Cross-validation is indispensable for assessing the robustness and generalizability of machine learning models, particularly in oncology where data variability is substantial. Effective cross-validation techniques help minimize bias and variance in performance estimates by ensuring that the model is rigorously tested against diverse data subsets (Schumacher et al., 1997). Key strategies are provided in the following sections.

6.2.1 *K*-Fold Cross-Validation

A widely utilized method involves dividing the dataset into kk equally sized folds. The model is trained on $k-1$ folds and validated on the remaining fold, with this process repeating kk times, each time using a different validation fold. The final performance metric is the average across all folds (Boughorbel et al., 2017):

$$\text{Average Metric} = \frac{1}{k}\sum_{i=1}^{k} M_i$$

where M_i represents the performance metric for the ith fold.

6.2.2 Stratified *K*-Fold Cross-Validation

To address class imbalances commonly present in oncology datasets, stratified k-fold ensures that each fold maintains the same class distribution as the entire dataset. This prevents skewed performance estimates and ensures that minority classes are adequately represented in each validation fold.

Leave-One-Out Cross-Validation (LOOCV) Particularly useful for small datasets, LOOCV includes utilizing one individual data point as the validation set, while the rest of the dataset is employed for training purposes. This process is conducted for each data point, maximizing data utilization and ensuring a comprehensive assessment of the model's efficacy:

$$\text{Average Metric} = \frac{1}{N}\sum_{i=1}^{N} M_i$$

where N represents the total number of data points.

6.2.3 Nested Cross-Validation

This advanced technique is essential for hyperparameter tuning while avoiding overfitting. It involves two layers of cross-validation:

1. *Outer loop:* divides the data into k outer folds to evaluate performance.
2. *Inner loop:* within each outer fold, the data is further split into \(m \) inner folds for hyperparameter optimization.

The overall performance metric is computed as:

$$\text{Metric} = \frac{1}{k}\sum_{j=1}^{k}\left(\frac{1}{m}\sum_{i=1}^{m} M_{ij}\right)$$

where M_{ij} is the performance metric for the ith inner fold within the jth outer fold.

6.2.4 Repeated Cross-Validation

To further enhance the reliability of performance estimates, cross-validation can be repeated multiple times with different random splits. The results are then averaged to provide a more stable performance metric (Schumacher et al., 1997).

6.3 External and Multicenter Validation for Enhanced Generalizability

External validation is paramount for confirming that deep learning models developed in oncology are applicable across different clinical settings and populations. This involves testing the model on independent datasets that were not involved in the training or initial validation phases. Multicenter validation extends this concept by evaluating the model across datasets from various institutions, thereby accounting for:

1. *Diverse imaging protocols:* Variations in imaging equipment and procedures across centers can influence model performance. Multicenter validation ensures that the model is robust to such differences.
2. *Heterogeneous patient demographics:* Differences in age, gender, ethnicity, and other demographic factors across centers can affect disease prevalence and presentation. Validating across multiple centers ensures the model performs well across diverse patient populations.
3. *Variable disease characteristics:* The prevalence and characteristics of specific cancers can vary between institutions. Multicenter validation helps ensure the model's applicability to various disease subtypes and stages.

The aggregated performance across multiple centers is typically calculated using a weighted metric to reflect the varying sizes of datasets from each center:

$$\text{Global Metric} = \frac{\sum_{i=1}^{N} n_i \cdot M_i}{\sum_{i=1}^{N} n_i}$$

where N is the number of centers, n_i is the samples total number from the i-th center, M_i is the performance metric (e.g., AUC-ROC) for the ith center.

6.3.1 Addressing Discrepancies with Domain Adaptation

Despite rigorous validation, models may exhibit performance discrepancies when applied to external datasets due to differences in data distributions. Domain adaptation techniques are employed to mitigate these issues by aligning the source (training) and target (external) distributions. One common approach involves adversarial learning, where the model is trained to minimize the discrepancy between domains while maintaining performance on the primary task. The domain adaptation loss can be formulated as:

$$\text{Domain Adaptation Loss} = \text{Task} - \text{Specific Loss} + \lambda \cdot \text{Discrepancy Loss}$$

where

- λ is a hyperparameter that balances the importance of the task-specific loss and the domain discrepancy loss.
- Task-specific loss: pertains to the primary objective, such as classification accuracy.
- Discrepancy loss: quantifies the difference between the source and target domain distributions.

Techniques such as gradient reversal layers (GRL) and maximum mean discrepancy (MMD) are commonly used to implement domain adaptation, enabling models to generalize better across different clinical settings.

6.4 Additional Considerations in Model Validation

6.4.1 Confidence Intervals and Statistical Significance

Beyond point estimates of performance metrics, it is crucial to report confidence intervals to convey the uncertainty associated with these estimates. Methods such as bootstrapping or Bayesian approaches can be utilized to calculate confidence intervals, offering a range within which the true metric value is expected to fall with a given probability (e.g., 95%).

6.4.2 Calibration Curves

Calibration assesses how well the predicted probabilities of a model reflect the actual probabilities of outcomes. A well-calibrated model ensures that, for instance, among all instances predicted with a 70% probability of being positive, approximately 70% are indeed positive. Calibration curves plot predicted probabilities against observed frequencies, and metrics such as the Brier score can quantify calibration quality.

6.4.3 Decision Curve Analysis (DCA)

DCA evaluates the clinical utility of a model by considering the net benefit across different threshold probabilities. It helps determine whether the model's predictions would lead to better clinical decisions compared to default strategies such as treating all or no patients.

6.4.4 Interpretability and Explainability

In clinical settings, comprehending the rationale behind a model's predictions is crucial for building trust and facilitating adoption. Techniques such as SHAP (SHapley Additive Explanations), LIME (Local Interpretable Model-agnostic Explanations), and saliency maps are utilized to offer insights into feature importance and model behavior, thereby aiding informed clinical decisions.

6.4.5 Reproducibility and Transparency

Ensuring that model development and validation processes are transparent and reproducible is critical. This involves:

- Detailed documentation of data preprocessing steps, model architectures, and hyperparameters.
- Sharing code and datasets where possible to allow independent verification.
- Utilizing standardized reporting guidelines, such as TRIPOD (Transparent Reporting of a multivariable prediction model for Individual Prognosis Or Diagnosis).

6.4.6 Ethical Considerations and Bias Mitigation

Models must be evaluated for potential biases that could lead to unfair treatment of certain patient groups. This involves:

- Assessing performance across different demographic subgroups.
- Implementing fairness metrics to detect and mitigate biases.
- Ensuring data diversity to enhance model fairness and equity.

6.5 Case Studies and Practical Applications

To illustrate the application of these metrics and validation techniques, consider the following case studies in oncology deep learning:

Case Study 1: Breast Cancer Detection
Deep learning models for image classification have found numerous applications in modern scenarios (Tiwari et al., 2020). One such model, developed to identify malignant tumors in mammography images, employs stratified ten-fold cross-validation, achieving an average AUC-ROC of 0.95, precision of 0.90, and recall of 0.92 (Shen et al., 2019). External validation on datasets from three different hospitals confirmed consistent performance, illustrating the model's robustness across diverse imaging protocols and patient demographics (Shen et al., 2019).

Case Study 2: Lung Cancer Prognosis
A model predicts the 5-year survival rates of lung cancer patients based on clinical and genomic data. Nested cross-validation reveals optimal hyperparameters that maximize the F1-Score. Multicenter validation shows an average MCC of 0.85, indicating strong predictive power. Domain adaptation techniques are applied to adjust for differences in genomic data distributions between centers, enhancing the model's generalizability (Singh & Singh, 2024; Singh et al., 2023).

Case Study 3: Skin Lesion Classification
A CNN classifies skin lesions as benign or malignant using dermoscopic images. Calibration curves indicate good probability estimates, and decision curve analysis shows a net benefit over existing diagnostic methods. Interpretability tools highlight key image features influencing the model's decisions, facilitating clinician trust and adoption.

Evaluating performance metrics and validating models are foundational steps in the establishment of applications of deep learning in oncology. By meticulously selecting appropriate metrics, employing robust cross-validation techniques, and conducting comprehensive external and multicenter validations, researchers can ensure that their models are both accurate and generalizable. Additionally, addressing interpretability, reproducibility, and ethical considerations further solidifies these advanced computational tools' clinical applicability and trustworthiness, ultimately contributing to improved patient outcomes and personalized cancer care (Khan et al., 2024).

7 Clinical Integration and Workflow in Oncology Deep Learning

7.1 Seamless Integration with PACS for Real-Time Clinical Applications

Deep learning algorithms in PACS will be seamlessly integrated in order to analyze medical image data instantly in clinical workflows. As the cornerstones of the modern medical infrastructure, PACS store, retrieves, and shares digital images from different settings of healthcare. Real-time automatic image interpretation by patients during assessment through AI-driven analytics in PACS can help hasten diagnosis with a higher degree of accuracy by doctors. The integration of systems is especially critical in acute care environments where swift decision-making significantly influences patient outcomes. Effective execution requires careful adherence to protocols related to data security, thereby safeguarding patient confidentiality, along with meeting interoperability standards to facilitate uninterrupted communication among various healthcare IT systems. Moreover, adherence to healthcare regulations, such as the Health Insurance Portability and Accountability Act (HIPAA), is essential for safeguarding sensitive medical information (Smith et al., 2021).

7.2 Enhancing Diagnostic Efficiency Through Real-Time Image Processing

The use of sophisticated real-time image processing, enabled through complex deep learning architectures, such as convolutional neural networks (CNNs), is highly enhanced diagnostic procedures that offer instantaneous and accurate analysis of intricate imaging datasets. These models can recognize complex trends and subtle abnormalities in radiologic images, as demonstrated through their ability to distinguish malignancy in tumors from benign lesions with a high degree of accuracy (Johnson et al., 2020). Thus, the ability to interpret real images accelerates not only diagnostic processes but maximizes know-how usage so that attention is focused only on requiring sensitive human judgment, and indeed, it means increased flows of patients and reduced diagnosis wait times so that an overall faster beginning of any treatment program occurs (Johnson et al., 2020). Further, it could help in early malignant lesions detection using real-time processing technologies that improve chances for survival by prompt intervention.

7.3 Navigating Regulatory Pathways and Ensuring FDA Compliance

The transition of deep learning applications from research to clinical practice in oncology calls for adherence to strict regulatory frameworks, especially those from the U.S. Food and Drug Administration (FDA) (FDA, 2022; Clarke & Hughes,

2022). The FDA's 1 guidelines for Software as a Medical Device (SaMD) detail the requirements for market authorization, including full clinical validation, rigorous risk management, and robust quality assurance processes. It is critical for developers to provide robust proof of the clinical effectiveness and safety of their artificial intelligence instruments, ensuring that these advances bring substantial benefits without unwarranted risks to patients. Clarity in AI decisions is also crucial because confidence among healthcare professionals and patients arises from it. Complying with these regulatory requirements thus requires continuous monitoring and after-market surveillance to address arising issues promptly and to be responsive to evolving clinical demands. This is important because compliance with FDA regulations promotes the lawful and ethical integration of AI in the healthcare sector and increases the trustworthiness and dependability of such technologies among the medical community (FDA, 2022; Clarke & Hughes, 2022).

7.4 Implementing Comprehensive Quality Assurance Protocols

Important preservation of integrity and efficacy can be achieved by comprehensive quality assurance protocols for AI-driven tools in clinical oncology settings. Such protocols include, among others, periodic checks on performance, which result in models that are persistently correct and reliable within long periods. Models will therefore need to be recalibrated since clinical data distributions change because of improvements in imaging technology or changes in patient demographics, among other factors. Besides, the newly acquired datasets used to validate artificial intelligence models are especially essential in identifying possible sources of bias or inaccuracy, which may emerge due to the fact that clinical practices change over time. Assessment of AI systems shall therefore be conducted periodically to determine whether they meet set requirements or operate outside their respective safety thresholds. The proper training of healthcare professionals in the effective use of artificial intelligence tools is crucial, as it maximizes the potential benefits of these technologies while simultaneously reducing the risk of misinterpretation or misuse. Proper quality assurance processes not only extend the lifespan and reliability of AI systems but also are critical to maintaining long-term patient trust and the successful integration of AI into everyday clinical practice (White et al., 2021).

8 Challenges and Future Directions

Deep learning algorithms in medical imaging for early cancer detection holds great promise but faces significant challenges that hinder widespread adoption and require ongoing research.

One major challenge is the reliance on large, high-quality labeled datasets, which are scarce in healthcare due to the labor-intensive and expertise-dependent annotation process (Gulum et al., 2021; Choi et al., 2022). Manual labeling by radiologists or oncologists is prone to interobserver variability, complicating dataset reliability.

While techniques such as transfer learning and synthetic data generation offer partial solutions, future research must focus on semi-supervised, unsupervised, and active learning strategies to reduce dependency on labeled data (Kumar & Alqahtani, 2022; Decherchi et al., 2021).

Device and protocol variability in imaging modalities (e.g., MRI, CT, PET) further complicates model deployment. Differences in hardware, imaging parameters, and post-processing algorithms lead to heterogeneous data, impacting model performance. Domain adaptation techniques, standardization of imaging protocols, and multi-institutional datasets can mitigate these issues. Approaches such as adversarial training and federated learning also enable adaptation to new environments without extensive retraining.

Data privacy is another critical concern, as imaging data contains sensitive personal information governed by regulations such as HIPAA and GDPR. Federated learning and differential privacy allow model training on distributed data without compromising patient privacy, while homomorphic encryption and blockchain-based systems enhance security (Jiang et al., 2023; Kumar & Alqahtani, 2022). Cybersecurity threats, such as adversarial attacks and data breaches, further necessitate robust, privacy-preserving frameworks for clinical AI (Decherchi et al., 2021).

In conclusion, addressing challenges such as data scarcity, device variability, and privacy concerns is essential for integrating deep learning into clinical workflows. Advances in labeling techniques, model adaptation, and privacy-preserving technologies will enable reliable and secure deep learning systems, facilitating earlier cancer detection and improved patient outcomes (Jiang et al., 2023; Decherchi et al., 2021).

9 Ethical and Legal Considerations in AI-Based Diagnosis

AI integration in cancer diagnosis offers transformative benefits, including timely detection and personalized treatment strategies, but also presents ethical and legal challenges. Key issues include ensuring transparency, accountability, informed consent, and addressing biases in AI models (Quazi et al., 2024). Biases stemming from algorithmic design, cultural limitations, or nonrepresentative datasets require fairness-aware algorithms and diverse, inclusive data. Regular bias audits and updates to accommodate societal changes are essential (Ricci et al., 2022).

Transparency is fundamental for ethical AI adoption. Ambiguities in responsibility for AI errors—whether attributed to developers, healthcare providers, or organizations—pose significant challenges, particularly in sensitive areas such as end-of-life care. Engaging stakeholders such as physicians, ethicists, and patient advocacy groups during AI system design ensures real-world relevance and ethical compliance.

AI systems also raise legal concerns, particularly regarding patient consent and autonomous decision-making. Practitioners must disclose potential biases, limitations, and AI's role in decision-making to ensure informed consent (Carter et al., 2020). Regulatory bodies such as EMA and FDA oversee medical AI, but legal frameworks need adaptation to address AI-specific nuances, ensuring safety, transparency, and accountability (Sharma & Manchikanti, 2024).

Ultimately, transparency and accountability in medical AI go beyond technical concerns to uphold patient welfare, fairness, and public trust. By prioritizing these principles, AI's potential can be harnessed responsibly, aligning with healthcare ideals and safeguarding patient rights.

10 Conclusion

The integration of deep learning in oncology has ushered in a transformative era, enhancing early cancer detection and enabling personalized treatment strategies. Deep learning techniques, such as U-Nets, CNNs, as well as transformers, have exhibited exceptional capabilities for analyzing diverse medical imaging modalities, from mammograms to MRI and CT scans. These innovations address critical challenges in cancer diagnostics, including low specificity, limited sensitivity, and imbalances in datasets.

Emerging techniques such as explainable AI, few-shot learning, and federated learning offer solutions to issues of interpretability, data scarcity, and privacy concerns. In spite of these advancements, difficulties remain, such as the necessity for standardized data, thorough validation across varied clinical environments, and adherence to ethical and regulatory standards. Addressing these challenges through collaboration among researchers, clinicians, and policymakers is essential for the widespread integration of AI-driven tools into clinical practice.

In conclusion, deep learning has significant potential to enhance diagnostic accuracy, improve patient outcomes, and transform the future of oncology. However, continuous efforts in research, validation, and ethical implementation are crucial to fully realize its potential (Fig. 1).

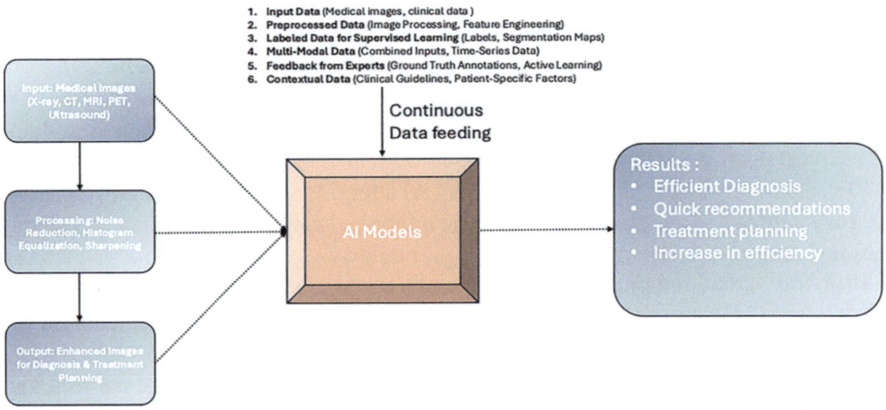

Fig. 1 Flowchart for medical imaging in AI-based cancer detection. It outlines the process from inputting medical images (X-ray, CT, MRI, PET, Ultrasound), through preprocessing techniques such as noise reduction and histogram equalization, to producing enhanced images for accurate diagnosis and effective treatment planning

References

Amri, Yessine, Amine Ben Slama, Zouhair Mbarki, Ridha Selmi, and Hedi Trabelsi. "Automatic glioma segmentation based on efficient U-net model using MRI images." Intelligence-Based Medicine 11 (2025): 100216.

Bahdanau, Dzmitry. (2014) "Neural machine translation by jointly learning to align and translate." arXiv preprint arXiv:1409.0473.

Black, P. M. (1991). Brain tumors. *New England Journal of Medicine, 324*(22), 1555–1564. https://doi.org/10.1056/NEJM199105303242205

Boughorbel, S., Jarray, F., & El-Anbari, M. (2017). Optimal classifier for imbalanced data using Matthews correlation coefficient metric. *PLoS One, 12*(6), e0177678.

Bray, F., Ferlay, J., Soerjomataram, I., Siegel, RL., Torre L.A, Jemal A. Global cancer statistics (2018): GLOBOCAN estimates of incidence and mortality worldwide for 36 cancers in 185 countries. CA Cancer J Clin. 2018 Nov;68(6):394–424. https://doi.org/10.3322/caac.21492

Carrington, A. M., Manuel, D. G., Fieguth, P. W., Ramsay, T., Osmani, V., Wernly, B., et al. (2022). Deep ROC analysis and AUC as balanced average accuracy, for improved classifier selection, audit and explanation. *IEEE Transactions on Pattern Analysis and Machine Intelligence, 45*(1), 329–341.

Chawla, N. V., Bowyer, K. W., Hall, L. O., & Kegelmeyer, W. P. (2002). SMOTE: Synthetic minority over-sampling technique. *Journal of Artificial Intelligence Research, 16*, 321–357.

Chen, R. J. et al. (2022). Pathomic Fusion: An integrated framework for fusing histopathology and genomic features for cancer diagnosis and prognosis. IEEE Trans. Med. Imaging, 41(4), 757–770. https://doi.org/10.1109/TMI.2020.3021387

Choi, Y., Garcia, M., Raman, S. S., Enzmann, D. R., & Brown, M. S. (2022). AI-human interactive pipeline with feedback to accelerate medical image annotation. In *Proceedings of SPIE 12033, medical imaging 2022: Computer-aided diagnosis, 120332U*. https://doi.org/10.1117/12.2611952

Clarke, T., & Hughes, M. (2022). Navigating FDA compliance for AI in oncology. *Medical Device Regulatory Affairs, 15*(1), 22–29.

Carter, S. M., Rogers, W., Win, K. T., Frazer, H., Richards, B., & Houssami, N. (2020). The ethical, legal and social implications of using artificial intelligence systems in breast cancer care. The Breast, 49, 25–32.

Decherchi, S., Pedrini, E., Mordenti, M., Cavalli, A., & Sangiorgi, L. (2021). Opportunities and challenges for machine learning in rare diseases. *Frontiers in Medicine (Lausanne), 8*, 747612. https://doi.org/10.3389/fmed.2021.747612

Dosovitskiy, Alexey, Lucas Beyer, Alexander Kolesnikov, Dirk Weissenborn, Xiaohua Zhai, Thomas Unterthiner, Mostafa Dehghani et al. (2020) "An image is worth 16x16 words: Transformers for image recognition at scale." arXiv preprint arXiv:2010.11929.

Do, Y. A., Jang, M., Yun, B. L., Shin, S. U., Kim, B., & Kim, S. M. (2021). Diagnostic performance of artificial intelligence-based computer-aided diagnosis for breast microcalcification on mammography. *Diagnostics, 11*(8), Article 8. https://doi.org/10.3390/diagnostics11081409

FDA. (2022). *Software as a medical device (SaMD): Clinical evaluation*. U.S. Food and Drug Administration.

Frid-Adar, M., Diamant, I., Klang, E., Amitai, M., Goldberger, J., & Greenspan, H. (2018). GAN-based synthetic medical image augmentation for increased CNN performance in liver lesion classification. *Neurocomputing, 321*, 321–331.

Ghosh, P., Azam, S., Hasib, K. M., Karim, A., Jonkman, M., & Anwar, A. (2021, July). A performance based study on deep learning algorithms in the effective prediction of breast cancer. In *2021 international joint conference on neural networks (IJCNN)* (pp. 1–8). IEEE.

Gulum, M. A., Trombley, C. M., & Kantardzic, M. (2021). A review of explainable deep learning cancer detection models in medical imaging. *Applied Sciences, 11*(10), 4573. https://doi.org/10.3390/app11104573

Girdhar N, Sinha A, Gupta S. DenseNet-II: an improved deep convolutional neural network for melanoma cancer detection. Soft comput. 2022 Aug 24:1-20. https://doi.org/10.1007/s00500-022-07406-z Epub ahead of print. PMID: 36034768; PMCID: PMC9400005.

Hussein, Sarfaraz, Robert Gillies, Kunlin Cao, Qi Song, and Ulas Bagci. "Tumornet: Lung nodule characterization using multi-view convolutional neural network with gaussian process." In 2017 IEEE 14th international symposium on biomedical imaging (ISBI 2017), pp. 1007–1010. IEEE, 2017.

He, K., Fan, H., Wu, Y., Xie, S., & Girshick, R. (2020). Momentum contrast for unsupervised visual representation learning. In *Proceedings of the IEEE/CVF conference on computer vision and pattern recognition* (pp. 9729–9738). IEEE.

Jiang, X., Hu, Z., Wang, S., & Zhang, Y. (2023). Deep learning for medical image-based cancer diagnosis. *Cancers (Basel), 15*(14), 3608. https://doi.org/10.3390/cancers15143608

Johnson, M., Davis, K., & Gupta, R. (2020). Real-time image processing in oncology: Enhancing diagnostic efficiency with deep learning. *Radiology Today, 40*(2), 34–39.

Jovčevska, I. (2020). Next generation sequencing and machine learning technologies are painting the epigenetic portrait of glioblastoma. *Frontiers in Oncology, 10*, 798. https://doi.org/10.3389/fonc.2020.00798

Kalaiyarasi, M., Dhanasekar, R., Ram, S. S., & Vaishnavi, P. (2020). Classification of benign or malignant tumor using machine learning. *IOP Conference Series: Materials Science and Engineering, 995*(1), 012028. https://doi.org/10.1088/1757-899X/995/1/012028

Kaushik, M., Joshi, R. C., Kushwah, A. S., Gupta, M. K., Banerjee, M., Burget, R., & Dutta, M. K. (2021). Cytokine gene variants and socio-demographic characteristics as predictors of cervical cancer: A machine learning approach. *Computers in Biology and Medicine, 134*, 104559.

Kousis, I., Perikos, I., Hatzilygeroudis, I., & Virvou, M. (2022). Deep learning methods for accurate skin cancer recognition and mobile application. *Electronics, 11*(9), Article 9. https://doi.org/10.3390/electronics11091294

Kumar, G., & Alqahtani, H. (2022). Deep learning-based cancer detection-recent developments, trend and challenges. *Computer Modeling in Engineering and Sciences, 130*(3), 1271–1307.

Kushwah, A. S., Gupta, M. K., Singh, R., & Banerjee, M. (2020). Cytokine gene variants and treatment outcome of cisplatin-based concomitant chemoradiotherapy in cervical cancer. *British Journal of Biomedical Science, 77*(2), 81–86.

Krizhevsky, A., Sutskever, I., & Hinton, G. E., ImageNet classification with deep convolutional neural networks. In Advances in Neural Information Processing Systems (NIPS),2012; 25, 1097 1105.

Kim, H.E., Cosa-Linan, A., Santhanam, N. et al. Transfer learning for medical image classification: a literature review. BMC Med Imaging 22, 69 (2022). https://doi.org/10.1186/s12880-022-00793-7

Koonce, B., & Koonce, B. Convolutional Neural Networks with Swift for TensorFlow: image recognition and dataset categorization. 2021. ISBN: 978-1-4842-6167-5.

Khan AA, Kumar R N, Chakma S, Das S. Sarcoma diagnosis by DNA methylation classifier: A systematic review, current status and future prospects. Pathol Res Pract. 2024 Nov;263:155634. https://doi.org/10.1016/j.prp.2024.155634 Epub 2024 Oct 4. PMID: 39383738.

LeCun, Y., Bengio, Y. & Hinton, G. Deep learning. Nature 521, 436–444 (2015). https://doi.org/10.1038/nature14539

Liu, Z., Ni, S., Yang, C., Sun, W., Huang, D., Su, H., Shu, J., & Qin, N. (2021). Axillary lymph node metastasis prediction by contrast-enhanced computed tomography images for breast cancer patients based on deep learning. *Computers in Biology and Medicine, 136*, 104715. https://doi.org/10.1016/j.compbiomed.2021.104715

Loizidou, K., Skouroumouni, G., Pitris, C., & Nikolaou, C. (2021). Digital subtraction of temporally sequential mammograms for improved detection and classification of microcalcifications. *European Radiology Experimental, 5*(1), 40. https://doi.org/10.1186/s41747-021-00238-w

Lundberg, S. (2017). A unified approach to interpreting model predictions. *arXiv preprint arXiv:1705.07874.*

Minna, J. D., Roth, J. A., & Gazdar, A. F. (2002). Focus on lung cancer. *Cancer Cell, 1*(1), 49–52. https://doi.org/10.1016/S1535-6108(02)00027-2

Milletari, Fausto, Nassir Navab, and Seyed-Ahmad Ahmadi. "V-net: Fully convolutional neural networks for volumetric medical image segmentation." 2016 fourth international conference on 3D vision (3DV). Ieee, 2016.

Mohsen, H., El-Dahshan, E.-S. A., El-Horbaty, E.-S. M., & Salem, A.-B. M. (2018). Classification using deep learning neural networks for brain tumors. *Future Computing and Informatics Journal, 3*(1), 68–71. https://doi.org/10.1016/j.fcij.2017.12.001

Niu, S., Huang, J., Li, J., Liu, X., Wang, D., Zhang, R., et al. (2020). Application of ultrasound artificial intelligence in the differential diagnosis between benign and malignant breast lesions of BI-RADS 4A. *BMC Cancer, 20*(1), 959. https://doi.org/10.1186/s12885-020-07413-z

N. Tajbakhsh et al., "Convolutional Neural Networks for Medical Image Analysis: Full Training or Fine Tuning?," in IEEE Transactions on Medical Imaging, vol. 35, no. 5, pp. 1299-1312, May 2016, https://doi.org/10.1109/TMI.2016.2535302

Oh, J.-H., Kim, H.-G., & Lee, K. M. (2023). Developing and evaluating deep learning algorithms for object detection: Key points for achieving superior model performance. *Korean Journal of Radiology, 24*(7), 698. https://doi.org/10.3348/kjr.2022.0765

Oktay, Ozan, Jo Schlemper, Loic Le Folgoc, Matthew Lee, Mattias Heinrich, Kazunari Misawa, Kensaku Mori et al. "Attention u-net: Learning where to look for the pancreas." arXiv preprint arXiv:1804.03999 (2018).

Park SY, Ayana G, Wako BD, Jeong KC, Yoon SD, Choe SW. Vision Transformers for Low-Quality Histopathological Images: A Case Study on Squamous Cell Carcinoma Margin Classification. Diagnostics (Basel). 2025 Jan 23;15(3):260. https://doi.org/10.3390/diagnostics15030260 PMID: 39941191; PMCID: PMC11817517.

Pranav, P., Patel, A., & Jain, S. (Eds.). (2024). *Machine learning in healthcare and security: Advances, obstacles, and solutions* (1st ed.). CRC Press. https://doi.org/10.1201/9781003388845

Prinzi, F., Orlando, A., Gaglio, S., & Vitabile, S. (2024). Interpretable radiomic signature for breast microcalcification detection and classification. *Journal of Imaging Informatics in Medicine, 37*(3), 1038–1053. https://doi.org/10.1007/s10278-024-01012-1

Punn, N. S., & Agarwal, S. (2022). Modality specific U-Net variants for biomedical image segmentation: A survey. *Artificial Intelligence Review, 55*(7), 5845–5889. https://doi.org/10.1007/s10462-022-10152-1

Quazi, F. (2024). Ethics & Responsible AI in Healthcare. International Journal of Global Innovations and Solutions (IJGIS). https://doi.org/10.21428/e90189c8.a16c4bb9

Rieke N, Hancox J, Li W, Milletarì F, Roth HR, Albarqouni S, Bakas S, Galtier MN, Landman BA, Maier-Hein K, Ourselin S, Sheller M, Summers RM, Trask A, Xu D, Baust M, Cardoso MJ. The future of digital health with federated learning. NPJ Digit Med. 2020 Sep 14;3:119. https://doi.org/10.1038/s41746-020-00323-1PMID: 33015372; PMCID: PMC7490367.

Ricci Lara, M.A., Echeveste, R. & Ferrante, E. Addressing fairness in artificial intelligence for medical imaging. Nat Commun 13, 4581 (2022). https://doi.org/10.1038/s41467-022-32186-3

Sheller MJ, Reina GA, Edwards B, Martin J, Bakas S. Multi-Institutional Deep Learning Modeling Without Sharing Patient Data: A Feasibility Study on Brain Tumor Segmentation. Brainlesion. 2019;11383:92–104. https://doi.org/10.1007/978-3-030-11723-8_9 Epub 2019 Jan 26. PMID: 31231720; PMCID: PMC6589345.

Schumacher, M., Holländer, N., & Sauerbrei, W. (1997). Resampling and cross-validation techniques: A tool to reduce bias caused by model building? *Statistics in Medicine, 16*(24), 2813–2827.

Selvaraju, R. R., Cogswell, M., Das, A., Vedantam, R., Parikh, D., & Batra, D. (2017). Grad-cam: Visual explanations from deep networks via gradient-based localization. In *Proceedings of the IEEE international conference on computer vision* (pp. 618–626). IEEE.

Shen, L., Margolies, L. R., Rothstein, J. H., Fluder, E., McBride, R., & Sieh, W. (2019). Deep learning to improve breast cancer detection on screening mammography. *Scientific Reports, 9*(1), 12495.

Shorten, C., & Khoshgoftaar, T. M. (2019). A survey on image data augmentation for deep learning. *Journal of Big Data, 6*(1), 1–48.

Singh, O., & Singh, K. K. (2024). Local binary pattern (LBP) and transfer learning based approach to classify lung and colon cancer. *SN Computer Science, 5*(6), 783.

Singh, O., Kashyap, K. L., & Singh, K. K. (2023). Mesh-free technique for enhancement of the lung CT image. *Biomedical Signal Processing and Control, 81*, 104452.

Singh, S. P., Wang, L., Gupta, S., Goli, H., Padmanabhan, P., & Gulyás, B. (2020). 3D deep learning on medical images: a review. Sensors, 20(18), 5097.

Skin Cancer Detection Based on Deep Learning—PMC. (n.d.). Retrieved November 15, 2024, from https://pmc.ncbi.nlm.nih.gov/articles/PMC9759648/

Snell, J., Swersky, K., & Zemel, R. (2017). Prototypical networks for few-shot learning. In *Advances in neural information processing systems* (p. 30). NIPS.

Sun, Y.-S., Zhao, Z., Yang, Z.-N., Xu, F., Lu, H.-J., Zhu, Z.-Y., et al. (2017). Risk factors and preventions of breast cancer. *International Journal of Biological Sciences, 13*(11), 1387. https://doi.org/10.7150/ijbs.21635

Smith M, Sattler A, Hong G, Lin S. From Code to Bedside: Implementing Artificial Intelligence Using Quality Improvement Methods. J Gen Intern Med. 2021 Apr;36(4):1061-1066. https://doi.org/10.1007/s11606-020-06394-w Epub 2021 Jan 19. PMID: 33469745; PMCID: PMC8041947.

Tang, Y., Liang, M., Tao, L., Deng, M., & Li, T. (2022). Machine learning–based diagnostic evaluation of shear-wave elastography in BI-RADS category 4 breast cancer screening: A multicenter, retrospective study. *Quantitative Imaging in Medicine and Surgery, 12*(2), 1223. https://doi.org/10.21037/qims-21-341

Tiwari, V., Pandey, C., Dwivedi, A., & Yadav, V. (2020, December). Image classification using deep neural network. In *2020 2nd international conference on advances in computing, communication control and networking (ICACCCN)* (pp. 730–733). IEEE.

T. -Y. Lin, P. Goyal, R. Girshick, K. He and P. Dollár, "Focal Loss for Dense Object Detection," 2017 IEEE International Conference on Computer Vision (ICCV), Venice, Italy, 2017, pp. 2999–3007, https://doi.org/10.1109/ICCV.2017.324

Vanguri, Rami S., Jia Luo, Andrew T. Aukerman, Jacklynn V. Egger, Christopher J. Fong, Natally Horvat, Andrew Pagano et al. "Multimodal integration of radiology, pathology and genomics for prediction of response to PD-(L) 1 blockade in patients with non-small cell lung cancer." Nature cancer 3, no. 10 (2022): 1151–1164.

White, A., Thompson, S., & Patel, D. (2021). Quality assurance protocols for AI in clinical oncology. *Journal of Healthcare Quality, 43*(3), 150–160.

Winkler, J. K., Sies, K., Fink, C., Toberer, F., Enk, A., Deinlein, T., et al. (2020). Melanoma recognition by a deep learning convolutional neural network—Performance in different melanoma subtypes and localisations. *European Journal of Cancer, 127*, 21–29. https://doi.org/10.1016/j.ejca.2019.11.020

Yala, A., Lehman, C., Schuster, T., Portnoi, T., & Barzilay, R. (2019). A deep learning mammography-based model for improved breast cancer risk prediction. *Radiology, 292*(1), 60–66. https://doi.org/10.1148/radiol.2019182716

Zhou Z, Siddiquee MMR, Tajbakhsh N, Liang J. UNet++: A Nested U-Net Architecture for Medical Image Segmentation. Deep Learn Med Image Anal Multimodal Learn Clin Decis Support (2018). 2018 Sep;11045:3-11. https://doi.org/10.1007/978-3-030-00889-5_1 Epub 2018 Sep 20. PMID: 32613207; PMCID: PMC7329239.

Ethical Consideration Regarding Patient Privacy and Data Security

Gayatri Mann, Anglina, and Pratibha Gaur

1 Introduction

Patients looking for medical services and information cannot find them anywhere else than on hospital websites. A patient visiting a hospital website can identify a doctor, make an appointment, see test results, and obtain medical information with just a few clicks. However, patients also face privacy dangers in conjunction with these benefits. In 2021, a US$ 18 million settlement was reached between Mass General Brigham and the Dana Farber Cancer Institute. The plaintiffs claimed that the hospital systems had utilized third-party tracking technologies on their public websites without obtaining users' consent. While the settlement was significant, further investigation has revealed that tracking technologies are frequently used on hospital websites.

Privacy policies are not always a perfect solution because they can be tedious to read and ambiguous. Users of hospital websites may find it time-consuming and challenging to comprehend privacy regulations, which makes them an inadequate answer for preserving their privacy. However, in the context of hospitals using tracking technologies, they play a vital role. Hospitals can be held responsible for their promises to preserve user privacy by using privacy policies as a means of

G. Mann · Anglina
Department of Biotechnology and Microbiology, School of Science and Humanities, SRM University, Sonepat, Haryana, India

P. Gaur (✉)
Department of Biotechnology and Microbiology, School of Science and Humanities, SRM University, Sonepat, Haryana, India

Centre for Drug Design Discovery and Development (C4D), SRM University, Sonepat, Haryana, India
e-mail: pratibhagaur@srmuniversity.ac.in

R. Nema et al. (eds.), *Advances in Cancer Detection, Prediction, and Prognosis Using Artificial Intelligence and Machine Learning*,
https://doi.org/10.1007/978-981-96-9346-7_17

holding them to account for violations that could result in regulatory attention or civil lawsuits. Researchers, writers, and consumer advocates can also find inconsistencies between declared and real privacy practices by using privacy policies. Lastly, although the privacy policy may not be read by the majority of hospital website visitors, its availability respects people's autonomy by enabling them to make more informed decisions concerning.

2 Patient Privacy and Data Security

By offering thorough, real-time access to patient health information, electronic health records (EHRs) play a pivotal role in improving care coordination by promoting more informed decision-making and care continuity. The digitization of health records, however, also brings up serious privacy issues, including the possibility of illegal access and data breaches. That being said, healthcare providers must put strong security measures in place to safeguard patient data while guaranteeing that authorized workers may access it when needed. The concept of security is "the technical and procedural measures required to prevent intentional denial of service attacks, prevent unauthorized access, modification, use, and dissemination of data stored or processed in a computer system, and protect the system from physical harm." Security aids in preventing unwanted access to medical records. A breach of security occurs when someone breaks into a computer system (and also potentially a breach of confidentiality). However, no security solution can stop those with access rights to the data from violating someone's privacy. The privacy of medical records (henceforth referred to as protected health information) and other personal information is a legitimate concern that is being raised more frequently. There are very serious consequences if this information is compromised. Negative consequences include losing personal health insurance, being fired from a job without cause, and having one's identity used illegally in a variety of ways, such as credit card charges or passport fraud.

3 Core Ethical Principles

The development of contemporary technology, such as cloud computing and sensors, has fundamentally altered traditional healthcare systems. Once paper-based medical records have been digitized, such systems can show the great promise of next-generation healthcare services. Medical devices can remotely sense a person's health, medical networks can broadcast that information, and cloud computing, fog, and the edge can process that information. Smart healthcare services including remote monitoring, diagnosis, treatment, and prescription based on personal electronic health (e-health) data will increasingly depend on innovative healthcare systems that can enhance quality of life. However, there are drawbacks to the current e-healthcare system. Even while it provides us with cutting-edge medical services, security issues are becoming more prevalent. Some of a person's most private information is e-health data. To improve the governance of healthcare data, privacy regulations such as the General Data Protection Regulation (GDPR) and the Health Insurance Portability and Accountability Act (HIPAA) were created; yet, e-health data breaches have been common. Additionally, e-health data's security attack vectors have been expanding in tandem with its increased usefulness and accessibility. Due to software flaws and wireless connections, 1.5 million medical devices have been compromised in the past 10 years, and cloud computing services that store and process e-health data have become targets for massive e-health data.

4 Legal and Ethical Framework

4.1 HIPAA

Health care organizations' operations have been impacted by the 1996 Health Insurance Portability and Accountability Act (HIPAA). There are five titles under HIPAA, and its rules are intricate. The parts of HIPAA that deal with safeguarding the confidentiality and integrity of patient medical records are well known. The adoption of electronic medical records is covered by new HIPAA regulations. HIPAA establishes guidelines for what should be secured and safeguarded, including protected health information (PHI). The privacy rule establishes guidelines that any organization handling health data must adhere to in order to safeguard patients' private medical information. It also governs the use and sharing of PHI. Laws pertaining to privacy and security are always changing to meet the demands of the digital age, new dangers, and emerging technology. The 1996 Health Insurance Portability and Accountability Act regulations are being implemented more rigorously than in the past. HIPAA rule infractions persist despite the need that personnel who handle or have access to patient data undergo periodic HIPAA training. HIPAA has been updated and modified recently, strengthening its requirements and expanding the list of those who must abide by it. A law that established national standards to safeguard the privacy of medical records was passed by the federal government in 1996. Title 45 of the Code of Federal Regulations, parts 160–164, contains information about HIPAA. The main goal of HIPAA, which was approved by Congress and signed into law by President Bill Clinton on August 21, 1996, was to guarantee millions of Americans' health insurance coverage would be portable and continuous. However, HIPAA encompasses more than just portability, privacy, and security. Title 45 of the Code of Federal Regulations, part 164, contains the five sections that make up HIPAA. The regulations are extensive and intricate.

4.2 Privacy Rule

The term "privacy" in relation to PHI refers to the duty of authorized users to maintain the confidentiality of such information. The privacy regulation establishes the minimal national standards that all covered entities must adhere to in order to safeguard patients' private medical information. It also governs the use and sharing of PHI. All of an individual's PHI must be kept private and secret in accordance with HIPAA regulations.

Health plans (such as Anthem, Medicare, and Cigna), clearinghouses for health care (such as billing firms), and healthcare providers (such as clinics, hospitals, and doctor's offices) as well as their business partners (such as consultants and attorneys) that engage in health care transactions are all subject to the HIPAA privacy rule. Patients also have the right to seek a copy of their medical records and their own radiographs under this provision.

4.3 Exception of Privacy Rule

PHI cannot be disclosed by an organization covered by HIPAA unless a patient gives written consent. HIPAA, however, specifies particular situations in which PHI may be disclosed without a person's express written consent. PHI may be shared without consent under several situations:

- *For general health care operations, payment, or treatment.* In order to get paid for their services, providers may also give billing businesses access to PHI. Additionally, PHI can be used to carry out compliance or system quality checks that enhance healthcare operations. Only two categories of clinical care information are prohibited by HIPAA from being shared between providers without the patient's express consent: written psychotherapy notes and substance-abuse records from a licensed abuse program. If the patient is able to consent to or reject a disclosure. For instance, a patient immediately consents that someone else can hear the PHI when they bring them into the room.
- *In the event of a natural disaster.* For instance, following the implementation of the disaster protocol in hospitals located in the Texas and Louisiana areas, certain infraction penalties were waived during Hurricane Harvey in 2017. The waiver is only applicable to hospitals in the disaster area during the emergency period specified in the public health emergency declaration and for up to 72 h following the execution of a hospital's disaster protocol. It is issued by the president of the United States or a state official.

4.4 GDPR

Before the General Data Protection Regulation (GDPR) was published, scientists were quite concerned about data privacy (Myers & Liskov, 2000; Priscakova & Rabova, 2013). The General Data Protection Regulation (EU GDPR), 2016—a comprehensive regulation with 99 articles—was published in 2018. All of the EU's member states are immediately impacted by this regulation, and it differs from earlier data privacy laws in that it also applies to any non-EU entity that handles the data of European individuals. In order to help privacy-conscious applications, create

Unified Modeling Language (UML) models that define privacy concepts and enhance the definition and use of privacy, Basso et al. (2016) provide a UML profile for data privacy modeling. A profile for privacy-aware data, known as the APDL, is also suggested by Alshammari and Simpson (2018) as an abstract model for the life cycles of personal data. They specifically make a distinction between the various actions that can be carried out on personal data throughout its lifecycle. The APDL profile does not currently follow the UML standard, despite their suggestion that it might be described in terms of UML. Interestingly, the GDPR is not specifically taken into account by these privacy data models. Some other works solely address certain parts of the GDPR. For example, Mougiakou and Virvou (2017) offer a model that examines GDPR requirements using an educational e-platform paradigm called Law Courses. It combines the GDPR, information privacy, and best practices with UML use case diagrams. A GDPR model and its auxiliary techniques are presented by Matulevičius et al. (2020) for the management of regulatory compliance in corporate operations. To model the various facets of the GDPR, including consent and data processing, they employ component diagrams. They do not, however, take data tracking into account or model the relationships between system roles.

The fact that legal specialists, not software or information engineers, defined the GDPR is one of the primary issues with its implementation in the field of information technology (Tamburri, 2020). As a result, a lot of the literature is devoted to attempting to assist software engineers with the GDPR's implementation. A lot of these projects make use of modeling approaches, which facilitate data management and give software engineers a broad perspective on systems. In an effort to standardize data protection laws among EU member states, the General Data Protection Regulation (GDPR) (General Data Protection Regulation (EU GDPR), 2016) went into effect on May 25, 2018. The Data Protection Directive, which was created in 1995 to harmonize data protection rules among EU member states in order to ease both internal and cross-border data transfers, was superseded by the GDPR in 2016. Since the GDPR is a regulation, recipients are directly covered by it; unlike the Data Protection Directive, no additional transpositions are necessary. The GDPR was implemented to raise the standard for the privacy of those impacted by the data protection regulations, create more legal certainty, and remove any potential barriers to the free movement of personal data. The GDPR's empowering of data subjects with specific rights to assist them in being confident of the privacy and safety of their personal data is another crucial component (General Data safety Regulation (EU GDPR), 2016). These rights include the following: the right to information, the right to access, the right to rectification, the right to erasure, the right to portability of data, the right to object, the right to prevent automated decision-making, and the right to restrict processing.

5 Patient Consent and Ethical Data Sharing

When a business uses consent to process personal data, it needs to be able to demonstrate that the person granted their consent. If the consent is a component of a wider agreement, it ought to be expressed in simple language, in a clear and distinct manner. Any legal provision that is violated by the agreement will be void. Individuals may withdraw their consent at any moment, in which case the processing completed prior to the withdrawal will not be impacted. Companies must educate individuals about their right to withdraw, and the process to take back consent should be as easy as granting it. The organization should lawfully process the child's personal data only if the child is at least 16 years old. Processing of data of those under the age of 16 is only permitted with agreement granted by the child's parent or legal guardian. While certain nations may have a lower age restriction, it cannot be less than 13 years old. The company must make reasonable measures to verify that consent is genuinely given by the parent or guardian, considering the technologies available.

Sharing genetic and sensitive data, especially in the context of research, presents various ethical problems. Consent and privacy are two of the main issues. Genetic data is unique and extremely personal, not only influencing the individual but also relatives as well. Because of this, it is crucial that researchers obtain consent authorization, making sure subjects are aware of the potential uses of their data. Getting family members to agree, particularly when genetic information affects them indirectly, is harder to accomplish. Furthermore, there are ethical concerns associated with sharing genetic data across family members because decisions taken by one individual may have an impact on other relatives. Data sharing regulations must therefore seek a compromise between the advancement of science and maintaining of family privacy, guaranteeing that individuals maintain control over the use of their genetic data.

It is a difficult ethical task to strike a balance between patient privacy and the advantages of data sharing for research and therapy. Sharing genetic and medical data can, on the one hand, improve treatment outcomes, boost personalized therapy, and hasten scientific discovery. However, protecting patient privacy is essential to preventing confidentiality violations and the exploitation of private data.

Patient control and permission: One of the most important ways to safeguard patients' privacy is to make sure they provide informed authorization. This includes describing in detail how their data will be used, shared, and kept. Patients can keep control over their data with the use of certain tools, which allows them to change their preferences or withdraw approval as research proceeds.

Limiting access and data anonymization: Research facilities frequently anonymize data to safeguard identities of patients. Reidentification is still possible, though, particularly with genetic data, since technological developments may make it possible to trace anonymized data back to specific people. Controlled access to data, where only authorized researchers can access sensitive information, can lessen these dangers. Tiered access is a common feature of data repositories, with more stringent guidelines for sensitive material such as genetic information.

Public trust: Preserving people's confidence in medical research is essential for maintaining this balance. Study indicates that the exploitation of genetic information or a lack of openness can erode confidence and discourage research participation. Models for ethical data-sharing that put patients' needs ahead of business interests foster trust and promote voluntary data sharing (Thorogood & Chokoshvili, 2023).

6 Anonymization Techniques and Tools

The process of converting personal information into a form that prevents direct or indirect identification of individuals while preserving the data's value for analysis or research is known as data anonymization. To do this, characteristics such as names, addresses, or other distinctive information are changed or removed. In some circumstances, anonymization can involve procedures such as data masking, pseudonymization, or the use of cryptographic techniques to further protect privacy. The procedure needs to be carefully planned, particularly when handling sensitive data that may provide additional difficulties in maintaining anonymity, such as genetic or medical records. Since people can occasionally be reidentified by connecting anonymized data with other datasets, true anonymization is difficult to achieve (Monteiro et al., 2024).

Anonymization is achieved through a variety of approaches. These are a few typical techniques:

Pseudonymization: Pseudonyms or codes are used in place of social security numbers or other personal identifiers. In the process, direct identifiers are obscured, yet data is still useful for analysis. When merged with other datasets, pseudonymized data can occasionally still be connected to specific people.

Data masking: Data masking involves altering data in a way that makes personally identifiable details difficult to see. For example, swapping out real names for names that are generated at random or jumbling personal characteristics. Test environments that require data to mimic real-world data while maintaining privacy standards are common uses for masked data.

Generalization: In order to make people less identifiable, this technique involves decreasing the precision of the data. For instance, an individual's age range (e.g., 30–40 years) can be displayed rather than their precise age. In the same manner, geographic data can be made broader by displaying areas rather than unique addresses.

Suppression: In some circumstances, it is safer to get rid of all personally identifiable information. This method works well for anonymizing exceptional data items, such as a set of unique traits that, even after anonymization, may be used to identify a specific person.

k-Anonymity: It is a statistical strategy in which each individual is made to be identical to at least "k" others through modification of the data. This indicates that the data in a dataset cannot be uniquely separated from the data of at least k−1 additional individuals. For instance, combining the records of several people to make sure they have similar traits.

Working with anonymized data complicates the ethics of consent. Although it is a common misconception that consent is not needed for anonymized data, this misconception can violate people's rights. Even with anonymized data, ethical guidelines emphasize the significance of gaining informed consent to make sure participants are aware of how their data might be used. This viewpoint emphasizes the necessity of ethical data governance frameworks by being consistent with the concepts of autonomy and respect for individuals. The risks associated with the reidentification and misuse of "anonymous" data are significant concerns in the field of data privacy and ethics. This is especially problematic for detailed datasets because identification can be made from seemingly simple information. A well-known instance involved linking people with publicly accessible voter registration databases to reidentify people in anonymized health information. By comparing anonymized medical information with publicly accessible records, researchers were able to determine the identity of the patients. Anonymized datasets can have patterns that can be exploited by techniques such as machine learning, data mining, and statistical analysis to derive identities or sensitive characteristics. Researchers have shown that the probability of a successful reidentification can be raised by combining different data points (such as demographics, geographical location, and health data) (Ferretti et al., 2021).

Misuse of anonymized data can occur in several ways, leading to ethical and legal implications. It is possible to modify even anonymized data to produce biased algorithms that discriminate against particular populations. For instance, if the underlying data reflects prejudiced cultural standards, gathered information utilized for machine learning could strengthen already-existing differences. Businesses may profit from anonymized datasets by exploiting them to create focused advertising campaigns or sell insights without the participants' knowledge or consent. This brings up moral concerns regarding data ownership and people's rights to their personal information. Although anonymized, datasets may still be subject to security breaches. Unauthorized access to anonymized data may arise from cyberattacks; this presents potential risks not only for people but also for corporations that manage such data (Florea & Florea, 2020).

Protecting individual privacy while preserving the usefulness of data has become a crucial challenge in a society that is becoming more and more data-driven. In order to exploit sensitive datasets in industries such as healthcare, banking, and the social sciences without jeopardizing individual privacy, data anonymization techniques are crucial.

ARX is a popular tool that offers strong and adaptable data anonymization techniques. The University of Luxembourg is the developer of the open-source data anonymization tool ARX. By using a variety of privacy-preserving strategies that lower the possibility of reidentification, it is intended to anonymize datasets. The software provides a complete solution for researchers and professionals dealing with sensitive data because it supports a variety of privacy models, algorithms, and risk assessment criteria. ARX's ability to strike a balance between privacy and data value is one of its main advantages. ARX assists businesses in adhering to privacy laws such as the GDPR, HIPAA, and others by anonymizing data in a way that minimizes the possibility of revealing people's identities while maintaining the dataset's statistical or analytical usefulness. Furthermore, the platform's high degree of customization enables customers to modify anonymization procedures to meet the unique privacy needs of their datasets (ARX, 2020).

ARX offers a number of proven anonymization methods, such as *Pseudonymization* that substitutes fictitious or pseudonym identifiers for direct identifiers such as names or social security numbers. Although pseudonymization aids in the protection of individual identities, it does not completely anonymize the data because it may be possible to relink the original information. *Generalization* aims to lessen identifiability; it entails converting particular data points into more inclusive groupings. For instance, age ranges could be used in place of precise ages, or certain zip codes could be extrapolated to broader geographic areas. Depending on how much generalization is used, this method may lead to some loss of information even while it preserves data utility and protects privacy. *Suppression* involves removing from the dataset particular values or complete records that present a high risk of reidentification. For example, ARX can suppress values in a dataset that have uncommon combinations of characteristics that identify individuals. Suppression is useful for getting rid of high-risk data, but it can make the dataset less complete overall.

ARX is compatible with the *k-anonymity* model, which guarantees that, with regard to the quasi-identifiers (such age, gender, or location) that could be exploited for reidentification, every record in the anonymized dataset is identical to at least $k-1$ other records. Although this method reduces the possibility of unique identification, it might not be impenetrable to all attacks, such as those that link other data sources. ARX additionally supports *l-diversity*, a privacy model that tackles scenarios in which sensitive features (such as disease information) could still be determined, even inside a k-anonymous dataset, in order to reinforce k-anonymity. L-diversity makes it more difficult to predict sensitive information by ensuring that there is adequate variation in the sensitive qualities within each equivalency class. By demanding that the distribution of sensitive attributes in any anonymized subset (equivalence class) be near to their distribution in the entire dataset, the *T-Closeness* privacy model goes one step further. This stops attackers from examining the statistical distribution of private attributes within a particular group in order to obtain information (Prasser et al., 2020).

ARX's strong risk assessment module is one of its primary features. By examining sensitive features and quasi-identifiers in the dataset, users can assess the possible danger of reidentification prior to implementing anonymization procedures. ARX offers a number of indicators, including uniqueness risk, to evaluate the likelihood that a person can be reidentified.

ARX also enables users to examine how anonymization affects data utility. Finding the ideal balance between minimizing privacy threats and preserving the dataset's analytical value is difficult because anonymization frequently necessitates some degree of data distortion. By evaluating various anonymization setups and quantifying the loss of data quality, ARX assists users in weighing trade-offs between privacy and utility (Prasser et al., 2014).

Like ARX, a number of data anonymization technologies provide a variety of methods to safeguard private data. These solutions protect privacy while preserving the usefulness of data in a variety of industries, including research, healthcare, and finance. Depending on the particular privacy requirements and the type of data, each tool uses a different technique.

sdcMicro, an R tool for statistical disclosure control (SDC) in microdata, is one of the popular substitutes. In addition to providing sophisticated SDC techniques such as microaggregation and post-randomization, sdcMicro provides techniques including data suppression, generalization, and noise addition. It is well-liked by statisticians and academics who work with microdata because it offers thorough risk analysis for disclosure management (Templ et al., 2015).

Another approachable tool that provides both k-anonymity and l-diversity approaches is *Amnesia*, which was created by the ATHENA Research Center. Through a desktop application or web-based interface, it may interactively anonymize data. Amnesia is perfect for businesses and institutions that need to quickly anonymize tabular datasets while also assessing reidentification threats because of its versatility, which includes both human and automatic anonymization procedures (Tomás et al., 2022).

Privacy Analytics is a commercial product that focuses in anonymizing datasets for adherence to laws such as the GDPR and HIPAA for organizations handling healthcare data. It incorporates real-time risk monitoring and provides the anonymization of data that is structured as well as unstructured. This tool's emphasis on safeguarding personal health information (PHI) makes it popular in the pharmaceutical and healthcare industries. Another technique for anonymizing microdata in official statistics is called *μ-Argus* (*Mu-Argus*), and it was created by Statistics Netherlands. To guarantee that data is still valuable while lowering the danger of reidentification, it uses techniques including local suppression, recoding, and k-anonymity in addition to a number of risk assessment procedures. National statistics agencies have adopted it, demonstrating its dependability in handling huge datasets containing sensitive data (Franconi & Polettini, 2004). A different strategy is offered by *Aircloak Insights*, which uses query-based anonymization and differential privacy to enable safe access to sensitive data. Even while analyzing aggregate data, it prevents individual reidentification by adding controlled noise to the answers. This tool is widely used in industries where sensitive data processing in real time is essential, such as financial services and telecommunications.

Anonimatron is an open-source application that substitutes sensitive fields with realistic, random data while preserving format integrity for database administrators wishing to anonymize huge datasets. It is a workable way to anonymize production databases for testing or development since it is simple to connect with relational database platforms such as MySQL or PostgreSQL. Another method employed in anonymization technologies is called *Format-Preserving Encryption* (*FPE*), which securely encrypts data while maintaining the original form of data, such as dates or credit card numbers. FPE is particularly helpful in sectors such as financial services and healthcare where data format maintenance is necessary for legacy systems or compliance with regulations. Lastly, *Aircloak's Diffix* provides query-based anonymization through differential privacy. It guarantees the anonymity of the people in the data while enabling analysts to query sensitive datasets. The tool's ability to avoid linkage attacks by providing noise to query results makes it especially useful for enterprises that need to guarantee secure data access without sacrificing privacy.

7 Misuse of Anonymous Data

7.1 Discrimination and Bias

Even anonymized data can be used to generate biased algorithms that unintentionally perpetuate discrimination against particular communities. Machine learning models trained on such data can reinforce existing inequities if the underlying data reflects biased societal norms or historical injustices. For instance, anonymized data related to healthcare, employment, or credit decisions might still capture patterns that disadvantage certain racial, gender, or socioeconomic groups. Without careful oversight and inclusion of fairness criteria, these biases can become entrenched in predictive models, leading to further marginalization and reinforcing systemic inequality. Additionally, bias in the data collection process itself—such as overrepresentation or underrepresentation of certain groups—can skew the outcomes of algorithmic decision-making (Ferretti et al., 2021).

7.2 Commercial Exploitation

Businesses may profit from the use of anonymized datasets by selling insights, creating targeted advertisements, or developing personalized services without the participants' knowledge or consent. Even though the data is anonymized, it still holds significant commercial value, which companies can exploit for financial gain. This raises moral concerns about data ownership and individuals' rights over their personal information, as participants may not be aware of how their data is being utilized or monetized. The lack of transparency in data use can erode trust and create an imbalance in power, where corporations benefit at the expense of individuals' privacy and autonomy. Furthermore, in some cases, anonymized data can be reidentified, amplifying these ethical concerns.

7.3 Data Breaches

Even after anonymization, datasets remain vulnerable to data breaches. Unauthorized access to anonymized data—whether through cyberattacks, internal misuse, or accidental exposure—poses significant risks. Advanced reidentification techniques can often reverse anonymization, allowing attackers to link seemingly anonymous data back to individuals. This can expose sensitive personal information, leading to identity theft, financial fraud, or harm to reputations. Not only does this endanger the individuals whose data is compromised, but it also exposes the companies handling the data to legal, financial, and reputational consequences. Such breaches highlight the ongoing challenges in ensuring both privacy and security, even with anonymized datasets (Florea & Florea, 2020).

8 Responsibilities of Healthcare Providers in Data Usage

Healthcare professionals have a critical obligation to protect sensitive patient information in the digital era. This obligation entails not only adhering to tight security procedures but also striking a delicate balance between safeguarding privacy and allowing sharing of information for research, treatment breakthroughs, and medical innovation. Healthcare providers have many responsibilities, including ensuring confidentiality and integrity of data while maintaining a duty of transparency. Transparency means that patients should be completely informed about how their data is gathered, stored, and utilized, as well as any third parties that may have access to it.

Thus, healthcare practitioners must strike a careful balance between the requirement for data confidentiality and the potential benefits of data sharing, especially in public health research, clinical trials, and precision medicine. Ethical frameworks highlight the importance of patient autonomy, beneficence, and justice, while legal requirements support this by enforcing clear policies and procedures that prioritize patients' rights. By carefully handling these commitments, healthcare providers help to develop trust, ensure data security, and advance the possibilities for healthcare innovations.

8.1 Transparency in Data Usage

A primary responsibility of healthcare professionals is to preserve transparency in the use of information concerning patients. This duty is based not only on ethical values but also on legal requirements established in privacy rules such as the European Union's General Data Protection Regulation (GDPR) and the United States' Health Insurance Portability and Accountability Act (HIPAA). These requirements require medical professionals to explicitly declare how they will manage patient data, including who will have ownership of it, the exact purposes for which it will be used, and any conditions under which it might be shared with other organizations.

Transparency entails providing patients with a clear awareness of the entire data lifecycle, from collection to prospective distribution. Under GDPR, patients have the right to find out what data is collected, why it is collected, and how long it will be maintained, as well as the ability to access, correct, or delete their personal information under specific conditions. HIPAA, on the other hand, requires identical requirements for disclosure in the United States, ensuring that patients are aware of their rights to privacy and how their medical data may be used or released, particularly if provided for purposes other than immediate care.

By adhering to these transparency principles, healthcare providers enable patients to take informed decisions about their data, thereby increasing patient trust and establishing an accountability culture within healthcare. It also helps to prevent the exploitation of confidential information and is consistent with broader ethical standards, including respect for the autonomy of patients and the duty to do what is in the patient's best interests. As healthcare organizations become more reliant on digital information, transparency is a critical component of ethical and compliance data stewardship.

8.2 Informed Consent

Obtaining patients' informed permission is a critical component of healthcare transparency. Informed permission ensures that patients understand and agree to the storage, use, and potential dissemination of their personal information. To make informed consent meaningful, healthcare professionals have to convey this information in straightforward, simple terms, avoiding technical terms that may confuse or mislead. This clarification enables patients to make informed decisions about their data and know the consequences of their agreement (Ferretti et al., 2021). Regulations such as the GDPR and HIPAA highlight the necessity of informed consent, making it a legal requirement for healthcare providers. The GDPR, for example, requires that consent be "freely given, specific, informed, and unambiguous," which means that patients want clear information to make an informed decision. Healthcare professionals follow ethical standards and promote trust by ensuring that patients understand how their data will be used, assuring them that their information will be treated with respect for their autonomy and rights.

Obtaining transparency in healthcare data procedures might be difficult in practice. The intricacy of data-sharing agreements, along with the technical jargon that commonly appears in consent forms and privacy rules, can make it difficult for patients to completely understand the ramifications of how their data is used. According to studies, many patients skim or do not read, let alone understand, privacy rules. This raises fundamental questions about whether informed consent is truly being gained. Informed consent requires patients to have a clear, accessible knowledge of data use, which is hampered by the complex language and extensive disclosures that are common in conventional consent agreements. For example, a patient may sign a consent form believing that their data will only be used for direct care, whereas in reality it may be shared with third parties for research or corporate interests. If the specifics are hidden in convoluted language, the patient's consent may be legally gained but does not reflect a true comprehension or voluntary agreement.

The discrepancy between legal conformity and genuine comprehension creates an ethical dilemma. Some healthcare providers and organizations are now using efficient methods to tackle this issue, such as utilizing simplified language, summarizing key points, and giving electronic consent tools that clarify concepts using visuals or interactive features. Healthcare practitioners can make great progress toward true transparency by emphasizing clarity and accessibility, ensuring that informed approval is both understandable and truly representative of patient autonomy.

9 Patient Control Over Personal Data

Beyond openness, medical professionals have a critical responsibility to empower patients by giving them the ability to manage their personal information. Patient autonomy is a key principle of medical ethics, based on the belief that people should have control over decisions impacting their medical care, including how their personal health information is handled. This autonomy extends beyond clinical care and includes the processing of patient data. Empowering patients entails giving them particular rights to view, update, and, in some situations, erase their personal information. Access to one's health data is critical for patients to make educated decisions and follow their progress over time. Furthermore, the capacity to update faulty information ensures the accuracy of medical records, lowering the possibility of errors that could harm treatment outcomes. Patients may have the right to seek the erasure of their data in certain circumstances, including as when it is no longer required or was collected improperly, as outlined in rules such as the GDPR.

Such rights are more than simply legal niceties; they are critical to protecting patients' privacy and autonomy in an increasingly digital healthcare context. Healthcare practitioners are increasingly utilizing user-friendly digital technologies and patient portals to help them maintain control. Such platforms not just make it convenient for patients to access and manage their health information but they also foster trust by emphasizing patients' active involvement in the management of their personal data.

9.1 Portability

The portability of data is especially significant in this era of increased mobility and telemedicine. Patients may switch healthcare providers or seek care from several specialists; therefore, it is critical that they may readily transmit their health data between providers while maintaining control over it. For example, electronic health record (EHR) systems that are interoperable across providers help ensure that patients retain control over their data while providing continuity of care.

9.2 Data Correction and Deletion

The capacity to fix mistakes in personal data is critical to providing safe and effective patient care. Inaccuracies in healthcare records can have major consequences, including misdiagnoses, ineffective therapies, and other negative health effects. As a result, healthcare providers are accountable for developing clear, accessible processes that allow individuals to examine and modify their information. An efficient repair method should be simple and user-friendly, accommodating individuals with varied backgrounds and levels of technical ability. This can be accomplished by providing several data correction channels, such as digital patient portals, documents, and telephone or personal support. Clear instructions and timely responses to rectification requests are critical for avoiding medical delays and establishing patient trust. Furthermore, healthcare professionals should prioritize alerting patients about their right to examine and update data, as many people may be unaware of this choice. Healthcare practitioners maintain a vital component of patient autonomy by providing patients with easily available tools for data verification and correction, thereby promoting accurate, high-quality care.

Additionally, patients should have the option to remove their data under certain instances. For example, under the GDPR's "right to be forgotten," individuals can request that their data be deleted if it is no longer required for the reason for which it was gathered or if they withdraw consent. However, in the healthcare setting, there may be constraints on this freedom because medical records must be preserved for a particular period of time for legal or regulatory reasons.

10 Emerging Privacy Challenges in Healthcare

The healthcare industry is undergoing a massive digital change, altering patient care, operational procedures, and medical research. Healthcare providers' efficiency and quality of care have significantly increased as they progressively incorporate new technology such as electronic health records (EHRs), telemedicine, artificial intelligence (AI), and the Internet of Things (IoT). However, this growth raises major privacy challenges that must be addressed in order to protect patient data and maintain the reliability of the healthcare system. Digital transformation within the healthcare sector is the use of diverse technologies to improve patient outcomes and operational efficiency. The integration of EHRs has transformed how patient information can be recorded, saved, and shared, allowing healthcare practitioners to obtain entire patient histories fast. Telemedicine helps make healthcare more accessible by allowing patients to consult with doctors remotely, and AI and machine learning are being used for predictive analytics, diagnosis, and individualized treatment regimens. IoT devices, such as wearable health monitors, constantly gather and send patient data, delivering real-time health measurements. Collectively, these developments have the ability to alter healthcare delivery and increase patient participation.

As healthcare systems move to more digital settings, they face additional privacy threats from data breaches, the integration of modern technologies, and the complexity of regulatory regulations. Data breaches, whether caused by cyberattacks or inadvertent exposure, pose significant risks to patient confidentiality, potentially resulting in unauthorized access to sensitive information. Furthermore, the employment of modern technologies, such as artificial intelligence and the Internet of Things, presents data security and ethical concerns. These devices frequently rely on large volumes of sensitive health information, necessitating effective data protection techniques. Some key challenges emerging are discussed below.

10.1 Digital Health Footprint and Identify-Ability

As more people use digital health apps and wearable devices, a massive quantity of data is continuously created, resulting in a comprehensive "digital health footprint." This digital footprint comprises not only standard health monitoring information but also data from online searches, social media interactions, and device usage habits. Such a vast amount of data offers a complete picture of a person's health habits and lifestyle choices, frequently giving insights that go beyond what is documented in normal medical records.

JAMA Network Open research emphasizes the potential ramifications of this occurrence, showing how data produced from these digital interactions can be utilized to infer sensitive health information. For example, app usage patterns or the types of content published on social media can indicate mental health issues, chronic illnesses, or even a susceptibility to specific diseases. This complicates the anonymization of data. Traditional approaches for preserving patient identity may be ineffective, as the rich context surrounding health-related data can assist reidentification, particularly when paired with other publicly available information (Grande et al., 2020).

These problems highlight the urgent need for broader regulatory approaches that include health data collected outside of traditional medical settings. Policymakers and healthcare institutions must evaluate all digital interactions that affect a patient's health. This could entail developing clearer criteria for data collection, usage, and sharing for digital health applications, as well as ensuring that patient permission is not only informed but also reviewed on a regular basis as technology advances.

11 AI-Enabled Health Apps and Data Privacy

The use of artificial intelligence (AI) in healthcare applications is drastically changing diagnostics and patient care, providing innovative options for individualized therapy and better patient outcomes. However, this technical innovation raises significant privacy concerns, which must be addressed in order to preserve sensitive patient information.

According to an IEEE study, AI systems used in healthcare apps frequently require large datasets to adequately train their models. The reliance on massive amounts of data raises serious worries about data misuse and unauthorized access. Sensitive patient data, such as health histories and personal identifiers, might be exposed to breaches, especially as AI systems are incorporated into numerous platforms and technologies (Wang & Zare, 2023).

The ramifications of these difficulties highlight the critical need for strong AI-specific privacy safeguards. Healthcare businesses must prioritize the creation and implementation of comprehensive frameworks that assure ethical AI use, including features such as increased data encryption, access controls, and ongoing vulnerability monitoring. Furthermore, explicit patient controls over data sharing are critical, allowing people to make informed decisions about how their data is used in AI applications. By proactively addressing these privacy concerns, the healthcare sector will be able to reap the benefits of AI while maintaining patient confidentiality. Establishing thorough criteria for data usage in AI applications would not only increase patient trust but also help regulatory compliance, resulting in a more secure and ethical healthcare landscape.

12 Challenges with Informed Consent and Data Control

Patients are typically unaware of the various ways in which their data is shared and used, especially when aggregated for research or commercial purposes. In today's healthcare context, sophisticated data-sharing protocols frequently enable the use of patient data in ways that go far beyond the patient's original agreement. This poses serious ethical questions about patient autonomy and trust, as people may not fully appreciate the implications of their agreement in a digital setting. Traditional consent models typically include a one-time agreement, which fails to account for the intricacies of data usage in the continuously changing digital health context. This static approach might result in circumstances in which patients believe they have lost control of their information, especially as healthcare technologies and procedures develop. The opaque nature of data sharing can erode trust, making patients unwilling to fully interact with healthcare systems, knowing that their information may be used in unexpected or unintended ways.

To address these challenges, new consent frameworks, such as dynamic consent, are being looked into. Dynamic consent gives patients continuous control over their data by allowing them to adjust their permission settings in real time as technology and healthcare practices change. Healthcare organizations can improve patient autonomy and confidence by implementing dynamic consent frameworks, which ensure that data use is more aligned with patient expectations. This approach protects patient rights while also promoting ethical research procedures and commercial data use. Finally, integrating these novel consent methods may result in a more ethical and patient-oriented approach to sharing data in healthcare, enabling increased engagement and collaboration while protecting individual privacy (Wang & Zare, 2023; Grande et al., 2020).

12.1 Lack of Comprehensive Regulatory Oversight for Health IoT Devices

The growing use of Internet of Things (IoT) devices for healthcare purposes, such as fitness watches and remote monitoring tools, has transformed patient care by delivering real-time health measurements and encouraging proactive health management. However, this technical innovation raises substantial privacy concerns, which must be addressed in order to preserve sensitive patient information. One of the biggest concerns with IoT devices is the lack of standardized legislative norms controlling their use and data collection. Unlike traditional healthcare data, which is normally governed by well-defined standards such as HIPAA, data collected by IoT devices frequently falls into legislative murky zones. This inconsistency exposes these devices to breaches and unwanted data exchange, raising worries about the security of private medical data (Wang & Zare, 2023).

To successfully address these privacy concerns, consistent data protection mechanisms tailored specifically to IoT devices in healthcare are needed. These protocols should establish specific requirements for data collection, storage, and sharing to ensure patient information is handled with care and transparency. Furthermore, promoting transparent sharing of data procedures would allow patients to understand how and by whom their data is being utilized, increasing trust in IoT devices.

Healthcare providers can better protect patient confidentiality while taking use of these novel technologies by creating comprehensive and uniform regulatory frameworks for IoT devices.

12.2 Cross-Border Data Flows and Privacy Compliance

Telemedicine and cross-border data transfers have transformed access to healthcare, allowing patients to get consultations and treatment from physicians all over the world. However, this poses major regulatory issues, owing to the way healthcare data travels between nations with different privacy regulations. The absence of consistency in data protection standards can create legal gaps and impede efforts to protect patient information when data is transported globally. Different countries and regions have evolved their own data privacy frameworks, which may cause conflicts and uncertainty for healthcare businesses. For example, Europe's General Data Protection Regulation (GDPR) mandates strong data protection and privacy obligations, with severe consequences for noncompliance. In contrast, US regulations, such as HIPAA, have distinct stipulations that may not be consistent with the GDPR's high standards. This disparity can provide major compliance issues for healthcare providers, especially those who operate in different jurisdictions.

Healthcare firms may struggle to ensure compliance with the different regulatory standards that apply to their operations, incurring legal liability and fines. Furthermore, patients' levels of privacy may vary depending on where their data is processed or stored, eroding their trust in telemedicine services.

To effectively manage these regulatory obstacles, healthcare providers should implement compliance strategies that emphasize adherence to the most stringent rules in their operations locations. This could include conducting rigorous audits of data management methods, investing in strong cybersecurity measures, and creating explicit cross-border data transfer regulations. By proactively addressing these difficulties and implementing strong compliance strategies, healthcare providers may not only protect patient data but also develop confidence with patients. This method will allow them to reap the benefits of telemedicine and cross-border data sharing while adhering to the greatest levels of patient privacy and data security (Grande et al., 2020).

12.3 Safeguarding Against Cybersecurity Threats

Implementing comprehensive security frameworks and protocols is critical for protecting healthcare systems from cyber-attacks. Healthcare data, particularly when stored and sent in cloud environments, is extremely sensitive, necessitating stringent security and privacy safeguards to prevent any intrusions. As a result, diverse approaches to data security are required, with a focus on cloud computing, medical imaging, and healthcare breach response.

12.4 Security and Privacy Requirements for Cloud Computing in Healthcare

Cloud computing is the delivery of numerous services via the internet, including as storage, computing power, databases, networking, software, and analytics. Instead of using a local server or personal computer, cloud computing allows users to access and store data, execute applications, and conduct other computing operations via remote servers located in data centers. Cloud computing offers various benefits to healthcare facilities, including scalability, resource efficiency, and cost reduction. However, healthcare data is especially vulnerable because of its sensitivity and regulatory confidentiality requirements. Healthcare data on the cloud requires enhanced security and privacy measures, including data confidentiality, integrity, and availability. Encryption, identity and access management (IAM), and continuous monitoring and auditing of cloud systems are among the most important security requirements (Ermakova et al., 2020).

Encryption is critical in cloud security because it transforms data into an unreadable format that can only be decoded with the correct decryption key, so protecting data during both storage and transport. Without effective encryption, hostile actors could intercept healthcare data, resulting in unlawful access and data leaks. IAM controls ensure that only authorized workers have access to sensitive data, reducing the likelihood of insider threats. Multi-factor authentication (MFA) can improve IAM by requiring additional layers of verification before access is given (Ermakova et al., 2020).

12.5 Comprehensive Security Framework for Medical Images

The storage and transmission of medical pictures are key components of healthcare systems that must be protected. Slik et al. (2003) present a complete security framework that focuses on data encryption, safe access protocols, and audit trails to prevent unwanted access and alteration. Medical imaging systems include enormous volumes of sensitive data, making them an appealing target for attackers. Slik et al. (2003) proposed a multi-layered security structure in which data is encrypted at several levels, ensuring that even if one layer is hacked, the remaining layers continue to provide protection (Slik et al., 2003).

Audit trails are especially important since they track every access and modification to medical pictures, allowing possible security breaches to be identified early on and traced back to their source. Furthermore, using secure communication protocols such as TLS (transport layer security) is suggested to protect data during transmission. The TLS protocol, which utilizes encryption to safeguard data as it flows from one system to another, is critical for maintaining the confidentiality and integrity of medical pictures in transit (Slik et al., 2003).

12.6 Data Breaches in Healthcare Systems

Despite certain safeguards, fraudsters have repeatedly attacked healthcare networks due to the enormous value of medical data. Common vulnerabilities in healthcare security systems, as well as the implications of data breaches, include patient privacy violations, financial losses, and reputational damage to healthcare companies. Phishing attempts, ransomware, and outdated software systems are some of the most typical causes of data breaches. Phishing attempts frequently trick healthcare workers into disclosing critical information, whereas ransomware encrypts data and demands a fee for its release (Reddy et al., 2021).

One method to reduce such risks is to provide frequent staff training on best practices in cybersecurity. Many healthcare breaches are the result of human error, such as clicking on malicious links or using passwords that are weak. Healthcare firms can lessen the chance of human error-related breaches by training personnel to recognize phishing emails and other cyber threats. Another important suggestion is to develop a solid data backup and disaster recovery plan. Data can be restored from backups, allowing healthcare institutions to quickly recover from ransomware attacks without having to pay a ransom. Regularly updating software and systems to patch vulnerabilities is also crucial, because obsolete software frequently has unpatched security holes that attackers can exploit (Reddy et al., 2021).

12.7 Emerging Solutions for Healthcare Data Security

Zero-trust security model in healthcare: The zero-trust model is a security framework based on the principle "never trust, always verify." Unlike standard security models, which assume that users on a network can be trusted, zero-trust views every access attempt as a risk. This concept is especially useful in healthcare, as data is dispersed across multiple platforms, networks, and devices, making it vulnerable to internal and external threats (Munjal & Bhatia, 2023). In healthcare, the zero-trust model uses multi-factor authentication (MFA), identity and access management (IAM), and network segmentation to ensure stringent access control. IAM guarantees that each user only has access to the data required for their position, reducing the risk of unauthorized access to critical patient information. Furthermore, continual monitoring within a zero-trust framework aids in the detection and response to unexpected behavior, such as accessing data from strange locations or at odd hours, which may suggest a potential breach (Munjal & Bhatia, 2023).

Adopting zero-trust is consistent with data protection laws such as the General Data Protection Regulation (GDPR) of the European Union, which mandate stringent controls over access to personal data. Adopting zero-trust models can help healthcare businesses better protect patient data, and guaranteeing privacy regulations are followed while lowering the possibility of data breaches (Mohammad Amini et al., 2023).

12.8 Homomorphic Encryption for Privacy-Preserving Data Processing

In the healthcare industry, homomorphic encryption has become a potent instrument for safe data processing. Homomorphic encryption enables computations on encrypted data, in contrast to conventional encryption techniques that need data to be decrypted before processing. This special quality is especially useful in the healthcare industry, where secrecy and data protection are crucial (Zhang et al., 2023).

Homomorphic encryption protects privacy while facilitating safe data analysis between various healthcare companies. For instance, homomorphic encryption enables data aggregation and analysis while maintaining encryption in research investigations involving patient data from several institutions. This guarantees that private patient data is protected even when it is being calculated (Boomija & Raja, 2023). Furthermore, federated learning—a cooperative machine learning technique that enables organizations to train models on local data while exchanging encrypted results—is supported by homomorphic encryption. In IoT-enabled healthcare systems, where medical equipment produces sensitive data that needs to be processed securely, this strategy is being utilized more and more (Zhang et al., 2023).

Although homomorphic encryption has advantages, it can slow down data processing and is computationally demanding. To increase its effectiveness and suitability for real-time applications in healthcare, research is still being done on encryption algorithm optimization (Munjal & Bhatia, 2023). With further development, homomorphic encryption could completely transform healthcare data processing while protecting privacy.

13 Blockchain Technology for Data Integrity and Security

Blockchain technology offers decentralized, transparent, and safe data storage, thus it has become very popular in the healthcare industry. Blockchain is a distributed ledger that is extremely difficult to tamper with since it saves data in permanent or immutable blocks or sections. In the healthcare industry, where data integrity and auditability are crucial, this technology is very beneficial (Adere, 2022). It is difficult for attackers to alter or delete records without detection.

Blockchain technology can help healthcare providers securely share patient data, improving interoperability and access control. For example, patient data can be maintained on a Blockchain network and accessed only by authorized parties via smart contracts that automatically enforce data-sharing laws. Blockchain's openness also allows patients to track who accessed their data, adding an extra layer of control and accountability (Ghosh et al., 2023).

Blockchain technology can also protect the healthcare supply chain, lowering the danger of counterfeit pharmaceuticals and fraudulent activity. By documenting each stage of the supply chain on a decentralized ledger, Blockchain secures the legitimacy of pharmaceuticals and medical supplies, which is critical for patient safety. Furthermore, the combination of Blockchain and homomorphic encryption has demonstrated potential for data protection in IoT healthcare applications. For example, Blockchain-based systems may securely store data from IoT devices, while homomorphic encryption allows data to be studied in its encrypted form, assuring end-to-end security (Ali et al., 2023).

Despite its benefits, Blockchain has scaling issues in healthcare because the technology can be resource-intensive. However, hybrid solutions are emerging to address these restrictions by storing only relevant metadata on the Blockchain and raw data in safe off-chain storage (Adere, 2022).

References

Adere, E. M. (2022). Blockchain in healthcare and IoT: A systematic literature review. In *Array* (Vol. 14). https://doi.org/10.1016/j.array.2022.100139

Ali, A., Al-rimy, B. A. S., Alsubaei, F. S., Almazroi, A. A., & Almazroi, A. A. (2023). HealthLock: Blockchain-based privacy preservation using homomorphic encryption in internet of things healthcare applications. *Sensors, 23*(15). https://doi.org/10.3390/s23156762

ARX. (2020). *ARX—Data anonymization tool*. ARX.

Alshammari & Simpson (2017), Towards a Principled Approach for Engineering Privacy by Design, APF 2017 (LNCS).
Alshammari & Simpson (2018), A model-based approach to support privacy compliance, Inf. Comput. Secur. 26(4):437–453.
Basso et al. (2016), PrivAPP: An approach for privacy-preserving reference architectures, Software Practice and Experience https://doi.org/10.1002/spe.2546.
Boomija, M. D., & Raja, S. V. K. (2023). A novel secure multiplicative homomorphic encryption for data security in healthcare. *Journal of Environmental Protection and Ecology, 24*(2)17–27.
Ermakova, T., Fabian, B., Kornacka, M., Thiebes, S., & Sunyaev, A. (2020). Security and privacy requirements for cloud computing in healthcare. *ACM Transactions on Management Information Systems, 11*(2). https://doi.org/10.1145/3386160
Ferretti, A., Ienca, M., Sheehan, M., Blasimme, A., Dove, E. S., Farsides, B., Friesen, P., Kahn, J., Karlen, W., Kleist, P., Liao, S. M., Nebeker, C., Samuel, G., Shabani, M., Rivas Velarde, M., & Vayena, E. (2021). Ethics review of big data research: What should stay and what should be reformed? *BMC Medical Ethics, 22*(1). https://doi.org/10.1186/s12910-021-00616-4
Florea, D., & Florea, S. (2020). Big data and the ethical implications of data privacy in higher education research. *Sustainability (Switzerland), 12*(20). https://doi.org/10.3390/su12208744
Franconi, L., & Polettini, S. (2004). Individual risk estimation in μ-argus: A review. In *Lecture notes in computer science (Including subseries lecture notes in artificial intelligence and lecture notes in bioinformatics), 3050*. https://doi.org/10.1007/978-3-540-25955-8_20
Ghosh, P. K., Chakraborty, A., Hasan, M., Rashid, K., & Siddique, A. H. (2023). Blockchain application in healthcare systems: A review. In *Systems* (Vol. 11, Issue 1). https://doi.org/10.3390/systems11010038
Grande, D., Luna Marti, X., Feuerstein-Simon, R., Merchant, R. M., Asch, D. A., Lewson, A., & Cannuscio, C. C. (2020). Health policy and privacy challenges associated with digital technology. *JAMA Network Open, 3*(7). https://doi.org/10.1001/jamanetworkopen.2020.8285
Matulevičius et al. (2020), GDPR business-process modeling & compliance method.
Mohammad Amini, M., Jesus, M., Fanaei Sheikholeslami, D., Alves, P., Hassanzadeh Benam, A., & Hariri, F. (2023). Artificial intelligence ethics and challenges in healthcare applications: A comprehensive review in the context of the European GDPR mandate. In *Machine learning and knowledge extraction* (Vol. 5, Issue 3). https://doi.org/10.3390/make5030053
Monteiro, S., Oliveira, D., António, J., Sá, F., Wanzeller, C., Martins, P., & Abbasi, M. (2024). Data anonymization: Techniques and models. *Smart Innovation, Systems and Technologies, 344*. https://doi.org/10.1007/978-981-99-0333-7_6
Mougiakou & Virvou (2017), Based on GDPR privacy in UML: Case of e-learning program, IISA, https://doi.org/10.1109/IISA.2017.8316456
Munjal, K., & Bhatia, R. (2023). A systematic review of homomorphic encryption and its contributions in healthcare industry. *Complex & Intelligent Systems, 9*(4). https://doi.org/10.1007/s40747-022-00756-z
Myers & Liskov (2000), Protecting Privacy Using the Decentralized Label Model, ACM TSEM, 9(4):410–442, https://doi.org/10.1145/363516.363526.
Priscáková & Rábová (2013), Model of Solutions for Data Security in Cloud Computing, Int. J. CS Eng & IT, arXiv:1307.3766.
Prasser, F., Kohlmayer, F., Lautenschläger, R., & Kuhn, K. A. (2014). ARX—A comprehensive tool for anonymizing biomedical data. In *AMIA … Annual symposium proceedings/AMIA Symposium*, pp. 984–993.
Prasser, F., Eicher, J., Spengler, H., Bild, R., & Kuhn, K. A. (2020). Flexible data anonymization using ARX—Current status and challenges ahead. *Software: Practice and Experience, 50*(7). https://doi.org/10.1002/spe.2812
Reddy, J., Elsayed, N., Elsayed, Z., & Ozer, M. (2021). *Data breaches in healthcare security systems*. School of Information Technology.
Slik, D., Montour, M., & Altman, T. (2003). Comprehensive security framework for the communication and storage of medical images. In *Medical imaging 2003: PACS and integrated medical information systems: Design and evaluation, 5033*. https://doi.org/10.1117/12.480475

Tamburri (2020), Design principles for the GDPR — FCA & evaluation, Information Systems 91:101469 https://doi.org/10.1016/j.is.2019.101469

Templ, M., Kowarik, A., & Meindl, B. (2015). Statistical disclosure control for micro-data using the R package sdcMicro. *Journal of Statistical Software, 67*(4). https://doi.org/10.18637/jss.v067.i04

Thorogood, A., & Chokoshvili, D. (2023). *Ethical and legal considerations in genomic data sharing: Evolution of the discourse and the road ahead.* https://doi.org/10.1007/978-3-031-29451-8_19

Tomás, J., Rasteiro, D., & Bernardino, J. (2022). Data anonymization: An experimental evaluation using open-source tools. *Future Internet, 14*(6). https://doi.org/10.3390/fi14060167

Wang, P., & Zare, H. (2023). A case study of privacy protection challenges and risks in AI-enabled healthcare app. In *Proceedings—2023 IEEE conference on artificial intelligence, CAI 2023.* https://doi.org/10.1109/CAI54212.2023.00132

Zhang, L., Xu, J., Vijayakumar, P., Sharma, P. K., & Ghosh, U. (2023). Homomorphic encryption-based privacy-preserving federated learning in IoT-enabled healthcare system. *IEEE Transactions on Network Science and Engineering, 10*(5). https://doi.org/10.1109/TNSE.2022.3185327

Further Research and Implementation of Deep Learning Technologies in Oncology for Enhanced Healthcare Delivery

Geeta Chhabra Gandhi, Roshmeet Chakraborty, Nishkarsh Kewlani, Dinesh Kumar Saini, and Kashif Zia

Abstract

Deep learning (DL) has emerged as a transformative technology in oncology, offering innovative solutions for cancer detection, prognosis, and personalized treatment. Artificial intelligence (AI)-driven methodologies, such as convolutional neural networks (CNNs), recurrent neural networks (RNNs), and transformer-based models, have demonstrated remarkable potential in analyzing medical imaging, genomic data, and electronic health records (EHRs). These advancements have significantly improved early cancer detection, tumor classification, and treatment planning by providing automated, high-precision insights. However, despite these breakthroughs, challenges remain in data availability, model interpretability, ethical concerns, and clinical integration. The need for large, diverse, and high-quality datasets poses a significant barrier, while the black-box nature of many DL models raises concerns about trust and adoption among clinicians. Additionally, regulatory compliance and infrastructure limitations further complicate AI deployment in oncology. This chapter highlights the importance of interdisciplinary collaboration among AI researchers, oncologists, bioinformaticians, and policymakers to develop explainable and clinically validated AI models. Future research should focus on integrating multimodal AI approaches, leveraging federated learning for data privacy, and developing

G. C. Gandhi · R. Chakraborty · N. Kewlani
Manipal University Jaipur, Jaipur, Rajasthan, India

D. K. Saini
Department of IoT and Intelligent Systems, Manipal University Jaipur, Jaipur, Rajasthan, India

K. Zia (✉)
University of Glasgow, Glasgow, UK
e-mail: Kashif.Zia@glasgow.ac.uk

R. Nema et al. (eds.), *Advances in Cancer Detection, Prediction, and Prognosis Using Artificial Intelligence and Machine Learning*,
https://doi.org/10.1007/978-981-96-9346-7_18

AI-driven personalized treatment strategies. Addressing these challenges will enable deep learning to revolutionize oncology, enhancing healthcare delivery, optimizing patient management, and ultimately improving survival rates.

1 Introduction

1.1 Overview of AI and Deep Learning (DL) in Oncology

Artificial intelligence (AI) and deep learning (DL) have become transformational forces in oncology, significantly advancing cancer diagnosis, treatment planning, and patient management. Deep learning, a subset of machine learning, employs artificial neural networks to analyze vast amounts of biomedical data, identifying complex patterns that may not be easily detected by human clinicians. AI models in oncology integrate data from diverse sources, including medical imaging modalities such as computed tomography (CT), magnetic resonance imaging (MRI), and positron emission tomography (PET) scans, as well as histopathology slides, genomic sequencing, and electronic health records (EHRs). By leveraging these datasets, AI-driven systems enhance diagnostic accuracy through automated tumor detection, segmentation, and classification, facilitating early cancer detection with higher precision.

Convolutional neural networks (CNNs) have proven particularly effective in processing medical images, enabling real-time analysis and identification of malignancies. Additionally, recurrent neural networks (RNNs) and transformer-based models contribute to predictive analytics, helping oncologists anticipate disease progression, treatment responses, and potential relapse risks. AI has also demonstrated remarkable potential in drug discovery by accelerating the identification of new therapeutic agents, optimizing drug repurposing strategies, and personalizing treatment plans based on a patient's molecular profile. This personalized approach to oncology allows for targeted therapies that minimize adverse effects and enhance treatment efficacy.

Moreover, AI-driven robotic-assisted surgeries and radiation therapy planning have improved surgical precision and minimized damage to surrounding healthy tissues. The increasing availability of high-performance computing, cloud-based platforms, and big data infrastructure further fuels AI's role in modern oncology research and clinical applications. However, despite these promising advancements, several challenges hinder the full-scale implementation of AI in oncology. Issues such as model interpretability, data privacy concerns, bias in AI algorithms, and regulatory barriers pose significant obstacles. Furthermore, the integration of AI into clinical workflows requires extensive validation through clinical trials and robust regulatory frameworks to ensure its safety and effectiveness. Moving forward, interdisciplinary collaboration among AI researchers, oncologists, bioinformaticians, and policymakers is crucial to overcoming these challenges. Future research should focus on enhancing AI transparency, adopting federated learning for secure data sharing, and developing AI-driven decision support systems that complement oncologists in delivering optimal patient care (Fig. 1).

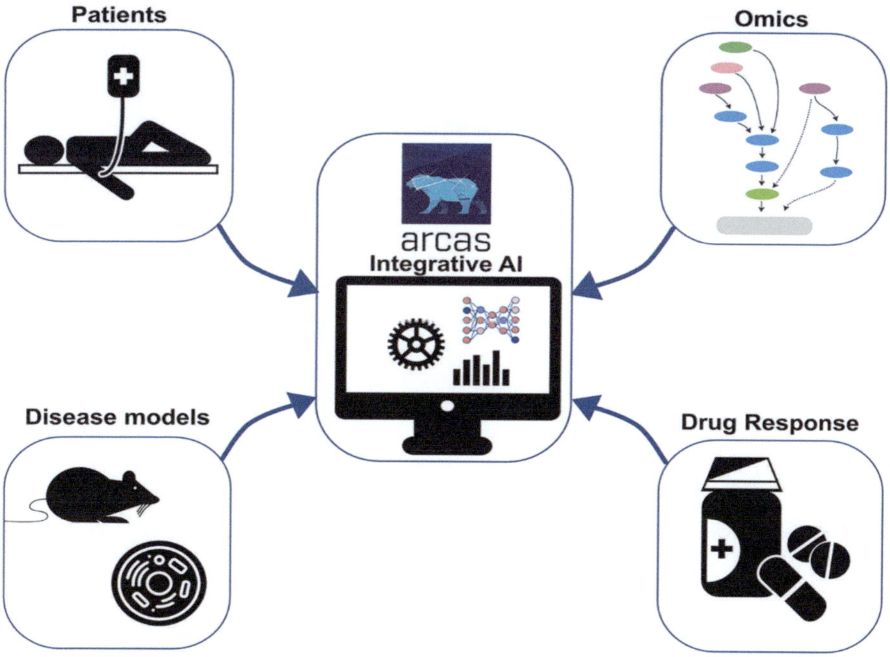

Fig. 1 A neural network processing CT or MRI scans to detect tumors. (https://static1.squarespace.com/static/5daed3278ea34a7aeca939a2/t/6052246b7fb49f6c6413a385/1615996011749/conceptVagueDHA_cancerDxOmics.png?format=1500w)

1.2 Importance of AI in Cancer Detection, Prediction, and Prognosis

Artificial intelligence (AI) has revolutionized cancer detection and prognosis, significantly improving diagnostic accuracy and reducing reliance on human interpretation. Traditional cancer diagnosis depends on radiologists and pathologists, whose interpretations can be subjective and vary based on expertise and experience. AI-driven models overcome these challenges by providing objective, highly consistent, and accurate analysis. These models can process vast amounts of medical data within seconds, detecting malignant lesions with greater sensitivity and specificity compared to conventional diagnostic methods. AI algorithms are particularly effective in analyzing medical imaging modalities such as mammograms, CT scans, MRIs, and histopathology slides, allowing for early detection of cancer at its most treatable stage (Fig. 2).

Beyond diagnosis, AI plays a crucial role in prognosis and treatment planning. AI-powered predictive models analyze past patient data, treatment responses, and clinical outcomes to forecast disease progression, recurrence risks, and survival rates. Machine learning techniques such as deep learning and reinforcement learning enable precise risk stratification, helping oncologists determine optimal treatment plans tailored to individual patients (Mennella et al., 2024). AI also contributes

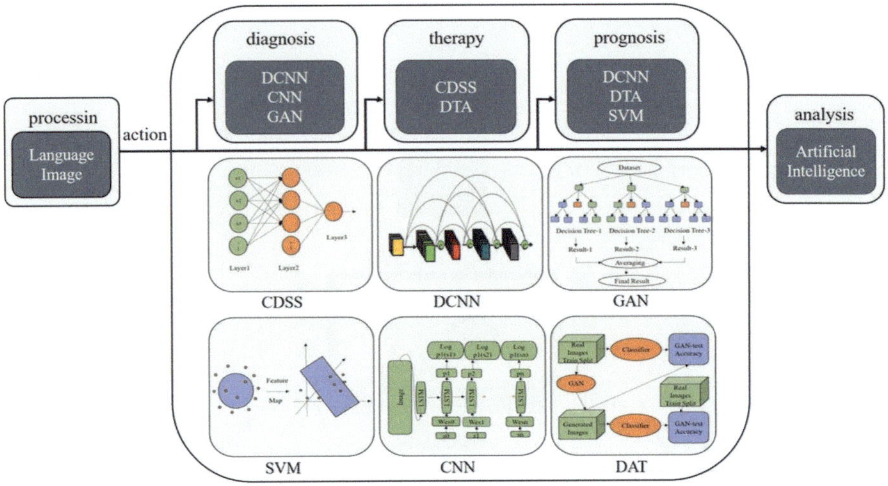

Fig. 2 A side-by-side comparison highlighting the accuracy differences between AI-assisted and traditional cancer detection techniques. (Pei et al., 2022)

to biomarker discovery, identifying genetic mutations and molecular signatures associated with specific cancer types. This facilitates the development of targeted therapies, advancing precision oncology by ensuring patients receive the most effective treatments based on their genetic profiles. The integration of AI with multimodal datasets—encompassing medical imaging, pathology, genomic data, and electronic health records (EHRs)—further enhances prognosis prediction and personalized treatment recommendations (Miotto et al., 2018; Sadeghi et al. (2024). AI-driven real-time monitoring of cancer patients allows treatment regimens to be dynamically adjusted based on patient response, minimizing adverse effects while maximizing therapeutic outcomes. Additionally, AI-assisted robotic surgeries and radiation therapy planning contribute to higher precision and reduced damage to healthy tissues. Despite these advancements, widespread clinical implementation of AI in oncology faces several challenges. Issues such as algorithmic bias, regulatory approval, data privacy concerns, and alignment with real-world clinical workflows must be addressed to ensure equitable and safe AI adoption. Moving forward, interdisciplinary collaboration between AI researchers, oncologists, regulatory bodies, and policymakers is essential to refine AI models, establish ethical AI guidelines, and seamlessly integrate AI into routine cancer care. If effectively implemented, AI has the potential to significantly reduce cancer-related mortality and improve the overall quality of life for patients worldwide.

1.3 The Gap in Research Breakthroughs and Actual Clinical Adoption

Despite the remarkable progress in AI-driven oncology, translating research advancements into real-world clinical applications remains a formidable challenge. One of the biggest obstacles is the lack of access to large, diverse, and high-quality datasets required to train robust AI Models (AI systems are often developed using datasets from specific populations, limiting their generalizability across diverse patient groups. Without broad representation, AI models may exhibit biases that reduce their reliability in different demographic or geographic settings. Another critical challenge is model interpretability, as clinicians need to understand how AI reaches its conclusions to trust AI-assisted decision-making. Many deep learning algorithms function as "black boxes," making it difficult for medical professionals to verify the reasoning behind AI-generated diagnoses or treatment recommendations (Ligero et al., 2025). The regulatory landscape also presents a significant hurdle, as AI-powered medical tools must undergo rigorous validation, including clinical trials and approvals from regulatory bodies such as the U.S. Food and Drug Administration (FDA) and the European Medicines Agency (EMA), before they can be widely implemented. These regulatory processes are often lengthy and complex, slowing down the adoption of AI-driven diagnostic and treatment solutions. Additionally, the integration of AI into clinical workflows is hindered by outdated healthcare infrastructure, as many hospitals and cancer treatment centers still rely on legacy systems that are not designed to support AI-driven analytics. Upgrading these infrastructures to accommodate high-performance AI applications requires substantial financial investment and technical expertise, which many institutions may lack (Lotter et al., 2024). AI systems also demand significant computational power and cloud-based resources, further complicating their deployment in resource-limited settings. Ethical concerns surrounding AI adoption in oncology further impede large-scale implementation, including data privacy, security risks, and biases in AI decision-making that raise concerns about patient confidentiality and equitable treatment. Ensuring compliance with stringent data protection regulations such as the Health Insurance Portability and Accountability Act (HIPAA) and the General Data Protection Regulation (GDPR) is crucial (Reis, 2025). Overcoming these challenges requires interdisciplinary collaboration between AI researchers, oncologists, bioinformaticians, and policymakers. Developing standardized protocols, regulatory guidelines, and explainable AI models will be key to ensuring safe and effective AI adoption in oncology. Future research should focus on creating scalable, interpretable, and clinically relevant AI solutions that seamlessly integrate into existing healthcare frameworks (Alum, 2025) ResearchGate, 2021b; Alsharif, 2024) (Fig.3).

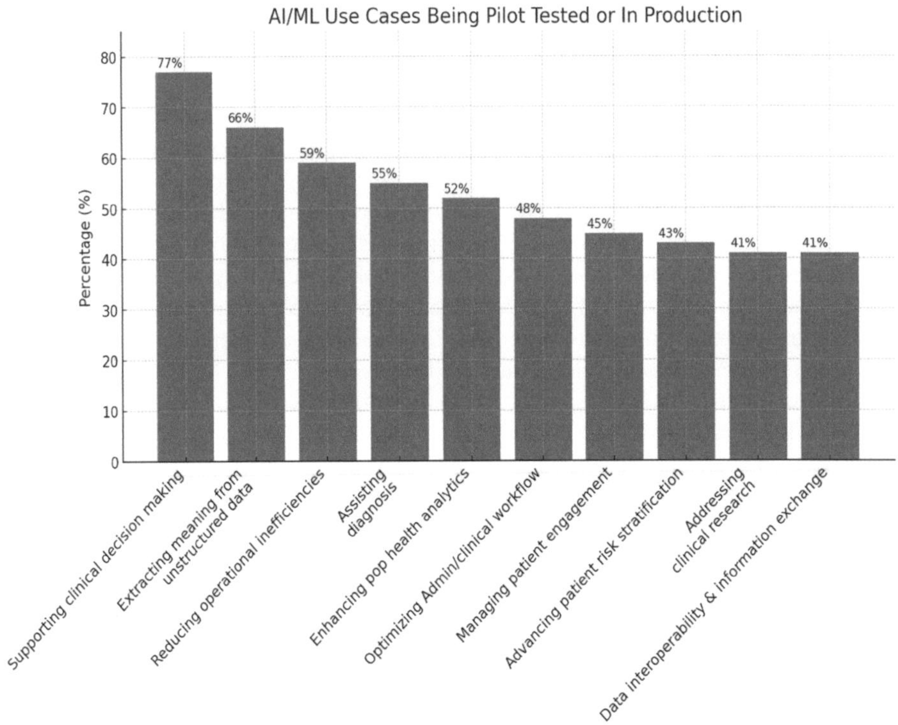

Fig. 3 AI/ML use cases being pilot tested or in production

2 The State of Deep Learning in Oncology

2.1 AI in Medical Imaging (CT, MRI, PET Scans, Histopathology Slides) (Fig. 4)

Medical imaging plays a crucial role in the early and accurate diagnosis of cancer, and recent advancements in artificial intelligence (AI) and deep learning have significantly improved the precision and efficiency of diagnostic processes. AI-driven models, particularly those based on deep learning, have revolutionized cancer detection by enhancing image analysis, automating tumor identification, and classifying cancerous tissues with high accuracy. Computer vision-based systems can analyze vast amounts of medical imaging data, such as computed tomography (CT), magnetic resonance imaging (MRI), and positron emission tomography (PET) scans, with greater speed and accuracy than traditional diagnostic methods. Convolutional neural networks (CNNs), in particular, have shown exceptional success in processing medical images, detecting abnormalities, and distinguishing between benign and malignant tumors. These AI models help radiologists by reducing diagnostic errors, minimizing variations in interpretation, and improving decision-making in complex cases (Liao et al. 2023). AI also enables early detection of cancer, which is

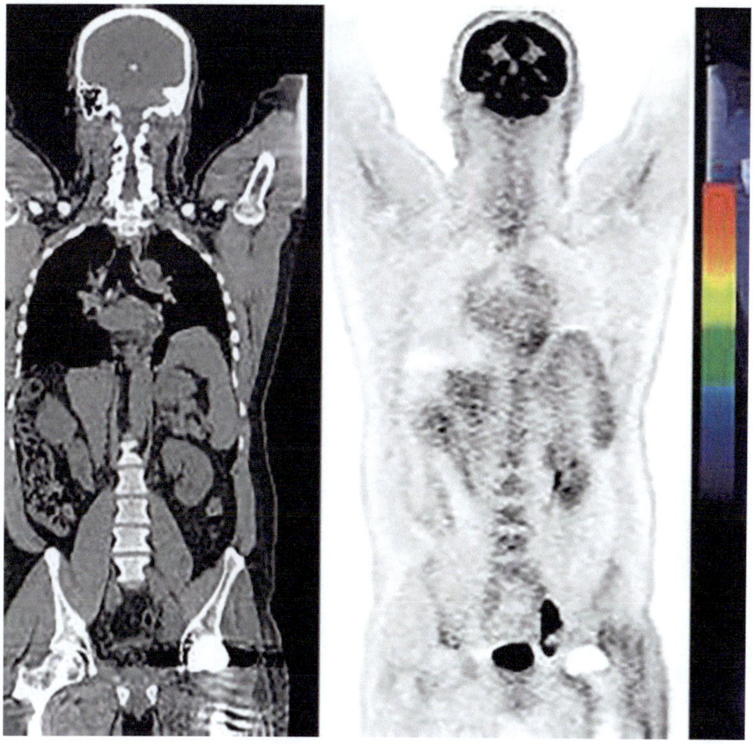

Fig. 4 AI-powered medical imaging: enhancing CT, MRI, PET, and histopathology analysis. (https://tse2.mm.bing.net/th?id=OIP.ag1dZMMw9NoqjxNxnrXLKAHaEd&pid=Api)

critical for improving patient prognosis and treatment outcomes. Additionally, AI-powered imaging systems can assist in tumor segmentation, aiding in targeted treatment planning and monitoring disease progression (Ramachandran, 2022). While these advancements offer significant benefits, challenges such as data privacy, model interpretability, and integration into clinical workflows remain barriers to widespread adoption in oncology (Olawade et al., 2024) (Fig. 5).

AI also enhances histopathological image analysis by enabling the automatic detection of microscopic cancer cells, assisting pathologists in identifying malignancies with precision comparable to human experts. Traditional histopathological analysis is often labor-intensive, prone to variability, and requires a highly trained specialist, but AI-driven models mitigate these challenges by providing consistent, objective, and highly accurate assessments (Sebastian and Peter, 2022). By analyzing digitized pathology slides, AI can differentiate between cancerous and noncancerous cells, detect subtle morphological changes, and identify rare or aggressive cancer subtypes that might be overlooked by human pathologists. Recent developments in AI-driven medical imaging have also led to the creation of automated tumor graders, which provide standardized and unbiased assessments of tumor severity, aiding oncologists in determining prognosis and treatment strategies. AI

a

	Cancer Type	Cox Regression (%)	CNN (%)	RNN (%)
●	Breast Cancer	72	85	88
●	Lung Cancer	68	80	84
●	Colorectal Cancer	70	82	86
●	Prostate Cancer	74	87	89
●	Ovarian Cancer	65	78	83

b

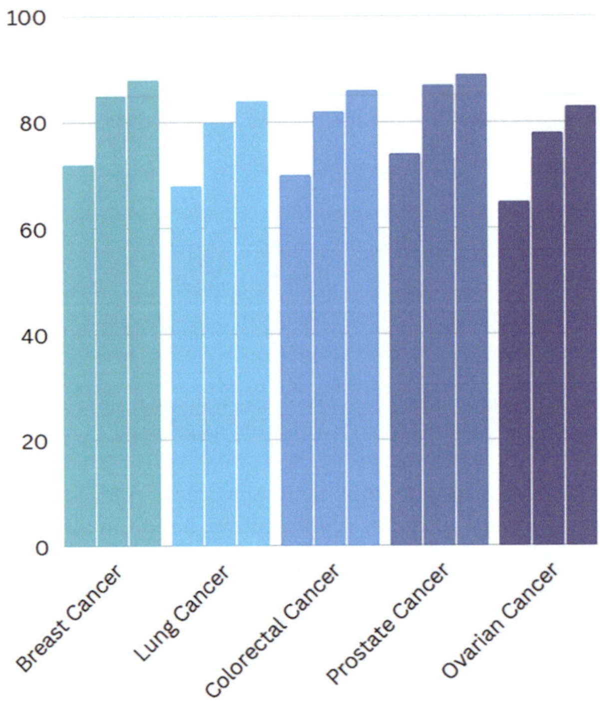

Fig. 5 (**b**) The efficacy of deep learning in survival rate prediction for cancer has been established in various studies. The graph compares the performance of Cox Regression, CNNs, and RNNs in survival prediction based on TCGA data, with deep learning methods clearly outperforming them

algorithms are being trained to assess multiple parameters such as tumor size, shape, and cellular composition to provide a comprehensive analysis of cancer progression (Truhn et al. 2024). Moreover, AI-integrated imaging systems can combine histopathological, radiological, and molecular data for a more holistic understanding of tumor biology. Future research aims to refine AI's ability to classify malignant and benign lesions with even greater accuracy, improve its adaptability across different populations, and facilitate seamless integration into routine clinical workflows. Efforts are also being made to ensure AI-driven diagnostic tools accessible to a broader range of healthcare institutions, including those in resource-limited settings, thereby democratizing advanced cancer diagnostics. Additionally, ethical considerations, regulatory compliance, and the development of explainable AI models are critical areas of ongoing research to enhance clinical trust and ensure safe and effective implementation.

2.2 Deep Learning in Genomics and Biomarker Discovery

AI-based genomics is transforming precision oncology by enabling the identification of genetic mutations and biomarkers that drive cancer progression. Deep learning algorithms analyze vast genomic sequencing datasets to uncover cancer-causing mutations, predict patient responses to targeted therapies, and assist in personalized treatment selection. By integrating transcriptomic and proteomic data, AI enhances biomarker discovery, leading to highly individualized treatment strategies that improve therapeutic efficacy and minimize adverse effects. AI-driven genomics also plays a crucial role in immunotherapy by identifying patients most likely to benefit from immune checkpoint inhibitors, thereby reducing the risk of ineffective treatments and unnecessary side effects. The ability of deep learning to process and interpret large-scale genomic data accelerates cancer research, enhances our understanding of tumor biology, and facilitates the development of novel precision therapies. Moreover, AI aids in detecting rare mutations that may be overlooked in traditional analysis, improving early cancer detection and prognosis prediction. However, challenges such as data heterogeneity, computational costs, and regulatory hurdles remain barriers to widespread clinical adoption. Standardizing genomic data, improving AI model interpretability, and integrating AI-powered genomic insights into routine oncology workflows are essential steps for realizing the full potential of AI in precision medicine. Addressing these challenges will require collaborative efforts between bioinformaticians, oncologists, AI researchers, and regulatory bodies to ensure AI-driven genomic analysis is accurate, ethical, and clinically viable (Fig. 6).

2.3 AI-Driven Predictive Models for Cancer Recurrence and Progression

Deep learning models are increasingly being employed to predict cancer recurrence and progression by analyzing longitudinal patient data. Advanced neural networks such as recurrent neural networks (RNNs) and long short-term memory (LSTM) networks are particularly effective in evaluating time-series data, allowing for continuous monitoring of tumor growth, treatment responses, and potential relapses. These models enable oncologists to track patient progress over time, offering insights into how an individual's cancer might evolve. AI-driven predictive

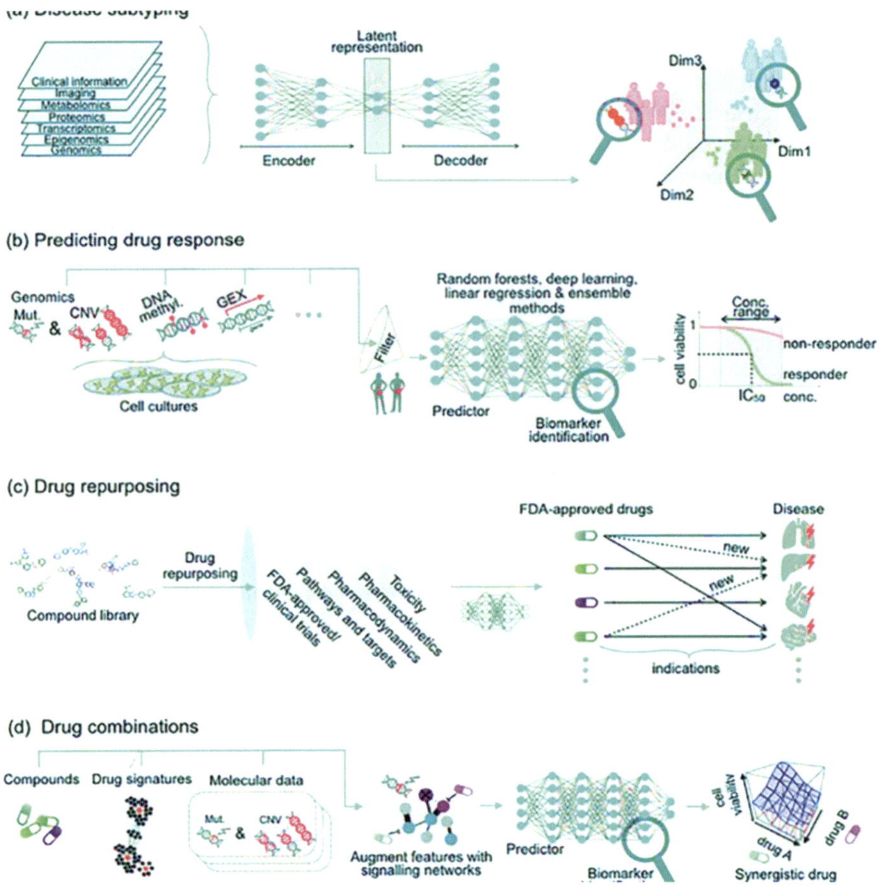

Fig. 6 Deep learning enhances precision medicine by integrating various data sources in early drug and biomarker discovery. (Bi et al. 2019), Ding et al., 2024)

analytics help in designing personalized follow-up regimens, ensuring that high-risk patients receive timely interventions to prevent recurrence (Li et al. 2025). Additionally, predictive models are extensively utilized in survival analysis, providing clinicians with the ability to estimate a patient's prognosis based on past cases with similar profiles. By leveraging historical patient data, AI-powered models enhance the accuracy of survival rate estimations, guiding oncologists in making informed decisions about treatment plans. The integration of AI in predictive oncology is also proving beneficial in identifying patterns of resistance to therapy, allowing for timely modifications to treatment approaches (Verlingue et al., 2024). These AI-driven insights enable clinicians to adopt proactive strategies, reducing the chances of disease progression and improving overall patient outcomes. Furthermore, AI models are increasingly incorporating multimodal data, including imaging, genomic, and clinical records, to provide a more holistic view of a patient's

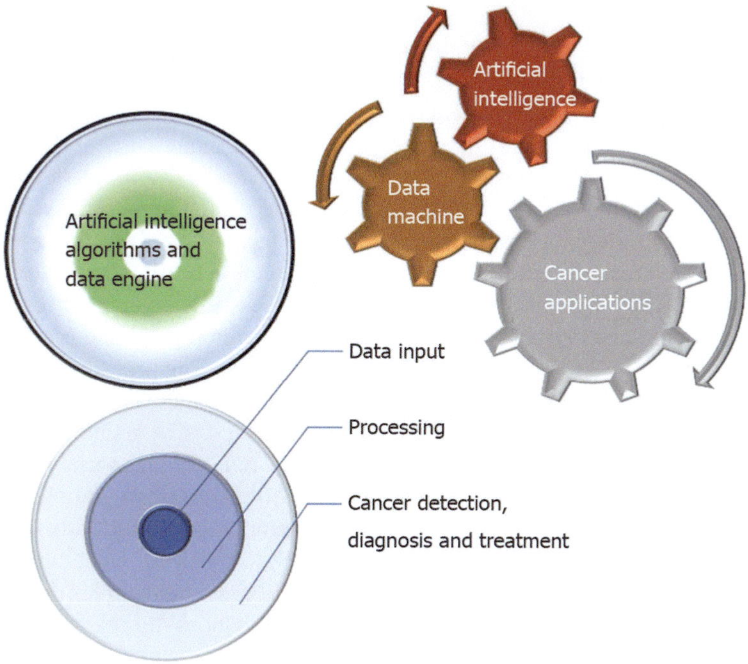

Fig. 7 AI predictive model for cancer recurrence. (ResearchGate, 2021a)

condition. This multidimensional approach enhances prediction accuracy and ensures that cancer management is tailored to individual patient needs. Despite these advancements, several challenges remain, including the need to improve model precision, address bias in training datasets, and integrate real-world patient data for more robust predictions. One of the critical areas of focus for future research is the validation of AI-driven predictive models through large-scale clinical trials, ensuring their reliability and generalizability across diverse populations. Ethical concerns such as patient data privacy and algorithm transparency must also be addressed to facilitate widespread clinical adoption. Moreover, interdisciplinary collaborations between AI researchers, oncologists, and healthcare institutions are essential to refine these predictive models and incorporate them into mainstream oncology practices. As AI technology continues to evolve, further advancements in predictive analytics will enhance cancer prognosis, enable early detection of relapses, and ultimately contribute to more effective, personalized cancer treatment strategies (Fig. 7).

2.4 Success Stories: AI Systems Already Assisting Oncologists

There are several AI-driven systems already proving their clinical utility in oncology, showcasing the transformative potential of artificial intelligence in improving cancer diagnosis, prognosis, and treatment. IBM Watson for Oncology, one of the most well-known AI-powered platforms, analyzes vast amounts of patient data, medical literature, and clinical trial findings to recommend evidence-based treatments tailored to individual patients (Lipkova et al., 2022). This system assists oncologists by providing data-driven insights, ensuring that treatment decisions are aligned with the latest research and best practices. Similarly, Google's DeepMind has developed AI models that identify breast cancer in mammograms with higher accuracy than radiologists, significantly reducing false positives and false negatives. These AI-driven diagnostic tools enhance early detection rates, which is crucial for improving patient outcomes. AI has also been successfully integrated into pathology labs, where deep learning models analyze histopathological slides to diagnose cancers such as lung and skin cancer. These models offer high precision in detecting malignancies, reducing human error and subjectivity in diagnosis. By standardizing the interpretation of biopsy samples, AI ensures more consistent and accurate cancer detection, aiding pathologists in making more reliable diagnoses (Wei et al., 2025). Beyond diagnostics, AI-driven systems contribute to treatment planning and personalized medicine. AI models analyze genomic sequencing data to identify mutations linked to cancer progression, enabling oncologists to select targeted therapies for individual patients. The ability of AI to process multi-omics data, including transcriptomic and proteomic information, enhances biomarker discovery, leading to better patient stratification and optimized therapeutic regimens. AI also plays a crucial role in predicting disease progression and recurrence. Recurrent neural networks (RNNs) and long short-term memory (LSTM) models analyze longitudinal patient data to track tumor growth, assess treatment responses, and forecast potential relapses. Predictive analytics allow clinicians to design personalized follow-up plans, ensuring timely interventions for high-risk patients. Additionally, AI models assist in survival analysis, providing insights into prognosis based on historical patient data. These capabilities empower oncologists to make data-driven decisions, improving the precision and effectiveness of cancer management strategies.

Despite these advancements, the widespread clinical adoption of AI in oncology faces several challenges. One of the biggest hurdles is the need for large, diverse, and high-quality datasets to train AI models effectively. Many AI systems are developed using data from specific populations, limiting their generalizability across different demographic groups. Additionally, model interpretability remains a significant concern, as clinicians must understand how AI systems reach their conclusions to trust AI-assisted decision-making. The regulatory landscape for AI in healthcare is complex, with AI-driven medical tools requiring rigorous clinical trials and FDA approvals before being integrated into standard practice. Furthermore, healthcare institutions often rely on legacy infrastructure, making the implementation of AI systems costly and technologically challenging. AI models also require substantial computational resources, which many hospitals and cancer centers may not have readily available. Ethical concerns, including data privacy, security, and algorithmic bias, further hinder large-scale adoption. AI models must be transparent and explainable to gain acceptance from both healthcare providers and patients. To address these challenges, interdisciplinary collaboration between AI researchers, oncologists, bioinformaticians, and policymakers is essential. Developing standardized protocols, regulatory guidelines, and open AI architectures will facilitate smoother integration into clinical workflows. Future research must focus on improving AI model explainability, ensuring fairness in algorithmic decision-making, and validating AI-driven predictions through large-scale clinical trials. Additionally, investing in AI education and training programs for healthcare professionals will enhance their ability to work alongside AI systems effectively. By overcoming these barriers, AI has the potential to revolutionize oncology, reducing diagnostic errors, personalizing treatment regimens, and ultimately improving cancer survival rates. The future of AI in oncology lies in its ability to seamlessly integrate into clinical practice, empowering oncologists with data-driven insights and enhancing patient care at every stage of the cancer treatment journey.

3 Challenges in the Use of Deep Learning in Oncology

3.1 Availability and Quality of Data: Limited Datasets, Class Imbalance, Privacy Issues

The unavailability and quality of data are among the biggest challenges in implementing deep learning in oncology. Deep learning algorithms rely on large, high-quality datasets to accurately and consistently make predictions. However, cancer-related datasets are often small, fragmented, and subject to significant class imbalance, where certain cancer types have far fewer samples than others (Williams et al., 2021). This scarcity of diverse and representative data limits the ability of AI models to generalize across different populations, potentially leading to biased predictions that do not perform well for underrepresented patient groups. Furthermore, cancer datasets are often heterogeneous, consisting of multiple data types such as imaging, histopathology slides, genomic sequences, and clinical records. Integrating these different data modalities into a single deep learning framework requires extensive preprocessing and standardization, which adds another layer of complexity to AI model development. Additionally, privacy regulations such as the Health Insurance Portability and Accountability Act (HIPAA) in the United States and the General Data Protection Regulation (GDPR) in Europe impose strict rules on the collection, storage, and sharing of patient data. While these regulations are crucial for protecting patient privacy, they also make it difficult to compile large, multi-institutional datasets necessary for training robust AI models. Many hospitals and research institutions are hesitant to share sensitive patient data due to legal and ethical concerns, further restricting the availability of comprehensive datasets. This lack of access to large-scale data slows down AI advancements in oncology and limits the development of more accurate and reliable predictive models. To address these challenges, federated learning has emerged as a promising solution. Federated learning is a decentralized approach that allows AI models to be trained across multiple institutions without sharing raw patient data. Instead of transferring data to a

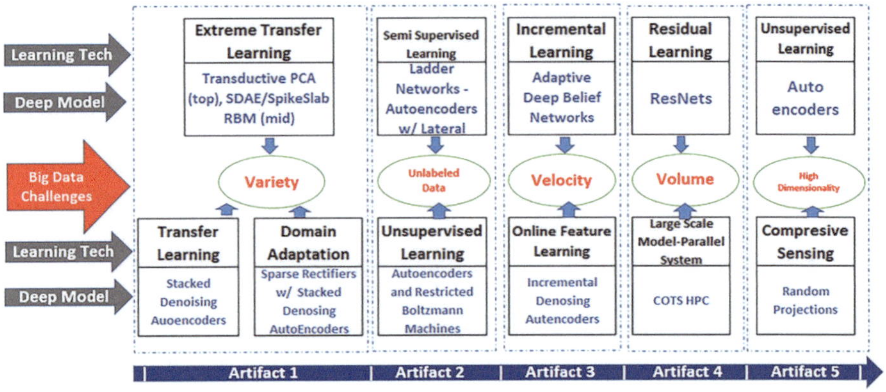

Fig. 8 Deep learning framework for big data challenges (World Journal of Gastrointestinal Oncology, 2021)

central repository, the model is trained locally on each institution's data, and only model updates (such as weight adjustments) are shared. This method preserves data privacy while enabling collaboration across multiple healthcare centers, ultimately helping AI systems learn from diverse datasets without violating privacy regulations. Despite its potential, federated learning in cancer research is still in its early stages and requires further development and validation. Future research should focus on improving the scalability and efficiency of federated learning techniques, ensuring data security, and developing standardized protocols to facilitate widespread adoption in oncology. By overcoming data accessibility barriers, AI can be more effectively integrated into cancer diagnostics, treatment planning, and personalized medicine (Fig. 8).

3.2 Model Interpretability and Trust: The Need for Explainable AI in Medicine

For deep learning models to gain widespread acceptance in clinical oncology, they must be interpretable and explainable. Currently, most AI-based models function as "black boxes," where their internal decision-making processes are not transparent to clinicians. This lack of interpretability is a significant barrier to adoption, as oncologists need to trust and understand AI-driven recommendations, particularly when making critical treatment decisions that can determine patient outcomes. If the reasoning behind an AI model's prediction is unclear, oncologists are unlikely to rely on it for cancer diagnostics, prognosis estimation, or treatment planning. Transparency is especially crucial in high-stakes fields such as oncology, where AI-generated insights must align with clinical reasoning to be useful in practice. To address this challenge, explainable AI (XAI) techniques have been developed to make AI-driven predictions more understandable. One such approach is SHapley Additive exPlanations (SHAP), which quantifies the contribution of each input feature to the model's prediction, providing oncologists with insights into why an AI system arrived at a particular conclusion. Similarly, attention mechanisms in deep learning models highlight which areas of medical images or genomic sequences are most influential in determining cancer diagnoses, enabling radiologists and pathologists to validate AI-generated results. These methods enhance transparency and help clinicians assess the reliability of AI outputs. Despite these advancements, there is still much to be done to make AI explanations more intuitive and actionable for clinical use. Many XAI techniques remain complex and technical, making them difficult for nontechnical medical professionals to interpret. AI explanations should not only be accurate but also presented in a way that aligns with clinical workflows, offering concise, context-specific insights that support medical decision-making. Efforts must focus on developing user-friendly interfaces, visual explanations, and standardized interpretability metrics that facilitate seamless integration of AI into oncology practice. Without sufficient interpretability, clinicians will be reluctant to rely on AI for diagnostics, risk prediction, and treatment recommendations, ultimately limiting its clinical applicability. Regulatory bodies such as the FDA and EMA also emphasize the need for transparency in AI-driven healthcare tools before they can be approved for clinical use. Therefore, future research should prioritize the simplification and standardization of XAI techniques, ensuring that AI-driven oncology solutions are both powerful and clinically meaningful. By enhancing AI interpretability, the medical community can build greater trust in AI systems, leading to more confident adoption and improved patient outcomes in cancer care.

3.3 Regulatory and Ethical Constraints: FDA Approvals, Legal Obstacles, Patient Consent

Regulatory approval for AI-driven medical devices is a lengthy and complex process, as agencies such as the U.S. Food and Drug Administration (FDA) and the European Medicines Agency (EMA) require extensive testing, clinical validation, and post-market surveillance to ensure that AI models used in oncology are safe, effective, and unbiased. AI-based diagnostic and treatment planning tools must undergo rigorous trials to demonstrate reliability across diverse patient populations, and any deviation from expected clinical outcomes could delay approvals significantly. Ethical considerations further complicate deployment, with concerns around patient consent, data privacy, and algorithmic fairness at the forefront. AI models trained on biased datasets may lead to disproportionate effects on specific demographic groups, resulting in unequal cancer diagnosis and treatment recommendations. Ensuring that AI models are developed using representative data, regularly audited for bias, and aligned with ethical AI principles is crucial to their responsible implementation. Additionally, legal liability remains a significant issue, as there is no clear framework to determine whether the software developer, the clinician, or the healthcare institution should be held accountable if an AI-based system makes an inaccurate or potentially harmful prediction. The ambiguity around responsibility and malpractice insurance for AI-driven medical decisions creates hesitancy among healthcare providers to adopt these technologies. Further, as AI systems continuously evolve through machine learning, regulators struggle to determine whether models should undergo new approvals for every update or modification, creating bottlenecks in AI deployment. Ethical dilemmas also arise in AI's role in treatment prioritization, especially in resource-limited settings where AI may influence decisions about which patients receive access to advanced therapies first. Transparency in AI decision-making is critical to fostering trust among oncologists, patients, and regulatory bodies, requiring developers to integrate explainable AI (XAI) methodologies that allow clinicians to understand and validate AI-generated recommendations. Moreover, collaboration between AI researchers, medical professionals, policymakers, and legal experts is necessary to create comprehensive regulatory frameworks that balance innovation with patient safety. Addressing these regulatory, ethical, and legal challenges is fundamental to the successful deployment of AI in oncology, ensuring that AI-driven medical tools contribute to equitable, high-quality cancer care while maintaining public trust and compliance with global healthcare standards.

3.4 Integration into Clinical Workflows: Compatibility with Current Hospital Systems

The integration of AI with existing hospital systems presents a significant challenge, as most healthcare organizations rely on legacy electronic health record (EHR) systems that are often incompatible with advanced AI-driven tools. Seamless embedding of AI into clinical workflows is critical to ensuring real-time decision-making support without disrupting well-established medical practices. Standardization of data formats, interoperability between different AI models and hospital infrastructure, and the development of user-friendly interfaces are essential for successful adoption. Many hospitals lack the technological infrastructure needed to support AI deployment at scale, making upgrades to IT systems a prerequisite before AI can be fully integrated. Moreover, AI-based recommendations must be easily interpretable by clinicians, necessitating training programs to familiarize medical professionals with AI-assisted diagnostics and decision-making. Without proper training, healthcare providers may either over-rely on AI without critical assessment or disregard AI insights altogether. Resistance to AI adoption is another major concern, as doctors and radiologists who have spent years perfecting traditional diagnostic methods may be hesitant to trust AI-generated outputs. Additionally, AI models require continuous updates to stay aligned with evolving medical research, making ongoing maintenance and compliance with regulatory standards a necessity. Ensuring that AI technologies are seamlessly incorporated into daily medical routines will require close collaboration between AI developers, hospital IT teams, and clinical staff. If poorly integrated, AI systems risk becoming underutilized or outright rejected by healthcare professionals who prefer conventional diagnostic methods. To address these challenges, AI tools must be designed with intuitive interfaces, minimal workflow disruptions, and clear justifications for their recommendations. Furthermore, strong cybersecurity measures must be in place to protect sensitive patient data

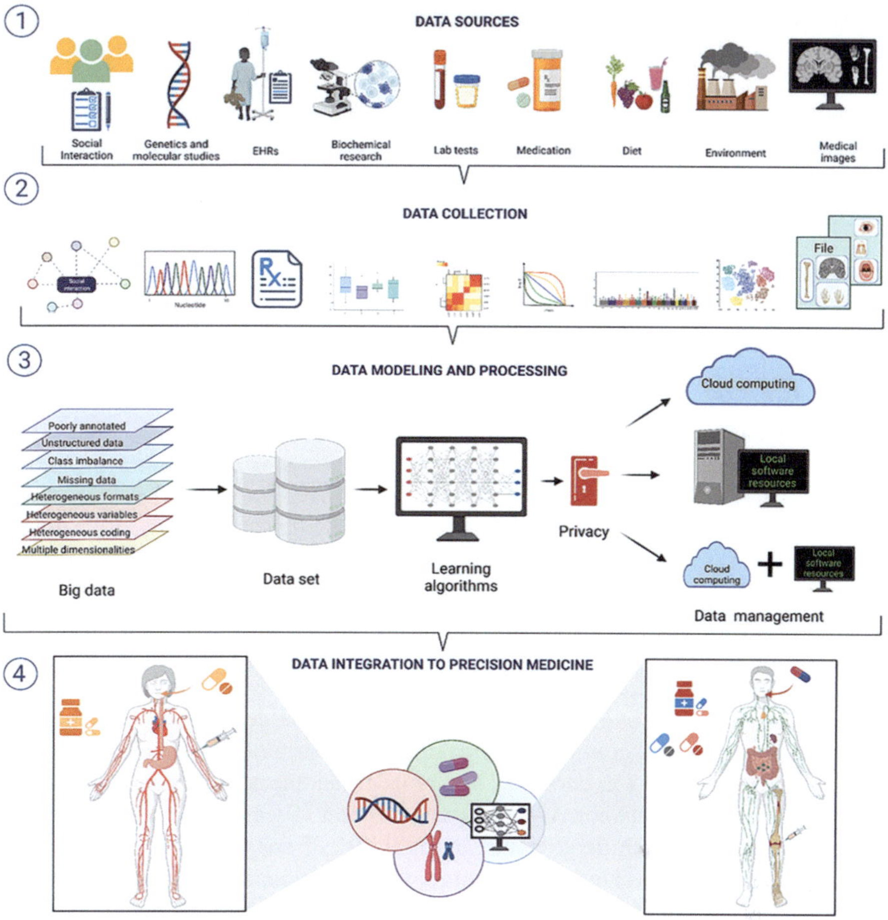

Fig. 9 Integration of artificial intelligence into the patient-care workflow of clinical oncology. (Syed, 2018)

when AI interacts with hospital networks. Without overcoming these technical, infrastructural, and human adoption hurdles, the transformative potential of AI in oncology and broader healthcare applications will remain unrealized, limiting its ability to enhance diagnostic accuracy, optimize treatment strategies, and improve patient outcomes (Fig. 9).

3.5 Computational Expense and Accessibility: Infrastructure Hurdles in Resource-Constrained Environments

The integration of AI with existing hospital systems remains a formidable challenge, as most healthcare institutions operate on outdated electronic health record (EHR) systems that are often incompatible with advanced AI-driven tools, creating barriers to seamless implementation. AI must be embedded within clinical workflows in a way that enables real-time decision-making support without disrupting established medical practices, requiring extensive standardization of data formats, interoperability between AI models and hospital infrastructure, and the creation of user-friendly interfaces that simplify adoption. A key obstacle is that many hospitals lack the necessary technological infrastructure to support large-scale AI deployment, necessitating significant IT upgrades before AI can be fully integrated into patient care. Moreover, AI-generated recommendations must be interpretable and trustworthy, making clinician training essential to ensure that healthcare providers understand how to work alongside AI-assisted diagnostics and treatment planning. Without adequate training, there is a risk that doctors may either over-rely on AI without critical evaluation or disregard AI insights altogether due to skepticism or lack of familiarity, thereby undermining its potential benefits. Resistance to AI adoption is another significant barrier, as experienced medical professionals—particularly radiologists and oncologists—who have spent years refining traditional diagnostic and treatment methods may hesitate to trust AI-generated outputs, especially if they are perceived as "black-box" solutions that lack clear explanations. Additionally, AI models require continuous updates to remain aligned with evolving medical research, necessitating ongoing maintenance, rigorous testing, and compliance with regulatory standards such as those set by the U.S. Food and Drug Administration (FDA) and the European Medicines Agency (EMA).(Beyond technical and regulatory hurdles, cybersecurity risks also pose a major concern, as AI systems handling sensitive patient data must be safeguarded against breaches and cyber threats that could compromise privacy and undermine patient trust. The successful integration of AI in oncology and broader healthcare applications demands a collaborative effort between AI developers, hospital IT teams, and medical professionals to ensure smooth deployment and usability. If not carefully integrated, AI tools risk being underutilized or outright rejected by clinicians who may prefer traditional diagnostic approaches, limiting their impact on improving patient outcomes. To address these adoption challenges, AI systems must be designed with intuitive, transparent interfaces that minimize workflow disruptions while providing clear justifications for recommendations, reinforcing clinician confidence in AI-assisted decision-making. Furthermore, AI-powered tools should complement rather than replace human expertise, enhancing the efficiency of healthcare professionals by automating routine tasks, reducing diagnostic errors, and personalizing treatment strategies based on patient-specific data. The long-term success of AI in oncology depends on overcoming technical, infrastructural, and human-related barriers through strategic planning, continuous model refinement, and comprehensive medical training programs that empower clinicians to harness AI effectively. Without resolving these challenges, the full potential of AI to revolutionize cancer diagnosis, optimize treatment plans, and improve survival rates will remain unrealized, hindering its ability to deliver meaningful advancements in precision oncology and personalized medicine.

4 Future Research Directions for Deep Learning in Oncology

4.1 Self-Supervised and Federated Learning to Enhance Data Privacy and Model Generalization

One of the major challenges in applying AI to oncology is the requirement for large, high-quality datasets, which are often difficult to access due to strict privacy regulations and data-sharing limitations. Cancer datasets are typically fragmented across multiple institutions, making it challenging to train AI models that are both accurate and generalizable across diverse populations. Traditional supervised learning approaches rely heavily on labeled data, but obtaining well-annotated medical datasets is expensive, time-consuming, and constrained by legal frameworks such as HIPAA (Health Insurance Portability and Accountability Act) in the United States and the GDPR (General Data Protection Regulation) in Europe. To address these challenges, advanced learning paradigms such as federated learning (FL) and self-supervised learning (SSL) have emerged as promising solutions that allow AI models to learn from vast amounts of decentralized data while maintaining patient confidentiality. Self-supervised learning reduces dependence on expensive human annotations by enabling AI models to extract meaningful patterns and representations from unlabeled data, making it particularly useful in oncology where labeled datasets are scarce. In contrast, federated learning enables multiple healthcare institutions to collaboratively train AI models without exchanging raw patient data, thus ensuring compliance with stringent data privacy laws. This decentralized approach allows AI to leverage a broader spectrum of real-world clinical data, improving its ability to generalize across different demographics and medical conditions. Despite these advantages, several challenges remain in deploying FL and SSL in oncology. Federated learning models must be secured against potential threats such as adversarial attacks and data poisoning, which could compromise patient safety (Fig. 10).

Communication overheads in FL also pose a significant challenge, as training AI models across multiple decentralized nodes requires efficient data transmission and synchronization without excessive computational costs. Moreover, SSL techniques need to be further refined to handle complex multimodal oncology data, which may include imaging, genomic sequences, and clinical notes. Addressing these challenges requires ongoing research into improving model security, optimizing FL frameworks for large-scale medical collaborations, and enhancing SSL algorithms to work effectively with heterogeneous oncology data. If successfully implemented, these advanced AI learning techniques have the potential to revolutionize cancer diagnosis, treatment planning, and prognosis prediction by making AI models more robust, interpretable, and widely deployable in real-world clinical settings. Future efforts should focus on integrating these technologies into existing healthcare infrastructures, fostering collaboration among AI researchers and oncologists, and ensuring that AI-driven cancer care remains ethical, unbiased, and aligned with patient-centered treatment approaches.

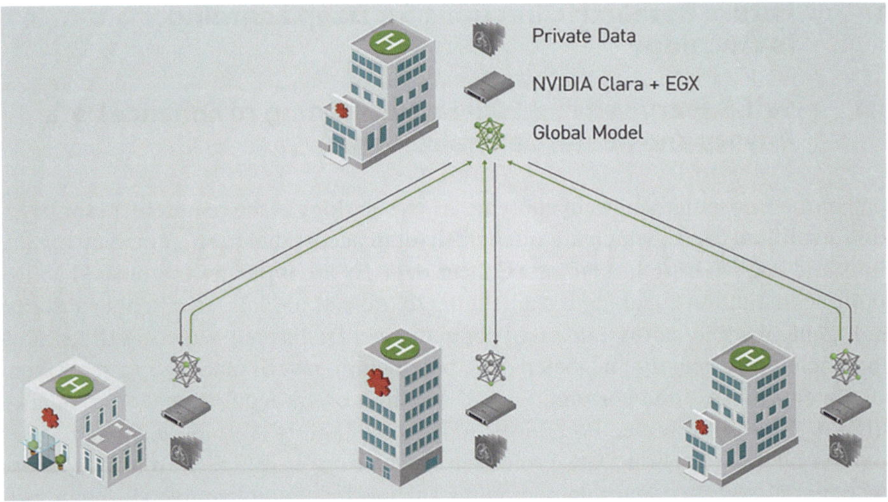

Fig. 10 Label-efficient self-supervised federated learning framework. (Adashek, 2019)

4.2 Multimodal AI: Convergence of Medical Imaging, Histopathology, and Genomics

Cancer diagnosis and treatment planning necessitate a comprehensive analysis of multiple data modalities, including medical imaging, histopathology slides, and genomic sequencing, yet current AI models are typically specialized in one domain, limiting their ability to provide a holistic understanding of a patient's condition. Future advancements in multimodal AI aim to integrate diverse data sources, enabling deep learning algorithms to deliver more accurate and precise predictions. By combining radiological imaging techniques such as CT, MRI, and PET scans with histopathological analysis and genomic biomarkers, multimodal AI can enhance diagnostic accuracy, facilitate rapid cancer subtype identification, and uncover novel treatment strategies. However, achieving seamless integration of heterogeneous medical data poses significant challenges, particularly in designing AI architectures that can efficiently process and interpret complex multimodal inputs. Researchers are exploring transformer-based models and cross-attention mechanisms to enable effective data fusion, ensuring that AI systems can extract meaningful relationships between different modalities while maintaining transparency and interpretability. The ultimate goal is to develop AI-driven oncology systems capable of delivering personalized treatment plans tailored to individual patients, improving both prognosis and therapeutic outcomes. Overcoming the technical and computational barriers associated with multimodal AI will require collaboration between oncologists, data scientists, and AI researchers to ensure that these systems are both clinically viable and aligned with real-world medical practices. As these technologies evolve, their integration into healthcare workflows could revolutionize oncology by enabling data-driven, precision medicine approaches that offer more targeted and effective cancer treatments.

4.3 AI-Driven Patient Profiling for Personalized Treatment Plans

Oncology therapy is heading toward personalized medicine wherein treatment is tailored based on an individual's genetic profile, tumor characteristics, and response to previous treatments. AI-powered patient profiling is at the forefront of this transformation, using deep learning to interpret genomic mutations, imaging reports, and clinical histories to prescribe tailored treatment regimens (Fig. 11).

Future studies will focus on creating better AI models to further improve precision oncology and make treatment recommendations more in line with the latest clinical guidelines and customized patient characteristics. Deep learning can potentially anticipate response to therapy, prevent injury by reducing side effects, and guide oncologists to therapy with the highest potential for success. The union of AI with electronic health records (EHRs) and real-world patient information will further boost the responsiveness of tailored treatment planning. The challenge remains that AI-generated recommendations need to be explainable, evidence-based, and updated with new clinical trial data and real-world experience on a regular basis.

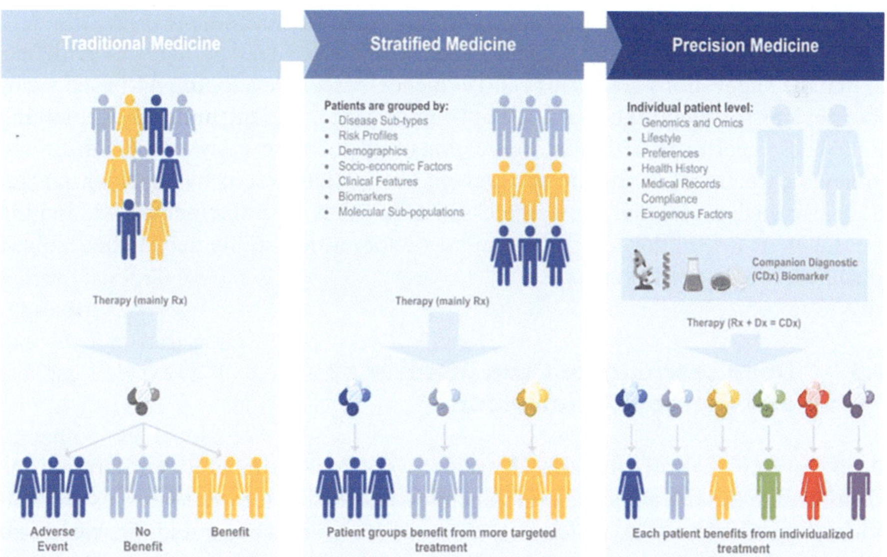

Fig. 11 AI for patient stratification and personalized treatment. (Yan et al., 2022)

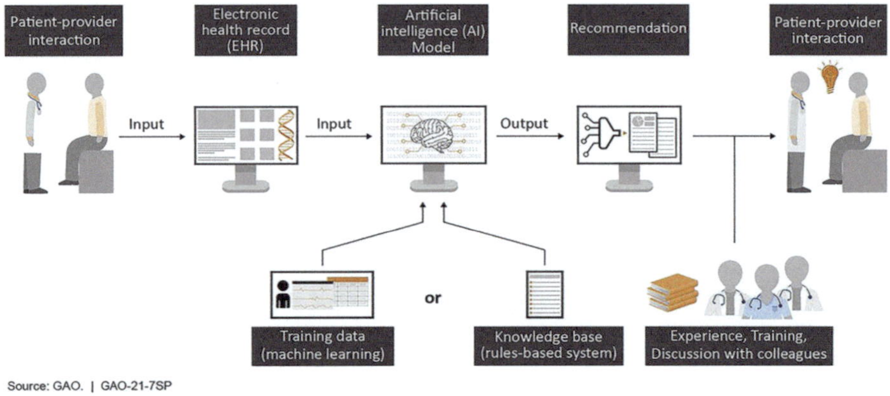

Fig. 12 AI-assisted medical diagnosis: increase assistance. (Penke, 2021)

4.4 Real-Time AI-Assisted Diagnostics and Decision Support Systems

The potential of AI in oncology lies in real-time diagnostics and decision support systems that provide clinicians with real-time information. Existing diagnostic workflows involve time-consuming manual assessments, leading to delays in treatment initiation. Decision support systems aided by AI can analyze imaging scans, histopathology slides, and patient histories in real-time and provide oncologists with real-time recommendations for diagnosis and treatment planning. Future work will focus on improving the reliability and efficiency of these AI-based systems through reinforcement learning and continuous learning techniques (Fig. 12).

AI systems will be made to adapt dynamically based on new medical evidence, so that the suggestions are current and evidence-based. Developing AI-based wearable devices for round-the-clock monitoring of cancer patients is another emerging area. These real-time AI-based applications will increase early cancer diagnosis, streamline treatment modifications, and enhance patient results by enabling quicker clinical decision-making. Nevertheless, issues such as implementing AI within present clinical workflows and ensuring regulatory acceptability need to be resolved to achieve widespread acceptance.

4.5 Deep Learning for Drug Discovery and Therapy Optimization

Deep learning is revolutionizing drug discovery by significantly accelerating the identification of potential cancer therapeutics, addressing the inefficiencies of traditional drug development, which is both costly and time-consuming. Conventional methods require years to bring a new drug to market, involving extensive laboratory testing, clinical trials, and regulatory approvals. AI-powered systems, however, can rapidly analyze vast chemical and biological datasets to predict drug–target interactions, select lead compounds, and repurpose existing drugs for novel oncological applications. By leveraging deep learning, pharmaceutical researchers can streamline drug screening processes, reducing both time and cost. Future advancements in AI-driven drug discovery will rely on generative models, such as variational autoencoders (VAEs) and generative adversarial networks (GANs), which are capable of designing novel molecular structures with high efficacy and low toxicity. These models can generate and refine drug candidates by learning latent patterns from existing molecular libraries, enabling the rapid synthesis of compounds with optimal pharmacological properties. Additionally, AI will play a crucial role in optimizing combination therapies by predicting drug synergies and minimizing adverse interactions, allowing for more effective multidrug treatment regimens tailored to individual patients. Deep learning models will further enhance clinical trial optimization by identifying the most suitable patient populations, predicting trial outcomes based on real-world data, and improving patient stratification for personalized medicine approaches. AI-driven insights can help refine trial designs, ensuring that new cancer drugs progress through the regulatory pipeline more efficiently. Furthermore, deep learning techniques integrated with computational biology and chemoinformatics will enable the discovery of precision oncology treatments, where drug regimens are customized based on a patient's genetic profile and tumor characteristics. The convergence of AI with computational drug design will facilitate a more targeted approach to cancer therapeutics, enhancing treatment efficacy while reducing side effects. As AI continues to evolve, its synergy with biotechnology and pharmaceutical sciences will transform cancer drug discovery into a more efficient, personalized, and economically viable process. However, challenges such as the need for high-quality labeled datasets, regulatory hurdles, and ethical concerns regarding AI-driven drug design must be addressed to maximize its potential. Close collaboration between AI researchers, oncologists, pharmacologists, and regulatory bodies will be necessary to ensure that deep learning technologies are harnessed effectively for clinical application. With continued advancements in AI, the future of cancer drug discovery holds promise for faster, more precise, and more accessible treatments, ultimately improving patient outcomes on a global scale.

5 Roadmap for Clinical Implementation

5.1 Approaches for Transitioning AI Models from Lab to Bedside

Model translation from the laboratory to real-world clinical application is a crucial step in ensuring the successful implementation of AI-driven oncology solutions. Before these models can be widely deployed in cancer treatment centers and hospitals, they must undergo rigorous validation, receive regulatory approvals, and seamlessly integrate into clinical workflows. Future research must prioritize enhancing AI models to improve their interpretability, generalizability, and alignment with established clinical best practices, ensuring that their recommendations are reliable and actionable for oncologists. One effective strategy for this transition is rolling deployment, where AI models are initially implemented in controlled clinical environments to assess their performance and impact before broader adoption. This phased approach allows for iterative improvements based on real-world feedback, mitigating potential risks and refining AI-driven decision-making. Additionally, AI solutions must be developed in close collaboration with oncologists and medical practitioners to ensure that they complement clinical workflows rather than disrupt established treatment protocols. Oncologists must be actively involved in training AI models to reflect nuanced medical judgment and real-world patient variability. Frameworks for AI integration, alongside standardized regulatory guidelines, will be critical in facilitating widespread adoption, helping healthcare institutions navigate compliance and ethical considerations. Interoperability with existing hospital information systems and electronic health records (EHRs) is another essential factor in successful deployment, enabling AI models to access and analyze patient data efficiently without creating additional burdens for clinicians. Seamless EHR integration will allow AI-driven insights to be readily available within standard diagnostic and treatment planning processes, improving efficiency and decision-making accuracy. Furthermore, ongoing monitoring and post-market surveillance will be necessary to ensure AI models continue to perform optimally as medical knowledge evolves and new treatment protocols emerge. Addressing these challenges requires a collaborative effort between AI researchers, healthcare providers, regulatory bodies, and hospital IT teams to create robust frameworks for AI adoption in oncology. Cybersecurity measures must also be in place to protect sensitive patient data and ensure compliance with privacy regulations such as HIPAA and GDPR. Without clear strategies for validation, integration, and long-term sustainability, AI oncology solutions risk underutilization or rejection by medical professionals. By focusing on regulatory alignment, user-centric design, and seamless interoperability, AI can effectively support oncologists in delivering more precise diagnoses, optimizing treatment plans, and ultimately improving patient outcomes (Fig. 13).

5.2 Interdisciplinary Collaboration: AI Researchers, Oncologists, Bioinformaticians, Policymakers

Success in oncology using AI relies on widespread collaboration across multiple disciplines, including AI researchers, oncologists, bioinformaticians, and policymakers. Effective AI-driven oncology solutions must be clinically relevant, explainable, and seamlessly integrated into real-world healthcare workflows, necessitating close cooperation between AI specialists and medical professionals. Oncologists play a critical role in defining the clinical objectives and ensuring that AI models align with medical best practices. Their expertise helps refine AI predictions, making them more interpretable and applicable in patient care. Bioinformaticians, on the other hand, are responsible for preprocessing and assembling large-scale genomic, histopathological, and clinical data, ensuring that AI models receive

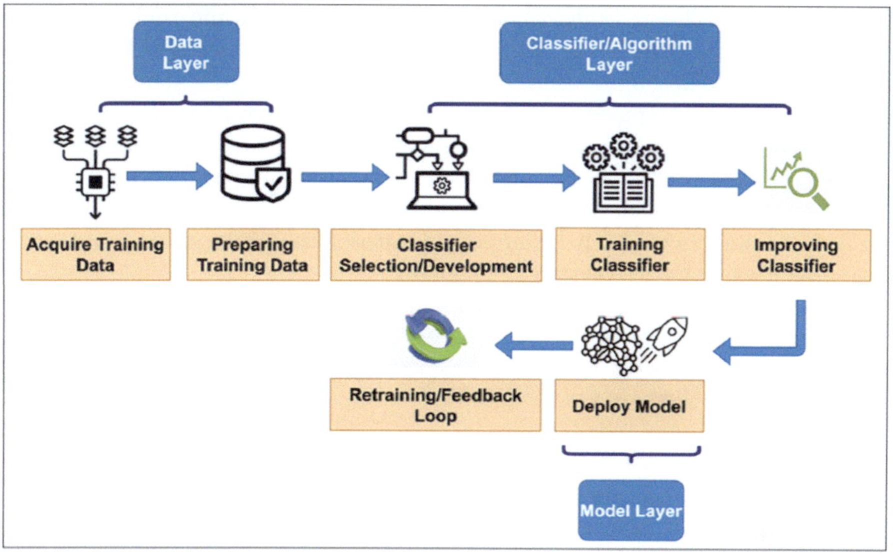

Fig. 13 Phases involved in deploying AI model. (U.S. Government Accountability Office (GAO), 2021)

high-quality input to aid cancer diagnosis and treatment. Without properly curated datasets, AI systems may generate unreliable results, emphasizing the need for rigorous data management and validation. Meanwhile, regulatory bodies have the responsibility of developing guidelines that foster ethical AI usage while safeguarding patient privacy and data security. Compliance with global privacy laws such as HIPAA and GDPR is essential to maintaining public trust and ensuring responsible AI deployment in oncology. Additionally, AI models must be continuously updated and validated against the latest medical knowledge to remain relevant and reliable in evolving clinical environments. Beyond the technical and regulatory aspects, patient advocacy groups play a vital role in ensuring that AI-enabled healthcare solutions remain patient-centered. Their involvement ensures that AI-driven recommendations consider not only clinical effectiveness but also patient experiences, preferences, and quality of life. Public trust in AI-powered oncology solutions will be strengthened when patients feel confident that these technologies are designed with their best interests in mind. Establishing interdisciplinary research initiatives and fostering international collaborations in AI-driven oncology can further accelerate progress. By sharing data, methodologies, and insights across borders, researchers can develop more robust AI models capable of generalizing across diverse patient populations. Initiatives such as open-source AI platforms and multi-institutional partnerships can democratize access to AI advancements, ensuring that innovations in AI-facilitated cancer therapy benefit patients worldwide. Moreover, cross-disciplinary collaboration can address critical challenges such as model interpretability, bias mitigation, and AI transparency, making these technologies more trustworthy and effective. Governments and funding agencies must support research programs that encourage AI-driven oncology advancements while establishing policies that promote responsible innovation. By creating an ecosystem where AI researchers, clinicians, bioinformaticians, policymakers, and patient representatives work together, the full potential of AI in oncology can be realized. This collaborative approach will not only improve diagnostic accuracy and treatment outcomes but also pave the way for more personalized and accessible cancer care. Ensuring that AI technologies are developed with inclusivity, transparency, and clinical utility in mind will ultimately lead to better patient experiences, optimized treatment strategies, and improved survival rates in oncology.

5.3 Education and Training: Preparing Clinicians to Work with AI Systems

To successfully integrate AI into oncology, it is crucial to equip clinicians with the knowledge and skills to interpret and effectively respond to AI-derived insights in clinical practice. AI literacy must become a fundamental component of medical education and continuing professional development to ensure that oncologists, radiologists, and pathologists can confidently utilize AI tools in their decision-making processes. Integrating AI education into medical school curricula will familiarize future healthcare professionals with the fundamental principles of artificial intelligence, including machine learning techniques, model interpretability, and ethical considerations. This foundational knowledge will help them critically evaluate AI-generated recommendations rather than accepting them at face value. Additionally, AI training should extend beyond theoretical instruction and include hands-on experience with AI-driven diagnostic and treatment technologies. This can be accomplished through specialized workshops, interactive simulations, and case-based learning that demonstrate real-world applications of AI in oncology.

For practicing clinicians, continuing education programs should incorporate AI-focused modules tailored to different specialties. Radiologists, for example, should receive training on AI-powered image analysis tools, while pathologists should learn about AI applications in digital pathology. Similarly, oncologists should understand AI's role in treatment planning, including predictive modeling for personalized therapies. These courses should also emphasize AI's limitations, potential biases, and strategies to mitigate them, preventing overreliance on AI-driven outputs. Beyond clinician training, AI developers must prioritize creating user-friendly interfaces that seamlessly integrate AI solutions into existing clinical workflows. The success of AI in oncology depends not only on model accuracy but also on ease of adoption. Reducing the learning curve through intuitive design, interactive dashboards, and automated result interpretation can enhance clinician engagement and encourage widespread use of AI tools in everyday practice. Furthermore, fostering interdisciplinary collaboration between AI developers and healthcare professionals will help ensure that AI solutions are tailored to meet the practical needs of clinicians. To further promote AI literacy, hospitals and healthcare institutions should implement structured AI training programs as part of professional development. Encouraging interdisciplinary team training—where clinicians, data scientists, and AI engineers work together—can improve understanding of AI's potential while fostering trust in AI-driven recommendations. Additionally, regulatory bodies and medical associations should establish standardized AI education requirements for oncologists and other specialists involved in cancer diagnosis and treatment. Ultimately, the successful application of AI in oncology will depend on the ability of clinicians to use AI tools confidently and responsibly. By prioritizing AI literacy in medical education, fostering hands-on learning experiences, and designing clinician-friendly AI interfaces, the healthcare industry can ensure that AI enhances—not replaces—clinical expertise. This balanced approach will lead to better patient outcomes, more precise cancer diagnoses, and improved treatment strategies, positioning AI as an indispensable tool in the fight against cancer.

5.4 Real-World Validation and Clinical Trials Before Large-Scale Rollout

Before the large-scale implementation of AI in oncology, rigorous real-world validation through prospective clinical trials is essential. These trials serve as a critical step in ensuring the validity, reliability, and safety of AI-driven tools across diverse patient populations. Without thorough clinical validation, AI models may demonstrate promising results in controlled research settings but fail to perform effectively in real-world scenarios where clinical variability is high. Prospective clinical trials provide AI models with the opportunity to be tested in real-world healthcare environments, assessing their generalizability across different hospitals, patient demographics, and treatment protocols. AI models must be trained and validated on datasets that reflect diverse patient populations, as cancer diagnosis and treatment can be influenced by a variety of genetic, environmental, and socioeconomic factors. By conducting trials in multiple healthcare institutions and under various clinical conditions, AI tools can be refined to ensure consistent accuracy and reliability, ultimately leading to better patient outcomes. Regulatory bodies such as the U.S. Food and Drug Administration (FDA) and the European Medicines Agency (EMA) require strong clinical evidence before granting clearance to AI-based diagnostic and therapeutic tools. Unlike traditional medical devices, AI models are dynamic—they continue to learn and evolve with new data, making continuous monitoring essential. As a result, regulatory frameworks must be updated to accommodate AI's iterative nature, ensuring that models maintain their accuracy and safety post-deployment. This requires ongoing performance evaluations, including real-world data monitoring and revalidation, to prevent model drift—a phenomenon where AI's predictive accuracy declines over time due to shifts in clinical data distributions.

In addition to regulatory oversight, establishing standardized evaluation criteria and procedures for AI validation is crucial. Unlike traditional medical devices that follow well-defined testing protocols, AI models often lack universally accepted benchmarks for clinical effectiveness. Future research should focus on developing standardized methodologies for assessing AI's impact in oncology, incorporating metrics such as sensitivity, specificity, and overall clinical utility. Collaborative efforts between AI developers, oncologists, and regulatory agencies will be necessary to define these standards and streamline the approval process, ensuring that AI-based solutions meet the highest levels of clinical efficacy. Moreover, post-deployment surveillance mechanisms should be implemented to continuously evaluate AI tools in practice. Hospitals and cancer treatment centers must integrate AI performance tracking systems that flag any potential degradation in accuracy, bias, or unintended consequences. AI-driven oncology solutions should also be subject to periodic retraining with fresh clinical data to keep up with advancements in cancer research and treatment methodologies. The future of AI in oncology depends on a robust regulatory and clinical validation framework that fosters trust and ensures patient safety. By prioritizing prospective clinical trials, continuous performance monitoring, and standardized evaluation criteria, the healthcare industry can pave the way for AI adoption in cancer diagnosis and treatment. Ultimately, the collaboration between AI researchers, oncologists, bioinformaticians, and regulatory authorities will determine the success of AI in transforming oncology, making it a powerful tool in the fight against cancer.

5.5 Public–Private Partnerships to Drive AI Adoption in Oncology

The successful adoption of AI in oncology demands substantial investment, infrastructure, and regulatory support, necessitating a multi-stakeholder approach to ensure its seamless integration into clinical practice. Public–private partnerships (PPPs) can play a crucial role in accelerating AI implementation by fostering cross-agency collaborations among research universities, government agencies, and private healthcare technology firms. Governments can provide financial support through research grants and funding programs, enabling the development of innovative AI applications in oncology, while the private sector can contribute by building sophisticated AI software tools, offering cloud computing resources, and training large-scale deep learning models. Pharmaceutical companies also stand to benefit from AI-driven advancements, utilizing these technologies for drug discovery, biomarker identification, and the personalization of cancer treatments, thus expanding the scope of precision oncology. Establishing AI research centers and innovation hubs that bring together industry experts, medical professionals, data scientists, and policymakers can drive AI-driven oncology solutions forward by promoting knowledge exchange and interdisciplinary collaboration. Moreover, global partnerships can enhance equitable access to AI-based cancer care, helping to bridge the gap between high-resource and low-resource healthcare settings, ensuring that advancements in AI-driven oncology are not limited to technologically advanced nations but reach underserved populations as well. Regulatory agencies must also work closely with AI developers and healthcare institutions to create standardized guidelines that facilitate AI's safe, ethical, and effective deployment in clinical workflows, addressing concerns such as model interpretability, patient data security, and real-world validation. By fostering an ecosystem where academia, industry, and government entities work in synergy, AI-driven oncology solutions can be scaled efficiently, ensuring their integration into routine cancer diagnosis, treatment planning, and drug discovery. Strategic collaborations and investments in AI infrastructure will ultimately determine the pace at which these cutting-edge technologies are adopted, influencing patient outcomes and reshaping the future of oncology worldwide.

6 Societal and Ethical Implications

6.1 AI Algorithm Bias and Impact on Cancer Diagnosis in Multicultural Populations

AI algorithm bias is a critical ethical issue in oncology, as biased AI models can lead to inaccurate cancer diagnoses and treatment recommendations, disproportionately affecting underrepresented groups. Many AI algorithms used in cancer diagnosis and treatment planning are trained on datasets predominantly composed of patients from specific racial, ethnic, or socioeconomic groups. As a result, these AI systems may be less accurate when diagnosing individuals from historically marginalized populations, leading to disparities in healthcare outcomes. If left uncorrected, algorithmic bias in oncology could contribute to delayed diagnoses, misdiagnoses, or inappropriate treatment plans, exacerbating existing healthcare inequities and worsening patient prognoses. One of the primary reasons for AI bias in oncology is the lack of diversity in medical datasets. Cancer data collected from hospitals and research centers often reflects a skewed demographic distribution, as clinical trials and imaging databases frequently overrepresent individuals of European descent while underrepresenting populations from African, Asian, Hispanic, and Indigenous backgrounds. Environmental and genetic factors influencing cancer progression vary significantly across different ethnicities, meaning that an AI model trained on a homogenous dataset may fail to recognize cancer patterns specific to underrepresented populations. For instance, breast cancer manifests differently in women of African descent than in Caucasian women, yet most AI-based mammography models are trained on datasets dominated by the latter. This imbalance can lead to increased false-negative or false-positive rates in minority populations, undermining trust in AI-driven oncology solutions. To address this challenge, AI researchers must focus on developing more representative and inclusive datasets by incorporating data from diverse populations. This requires collaboration between hospitals, research institutions, and public health agencies to ensure that AI models are trained on a wide spectrum of genetic, environmental, and socioeconomic data. Large-scale international collaborations can facilitate the aggregation of diverse datasets, ensuring that AI algorithms generalize well across different patient populations. Moreover, synthetic data generation techniques, such as generative adversarial networks

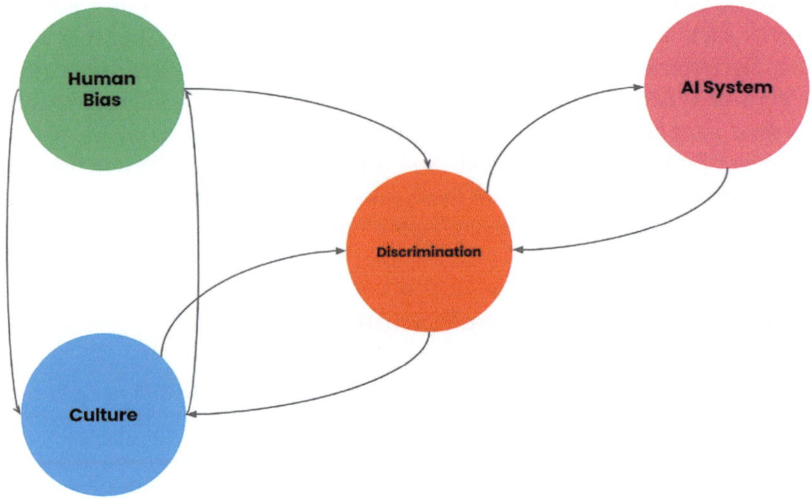

Fig. 14 Types of AI bias. (Labrecque Langlais et al., 2022)

(GANs), can be employed to augment training datasets with underrepresented demographic characteristics, reducing bias in AI predictions (Fig. 14).

Beyond dataset diversification, explainability techniques and bias auditing tools can help identify and mitigate algorithmic bias in oncology. Explainable AI (XAI) methods, such as SHapley Additive exPlanations (SHAP), Local Interpretable Model-agnostic Explanations (LIME), and saliency mapping, enable researchers and clinicians to interpret AI decision-making processes, making it possible to detect biases in model outputs. Bias auditing tools, such as fairness metrics and adversarial debiasing techniques, can be used to systematically assess whether AI models produce significantly different predictions based on race, ethnicity, gender, or socioeconomic status. By integrating these tools into the AI development pipeline, researchers can proactively address bias before deploying AI systems in real-world clinical settings.

Regulatory bodies, including the U.S. Food and Drug Administration (FDA), the European Medicines Agency (EMA), and the World Health Organization (WHO), must prioritize fairness and equity in AI-driven oncology solutions. Establishing guidelines for fairness-aware AI models will ensure that deep learning systems undergo rigorous bias assessments before being approved for clinical use. Additionally, healthcare institutions should implement continuous monitoring programs to evaluate AI performance in diverse clinical settings, ensuring that biases do not emerge over time as models encounter new patient populations.

Future research in AI for oncology should focus on designing fairness-aware models with built-in bias detection and correction mechanisms. Techniques such as adversarial training, counterfactual fairness modeling, and reweighting algorithms can help develop more equitable AI models. Moreover, interdisciplinary collaboration between AI researchers, oncologists, ethicists, and policymakers will be essential in ensuring that AI-driven cancer diagnosis and treatment solutions benefit all patient populations equitably. By diversifying training datasets, leveraging explainability techniques, integrating bias detection mechanisms, and enforcing fairness regulations, the healthcare industry can mitigate AI bias and promote equitable access to advanced cancer diagnostics and therapies. A commitment to fairness, transparency, and inclusivity in AI development will be key to realizing the full potential of AI in oncology and improving cancer outcomes for all populations worldwide.

6.2 Data Security and AI Transparency in Handling Sensitive Patient Data

With the increasing integration of AI in cancer diagnosis and treatment planning, safeguarding patient privacy and data security is paramount. AI-based oncology relies on vast amounts of sensitive patient information, including genomic sequences, medical imaging, and electronic health records (EHRs). Ensuring the confidentiality, integrity, and availability of this data is critical to maintaining trust in AI-driven healthcare solutions. As AI continues to reshape oncology, implementing robust security measures, ethical AI frameworks, and regulatory compliance will be essential to addressing concerns around patient privacy. One of the main challenges in AI-driven oncology is the risk of data breaches and unauthorized access. Patient health data is a prime target for cyberattacks, given its high value on the black market. Traditional centralized data storage systems pose a vulnerability, as a single point of failure can lead to massive breaches. To mitigate these risks, emerging techniques such as federated learning and homomorphic encryption provide new ways to train AI models while preserving patient anonymity. Federated learning enables AI models to learn from decentralized datasets without transferring raw patient data to a central repository. Instead, the AI model is trained locally at multiple institutions, and only the learned parameters are shared with a central server. This approach reduces the exposure of patient data and minimizes privacy risks while still enabling AI models to benefit from diverse medical datasets. Homomorphic encryption, on the other hand, allows AI algorithms to perform computations on encrypted data without decrypting it, ensuring that sensitive patient information remains protected throughout the processing pipeline. Implementing these privacy-preserving techniques can significantly enhance data security in AI-driven oncology.

Beyond technical safeguards, transparency in AI decision-making is crucial for building trust among clinicians and patients. Black-box AI models—those that produce predictions without explaining their reasoning—often generate skepticism, particularly in high-stakes fields such as oncology. If oncologists cannot interpret or validate an AI system's recommendations, they may be reluctant to rely on its insights for treatment planning. This is where explainable AI (XAI) frameworks play a crucial role. XAI techniques, such as SHapley Additive exPlanations (SHAP), Local Interpretable Model-agnostic Explanations (LIME), and saliency maps, make AI decision-making processes more interpretable. By providing visual or numerical explanations for AI-generated predictions, these techniques empower oncologists to validate AI-driven insights and integrate them into clinical workflows more confidently. This enhances accountability and ensures that AI recommendations align with established medical knowledge and best practices.

Policymakers and healthcare organizations must also implement strong data governance frameworks to regulate AI-driven oncology applications. Compliance with data protection laws such as the Health Insurance Portability and Accountability Act (HIPAA) in the United States and the General Data Protection Regulation (GDPR) in the European Union is necessary to prevent unauthorized data access and ensure ethical AI deployment. These regulations establish guidelines for data storage, processing, and sharing, helping to mitigate risks related to data misuse and patient privacy violations. Additionally, ethical considerations must be embedded in AI development and deployment. AI models should be designed to prioritize patient welfare, fairness, and accountability. Institutions must establish strict guidelines on informed consent, ensuring that patients understand how their data will be used in AI-driven research and clinical applications. Regular audits of AI models should also be conducted to assess their performance, security, and compliance with ethical standards.

Federated learning and homomorphic encryption offer innovative ways to enhance data protection, while XAI frameworks promote transparency and trust in AI-driven decisions. Healthcare organizations, AI developers, and policymakers must work together to establish strong regulatory frameworks, enforce data protection standards, and prioritize ethical considerations. By integrating robust security measures and transparent AI practices, the healthcare industry can harness the full potential of AI in oncology while safeguarding patient rights and trust.

6.3 Healthcare Access Disparities: Ensuring AI-Driven Oncology Solutions Are Available to All Population Groups

Artificial intelligence has the potential to revolutionize cancer care by improving diagnostic accuracy, treatment personalization, and overall patient outcomes. However, for AI to benefit all populations equitably, its integration into healthcare must address disparities in access, particularly in low-income and underserved communities. Currently, AI-driven oncology solutions are primarily developed and piloted in high-resource settings with well-established healthcare infrastructures, leading to concerns about their applicability in regions with limited medical resources. The availability of high-quality medical images, genomic sequencing, and AI-driven diagnostics is scarce in low-income settings, which could widen existing health disparities rather than reduce them. If left unaddressed, AI could reinforce inequalities by enhancing cancer care only for those who already have access to advanced medical technology while leaving vulnerable populations behind. To ensure global and equitable access to AI-based oncology solutions, governments, health institutions, and technology providers must invest in cost-effective

AI models that are accessible in low-resource settings (Kumari et al., 2025). This includes the development of lightweight AI algorithms that can function on lower computational power, cloud-based diagnostic platforms that allow remote healthcare providers to access AI-driven insights, and mobile health (mHealth) applications that enable patients in remote areas to receive AI-assisted cancer screening and early detection services. Such innovations would not only make AI-powered oncology tools more widely available but also significantly improve cancer detection rates in underserved regions where specialist healthcare services are limited. Moreover, AI literacy training programs must be instituted to empower healthcare workers in resource-limited settings with the knowledge and skills needed to integrate AI into their clinical workflows. Without adequate training, frontline health workers may struggle to interpret AI-generated recommendations, diminishing the technology's potential impact. By equipping doctors, nurses, and community health workers with AI education, these tools can be better utilized to improve cancer diagnosis and treatment outcomes. Additionally, fostering public–private collaborations is crucial for democratizing AI-driven healthcare by underwriting initiatives that increase access to AI-based oncology solutions across all socioeconomic tiers. Governments can provide funding for AI research and healthcare infrastructure development, while private technology companies can contribute by designing AI solutions tailored for low-resource environments. Nongovernmental organizations (NGOs) and academic institutions can further support these efforts by developing training programs and conducting implementation research to refine AI integration in diverse settings. International partnerships can also play a key role in making AI-driven oncology tools more accessible across countries with varying levels of healthcare infrastructure, ensuring that AI's benefits are not confined to wealthy nations. Ethical considerations should also be central to AI deployment, prioritizing fairness, inclusivity, and transparency to prevent bias and ensure that all patients, regardless of socioeconomic status, receive high-quality cancer care. Addressing these inequities will ensure that AI enhances healthcare equity rather than exacerbating existing disparities. By strategically investing in affordable AI-driven oncology solutions, enhancing AI literacy among healthcare professionals, and fostering multi-sector collaborations, the global healthcare community can work toward a future where AI in cancer care is universally accessible, improving patient outcomes for all populations regardless of geographic or economic barriers (Fig. 15).

7 Conclusion and Call to Action

7.1 Summary of Key Insights from the Chapter

Deep learning has the potential to revolutionize oncology by enhancing cancer detection, prognosis, and personalized treatment planning, significantly improving patient outcomes. AI models can analyze vast volumes of medical data, including imaging scans, genomic sequences, and electronic health records, to detect patterns that may not be visible to human clinicians. By improving diagnostic accuracy,

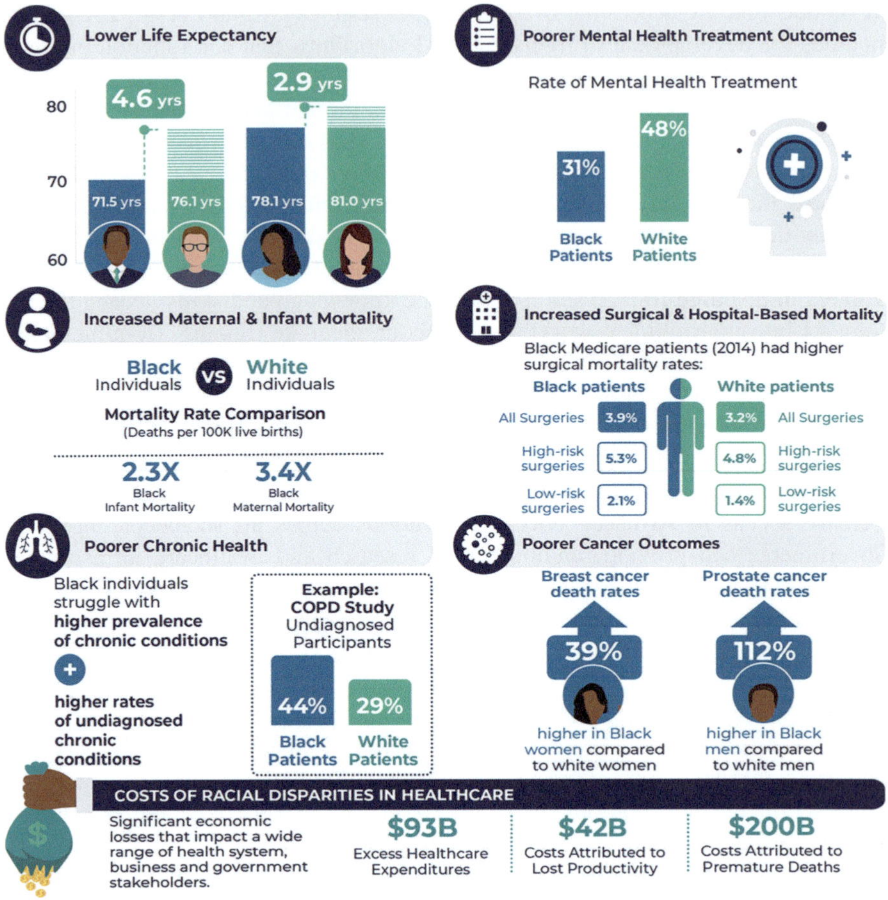

Fig. 15 Disparities in health care. (DataCamp 2020)

predicting disease progression, and optimizing therapeutic strategies, AI has the capability to streamline cancer care and reduce the burden on healthcare professionals. However, several challenges must be addressed to ensure the successful integration of AI in oncology. One of the primary issues is data availability, as AI models require large, diverse, and high-quality datasets to generalize effectively across different patient populations. Many existing datasets are biased toward specific demographic groups, leading to disparities in AI performance across different racial, ethnic, and socioeconomic backgrounds. Model interpretability is another critical concern, as deep learning algorithms often function as "black boxes," making it difficult for oncologists to understand the reasoning behind AI-generated recommendations. Without clear explanations, clinicians may struggle to trust AI-driven decisions, hindering adoption in clinical practice. Additionally, regulatory approval for AI-based oncology solutions remains a challenge, as governing bodies such as the FDA and EMA require extensive clinical validation to ensure safety, efficacy, and generalizability before deployment. Ethical considerations must also be prioritized, including issues of patient privacy, informed consent, and bias mitigation to prevent AI from exacerbating healthcare disparities. To overcome these obstacles, interdisciplinary collaboration between AI researchers, oncologists, regulatory authorities, and policymakers is crucial. AI solutions must undergo rigorous field testing through prospective clinical trials to validate their real-world applicability. Furthermore, robust data stewardship policies must be established to ensure patient data security and compliance with legal frameworks such as HIPAA and GDPR. By addressing these challenges through strategic research, transparent AI model development, and equitable access to AI-driven oncology tools, the healthcare industry can unlock the full potential of deep learning to improve cancer diagnosis, treatment, and patient care on a global scale.

7.2 The Need for Further Research and Development in DL-Oncology

Though significant progress has been made, deep learning in oncology is still in its early stages, requiring continuous research and development to refine AI models, improve their reliability, and make them clinically viable. AI algorithms must be regularly updated to integrate the latest medical discoveries, enhance their generalizability across diverse populations, and minimize biases that can lead to disparities in cancer diagnosis and treatment. One promising avenue for advancing AI in oncology is research into federated learning, which enables AI models to learn from decentralized datasets while preserving patient privacy. This approach can help overcome data-sharing restrictions and ensure that AI models are trained on diverse, representative data. Additionally, multimodal AI, which combines various data sources such as medical imaging, genomic sequencing, and electronic health records, has the potential to provide a more comprehensive understanding of cancer and improve diagnostic accuracy. Self-supervised learning techniques are also gaining traction, allowing AI models to learn from vast amounts of unannotated medical data, reducing the dependence on manually labeled datasets and enhancing AI's scalability. Furthermore, real-time diagnostic AI software and AI-driven clinical decision support systems are being developed to facilitate early cancer detection, optimize treatment planning, and improve patient outcomes. These tools can assist oncologists by providing precise, data-driven recommendations, reducing diagnostic errors, and accelerating treatment decisions. However, for AI to reach its full potential in oncology, significant investment is required to expand research, enhance AI literacy among medical professionals, and establish regulatory frameworks that ensure the safe and ethical deployment of AI technologies. Given the rising global cancer burden, the need for more efficient and scalable diagnostic and treatment methods has never been greater. AI-powered oncology solutions can help bridge the gap between resource-rich and resource-limited healthcare settings by enabling cost-effective, automated cancer screening and diagnosis. Collaboration between AI researchers, healthcare professionals, policymakers, and industry leaders will be

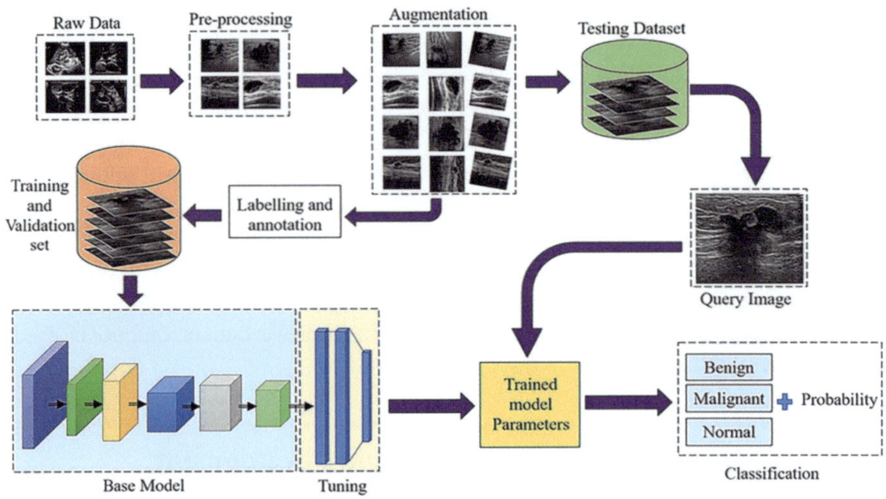

Fig. 16 Block diagram of proposed deep learning-based automatic breast tumor detection. (Wirelist Solfatara, 2023)

crucial to driving innovation, ensuring AI's responsible integration into clinical practice, and ultimately improving cancer care worldwide. With sustained research efforts and strategic implementation, AI has the potential to revolutionize oncology and make precision cancer care more accessible and effective for all patients (Fig. 16).

7.3 A Call for Cooperation Between Researchers, Healthcare Providers, and Policymakers

The successful implementation of AI in oncology depends on strong collaboration among AI researchers, clinicians, bioinformaticians, regulatory bodies, and policymakers. AI scientists must work closely with oncologists to develop clinically relevant and interpretable models that meet the demands of real-world healthcare. For AI to be truly useful in cancer diagnosis and treatment, it must align with medical workflows, integrate seamlessly with electronic health records, and provide insights that clinicians can trust and act upon. Close collaboration between AI experts and medical professionals ensures that AI models are not only technically sophisticated but also practically useful in clinical settings. Regulatory bodies play a crucial role in ensuring patient safety and ethical AI deployment by establishing clear guidelines for AI approval, validation, and continuous monitoring. Organizations such as the FDA and EMA must set standardized evaluation metrics to assess AI performance across diverse patient populations and clinical environments. This will promote transparency, build trust in AI-driven solutions, and prevent the deployment of unreliable models that could compromise patient care. Policymakers also have a vital role in driving AI adoption in oncology by funding AI-based healthcare initiatives, supporting research grants, and creating policies that facilitate AI integration into healthcare infrastructure. Government-backed initiatives and funding programs can help overcome financial barriers, enabling hospitals and research institutions to implement AI-driven technologies more effectively. Furthermore, international AI research collaborations and knowledge exchange programs can significantly accelerate advancements in AI-driven oncology. By sharing datasets, research methodologies, and technological innovations across borders, researchers can develop more robust AI models that work across diverse populations. Collaboration between high-resource and low-resource healthcare settings can also ensure that AI-driven cancer diagnosis and treatment tools are equitably accessible, reducing disparities in cancer care. Additionally, interdisciplinary research hubs that bring together AI developers, oncologists, bioethicists, and healthcare administrators can foster innovation, addressing key challenges such as algorithmic bias, data privacy, and model interpretability. The integration of AI into oncology is a complex but promising endeavor that requires sustained investment, regulatory support, and interdisciplinary cooperation. With strategic collaboration among all stakeholders, AI has the potential to enhance cancer diagnosis, treatment planning, and patient outcomes, ultimately transforming oncology into a more efficient, precise, and equitable field of medicine.

7.4 The Role of Funding Agencies and Global Initiatives in Enhancing AI for Cancer Care

International healthcare organizations and funding agencies play a crucial role in advancing AI-driven cancer diagnosis, treatment, and research. Governments, private hospitals, and nongovernmental organizations (NGOs) must invest in AI research projects, infrastructure development, and clinician training to facilitate the seamless integration of AI into oncology. Financial support from these stakeholders can help create large, high-quality datasets for training AI models, establish AI-powered diagnostic centers, and enhance medical professionals' ability to work with AI-driven tools. Public–private collaborations can further drive innovation by promoting knowledge exchange and funding cutting-edge AI projects focused on cancer detection, personalized treatment planning, and drug discovery. By fostering partnerships between technology companies, research institutions, and healthcare providers, these collaborations can accelerate AI development while ensuring that solutions address real-world clinical challenges. International cooperation among leading AI research institutes and cancer treatment centers is essential for developing standardized, high-quality AI models that can be applied across different populations and healthcare systems. AI models trained on diverse, global datasets can improve diagnostic accuracy and minimize algorithmic biases that often arise due to region-specific data limitations. This approach ensures that AI-driven oncology tools work effectively in both high-resource and low-resource settings, reducing disparities in cancer care. Additionally, funding agencies and policymakers must prioritize AI-driven cancer care as a global health issue. By allocating resources to AI research and deployment, they can support initiatives that bring AI-powered diagnostics and treatments to underserved communities. Investment in cloud-based AI platforms and mobile health applications can expand access to cancer care in remote and low-income areas, where specialized oncologists and advanced diagnostic tools may be scarce. Policymakers must also establish regulatory frameworks that promote ethical AI use, safeguard patient data, and ensure AI models' clinical reliability (Wang et al., 2023a; Khalifa and Albadawy, 2024). Encouraging cross-border collaborations and research exchange programs will further accelerate AI innovation in oncology, enabling rapid advancements in cancer detection and treatment. With strategic investments and global cooperation, AI has the potential to revolutionize oncology by improving cancer diagnosis, treatment personalization, and overall patient outcomes, making high-quality cancer care more accessible and equitable worldwide.

References

Adashek, J. (2019). *Integration of artificial intelligence into the patient-care workflow of clinical oncology*. Retrieved from https://www.researchgate.net/figure/Integration-of-artificial-intelligence-into-the-patient-care-workflow-of-clinical_fig1_334640630

Alsharif, F. (2024). Artificial intelligence in oncology: Applications, challenges and future Frontiers. *International Journal of Pharmaceutical Investigation [Internet]., 14*(3), 647–656. https://doi.org/10.5530/ijpi.14.3.76

Alum, E. U. (2025). AI-driven biomarker discovery: Enhancing precision in cancer diagnosis and prognosis. *Discover Oncology, 16*(1), 313. https://doi.org/10.1007/s12672-025-02064-7. PMID: 40082367; PMCID: PMC11906928.

Bi, W. L., Hosny, A., Schabath, M. B., et al. (2019). Artificial intelligence in cancer imaging: Clinical challenges and applications. *CA: A Cancer Journal for Clinicians, 69*(2), 127–157. https://doi.org/10.3322/caac.21552. PMID: 30720861; PMCID: PMC6403009.

DataCamp(2020). *Data demystified: The different types of AI bias.* Retrieved from https://www.datacamp.com/blog/data-demystified-the-different-types-of-ai-bias

Ding, Y., Wang, L., Yu, J., et al. (2024). Advancements in AI-driven medical imaging: CT, MRI, and PET scans in oncology. *Journal of Medical AI*, PMC10740686.

Khalifa, M., & Albadawy, M. (2024). AI in diagnostic imaging: Revolutionising accuracy and efficiency. *Computer Methods and Programs in Biomedicine Update, 5*, 100146. https://doi.org/10.1016/j.cmpbup.2024.100146

Kumari, D., Naidu, M. V. S. S., Panda, S., et al. (2025). Predicting breast cancer recurrence using deep learning. *Discover Applied Sciences, 7*, 113. https://doi.org/10.1007/s42452-025-06512-5

Labrecque Langlais, É., Thériault-Lauzier, P., Marquis Gravel, G., & Avram, R. (2022). Novel artificial intelligence applications in cardiology: Current landscape, limitations, and the road to real-world applications. *Journal of Cardiovascular Translational Research*. Retrieved from https://www.researchgate.net/figure/Phases-Involved-in-Deploying-AI-Model_fig6_361253530

Li, M., Xu, P., Hu, J., Tang, Z., & Yang, G. (2025). From challenges and pitfalls to recommendations and opportunities: Implementing federated learning in healthcare. *Medical Image Analysis, 101*, 103497. https://doi.org/10.1016/j.media.2025.103497

Liao, J., Li, X., Gan, Y., Han, S., Rong, P., Wang, W., Li, W., & Zhou, L. (2023). Artificial intelligence assists precision medicine in cancer treatment. *Frontiers in Oncology, 12*, 998222. https://doi.org/10.3389/fonc.2022.998222. PMID: 36686757; PMCID: PMC9846804.

Ligero, M., El Nahhas, O. S. M., Aldea, M., & Kather, J. N. (2025). Artificial intelligence-based biomarkers for treatment decisions in oncology. *Trends Cancer., 11*(3), 232–244. https://doi.org/10.1016/j.trecan.2024.12.001

Lipkova, J., Chen, R. J., Chen, B., Lu, M. Y., Barbieri, M., Shao, D., Vaidya, A. J., Chen, C., Zhuang, L., Williamson, D. F. K., Shaban, M., Chen, T. Y., & Mahmood, F. (2022). Artificial intelligence for multimodal data integration in oncology. *Cancer Cell, 40*(10), 1095–1110. https://doi.org/10.1016/j.ccell.2022.09.012

Lotter, W., Hassett, M. J., Schultz, N., Kehl, K. L., Van Allen, E. M., & Cerami, E. (2024). Artificial intelligence in oncology: Current landscape, challenges, and future directions. *Cancer Discovery, 14*(5), 711–726. https://doi.org/10.1158/2159-8290.CD-23-1199. PMID: 38597966; PMCID: PMC11131133.

Mennella, C., Maniscalco, U., De Pietro, G., & Esposito, M. (2024). Ethical and regulatory challenges of AI technologies in healthcare: A narrative review. *Heliyon., 10*(4), e26297. https://doi.org/10.1016/j.heliyon.2024.e26297. PMID: 38384518; PMCID: PMC10879008.

Miotto, R., Wang, F., Wang, S., Jiang, X., & Dudley, J. T. (2018). Deep learning for healthcare: Review, opportunities and challenges. *Briefings in Bioinformatics, 19*(6), 1236–1246. https://doi.org/10.1093/bib/bbx044. PMID: 28481991; PMCID: PMC6455466. https://pmc.ncbi.nlm.nih.gov/articles/PMC6455466/#sec2.

Olawade, D. B., David-Olawade, A. C., Wada, O. Z., Asaolu, A. J., Adereni, T., & Ling, J. (2024). Artificial intelligence in healthcare delivery: Prospects and pitfalls. *Journal of Medicine, Surgery, and Public Health, 3*, 100108. https://doi.org/10.1016/j.glmedi.2024.100108

Pei, Q., Luo, Y., Chen, Y., Li, J., Xie, D., & Ye, T. (2022). Artificial intelligence in clinical applications for lung cancer: Diagnosis, treatment, and prognosis. *Clinical Chemistry and Laboratory Medicine, 60*(12), 1974–1983. https://doi.org/10.1515/cclm-2022-0291

Penke, V. B. (2021). *AI for patient stratification and personalized treatment: The future of AI-informed drug repurposing*.

Ramachandran, V. (2022). *Using large-scale dataset to identify opportunity for implementing technology-based intervention to improve cancer care in India: Innovation report*. Doctoral dissertation, University of Warwick.

Reis, T. C. (2025). Deep learning in oncology: Transforming cancer diagnosis, prognosis, and treatment. *Emerging Trends in Drugs, Addictions, and Health, 5*, 100171. https://doi.org/10.1016/j.etdah.2025.100171

ResearchGate. (2021a). *AI in early drug and biomarker discovery*. ResearchGate. https://www.researchgate.net/figure/AI-in-early-drug-and-biomarker-discovery-a-Deep-learning-empowers-precision-medicine_fig1_352084247

ResearchGate. (2021b). *Block diagram of proposed deep-learning based automatic breast tumor detection*. Retrieved from https://www.researchgate.net/figure/Block-diagram-of-proposed-deep-learning-based-automatic-breast-tumor-detection_fig2_353757886

Sadeghi, Z., Alizadehsani, R., Cifci, M. A., Kausar, S., Rehman, R., Mahanta, P., Bora, P. K., Almasri, A., Alkhawaldeh, R. S., Hussain, S., Alatas, B., Shoeibi, A., Moosaei, H., Hladík, M., Nahavandi, S., & Pardalos, P. M. (2024). A review of explainable artificial intelligence in healthcare. *Computers and Electrical Engineering, 118*(Pt A), 109370. https://doi.org/10.1016/j.compeleceng.2024.109370

Sebastian, A. M., & Peter, D. (2022). Artificial intelligence in cancer research: Trends, challenges and future directions. *Life (Basel)., 12*(12), 1991. https://doi.org/10.3390/life12121991. PMID: 36556356; PMCID: PMC9786074.

Syed, T. (2018). *Deep learning framework for big data challenges*. Retrieved from https://www.researchgate.net/figure/Deep-learning-framework-for-Big-data-challenges_fig12_327211800

Truhn, D., Tayebi Arasteh, S., Saldanha, O. L., et al. (2024). Encrypted federated learning for secure decentralized collaboration in cancer image analysis. *Medical Image Analysis, 92*, 103059. https://doi.org/10.1016/j.media.2023.103059

U.S. Government Accountability Office (GAO). (2021). *Artificial intelligence in health care: Benefits and challenges of machine learning in drug development*. Retrieved from https://www.gao.gov

Verlingue, L., Boyer, C., Olgiati, L., Brutti Mairesse, C., Morel, D., & Blay, J. Y. (2024). Artificial intelligence in oncology: Ensuring safe and effective integration of language models in clinical practice. *The Lancet Regional Health—Europe, 46*, 101064. https://doi.org/10.1016/j.lanepe.2024.101064

Wang, L., Chen, X., Zhang, L., Li, L., Huang, Y., Sun, Y., & Yuan, X. (2023a). Artificial intelligence in clinical decision support systems for oncology. *International Journal of Medical Sciences, 20*(1), 79–86. https://doi.org/10.7150/ijms.77205

Wei, Q., Pan, S., Liu, X., Hong, M., Nong, C., & Zhang, W. (2025). The integration of AI in nursing: Addressing current applications, challenges, and future directions. *Frontiers in Medicine (Lausanne)., 12*, 1545420. https://doi.org/10.3389/fmed.2025.1545420. PMID: 40007584; PMCID: PMC11850350.

Williams, D., Hornung, H., Nadimpalli, A., & Peery, A. (2021). Deep learning and its application for healthcare delivery in low and middle income countries. *Frontiers in Artificial Intelligence, 4*, 553987. https://doi.org/10.3389/frai.2021.553987

Wirelist Solfatara. (2023). *Disparities in health care*. Retrieved from https://wirelistsolfatara.z21.web.core.windows.net/disparities-in-health-care.html

World Journal of Gastrointestinal Oncology. (2021). *Applications of artificial intelligence in early detection of cancer*. Retrieved from https://www.wjgnet.com/2644-3228/figures/v1/i2/39.htm

Yan, R., Qu, L., Wei, Q., Huang, S., Shen, L., Rubin, D., Xing, L., & Zhou, Y. (2022). Label-efficient self-supervised federated learning for tackling data heterogeneity in medical imaging. *arXiv preprint arXiv:2205.08576*. Retrieved from https://arxiv.org/abs/2205.08576

MIX
Papier aus verantwortungsvollen Quellen
Paper from responsible sources
FSC® C105338

If you have any concerns about our products,
you can contact us on
ProductSafety@springernature.com

In case Publisher is established outside the EU,
the EU authorized representative is:
Springer Nature Customer Service Center GmbH
Europaplatz 3, 69115 Heidelberg, Germany

Printed by Libri Plureos GmbH
in Hamburg, Germany